1st International Conference on Industrial, Electrical and Electronics (ICIEE 2018)

The Opportunities, Challenges, and Future Impacts of Industry 4.0

MATEC Web of Conferences Volume 218 (2018)

Anyer, Indonesia
4 - 5 September 2018

Editors:

W. Martiningsih
R. Wiryadinata
S. Praptodiyono
M.I. Santoso
I. Saraswati

ISBN: 978-1-5108-7404-6

Printed from e-media with permission by:

Curran Associates, Inc.
57 Morehouse Lane
Red Hook, NY 12571

Some format issues inherent in the e-media version may also appear in this print version.

Printed by Curran Associates, Inc. (2018)

For additional information, please contact EDP Sciences – Web of Conferences
at the address below.

EDP Sciences – Web of Conferences
17, Avenue du Hoggar
Parc d'Activité de Courtabœuf
BP 112
F-91944 Les Ulis Cedex A
France

Phone: +33 (0) 1 69 18 75 75
Fax: +33 (0) 1 69 28 84 91

contact-edps@webofconferences.org

Additional copies of this publication are available from:

Curran Associates, Inc.
57 Morehouse Lane
Red Hook, NY 12571 USA
Phone: 845-758-0400
Fax: 845-758-2633
Email: curran@proceedings.com
Web: www.proceedings.com

TABLE OF CONTENTS

1st International Conference on Industrial Electrical and Electronics (ICIEE)

"The opportunities, challenges, and future impacts of industry 4.0"
Anyer, Banten, Indonesia, September 4-5th, 2018

The 1st International Conference on Industrial Electrical and Electronics (ICIEE) aims to bring together leading academic scientists, researchers and research scholars to exchange and share their experiences and research results on all aspects of Industry 4.0 and Internet of Things. It also provides a premier interdisciplinary platform for researchers, practitioners and educators to present and discuss the most recent innovations, trends, and concerns as well as practical challenges encountered and solutions adopted in the fields of Industry 4.0 and Internet of Things.

International Board and Program Committee
of the 1st International Conference on Industrial Electrical and Electronics (ICIEE)

- ✓ Prof. Dr. Tumiran
  (Gadjah Mada University, Indonesia)
- ✓ Prof. Dr. Samsir Abduh
  (Universitas Trisakti, Indonesia)
- ✓ Prof. Dr. Ir. Gamantyo Hendrantoro, Ph.D.
  (Institut Teknologi Sepuluh November, Indonesia)
- ✓ Prof. Ir. Sunarno, M.Eng., Ph.D.
  (Universitas Gadjah Mada, Indonesia)
- ✓ Gunawan Wibisono, M.Sc., Ph.D
  (Universitas Indonesia, Indonesia)
- ✓ Dr. Iznan H. Hasbullah
  (Universiti Sains Malaysia, Malaysia)
- ✓ Muhammad Reza, Ph.D.
  (Telkom University, Indonesia)
- ✓ Dr. Ir. Yuyu Wahyu, MT.
  (Indonesian Institute of Science, Indonesia)
- ✓ Bachtiar Nappu, Ph.D.
  (Universitas Hasanuddin, Indonesia)
- ✓ Cahyo Mustiko, Ph.D.
  (Universitas Mataram, Indonesia)
- ✓ Dr. Ir. Sholeh Hadi Pramono
  (Universitas Brawijaya, Indonesia)
- ✓ Dr. Ali Hanafiah Rambe, ST., MT.
  (Universitas Sumatera Utara, Indonesia)
- ✓ Dr. Hermawan, DEA.
  (Universitas Diponegoro, Indonesia)
- ✓ Dr. Ir. Yuli Kurnia Ningsih, MT.
  (Universitas Trisakti, Indonesia)
- ✓ Dr. Wahyudi
  (Universitas Diponegoro, Indonesia)
- ✓ Dr. Azlan Osman
  (Universiti Sains Malaysia, Malaysia)
- ✓ Dr. Muhammad Haikal Satria
  (Universiti Teknologi Malaysia, Malaysia)
- ✓ Dr. Arbai Yusuf
  (C-Tech Labs Edwar Technology, Indonesia)
- ✓ Dr. Marlin Baidillah
  (Chiba University, Japan)
- ✓ Dr. Victor Odedo
  (University of Bath, United Kingdom)
- ✓ Dr. Wahyu Susihono
  (Universitas Sultan Ageng Tirtayasa)
- ✓ Dr. Ing. Asep Ridwan
  (Universitas Sultan Ageng Tirtayasa)

Organizing Committee

of the 1st International Conference on Industrial Electrical and Electronics (ICIEE),

❖ Honorary Chair

Prof. Dr. H. Sholeh Hidayat, M.Pd.
Rector – University of Sultan Ageng Tirtayasa

❖ General Chair

Dr. Ir. Supriyanto, M.Sc., IPM
Head of Electrical Engineering Department. University of Sultan Ageng Tirtayasa.

❖ Technical Chair

H. M. Hartono, M.T. (ICIEE)
Heri Haryanto, M.T. (NCIEE)

❖ Co. Chair

1. Dr.-Ing. M. Iman Santoso, M.Sc.
2. Dr. Romi Wiryadinata, M.Eng.
3. Dr. Irma Saraswati, M.T.
4. Dr. Wahyuni Martinginsih, M.T.
5. Dr. Alimuddin, M.T.

❖ Members

1. Ir. Ri Munarto, M.Eng.
2. Rian Fahrizal, M.Eng.
3. M. Sadikin, M.T.
4. M. Otong, M.T.
5. Teguh Firmansyah, M.T.
6. Herudin, M.T.
7. H. Alif Maulana, M.T.
8. Anggoro Suryo Pramudyo, M.Kom.
9. Imamul Muttakin, M.Eng.

Keynote Speakers 1

Prof. Dr. Rosni Abdullah
(Universiti Sains Malaysia, Malaysia)

The opportunities and challenges of industry 4.0

Abstract

The Internet of Things (IoT) has not only revolutionised business processes, but has strong potential to profoundly impact the life of many citizens. Many sectors such as healthcare, transportation and manufacturing among others, are starting or have started to adopt IoT by connecting devices and creating innovation in the respective sectors. One sector that will particularly benefit is the industrial sector. In Industry 4.0, the connected ecosystem of the IoT, will provide manufacturers and consumers increased automation, improved communication and monitoring, along with self-diagnosis and new levels of analysis to improve productivity. This talk will introduce the framework of IoT in the context of industry 4.0 and explore the opportunities and challenges that lay ahead.

Keynote Speakers 2

Prof.Dr.Yilmaz Uygun, M.Sc. (Dept. Mathematics & Logistics, Jacob Univ. Bremen, Germany)

Human Resources Requirements for Industry 4.0

Abstract
Industry 4.0 is currently changing industrialized countries´ economies. Previously uncoupled items will connect to each other to develop self-controlled production systems by using the latest software and hardware in information and communication technology. This rapidly increasing technologies not only have the potential to substitute repetitive jobs but also pose a threat to more sophisticated ones. Against the backdrop of this, important questions arise which will be addressed with this speech; such as the type of jobs that are affected, the qualifications and skill sets that are needed in Industry 4.0. Other specific queries that appear are; type machines/robots that will be dominant, the policy taken in the factory, new communication of work (e.g., man-machine interaction) and the operating procedure for the machines. All of these terms should be addressed and mapped based on the locus of the industries which vary depending on their technological maturity level. Afterwards, set of cognitive readiness competencies should be prepared. Finally, the relevant vocational educations, trainings and certifications can be derived from those set of competencies.

Keynote Speakers 3

Prof. Poki Chen (National Taiwan University of Science and Technology, Taiwan)

Abstract
Industrial production is one of the key competitions among global players. After Industry 4.0 introduced by German government, most of the international big players invest tons of money to offer their production system more flexibility, faster adaptationforboth individualizedand customized products through digitalization and new technologies. For implementation, Industry 4.0 needs decentralized control and advanced connectivity to collect and exchange real-time information for production processesoptimization.Expected as the newest industrial revolution, it opens a huge market for both software and hardware engineers. To generate big enough data for optimizing the production automation, it needs a lot of sensors to identify, locate, track and monitor the equipment and the products. To connect all the machines, products and factories, a market booming for sensors not only can be expected but also is already undergoing.In this speech, feasible solutions for Industry 4.0 from micro-scale MEMS (Micro Electro Mechanical System)accelerometers, humidity, pressure,acoustic,gas sensors all the way to nano-scale image, illumination, position and temperature sensors will be highlighted to point out the possible ways for both research and commercialization.

Preface

We would like to welcome all participants from academicians, government, and industry to attend the International Conference on Industrial, Electrical and Electronics (ICIEE 2018) in Aston Beach Hotel, Anyer, Indonesia. ICIEE 2018 is organized by Department of Electrical Engineering, the Universitas Sultan Ageng Tirtayasa, supported by FORTEI (Electrical Engineering Higher Education Forum of Indonesia).

This conference is a great event in the field of Industrial, Electrical and Electronics. World Class Speakers are scheduled to deliver speech on hot topics in industry 4.0 so as it can be beneficial to all delegates. The committee noted that 143 abstracts submitted to the conference systems. However, after go through a peer-review process by the program (editorial) committee and international boards, the final accepted papers are 76.

On behalf of the organizing committee, the editors would like to express our sincere gratitude and appreciation to all participants for coming down to Anyer Beach to share and present your research findings. We are also indebted to all the international reviewers for helping us in reviewing all the papers for ensuring high quality of all the accepted papers. I would like to thanks all sponsorships for their valuable supports. We believe that the conference will be a key stage to improve our research and show great development to the world in the field of industrial, electrical, electronics and information technology

I would also like to extend my thanks to all the organizing committee for working very hard to make this conference as today and record my personal apology for any shortcomings. Any recommendations and suggestions for improvement are very much appreciated and most welcome.

The ICIEE 2018 is one step to reach our vision towards an excellent department in 2020. I believe this international event is able to encourage our spirit to move on a better condition as well as expand our collaboration and networking.

Editor of ICIEE

edp sciences

Web of Conferences

Statement of Peer review

In submitting conference proceedings to *Web of Conferences*, the editors of the proceedings certify to the Publisher that

1. They adhere to its **Policy on Publishing Integrity** in order to safeguard good scientific practice in publishing.
2. All articles have been subjected to peer review administered by the proceedings editors.
3. Reviews have been conducted by expert referees, who have been requested to provide unbiased and constructive comments aimed, whenever possible, at improving the work.
4. Proceedings editors have taken all reasonable steps to ensure the quality of the materials they publish and their decision to accept or reject a paper for publication has been based only on the merits of the work and the relevance to the journal.

Title, date and place of the conference

The 1st International Conference on Industrial, Electrical and Electronics (ICIEE) 2018
September 4 - 5, 2018
Anyer, Indonesia

Proceedings editor(s):

1. Dr. Wahyuni Martiningsih
2. Dr. Romi Wiryadinata
3. Dr. Supriyanto
4. Dr. -Ing. M. Iman Santoso
5. Dr. Irma Saraswati

Date and editor's signature 27/09/2018

Techno Economic Analysis of Hybrid Sea Floating Photovoltaic and Wind Turbine for Fish Cold Storage At Remote Island In Indonesia (Case Study: Small Kei Island, Maluku, Indonesia)

Fidel Rezki Fajry[1], *Iwa* Garniwa*[1], *Achmad* Fajar Tofani[1]
[1]Department of Electrical Engineering, Universitas Indonesia, Jakarta, Indonesia

Abstract. Most of fisherman in Small Kei Island, Indonesia, are difficult to keep the good quality of fish for several days because unavailability of Fish Cold Storage. Due to the area is also far from grid, this paper is aimed to analyse the most optimized sizing of integrating the new technology of sea floating photovoltaic and wind turbine to achieve the minimal levelized cost of energy for powering 100 tons of fish cold storage. Ten scenarios with implemented off-grid scheme by dividing battery bank into two equal parts that will be operated interchangeably to meet 24 hours load requirement on hybrid sea floating photovoltaic and wind turbine system are performed. As result, 100 kW of wind turbine and 803kWp of sea floating photovoltaic are chosen with the most minimal levelized cost of energy of USD 0.52/kWH by life cycle cost method of 20 years. With define electricity tariff of USD 0.81/kWH and interest loan of 9%, the economic analysis result fulfil of four indicators requirement with payback period of 6 years 6 months 20 days, net present value of USD 10,981, internal rate of return of 9.10%, and profitability index of 1.60.

1 Introduction

Fish cold storage can preserve fish for a long period and ensure the fish quality is in a good quality. In 2016, the installed fish cold storage in Indonesia is only reached 11.49% of total demand [1]. One of the main problem is due to the electricity which is not distributed. Hence, those areas need standalone power generation for fish cold storage. Nowadays, the type of power generation from renewable energy is proposed to be developed. It is appropriate with PT. Perusahaan Listrik Negara that prioritizes to utilize renewable power generation to supply electricity in remote areas, mainly for those are far from grid.

Studies related to cold storage powered by integrating renewable power generation are quite scarce in open literature. The two tons of Fish Cold Storage powered Photovoltaic (COSPV) is proposed in [2]. The PV system scheme is on-off grid and have focused to obtain the estimation cost of COSPV. A rooftop solar PV – diesel generator hybrid mini

*Corresponding author: garniwa@gmail.com

cold storage for rural area is proposed in [3]. The PV system scheme is on grid. The running concept of cold storage is by consuming power from the roof top solar PV panel, and the diesel generator will be only absorbed during night time. The consideration focused on financial analysis and costing. Then, a conceptual model of a cold storage warehouse powered by on grid PV and grid in smart grid setting is presented in [4]. The system allows the cold storage to withdraw power from the grid and feed power to the grid. A formula with linear programming is used as an optimization to investigated optimal energy management strategies under different tariff scenario.

On recent technology, sea floating photovoltaic (SFPV) has been developed [5]. SFPV system can increase efficiency due to reflection of light from the water and the PV panel temperatures is maintained lower by the natural evaporative cooling from waterbody [6].

On this paper, the objective is to analyse the optimum size of an off-grid hybrid sea floating photovoltaic (SFPV) and wind turbine by setting the power capacity of both renewable generation to achieve the minimum levelized cost of energy (LCOE) for powering DC Fish Cold Storage in Small Kei Island, Indonesia. However, according to economic analysis, SFPV and Wind Turbine with off-grid concept for powering DC fish cold storage are feasible to be developed.

2 Energy Resource and Load Profile Data

The load profile of one hundred tons of Fish Cold Storage is shown on **Fig. 1**. The peak load of 49kW is occurred almost eighteen hours during the main freezing load of compressor, condenser, and evaporator running from 6:00AM to 11:59PM. The six (6) hours remain of 16.7 kW is the recovery period when defrost running from 00:00 to 5:59AM. Another load is consisted of lighting, heating, mechanical handling equipment, etc.

The annual of both daily solar radiation and wind speed data in Small Kei Island are 5.57 kWh/m2/day (232 W/m2) and 4.0 m/s, which were collected from Atmospheric Science Data Center that refer to NASA **[7]**.

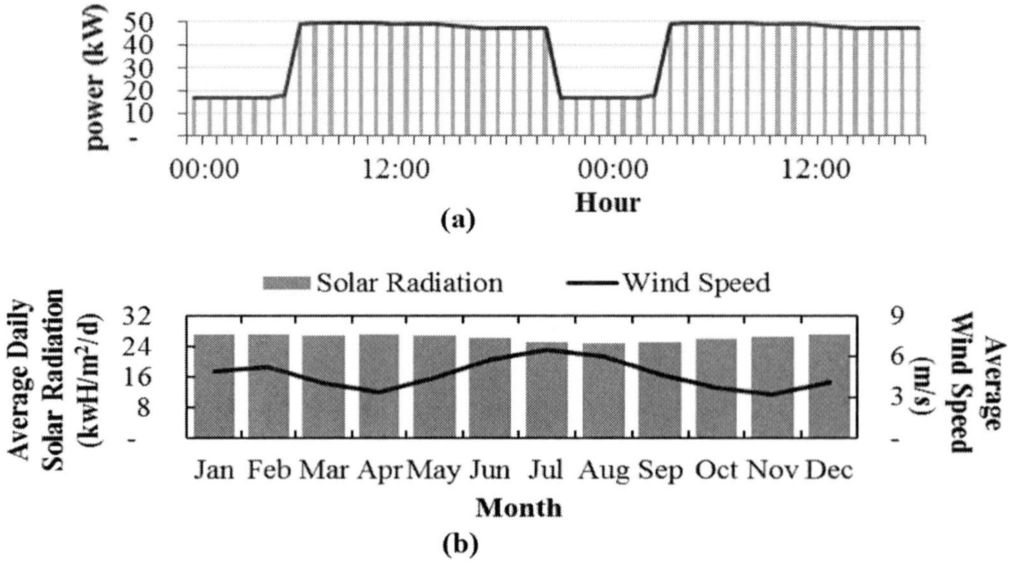

Fig. 1. (a) daily load curve of 100 tons of fish cold storage, (b) monthly average daily solar radiation and wind speed at Small Kei Island

3 Component Sizing of Hybrid System and LCOE Calculation

The proposed hybrid SFPV and wind turbine system uses off-grid scheme as shown on **Fig 2.** The off-grid system configuration uses two equal battery banks that will be operated interchangeably with the strategy schedule of operation as shown on **Table 2**. Each of battery bank will be charged and discharge for 24 hours. The purposed is to make sure that the battery can be fully operated when the weather is unpredictable and meet load requirement for daily 24 hours basis. During discharged, the battery banks will be cut-off if the remain of capacity achieve 30% from total capacity. After obtained the load requirement, the sizing of hybrid system is started from battery bank, wind turbine, and SFPV module as mentioned on below.

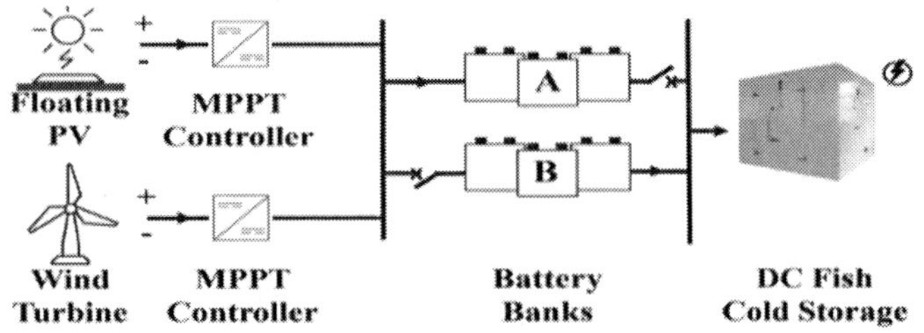

Fig. 2. Block diagram of the proposed hybrid SFPV and wind turbine

3.1 Sizing Battery Banks

The total capacity of battery banks per unit (Cap) can be calculated from the total load demand (TLD), Depth of Discharge (DOD), and the day without sun (dws) as derived:

$$Cap_{(VAH)} = \frac{TLD_{(kwh/day)} * dws_{(day)}}{DOD} \tag{1}$$

3.2 Sizing Wind Turbine

A wind turbine generator can start generating power if the wind speed (v) exceeds the cut-in value (V_i). Then, it will generate the rated power (P_{rated}) if the wind speed exceed the rated speed (V_r) of the wind turbine. It will stop if the wind speed exceeds the cut-out (V_o) value to protect generator **[3]**. The power output of Wind Turbine (P_{WT}) is expressed as **[8]**:

$$\begin{aligned} P_{wind} &= 0 && if \;\; v \le V_i \;\; or \;\; v \ge V_o \; ; \\ P_{wind} &= P_{rated}\eta_{turbine}\frac{v - V_i}{V_r - V_i} && if \;\; V_i < v < V_r ; \\ P_{wind} &= P_{rated} && if \;\; V_r < v < V_o ; \end{aligned} \tag{2}$$

Table 2. The strategy schedule scenario of battery banks operation

Day	**1st**		**2nd**		**3rd**		**4th**	
Time								
Battery A	X	X	V	V	X	X	V	V
Battery B	V	V	X	X	V	V	X	X

Table 3. Small Kei Island Site & Environmental Data

Description	Requirement of Floating PV	Small Kei Island Data [9]
Water depth (low tidal)	Minimum 1 m, optimal 3-10 m, maximum 25m	2.75 m
Waves	At 1-3 m depth maximum 0.5 m waves, At depth >3 m maximum 1.5 m waves	0.5 m
Wind Speed	Maximum 100-120 km/h	14.4 km/h
Water Current	Maximum 8 km/h	0.54 km/h
Seabed	Preferably Sand, mud, or rubble.	Sandy

3.3 Sizing Solar Floating Photovoltaic (SFPV) Module

SFPV system concept removes big land requirement. If inland PV requires requirement on civil and trenching, SFPV requires platform, mooring system, and under-water cable as described on **[6]**. On this paper, the reason to utilize SFPV is due to the site & environmental data of Small Kei Island fulfil the requirement to install SFPV as shown on **Table 3**, and have better efficiency of module compare to inland PV due to water cooling effect.

Each of SFPV module has the capacity of 24kWp that consists of 96 units of 250kW panels. To calculate the required power generation of SFPV shall consider the average effective time (E_{ff}) of solar radiation in Small Kei Island, which is 4.5 hours per day **[10]**, the average of PV panel efficiency ($\eta_{PVmodule}$) according to recent research, which is 17%, and charge controller efficiency (η_{MPPT}). The real capacity of SFPV (Cap_{SFPV}) can be expressed as follows:

$$Cap_{SFPV(kWp)} = \left[\left(\frac{Cap_{(VAH)}}{dws_{(day)}}\right) - W_{T(kWH/day)}\right] * E_{ff(h/day)} * \eta_{\text{PVmodule}} * \eta_{MPPT} \quad (3)$$

3.4 LCOE Calculation

To compare different capacities of electricity generation on a comparable basis, LCOE **[11]** with 20 years lifetime project is calculated as follows:

$$LCOE_{\left(\frac{USD}{kWH}\right)} = \frac{Life\ Cycle\ Cost\ (LCC)_{(USD)}}{Lifetime\ energy\ production_{(kWH)}} \quad (4)$$

The LCC is obtained refer to unit cost as shown on **Table 5** on below.

Table 5. Unit cost of hybrid SFPV & wind turbine system

Description	Unit Cost		
	SFPV (USD / kWp)	Wind Turbine (USD / kW)	Battery (USD/kVAH)
Material	270.00	2,274.00	220.00
Grid Connection	150.00	350.00	Included
Construction	236.00	560.00	39.65
Other Cost	410.00	350.00	N/A
Land Cost	4.00	60 .00	N/A
O&M Cost	10.72	898.00	545.79

4 Result And Discussion

The results of the optimal sizing simulation are presented according to various combination power of SFPV and wind turbine with 10 scenarios as shown on **Table 4**. Increasing the capacity of wind turbine creates impact on SFPV capacity that is decreased as the compensation to meet total load requirement. Then, the LCOE is reduced when the capacity of wind turbine is increased and the capacity of SFPV is decreased. It can be understood that SFPV has both LCC and LCOE greater than wind turbine because it has greater capacity to compensate the effect from effective time of radiation solar which is only 4-5 hours per day. Then with the concept of using two equal battery banks that operated interchangeably, using typical battery bank of flooded lead acid with cycle of 2430, the battery banks has lifetime of 6.67 years and need additional 2 times replacement during project lifetime of 20 years. It makes the O&M cost of Battery banks become greats and make the LCOE of off-grid hybrid renewable power generation become expensive. As result, the hybrid system of 100kW wind turbine and 803kWp SFPV is chosen with the most minimal LCOE of USD 0.52/kWH.

Table 4. Simulation result of optimization hybrid SFPV & Wind Turbine

Scenario	Total Load (kWh/ day)	Battery Bank Capacity per Unit			Wind Turbine		SFPV		LCOE (USD/ kWh)
		V	AH	DOD	Rated (kW)	Output (kWh/day)	Rated (kWp)	Output (kWh/day)	
1	943.88	48	28,092	70%	10	76.50	1,751	1,271.89	0.616
2	943.88	48	28,092	70%	20	153.01	1,645	1,195.39	0.606
3	943.88	48	28,092	70%	30	229.51	1,540	1,118.88	0.596
4	943.88	48	28,092	70%	40	306.01	1,435	1,042.38	0.585
5	943.88	48	28,092	70%	50	382.52	1,330	965.88	0.575
6	943.88	48	28,092	70%	60	459.02	1,224	889.37	0.564
7	943.88	48	28,092	70%	70	535.52	1,119	812.87	0.553
8	943.88	48	28,092	70%	80	612.03	1,014	736.37	0.542
9	943.88	48	28,092	70%	90	688.53	908	659.86	0.531
10	943.88	48	28,092	70%	100	765.04	803	583.36	0.520

According to data of scenario chosen, we can see that the percentage of energy output (kWh/day) of both wind turbine and SFPV are 57% and 43% from total energy output and meet load requirement for 24 hours as shown on **Fig 3**.

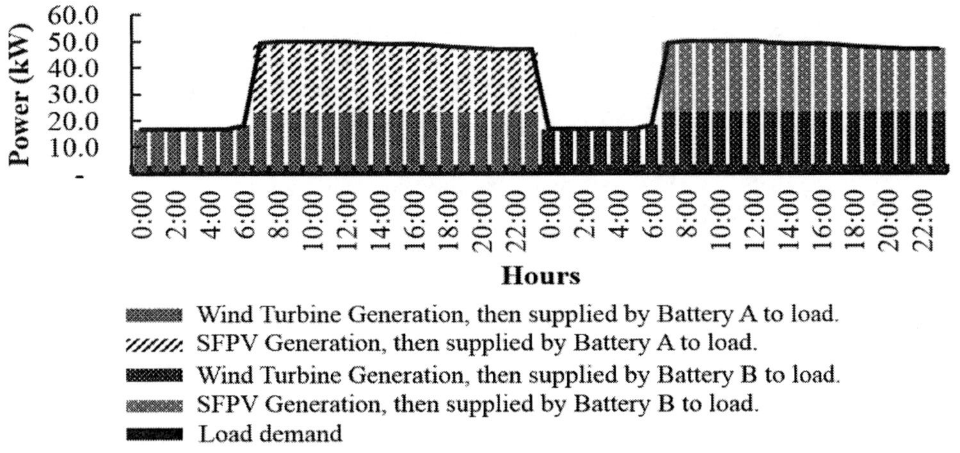

Fig. 3. Power generation curve of hybrid SFPV and wind turbine

Define ideal electricity tariff of 0.81/kWH, then interest loan (i) of 9%, the economic analysis on **Table 6** shown that hybrid SFPV and wind turbine is feasible to be developed.

Table 6. Economic analysis result of four indicators

Indicator	Requirement	Result
Payback Period (PBP)	less than project lifetime	6.56 years
Net Present Value (NPV)	> 0	USD 10,981
Internal Rate of Return (IRR)	> i	9.1%
Profitability Index (PI)	≥1	1.6

5 Conclusion

The hybrid system of 100kW wind turbine and 803kWp SFPV with off-grid scheme is feasible to be developed for powering fish cold storage in Small Kei Island during 24 hours continuously with minimum LCOE of USD 0.52/kWH. According to scenarios, PV is not recommended to be developed as standalone power generation due to have great capacity and LCOE. However, lower trend price of PV module and battery technology in the future can change the condition and the evaluation of this research results is still required.

The authors gratefully acknowledge the financial support provided by Universitas Indonesia through the 2018 PITTA funding scheme managed by the Directorate of Research and Community Engagement (DRPM UI).

References

1. Ministry of Marine Affairs and Fisheries Republic of Indonesia, "Indonesia Marine And Fisheries Business And Investment Opportunities", 2017.
2. B.Sena, Fauzun, Indarto, "Cost Estimation in Designing Cold Storage Powered Photovoltaic System (COSPV) for Two Tons of Fish in Indonesia", *TEKNOSAINS*, vol.**12**, No.2, 2016.
3. M. R. Khan and S. Iqbal, "Solar PV-diesel hybrid mini cold storage for rural Bangladesh," *2014 3rd International Conference on the Developments in Renewable Energy Technology (ICDRET)*, Dhaka, 2014, pp. 1-3.
4. R. A. Verzijlbergh and Z. Lukszo, "Conceptual model of a cold storage warehouse with PV generation in a smart grid setting," *2013 10th IEEE International Conference On Networking, Sensing And Control (ICNSC)*, Evry, 2013, pp. 889-894.
5. SWIMSOL GmbH, "Solar Energy for Tropical Islands", 2015.
6. A. Sahu, N. Yadav, K. Sudhakar, "Floating photovoltaic power plant: A review", *Renewable and Sustainable Energy Reviews*, vol. **66**, 2016, pp. 815-824.
7. Atmospheric Science Data Center. (2018, 3 February 2018). *Surface meteorology and Solar Energy - A renewable energy resource web site (release 6.0)*. Available from: https://eosweb.larc.nasa.gov
8. M. R. Ranjbar, S. Kouhi, "Sources' Response for supplying energy of a residential load in the form of on-grid hybrid systems", *International Journal of Electrical Power & Energy Systems*, vol. **64**, 2015, pp. 635-645.
9. BMKG Indonesia. (2018, 3 February 2018). *Pusat Meteorologi Maritim BMKG - Ocean Forecast System (OFS)*. Available from: http://peta-maritim.bmkg.go.id/static/
10. Lee, E., "A study of solar installation in Singapore", *Schneider Electric - Facilities Management Seminar*, 5 August 2011.
11. C. S. Lai and M. D. McCulloch, "Sizing of Stand-Alone Solar PV and Storage System With Anaerobic Digestion Biogas Power Plants," in *IEEE Transactions on Industrial Electronics*, vol. **64**, no. 3, pp. 2112-2121, March 2017.

Comparison methods of short term electrical load forecasting

Hartono[1*], *Arif* Marifa Ahmad[2], M.Sadikin[3]

Abstract. The supply of electricity that exceeds the load requirement results in the occurrence of electrical power losses. To provide the appropriate power supply to these needs, there must be a plan for the provision of electricity by making prediction or estimation of electrical load. Therefore the issue of electrical load forecasting becomes very important in the provision of efficient power. In this study, the author tries to build a model of short-term electrical load prediction using artificial neural network (ANN) with learning algorithm levenberg-marquardt (Trainlm), Bayesian regularization (Trainbr) and scaled conjugate gradient (Trainscg). Scope of research data retrieval is limited electrical load on the work area of Serang City. The results of this study show that the JST prediction of levenberg-marquardt (Trainlm) algorithm is better than the calculated prediction using Bayesian regularization (Trainbr) and scaled conjugate gradient (Trainscg) algorithms. The electric load prediction shows that the average error (Trainlm) is 3.37. Thus, it can be concluded that the electrical load prediction using the levenberg-marquardt (Trainlm) JST algorithm is more accurate than that of the Bayesian regularization (Trainbr) JST algorithm and the scaled conjugate gradient (Trainscg)

1 Introduction

The supply of electricity that exceeds the load requirement results in the occurrence of electrical power losses. Conversely, in case of supply shortage (supply) electricity, will result in the occurrence of blackouts. To provide the appropriate power supply to these needs, there must be a plan for the provision of electricity by making prediction or estimation of electrical load. Forecasting is the activity of predicting or predicting what will happen in the future. Forecasting is usually divided into two, namely qualitative meaning of forecasting based on data that is not in the form of numbers and quantitative means forecasting based on data in the form of numbers. Forecasting is done based on data or variables that have occurred previously systematically combined with a particular method and processed to obtain estimates in the future. Good forecasting will reduce unnecessary and unnecessary development risks in the supply of electricity to customers [1].

The best algorithm at the time to forecast the load and in the distribution of electrical energy is not deficient or overloaded so there is no loss on the provider of electrical energy. The reason of the use of artificial neural network (JST) is because ANN is able to learn the pattern of data or sample input or can do generalization, abstraction and extraction of

[1] Corresponding author: sallam.hartono@yahoo.com

property statistics data historical data electrical load then can make prediction or forecast of big load which will come.
Neural Network or Neural Network (ANN) is an information processing system that has characteristics resembling a biological neural network. [4] Artificial neural networks are used for electric load forecasting because it can easily formulate a very flexible experience of forecasting and knowledge. Artificial neural networks have a good approximation to nonlinearity so that it can produce better error response. [3]
Figure 2.1 is an illustration of a neuron.

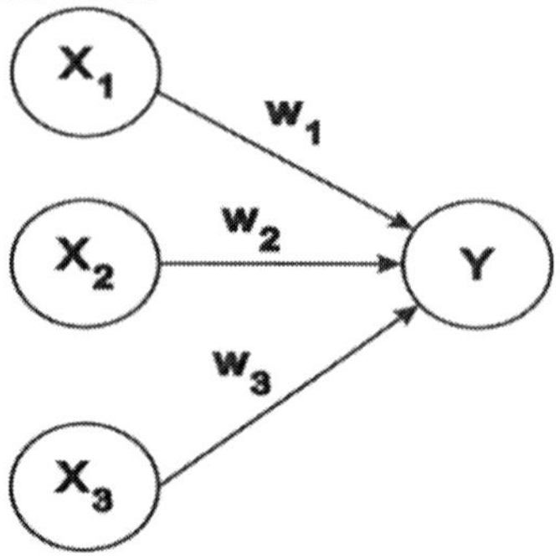

Fig. 1. Neural network 1. neuron Y artificial [4]

Y receives input from neurons x1, x2, x3 with the weights of each relationship are w1, w2, w3 the three implants of neurons present when summed will yield the equation:

$$net = x_1w_1 + x_2w_2 + x_3w_3 \quad (1)$$

the amount of input received by Y follows the activation function y = f (net). [5]

1.1 Backpropagation Learning Algorithm

Artificial neural network backpropagation was first introduced by Rumelhart, Hinton and William in 1986, then Rumelhart and Mc Clelland developed it in 1988. The backpropagation algorithm for Neural network is generally applied to multi-perceptron perceptrons. Perceptron has at least an input section, an output section and several layers that are between input and output. This middle layer, also known as hidden layers, can be one, two, three and so on. In practice, the number of hidden layer at most is three layers. With these three layers almost all the problems in the industrial world have been resolved. The output of the last layer of the hidden layer is directly used as the output of the Neural network.

1.1.1 Bayesian regularization Learning Algorithm (Trainbr)

Bayesian regularization minimizes the linear combination of quadratic errors and weights. It also modifies the linear combination so that at the end of the training, the resulting network has good generalization qualities. Training stops when one of these conditions occurs:

1. The maximum number of epochs (repetitions) is reached.
2. Maximum amount of time exceeded.
3. Performance is minimized to destination.
4. Gradient performance drops below min_grad.
5. you exceed mu_max.

1.1.2 Learning Algorithm scaled conjugate gradient (Trainscg)

The scaled conjugate gradient algorithm is a modification of backpropagation. Training stops when one of these conditions occurs:

1. The maximum number of epochs (repetitions) is reached.
2. Maximum amount of time exceeded.
3. Performance is minimized to destination.
4. Gradient performance drops below min_grad.

1.1.3 Algoritma Pembelajaran Levenbreg-Marquardt (Trainlm)

Like the quasi-newton method, the levenbreg-marquardt algorithm is designed using a second derived approximation without having to compute a hessian matrix. Newton method is very fast and accurate to get the minimum error, therefore the algorithm is expected as soon as possible to change the value to be equal to 0. For that, after some iteration, this algorithm will decrease the value of m, the increase of m value is only done when needed a step temporary) to decrease performance function. [7]

2 Research methods

The whole series for the completion of this research are as follows:

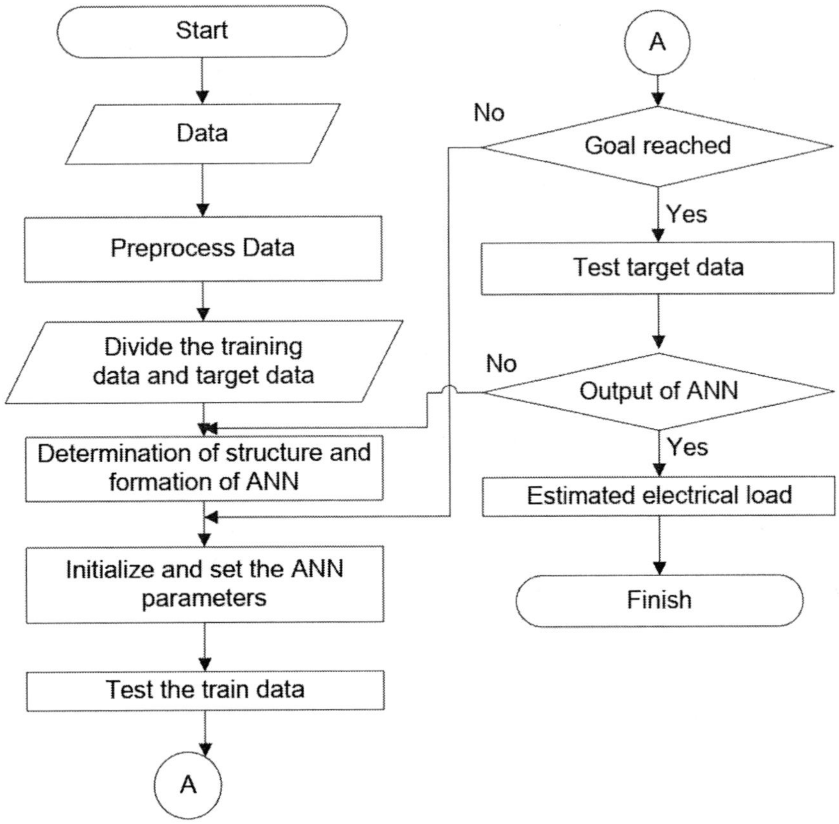

Fig. 2. Flowchart research

3 Results and discussion

The following test results can be seen in the algorithm levenberg-marquardt, Bayesian regularization and scaled conjugate gradient in the figure below:

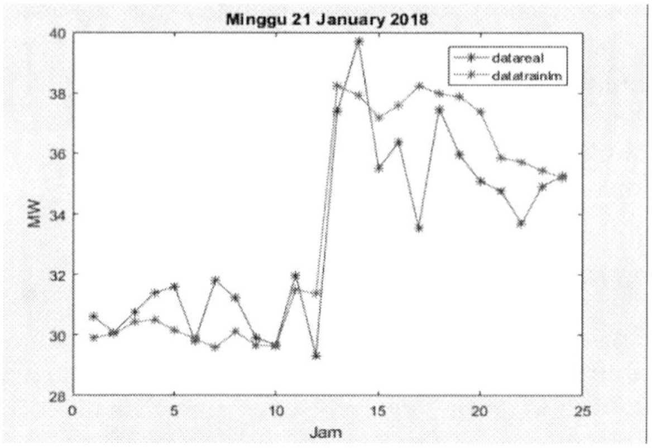

Fig. 3. Comparison of electrical trainlm expenses sunday 21 january 2018

Figure 3 shows the comparison between actual loads compared with the results of JST forecasting. Based on Figure 3.1 it is known that the largest difference between the real load and the trainlm prediction load occurs at 16.00 with a difference of -4.6888 MW minus sign (-) indicates that the predicted result is greater than the real load while the unmarked () is a plus which means the opposite the real burden is greater than the load of the predicted result, while thesmallest difference occurs at 23:00 which is a difference of 0.0538 MW.

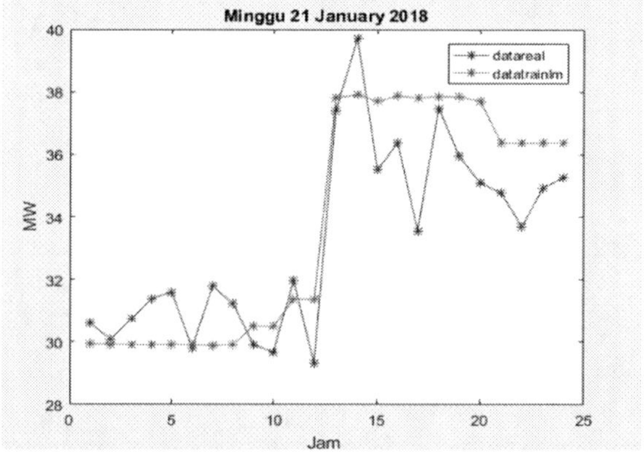

Fig. 4. Comparison of trainbr electrical weights week 21 january 2018

Based on Figure 4 it is known that the largest difference between real load and trainbred prediction load occurs at 16.00 with a difference of -4.2442 MW minus sign (-) indicates that the predicted result is greater than the real load while the unmarked () is a plus which means the opposite is the real burden is greater than the load of the predicted results, while the smallest difference occurs at 05.00 that is the difference of -0.0977 MW.

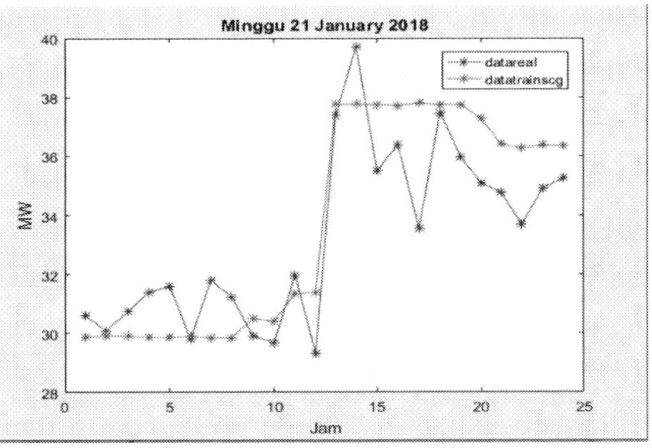

Fig. 5. Comparison of trainscg electrical expenses week 21 january 2018

Based on Figure 5 it is known that the greatest difference in real load and predicted load trainscg occurs at 16.00 with a difference of -4.2601 MW minus sign (-) indicates that the predicted result is greater than the real load while the unmarked () is a plus which means the opposite is the real burden is greater than the load of the predicted result, while the smallest difference occurs at 05.00 that is the difference of -0.1048 MW

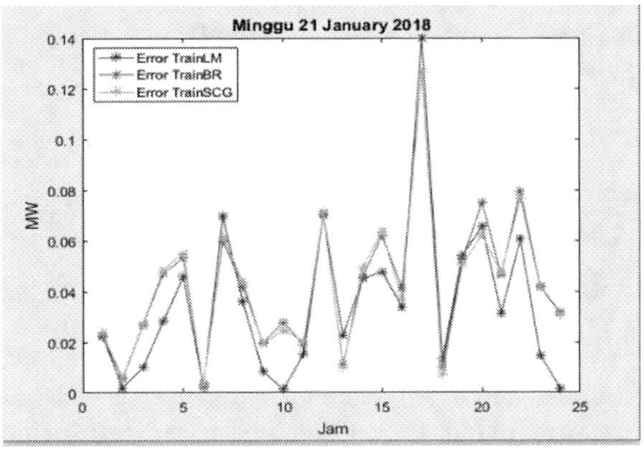

Fig. 6. Comparison of error values week 21 january 2018

Figure 6 shows the difference in load if trained in trainlm is 13.97% for the largest load at 16:00 hours while for the smallest load is 0.152% at 23:00, the trainbr is 12.64% for the largest load at 16:00 hours while for the smallest load is 0.32% at 05:00, and on trainscg is 12.69% for the largest load at 16:00 hours while for the smallest load is 0.0152% at 05:00. The average error in a day (24 hours) is the number of errors per hour divided by 24 so the results can be viewed in the following table 1.

Table 1. Comparison of average error sunday 21 january 2018

Method	ANNTrainlm	ANNTrainbr	ANNTrainscg
Average error	3,503	4,2453	4,2159

From the results of the average error in Table 1 each learning algorithm ANN on January 21, 2018. Apparently the prediction algorithm learning algorithm imitation levenberg-marquardt (trainlm) better than 2 learning algorithms Bayesian regularization (trainbr) and scaled conjugate gradient (trainscg).

The following are the daily power load estimation results of Serang City on transformer 3. Using 3 artificial neural network learning algorithms namely levenberg-marquardt (trainlm), Bayesian regularization (trainbr) and scaled conjugate gradient (trainscg compared to actual loads) real) for January 22 - January 27, 2018 presented in tabular form

Table 2. Average error comparison dated monday 22 january 2018

Method	ANN Trainlm	ANN Trainbr	ANN Trainscg
Average error	3.41%	3.49%	3.39%

In table 2 for Monday, January 22, 2018, the average error of prediction of neural network model of levenberg-marquardt (trainlm) is 3.41%, and Bayesian regularization (trainbr) 3.49% and scaled conjugate gradient (trainscg) of 3.39 %. Predicted results today show that the prediction of a scaled conjugate gradient (trainscg) JST learning algorithm is better than 2 other learning algorithms

Table 3. Average comparison of errors dated 23 tuesday january 2018

Method	ANNTrainlm	ANNTrainbr	ANNTrainscg
Average error	3.2%	3.6%	3.4%

In Table 3 for Tuesday, January 23, 2018, the average error of predicted neural network model of Levenberg-Marquardt (trainlm) is 3.29%, and Bayesian regularization (trainbr) 3.69% and scaled conjugate gradient (trainscg) of 3.46 %. Today's prediction results show that the learning algorithm of levenberg-marquardt (trainlm) JST learning is better than 2 other learning algorithms.

Table 4. Comparison average error date wednesday january 24, 2018

Method	ANNTrainlm	ANNTrainbr	ANNTrainscg
Average error	3.43%	3.98%	3.96%

In Table 4 for Wednesday, January 24, 2018, the average error of prediction of neural network model of Levenberg-Marquardt (trainlm) is 3.43%, and Bayesian regularization (trainbr) 3.98% and scaled conjugate gradient (trainscg) is 3.96 %. Today's prediction results show that the learning algorithm of levenberg-marquardt (trainlm) JST learning is better than 2 other learning algorithms.

Table 5. Comparison average error of 25 thursday january 2018

Method	ANNTrainlm	ANNTrainbr	ANNTrainscg
Average error	3.24%	3.94%	3.25%

Table 5 for Thursday, January 25, 2018, the average error of prediction of neural network model of levenberg-marquardt (trainlm) is 3.24%, and Bayesian regularization (trainbr) 3.94% and scaled conjugate gradient (trainscg) of 3.25% . Today's prediction results show that the learning algorithm of levenberg-marquardt (trainlm) JST learning is better than 2 other learning algorithms

Table 6. Average comparison of errors dated 26th of january 2018

Method	ANNTrainlm	ANNTrainbr	ANNTrainscg
Average error	3.12%	4.5%	4.24%

Table 6 for Friday, January 26, 2018, average error of prediction of neural network model of levenberg-marquardt (trainlm) is 3.14%, and Bayesian regularization (trainbr) 4.5% and scaled conjugate gradient (trainscg) of 4.25%. Today's prediction results show that the learning algorithm of levenberg-marquardt (trainlm) JST learning is better than 2 other learning algorithms

Table 7. Comparison average error of 27 saturday 2018

Method	ANNTrainlm	ANNTrainbr	ANNTrainscg
Average error	3.56%	5.09%	5.14%

In Table 7 for Saturday, January 27, 2018, the average error of prediction of neural network model of levenberg-marquardt (trainlm) is 3.564%, and Bayesian regularization (trainbr) 5.09% and scaled conjugate gradient (trainscg) of 5.14% . Today's prediction results show that the learning algorithm of levenberg-marquardt (trainlm) JST learning is better than 2 other learning algorithms

4 Conclusion

1. The result of electrical power forecast for one week starting from January 21, 2018 to January 27, 2018, gives the following results:
   a. The results of learning algorithm of neural network of levenberg-marquardt (trainlm) is better for January 21, January 23, January 24, January 25, January 26 and January 27, and the predicted result of scaled conjugate gradient (trainscg) is better only for the 22nd January.
   b. Average error of prediction of neural network for one week (trainlm) 3,37%, trainbr 4.13% and train scg 3.95%
2. predictions using three learning methods of learning algorithm of learning algorithm of artificial neural network levenberg-marquardt (trainlm), Bayesian regularization (trainbr) and scaled conjugate gradient (trainscg) have quite accurate results. But trainlm get results faster when simulating using MATLAB.

Reference

1. D. A. Sari, "Peramalan Kebutuhan Beban Jangka Pendek Menggunakan Jaringan Syaraf Tiruan Backpropagation," *Universitas Diponegoro*, Semarang, (2007).
2. Fausett L. *Fundamentals Of Neural Networks Architectures, Algorithmsand Applications* , London: Prantice-Hall, Inc (1994).
3. I. Handayani, "Peramalan Beban Puncak Listrik Jangka Pendek Menggunakan Metode Jaringan Syaraf Tiruan," *J. Sist. Kendali Tenaga Elektron. Telekomun. Komput.*, Vol. 1, No. 4, Pp. 2337–439, (2013).
4. Lee Ky, Park Jh. Short Term Load Forecasting Using An Artificial Neural Network, Ieee (1992)
5. Marsudi D. *Operasi Sistem Tenaga Listrik*, Yogyakarta: Penerbit Graha Ilmu (2006)
6. R. Dwisatya, M. R. Kirom, M. Si, A. G. Abdullah, And M. Si, "Prediksi Beban Listrik Jangka Pendek Berbasis Algoritma Feed Forward Back Propagation Dengan Mempertimbangkan Variasi Tipe Hari," *Universitas Telkom*, Vol. 2, No. 3, Bandung, Pp. 7315–7322, (2014)
7. S. Ruliah And R. Rolyadely, "Prediksi Pemakaian Listrik Dengan Pendekatan Back Propagation," *Sekolah Tinggi Manajemen Informatika Dan Komputer Banjarbaru*, Vol. 3, No. 1, Kalimantan Selatan, P. 466, (2014)
8. Triwulan Yayu, Nasrun Hariyanto And Sabat Anwari, "Peramalan Beban Puncak

Listrik Jangka Pendek Menggunakan Metode Jaringan Syaraf Tiruan", Institut Teknologi Nasional, Vol. **1**, No 4

9. Misra Soumya, Rajasree Prusty And Prados Kumar Huta. " Analysis Of Levenberg-Marquardt And Scaled Conjugate Gradient Training Algorithms For Artificial Neural Network Based Ls And Mmse Estimated Channel Equalizers", (Ieee 2015 International Conference On Man And Machine Interfacing (Mami))
10. Sahay Bhushan Kishan, Sunet Sahuu And Pragya Singh " Short-Term Load Forecasting Of Toronto Canada By Using Different Ann Algorithms" , (978-1-5090-0128-6/16/$31.00 ©2016 Ieee)
11. Sujito, "Perhitungan Life Time Transformator Jaringan Distribusi 20 Kv Di Apj Malang," *Univeritas Negeri Malang*, Malang, Pp. 63–75 (2009)
12. H. G. Nugraha And Azhari, "Pelatihan Bobot Jaringan Syaraf Tiruan Menggunakan Particle Swarm Optimization Untuk Peramalan Tingkat Inflasi," *Universitas Gajah Mada*, Yogyakarta, Pp. 292–302 (2014)

Power quality improvement using dynamic voltage restorer in distribution system PT. DSS Power Plant

Wahyuni Martiningsih[1,*], *Untung* Yudho Prakoso[1] , and *Herudin*[1]

[1]Departement Electrical Engineering, Universitas Sultan Ageng Tirtayasa, Banten, Indonesia

Abstract. In the power plant power system, voltage drops often occur and one of the contributing factors is a short circuit that affects the performance of motor drivers on the mills in the generator. In this research proposed the use of compensation dynamic voltage restorer (DVR) to damped voltage sag. The DVR using a three-phase voltage source inverter (VSI) with voltage loop control (PI). To detect the voltage drop system using park transform and apply it in the voltage regulator as a control system function that detects the voltage amplitude at the sensitive load continuously. The result of simulation in system distribution with active load 280 kVA and 50 kVAR reactive, before a voltage compensation system 0.67 pu and after voltage compensation system becomes 0.99 pu on nominal voltage Of 380V. The performance of DVR is reliable because it is able to compensate quickly to changes in the value of the voltage

1 Introduction

The quality of electric power becomes an important issue because the electricity that is channeled to the consumer must be really good quality. Disturbances to the electrical system can occur at any time and can't be avoided. The disturbances can come from both internal and external of the power plant system and can cause the problem of voltage drop. Voltage drop can impact the power quality of the power system.

Power quality problems are: voltage drop, flicker, voltage imbalance, disconnection and harmonic problem [1][2]. Sensitive equipment such as computers, releases, programmable logic controllers (PLC), electric motor drivers and so on, are extremely sensitive to voltage changes caused by interference with other parts of the system[3]. The voltage drop is defined as a decrease in the rms value of the voltage that can occur from 10 ms to 1 min with the depth of the falling voltage of 0.9 pu to 0.1 pu of a nominal pu based on the IEEE standard 1159-1995 [4][5].

In the electrical system of PT. DSS power plant, often there is a decrease of voltage and one of the causal factor is a short circuit that resulted in the performance of motor drivers on the mills in generator unit 6. In this research will be modeling three-phase system that experienced a decrease in voltage, from the results modeling will be analyzed the voltage drop recovery that occurs with the sensitive load to be protected by the DVR from the

* Corresponding author: y_martiningsih@untirta.ac.id

effect of the voltage drop, so that a large pattern of DC PWM inverter voltage required by the DVR in accordance with the needs of sensitive load to be protected.

1.1 DVR

DVR is a power electronic switching device which is connected in series to the load voltage bus to inject a dynamically controlled voltage. This voltage can eliminate effects of fault of voltage bus on a sensitive load [6].

DVR is equipment used to recover a voltage or improve the voltage quality on the load side and its position is mounted in series between the source and the load. DVRs are coupled in series with distribution systems to protect sensitive equipment against the occurrence of voltage drops [7],[8]. The basic function of the DVR is to detect the occurrence of voltage drops that occur on the power system channel, and then inject the voltage to compensate for the voltage drop that occurs. Therefore the DVR is placed close to the sensitive load that is protected.

The DVR works depending on the type of interference or an event occurring in the system, generating the injected voltage obtained from the DC energy storage unit and then converted to AC voltage by the voltage source inverter (VSI) [9]. To set the controller on the DVR is used dq0 transformation or Park transformation. The dq0 method will provide information on the depth of the voltage drop and the phase shift with the starting point and end point of the voltage drop. The transformation process from three phase system to dq0 system is shows at Equation 1[6].

$$\begin{bmatrix} V_d \\ V_q \\ V_0 \end{bmatrix} = \frac{2}{3} \begin{bmatrix} \cos(\omega t) & \cos(\omega t - 120) & \cos(\omega t + 120) \\ -\sin(\omega t) & -\sin(\omega t - 120) & -\sin(\omega t + 120) \\ \frac{1}{2} & \frac{1}{2} & \frac{1}{2} \end{bmatrix} \begin{bmatrix} V_a \\ V_b \\ V_c \end{bmatrix} \quad (1)$$

The DC source is a very important element for the DVR. The DC source will be converted to AC voltage by the inverter. The magnitude of the DC source will affect the magnitude of the output voltage as described in equation 3

$$M_a = \frac{A_r}{A_c} \quad (2)$$

$$V_{DC} = \frac{V_{LL}}{0{,}707\, Ma} \quad (3)$$

where:

M_a = amplitudee

A_r = amplitude reference

A_c = amplitude of carrier

V_{DC} = voltage source VSI

V_{LL} = rms voltage line to line

2 Metodology

This research was conducted on a distribution channel of PT. DSS Power Plant. The research process of simulating the recovery of drop voltage can be explained through the block diagram in Figure 1.

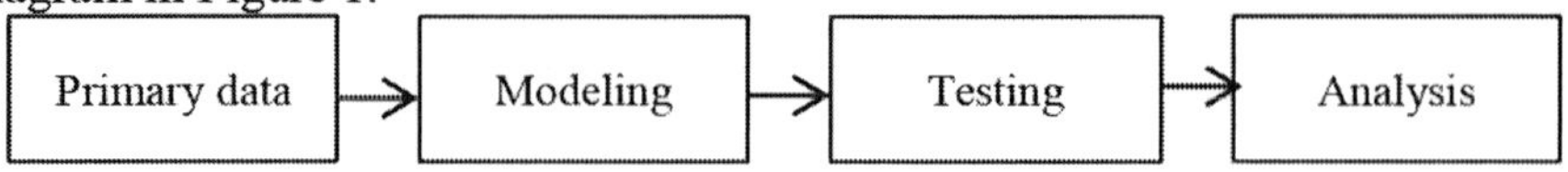

Fig. 1. Block diagram of research

Explanation of the steps of the block diagram in Figure 1:

1. Primary data, carried out at the beginning of the study which aims to support of the research, while the data taken include voltage, current, power, frequency, and transformer specifications.
2. Modeling, that is making the design of system design as a whole.
3. Testing the simulation is to ensure that the designed DVR can work properly.
4. Analysis of the results of the simulation, data which obtained of simulation and then analyzed so that conclusions achieved.

2.1 Parameter Model

The parameters used in the DVR model for the distribution system of PT. DSS Serang Power Plant as shown in Table 1.

Table 1. Parameter of modeling

Parameter	Nilai
Source voltage	20 kV
Transformer Distribution	2500 kVA
Frequency	50 Hz
Primary voltage	20 kV
Primary winding resistance	2.16 Ω
Primary winding inductance	8.6 mH
Secondary voltage	380V
Secondary winding resistance	0.02 Ω
Secondary winding inductance	0.08 mH
Magnetization resistance and reactance	577 Ω
P	280 kVA
Power of Inductive load	40 kVAR
Power of Capasitive load	10 kVAR

Modeling a voltage source uses a three phase system controlled voltage source. The source voltage is delivered to the load through a distribution transformer of 20 kV / 380 V. The load used in the simulation has a capacity of 280 kVA and 50 kVAR.

2.2 Research Design

In this study, DVR system modelling consists of several block diagrams, including load systems, controls, inverter voltage sources, filters, and injection transformers. The research design of the DVR can be seen in Figure 2.

The inverter used is a unit of three-phase inverter circuit with the ignition switch using the SPWM method. The output voltage generated by the inverter depends on the modulation index of the SPWM and the DC source voltage rating. The inverter converts DC voltage to AC voltage. Determination of DC source capacities can be seen in equation 3. Because the rating of the primary voltage on the injection transformer is 380 volt rms and if it is known the maximum M_a is 0.8, then the required DC source is 672V.

Passive filters at Figure 3. are composed of capacitors and inductors with a frequency that is set to the harmonic voltage frequency to be removed. Inductance and capacitance in this filter are 1 mH and 1 μF respectively.

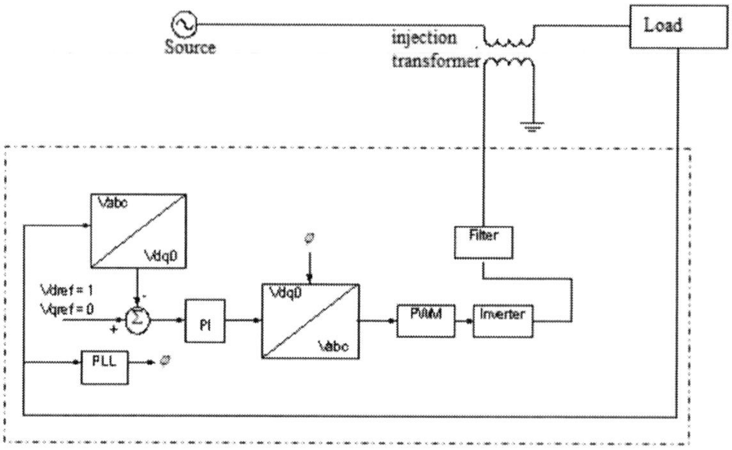

Fig.2. Block diagram of DVR design

SPWM is a comparison circuit between 3 phase sinusoidal signal and carrier signal. If a sinusoidal signal > signals the carrier, it will generate a pulse 1 and if the sinusoidal signal < carrier signal, it will generate a pulse 0. The pulse is by this SPWM which will activate the inverter. The carrier signal amplitude in this simulation is set to 1.25.

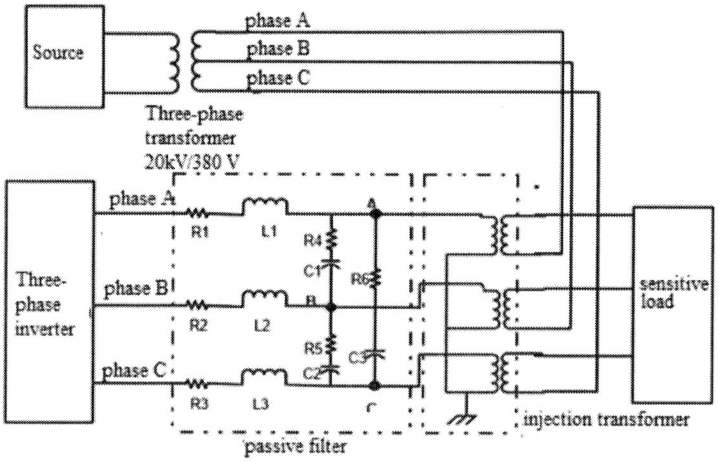

Fig.3. Design of three phase LC filter circuit

3 Results and Discussion

In this study, the distribution system of PT DSS power plant installed by the DVR is as shown in Figure 4. The DVR conneted in series with distribution system as compensator voltage sag that occurs.

The control system in this study serves to detect voltage disturbances. Load voltage is measured continuously which is then compared with the reference voltage using voltage loop control (PI). This controller has two constants, namely PC (Proportional Constant) and IC (Integral Constant). PC and IC are tuned until they reach the desired response.

The simulation design is modeled in the form of a distribution system of 20 kV/380 V. When there is no one phase disturbance to the ground on a three phase supply voltage feeder (V_A, V_B, and V_C) in a stable condition.

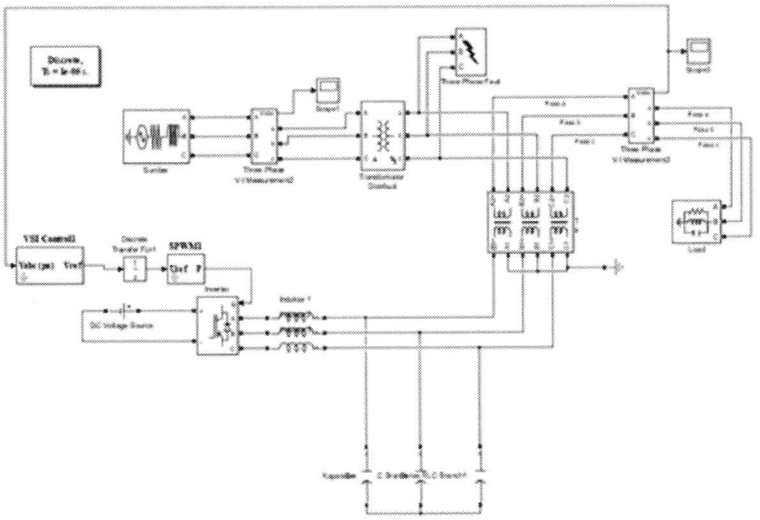

Fig. 4. System distribution with DVR

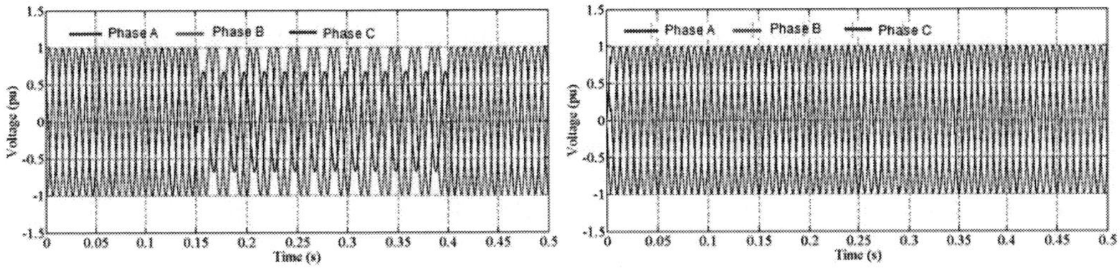

(a)

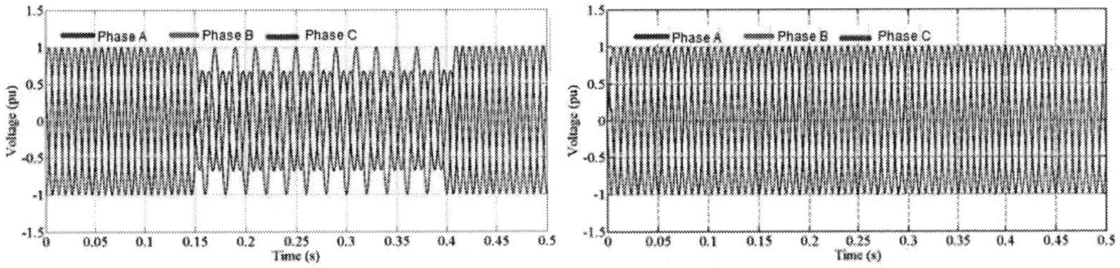

(b)

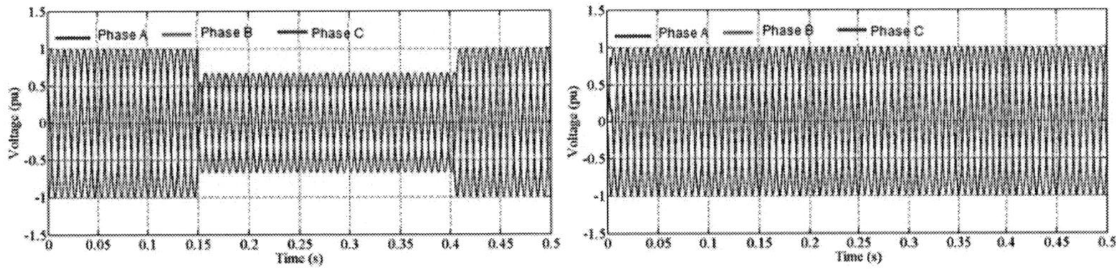

(c)

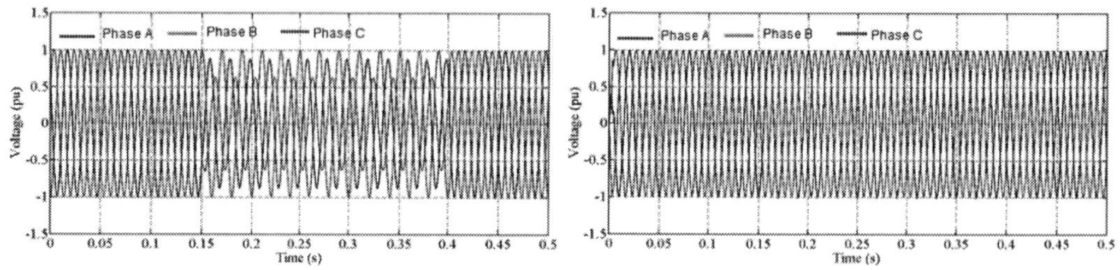

(d)

Fig.5. Simulation results with disturbances (a) single phase to the ground, (b) two phase to ground (c) three phase to ground and (d) phase to phase

Figure 5 (a) on the left is a simulation on a distribution system that experiences a single phase disturbance to the ground. There is voltage drop in the phase (0.6779 pu) which disturbanced. The voltage drop is handled by the DVR with injecting voltage, and then occurred a recovery of power quality in the system like shown Fig.5.(a)on the right.

Figure 5 (b) the disturbance that occurs is a two-phase disturbance to the ground (phase A and phase B). There was a voltage decrease in phase A (0.6786 pu) and phase B (0.6645 pu). The installed DVR can cause recovery of voltage stability to 0.9954 pu.

Three-phase disturbance that occurs in the distribution system causes a voltage drop in all three phases. In this case there is a voltage drop in phase A to 0.6714, phase B (0.6698 pu) and phase C (0.6707 pu). The DVR can handled that disturbance and recovery the voltage at 0.9967 pu, like seen at Figure 5 (c).

Figure 5. (d) the last simulation is a simulation for phase-phase disturbance. Disturbance occurs between phase A and phase B. The voltage on phase A drops to 0.8778 pu and phase B drops to 0.6318 pu. After the DVR is installed on distribution system, the voltage recovery occurs at the point 0.9921 pu.

4 Conclusion

The simulation results show that the proposed DVR is capable of repairing power quality interference. The DVR control block will detect the disturbance of voltage that occurs and the DVR functions as a compensator. PT. DSS Power Plant often occurred short circuit fault. DVR works to inject voltage when the systems undergo voltage sag. DVR is capable of repairing voltage sag that occurs from 0.67 pu to 0.99 pu with recovery time of 0.25 seconds.

References

1. K. Sandhya, A. J. Laxmi, and M. P. Soni, "Design of PI and Fuzzy Controllers for Dynamic Voltage Restorer (DVR)," *AASRI Procedia*, vol. **2**, pp. 149–155, (2012).
2. D. Tien, R. Gono, and Z. Leonowicz, "A Multifunctional Dynamic Voltage Restorer for Power Quality Improvement," *Energies*, vol. **11**, no. 6, p. 1351, (2018).
3. K. Bhumkittipich and N. Mithulananthan, "Performance enhancement of DVR for mitigating voltage sag/swell using vector control strategy," *Energy Procedia*, vol. **9**, pp. 366–379, (2011).
4. S. Khalid and B. Dwivedi, "Power Quality Issues , Problems , Standards & Their Effects in Industry with Corrective Means," *Int. J. Adv. Eng. Technol.*, vol. **1**, no. 2, pp. 1–11, (2011).
5. S. P. Mishra, B. Biswal, J. P. Roselyn, and D. Devaraj, "Simulation and analysis of DVR for mitigating voltage sags and swells," *Procedia Eng.*, vol. **64**, pp. 341–350, (2013).
6. A. M. Saeed, S. H. E. Abdel Aleem, A. M. Ibrahim, M. E. Balci, and E. E. A. El-Zahab, "Power conditioning using dynamic voltage restorers under different voltage sag types," *J. Adv. Res.*, vol. **7**, no. 1, pp. 95–103, (2016).

7. R. Ibrahim, A. M. A. Haidar, M. Zahim, and H. Iu, "The effect of DVR location for enhancing voltage sag," *2010 Int. Conf. Intell. Adv. Syst. ICIAS 2010*, pp. 1–4, (2010).
8. A. J. Laxmi, K. U. Rao, M. Sushama, and N. T. Devi, "Hardware Implementation Of Single Phase Dynamic Voltage Restorer," pp. 204–209, (2010).
9. C. Benachaiba and B. Ferdi, "Power Quality Improvement Using DVR," *Am. J. Appl. Sci.*, vol. **6**, no. 3, pp. 396–400, Mar. (2009).

Control design variable speed drive three-phase induction motor based cascaded H-Bridge seven level

Muhamad Otong[1,*], Ceri Ahendyarti[2]

[1]Universitas Sultan Ageng Tirtayasa, Power Systems Department, 42435 Banten, Indonesia
[2]Universitas Sultan Ageng Tirtayasa, Instrumentation Department, 42435 Banten, Indonesia

Abstract. The design variable speed drive induction motor is based on the power supply system of cascaded H-bridge seven level inverter. The variable speed regulation of the induction motor is performed by controlling the variable magnitude and output voltage frequency of the CHB circuit. The output voltage of the CHB circuit is generated by synthesizing the reference voltage signal derived from the field oriented control algorithm model of the induction motor. Testing and circuit operation mode using switching phase-switch technique on variable speed drive system of induction motor is done to know system performance in transient and steady-state by using computer software. The response of the simulation shows the conformity of the proposed concept. From this research is expected to improve the quality and capacity of output power in modular way so that it can be applied to the electric drive in general.

1 Introduction

Controlling the induction motor speed variable requires a controlled DC to AC voltage converter circuit or also called the "inverter" [1] - [3]. The form of cascaded H-bridge (CHB) multilevel inverter is a power electronics converter circuit that is very popular and suitable for applications that require high power. The advantage of using the CHB multilevel inverter circuit topology is to reduce harmonic distortion, reduce electromagnetic interference, and can be manufactured in a modular form [3] - [5]. The expected goal in this paper is to design a controlled AC voltage source that can be used to supply and control the speed of the induction motor and design an induction motor variable speed control algorithm using the field oriented control (FOC) method.

2 CHB Multilevel inverter circuit topology

As the name implies, the CHB multilevel inverter circuit is formed by connecting the basic module block of the H-bridge inverter in series or multilevel to get a higher output voltage and also to reduce harmonic defects. Figure 1 is an example of a topology of a 5-

* Corresponding author: muhamadotonguntirta@yahoo.com

level multilevel CHB series inverter consisting of 2 block modules of the H-bridge inverter circuit and supplied by an isolated DC voltage E. In figure 1, the multilevel inverter CHB circuit can produce 5- voltage level in each phase. When the switch S_{11}, S_{21}, S_{12} , and S_{22} is on conduction, then the output voltage of the inverter for the H1 module and the H2 module is $v_{H1} = v_{H2} = E$, and the phase voltage phase output CHB multilevel inverter is $v_{AN} = v_{H1} + v_{H2} = 2E$. In the same way, when the switch S_{31}, S_{41}, S_{32} and S_{42} is on conduction, , then the output voltage of the inverter for the H1 module and the H2 module is $v_{AN} = v_{H1} + v_{H2} = -2E$.

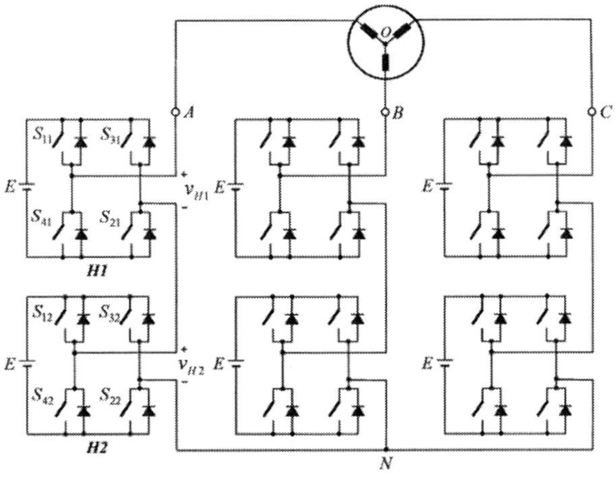

Fig. 1. CHB multilevel 5-level inverter circuit.

In general, the number of voltage levels generated by the multilevel inverter CHB circuit output can be formulated as follows: $m = 2H + 1$

With H is the number of H-bridge modules per phase, and m is the number of voltage levels generated by the multilevel inverter CHB circuit whose value is always in the form of an odd number

3 CHB Multilevel inverter control using phase-shift SPWM

The basic principle of the Phase-shifted SPWM (sinusoidal pulse width modulation) scheme in a CHB multilevel inverter circuit with m-level voltage can be developed using a unipolar modulation technique that requires a number of (m-1) carrier triangle waves with the same magnitude and frequency but having the difference-phase angle between 2 pieces of carrier wave adjacent triangle is shifted (phase-shifted) by : $\phi_{cr} = \frac{360^0}{(m-1)}$.

The signal to be modulated is usually a three-phase sinusoidal wave that can be adjusted either magnitude or frequency or both. The gate signal generated is obtained by comparing the signal to be modulated with the triangular carrier signal.

Figure 2 is an example of a voltage output waveform in a 7-level multilevel CHB circuit generated by using the principle of the Phase-shifted SPWM scheme which has 6 carrier waves and a different phase of 600 (only phase-A is shown).

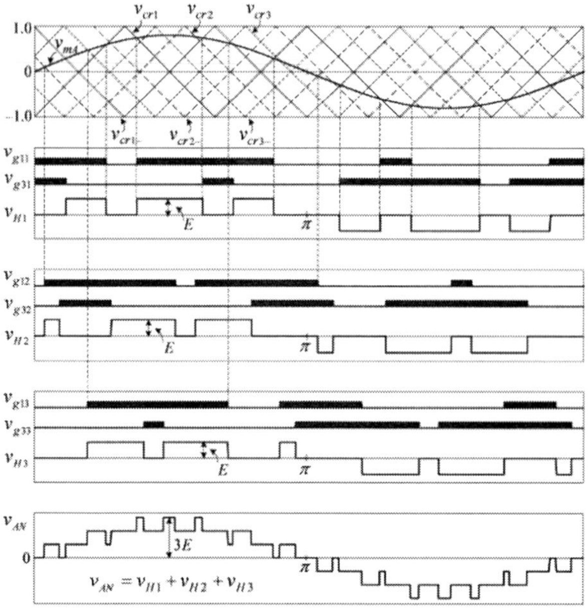

Fig. 2. Seven level multilevel CHB voltage output form uses Phase-shifted SPWM.

4 Dynamic modelling of induction motor

Making an induction motor modelling is the first step that must be done in designing a control system. Equation (1) through equation (6) is a dynamic model of an induction motor given in the form of a differential equation using the transformation of the coordinate axis dq (Park Transformation) which is placed parallel to the direction of the rotor magnetic field flux (rotor flux field oriented) [1], [4].

$$T_e = \frac{3}{2} Z_p \frac{L_h}{L_r} \Psi_{rd}(t)\, i_{sq}(t) \tag{1}$$

$$\Psi_{rd}(t) + \tau_r \frac{d\Psi_{rd}(t)}{dt} = L_h\, i_{sd}(t) \tag{2}$$

$$\omega_s(t) = \omega_e(t) + \frac{L_h i_{sq}(t)}{\tau_r \Psi_{rd}(t)} \tag{3}$$

$$\frac{d\omega_m(t)}{dt} = -\frac{f_d}{J_m}\omega_m(t) + \frac{3}{2}\frac{Z_p L_h}{L_r J_m}\Psi_{rd}(t) i_{sq}(t) - \frac{T_L}{J_m} \tag{4}$$

$$i_{sq}(t) + \tau_\sigma' \frac{di_{sq}(t)}{dt} = -\omega_s(t)\,\tau_\sigma'\, i_{sd}(t) - \frac{k_r}{r_\sigma}\Psi_{rd}(t) + \frac{1}{r_\sigma} u_{sq}(t) \tag{5}$$

$$i_{sd}(t) + \tau_\sigma' \frac{di_{sd}(t)}{dt} = \omega_s(t)\,\tau_\sigma'\, i_{sq}(t) + \frac{k_r}{r_\sigma \tau_r}\Psi_{rd}(t) + \frac{1}{r_\sigma} u_{sd}(t) \tag{6}$$

5 Design induction motor speed control systems

From the point of view of the control system design, it can be seen in the dynamic equation (5) and (6) that the variables that are manipulated to formulate the problem of the induction motor control system are determining the stator voltage signal $u_{sd}(t)$ and

$u_{sq}(t)$, in this case it is a reference voltage signal to the input signal to the PWM inverter. Therefore, to determine and manipulate the stator voltage signal. Therefore, to determine and manipulate the stator voltage signal $u_{sd}(t)$ and $u_{sq}(t)$ then we need a current controller which functions to regulate the amount of stator current $i_{sd}(t)$ dan $i_{sq}(t)$ in accordance with the motor operating mode as shown in Figure 3.

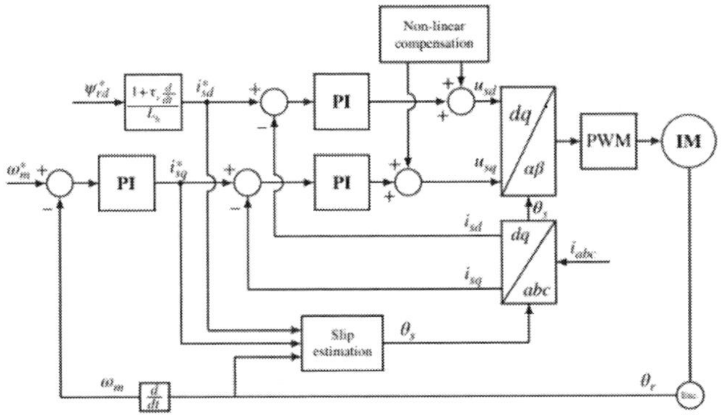

Fig. 3. Block diagram of the induction motor speed control system

Because the purpose of the induction motor control system design that is desirable is to regulate its speed, it can be seen also in equations (4) and (6) that the goal will be achieved by manipulating and controlling the current $i_{sq}(t)$ and maintain the magnitude of the rotor flux $\Psi_{rd}(t)$ constantly at nominal value.

6 Simulation and analysis

Table 1 are some of the system parameter data needed for simulation on an induction motor variable speed drive (VSD) system.

Figure 5 is the response of the actual output speed of the induction motor rotor. Seen in Figure 5 when t = 1 second the speed is inputted with the step 150 rad / second function, and the system responds with a speed that continues to rise linearly from the starting point, then the speed reaches a steady-state response of 150 rad / seconds at t = 1.63 seconds.

Table 1. Data of 3 phase induction motor and CHB seven level inverter

Power (Hp)	HP = 50		Cascaded H-*bridge*	n = 3
Voltage (volt)	Vll = 460		Level inverter m	2.n + 1 = 7 *level*
Frequency (Herz)	f = 60 Hz		Switch Type	IGBT and Diode
Pole	z_p = 2		Switching frequency	f_c = 12 KHz
Inertia (Kg.m^2)	J = 0,4		DC Bus H- *bridge*	E = 125,2 Volt
Friction (N.m.s)	F = 0,02187			
Stator resistance Ohm	r_s = 0,09961			
Stator inductance (H)	l_s = 0,031257			
Rotor resistance Ohm	r_r = 0,05837			
Rotor inductance Henry	l_r = 0,031257			
Mutual induktansi Henry	l_h = 0,03039			

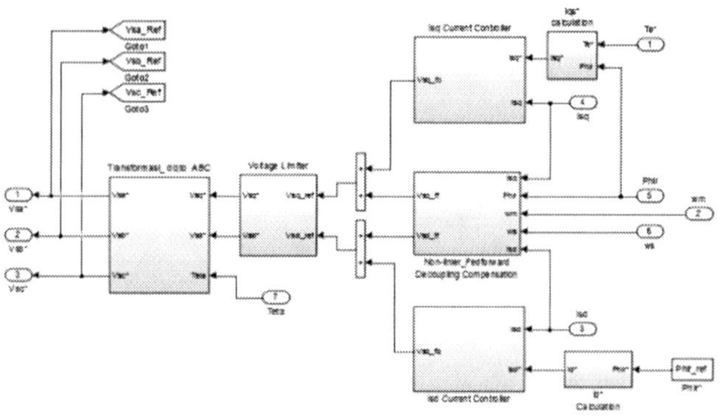

Fig. 4. Block variable speed drive induction motor system

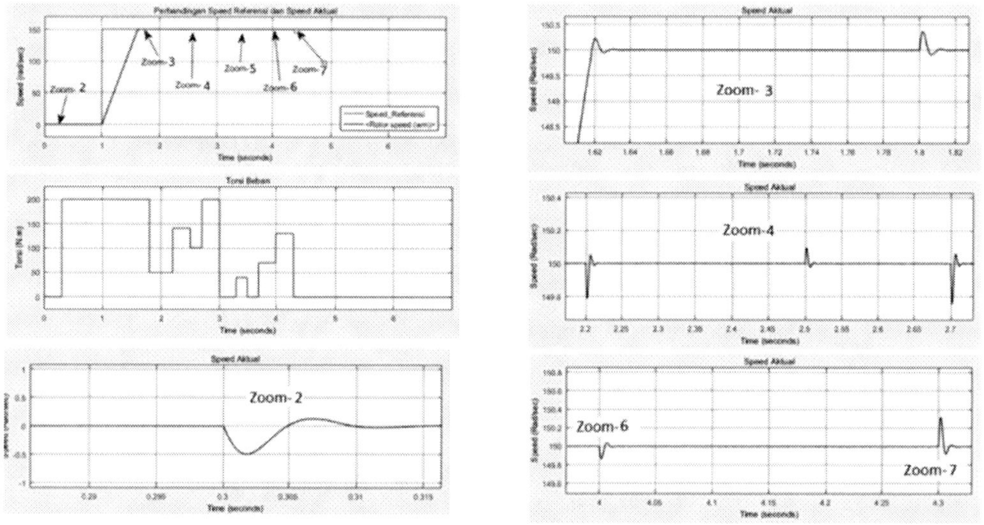

Fig. 5. Speed responds with load torque disturbance.

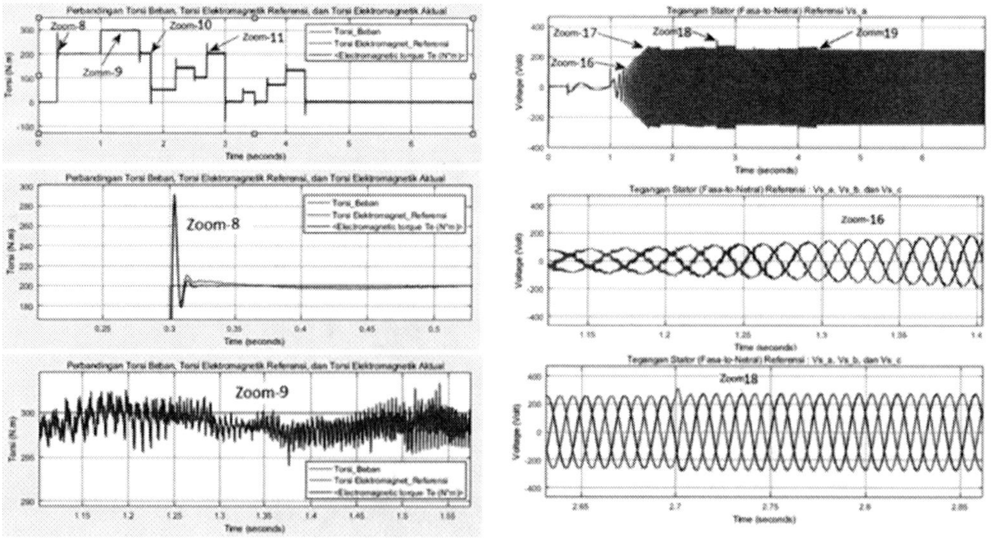

Fig. 6. Respond electromagnetic torque, and synthesizing the reference stator voltage signal.

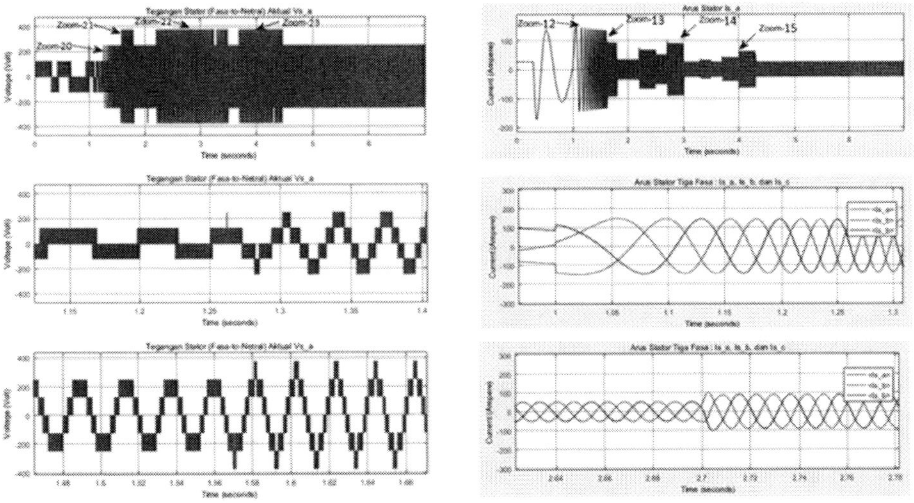

Fig. 7. Output voltage from CHB seven level inverter, and respond stator current

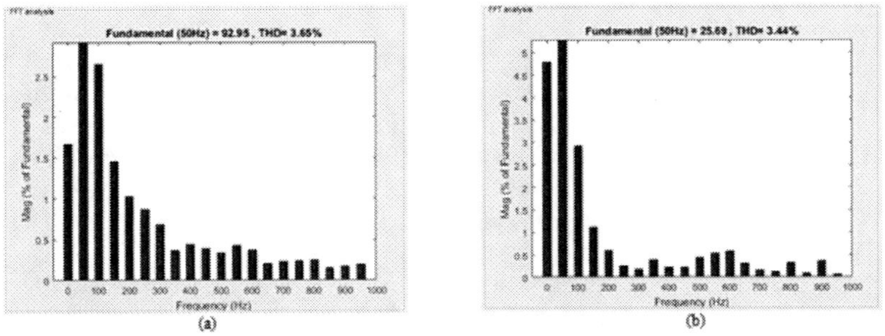

Fig. 7. (a) Harmonic spectrum of stator current at 1.7 sec, speed 150 rad/sec, load torque 200 N.m.
(b) Harmonic spectrum of stator current at 6.0 sec, speed 150 rad/sec, no load torque.

7 Conclusion

From the results simulation and analysis of the discussion in Section 6, it can be concluded that the design using the CHB multilevel inverter circuit topology has fulfilled the requirements to be used as a controlled voltage sources in supplying induction motor loads. This is indicated by the occurrence of a process of improving the quality of power with the form of a stator current that is almost sinusoidal by producing total harmonic distortion (THD) of 3.65%. The simulation results also show that, the FOC induction motor algorithm can control and track the rotational speed of the induction motor rotor according to the desired reference speed signal by producing a speed overshoot of 0.17% in 0.62 seconds and the system reaches a steady state-state within 0.02 seconds after transient occurs.

References

1. I.G. Christopher Raj, P. Renuga, M. Arul Prasanna, IJEPE, **3**, (2010)
2. J. Gholinezhad, R. Noroozian, JEET, **8**, 304-315, (2013)
3. S. Vazquez, J. I. Leon, L. G. Franquelo, J. J. Padilla, J. M. Carrasco, IEEE, **56**, (2009)
4. A. Mishra, P. Choudhary, IJETAE, **2**, (2012)

Stability issues in presence variable distributed generation into radial distribution network

Yuli Asmi Rahman[1]*, *Agus* Siswanto[2] , and *Irwan* Mahmudi[1]

[1]Electrical Engineering Department, Tadulako University, 94119 Palu, Indonesia
[2]Electrical Engineering Department, 17 Agustus 1945 University, 45135 Cirebon, Indonesia

Abstract. Related to environmental issues resulting from the use of traditional energy sources, drive usage of renewable energy is increasing. Changes in the structure of the network will certainly affect the changes in voltage stability. In this paper, discuss the impact of the stability after distributed generation penetration whose its output intermittent relatively. The simulation based PSAT software and tested into IEEE 30 bus system. Observation of voltage deviation and SVSI-index on some load buses in conditions before and after integration of wind generation. Load condition is a very determining factor of bus voltage stability index. This relates to the active power and reactive power needs that must be injected by DG penetration.

1 Introduction

An increase in the demand for loads in electrical power systems is a problem faced mainly in direct distribution networks with consumers. The need for additional power generation capacity, progress in power deregulation and utility restructuring and tight constraints on the construction of new transmission lines encourage interest in Distributed Generation (DG).

DG can be derived from both conventional and renewable energy sources. Currently, promising renewable energy sources for DG applications are photovoltaic (PV), micro-hydro, wind energy conversion system, geothermal energy, and so on. DG has a generating power range of 1-300 MW [1]. DG placement of pair size-location properly of the impact on the reduction of operating loss and operating cost at peak load, improvement of voltage profile and load factor, and increased reliability and stability of the system [2, 3]. Heavy demands and changed system configuration include integration of DG, and they were taken the role in system stability. Stability analysis is necessary to be pondered in the expansion and operation of electrical power systems.

In this study, to observe the impact of stability resulting from the integration of DG renewable energy, the study conducted by wind generation with different level penetration, DG location and load fluctuation. Stability analysis in this paper is done by using a static analysis approach. Various indicators for analyzing voltage stability in the literature are divided into three main groups [4]: line VSIs, bus VSIs, and VSIs overalls. In this paper,

[1] Corresponding author: asmi_yuli81@yahoo.co.id

voltage stability margin (VSM) and Simplified Voltage Stability Index (SVSI) indicators. VSM use to identify the weak bus as location DG placement because it calculates the weakest load bus of the power system properly [5]. The simplicity of SFVSI formula[6], indeed, computation only need voltage bus magnitude.

The proposed method was implemented in the IEEE 30 bus radial system using PSAT software. PSAT software display the bus voltage profile information before and after DG placement, as a result, optimal power load flow.

2 Voltage stability assesment

The ability of an electric power system to remain in equilibrium condition both during normal operation and after the interruption is defined as the stability of the electric power system [7]. The generator runs at synchronous speed when the condition is balanced. Load fluctuations include an increase and decrease will affect the electrical power output that simultaneously impacts the change of mechanical power prime movers. If prime mover mechanical power cannot adjust to load conditions then this condition is called an unstable system. Simulated interference on the system is done to analyze the voltage stability, especially the analysis of the system's ability to change the load. This analysis is categorized as steady-state stability with small disturbance stimulation.

The measurement of the proximity of the operating point to the voltage instability is validated by the value of the voltage stability index [8]. It indicates a possible limit for the system to become unstable under steady state conditions. Conventionally, to know the condition of the bus is observed through the P-V curve and the Q-V curve which depicted in Figure 1. The plot of the curve illustrates the relationship between power and voltage when subjected to load changes.

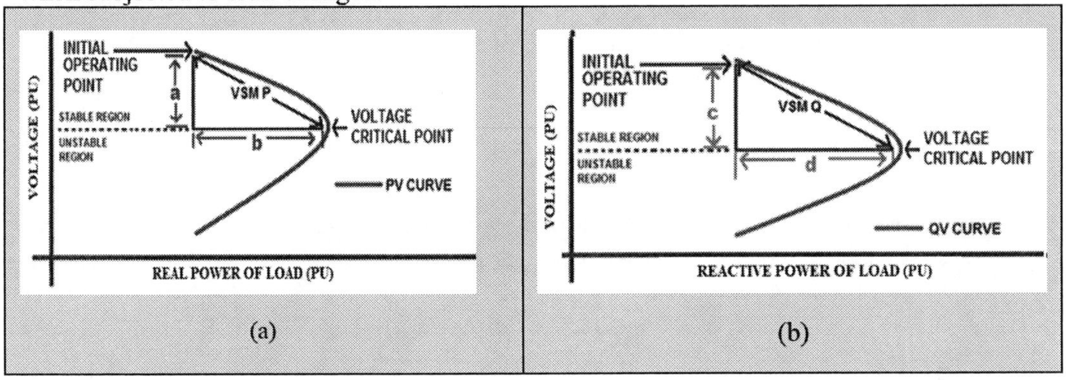

Fig. 1. (a) P-V Curve (b) Q-V Curve

Load changes affect the value of active or reactive power In the plot of the P-V curve and the Q-V curve, there is a knee point that identifies the stability limit value known as the Voltage Stability Margin (VSM). High or low VSM indicator value becomes a measure of the security level of the bus, where the higher the VSM value the bus is safe and vice versa[8].

2.1 Voltage stability margin (VSM)

VSM is divided into VSM (P) and VSM (Q). VSM (P) states the stability index on active power load and VSM (Q) for stability index at reactive load (Q). In Figure 1, let $P_{initial}$ and $Q_{initial}$ is the values of each active and reactive load power at the initial operating point. P critical and Q critical load states are active and reactive power at the critical voltage points, respectively. The critical voltage in the system is the value of the bus voltage at the point of collapse voltage. The system will experience a collapse voltage when its value is less than zero, where the critical voltage value of VSM is zero. The equation (1) use to determine VSM:

$$VSM = \frac{V_{initial} - V_{critical}}{V_{critical}} \tag{1}$$

where $V_{initial}$ is bus voltage when normal operation and $V_{critical}$ is the bus voltage when voltage collapse like depicted in Figure 1.

2.2 Simplified voltage stability index (SVSI)

Some of voltage bus stability indicators in the reference are: voltage collapse prediction index ($VCPI_{bus}$) [9], L-index [10], voltage stability index (VSI_{bus}) [11], ZL=ZS ratio [12], and SVSI [13] .

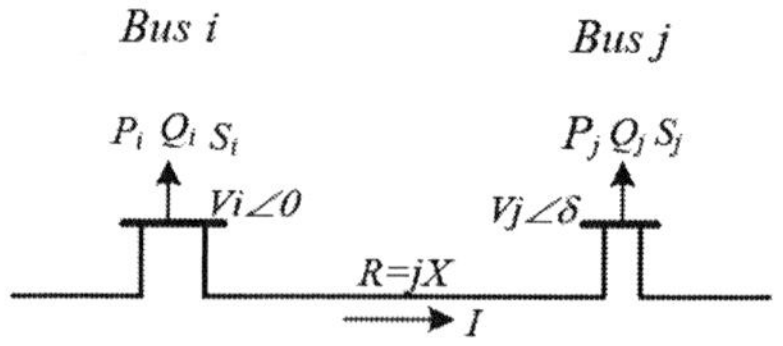

Fig.2. Representation of two bus power system

SVSI is a derivative of VSI index function which used to specify weak bust in the power system. Based on model in figure 2, definition of SVSI can be expressed as:

$$SVSI_i = \frac{\Delta V_i}{\beta * V_i} \tag{2}$$

Voltage deviation (ΔV_i) is difference between the nearest generator with load buses. The correction factor β[9] is expressed as :

$$\beta = 1 - Max(|V_m| - |Vl|)^2 \tag{3}$$

The SVSI threshold value for keeping the system stable refers to inequality:

$$Max\ (SVSI) < 1 \tag{4}$$

The advantage of this indicator is applied to a large network. The calculation requires the parameters of the bus voltage of the sender and receiver side.

3 Proposed Methods

To observe the stability impact of DG, this used the IEEE 30 bus test system. This test case consists of one swing bus, five generator buses, and 24 load buses. Figure 3 is the layout of the schematic IEEE 30 bus system which built using the PSAT software. This is a Matlab toolbox for analysis and simulation of the electric power system.

The DG used is a 50 MW wind generator. Model of wind generator using doubly fed induction generator [14] with the penetration level of 15% and 30%. The load fluctuations are considered as small disturbances in the system that affect the power system stability. Loads in variation overload by 40% and under load by 35% [13,15]. Load flow is varied on active power load (P), reactive power (Q) and P + Q load simultaneously.

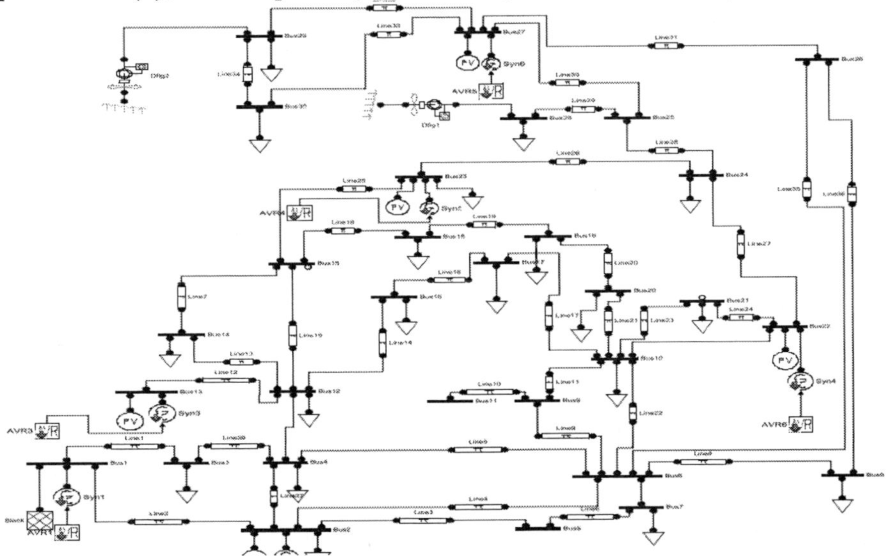

Fig. 3. PSAT model of IEEE 30 bus system connected wind generation

4 Simulation and Results

Using PSAT software, the simulation was tested on the IEEE 30- bus power system. This simulation aims to obtain the influence of DG integration on system stability using SVSI indicators. Active and reactive power load fluctuations provide a different SVSI response due to the bus voltage deviation generated. In testing, only three buses were selected for load fluctuations as shown in table 1. Table 1 shows bus voltages, before and after penetration DG about 15% and 30% levels.

Table 1. Running load flow results for IEEE 30 bus system

Bus	Voltage deviation		
	Initial condition	Wind integration	
		15%	30%
26	0.026	0028	0.035
29	0.052	0.073	0.116
30	0.054	0.081	0.128

Based on the results of running flow by PSAT, the β value obtained by 0.99374 calculated using equation (3). The calculation of voltage stability under different load conditions for different level penetration conditions is shown in figure 4.

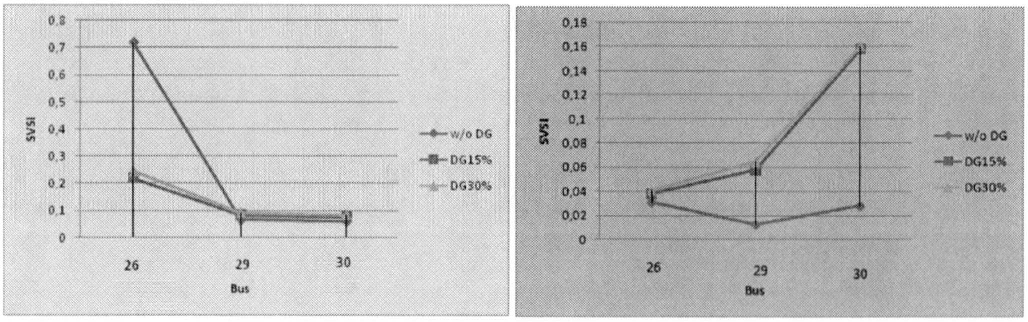

Fig. 4. (a) Voltage stability indices of heavy load **(b)**Voltage stability indices of light load

The results show that the SVSI ratio on the conditions of base loading and heavy loading (40%) looks the most critical on the bus 30 when 15 % penetration of wind generator on buses 26 and 29. This value corrected due to active power contribution and reactive power of DG integration[16-17]. The addition of DG penetration level by 30% affects the shift in SVSI value which is increasingly towards a stable direction. Penetration DG improves the voltage deviation that corrects the value of β. Different conditions are obtained during light loads; the SVSI index has increased 1.33%. This increase means that the condition of the bus leads to critical conditions. Increasing DG penetration level makes the bus voltage stability index does not make the system to the stable direction. This may be due to the active power and reactive power supplied by the grid-connected plant is sufficient. As a result, DG injected power causes an increased power loss effect resulting in a voltage drop across the load bus. This indicates that DG integration on the grid does not guarantee system stability will be better. There are load condition factors and DG sizes which determine the stability index as a measure of the system bus voltage stability.

5 Conclusion

In this paper, stability issues due to DG penetration on IEEE 30 bus systems are measured using SVSI and VSM. This study shows that DG integration affects the value of voltage stability. The load changes and the DG penetration level determine the critical value of the load bus. Need to assess DG size and location properly, because this determines active power and the reactive power injected into the system.

The acknowledgements

The authors are very grateful to Tadulako university through the technology research and higher education ministry to provide research grants for doctoral dissertation (097 / SP2H / LT / DRPM / 2018).

References

1. Yusran, A. Mochamad, and S. Adi, Optimization scheme of distributed generation installation growth considering network power quality, Journal of Theoretical and Applied Information Technology,vol.53, pp. 30-39 (2013)

2. Wang, Z. and J. Wang, Review on Implementation and Assessment of Conservation Voltage Reduction. IEEE Transactions on Power Systems, 29, 3 (2014)
3. Rahman Y.A. et al, Evaluating the effect placement capacitor and distributed photovoltaic generation for power system losses minimization in radial distribution system, Engineering International Conference (EIC 2017), vol. 1941, p. 6 (2018)
4. Modarresi, J., E. Gholipour, and A. Khodabakhshian, A comprehensive review of the voltage stability indices. Renewable and Sustainable Energy Reviews, 63 (2016)
5. Ahmad Fateh Mohamad Nor, M.S., Aida Fazliana Abdul Kadir and Rosli Omar, Determining Voltage Stability Margin Values by Measuring the Hypotenuse Under PV and QV Curves, International Journal of Electrical Engineering and Applied Sciences, 1,1 (2018)
6. Oukennou, A. and A. Sandali. Assessment and analysis of Voltage Stability Indices in electrical network using PSAT Software. in 2016 Eighteenth International Middle East Power Systems Conference (MEPCON) (2016)
7. F. B. K. Mahmood. et al, Weakest location exploration in IEEE-14 bus system for voltage stability improvement using STATCOM, synchronous condenser and static capacitor International Conference on Electrical, Computer and Communication Engineering (ECCE) (2017)
8. Amjady, N. and M.R. Ansari, Improving voltage security assessment and ranking vulnerable buses with consideration of power system limits. International Journal of Electrical Power & Energy Systems, vol.25, p. 10 (2003)
9. Balamourougan V, S.T., Sachdev MS, Technique for online prediction of voltage collapse IEE Proceeding Generation Transmission Distribution, 151, 453 (2004)
10. Kessel P, G.H., Estimating the voltage stability of a power system. IEEE Transaction Power Delivery. 346 (1986)
11. MH, Haque, Use of local information to determine the distance to voltage collapse, International Power Engineering Conference (IPEC) p. 407-419 (2007)
12. A. Wiszniewski, New criteria of voltage stability margin for the purpose of load shedding ,IEEE Transaction Power Delivery, **22**, 1367 (2007)
13. S. Pérez-Londoño , L.F.R., G. Olivar, A Simplified Voltage Stability Index (SVSI), Journal of electrical power and energy system, **83** (2014)
14. B. Babypriya , R.A., Modelling, Simulation and Analysis of Doubly Fed Induction Generator for Wind Turbine, Journal of electrical engineering, 60, 2 (2009)
15. Panda, R., et al. Impact analysis of wind power integration in existing power systems for study of voltage stability conditions, Eighteenth National Power Systems Conference (NPSC) (2014)
16. Siswanto, A., et al., Stability improvement of wind turbine penetrated using power system stabilizer (PSS) on South Sulawesi transmission system. Engineering International Conference (EIC2017), vol. 1941, p. 6 (2018)
17. Gunadin, I.C., Z. Muslimin, and A. Siswanto. Transient stability improvement using allocation power generation methods based on moment inertia, International Conference on Electrical Engineering and Informatics (ICELTICs). (2017)

Implementation of talis and dc house system for rural areas in indonesia

Dwi Riana Aryani[1] , *Faiz* Hunsnayain[1], *Andres Pramana* Edward[1], *Hwachang* Song[2], *Yoyok Dwi Setyo* Pambudi[1], *Anne* Zulfia[3,4], and *Chairul* Hudaya[1,4*]

[1]Dept. of Electrical Engineering, Universitas Indonesia, Depok, Jawa Barat, 16424, Indonesia
[2]Dept. of Electrical and Information Engineering, Seoul National University of Science and Technology, 232 Gongneung-ro, Nowon-gu, Seoul, 01811, Republic of Korea
[3]Dept. of Metallurgy and Materials Engineering, Universitas Indonesia, Depok, Jawa Barat, 16424, Indonesia
[4]Energy System Engineering, Faculty of Engineering, Universitas Indonesia, Depok, Jawa Barat, 16424, Indonesia

Abstract. In 2017, the aggregate electrification ratio in Indonesia has been achieving 92.8%. However, some rural areas such as in Maluku and Papua Islands still have low electrification ratio (~70%). One of the main problems in supplying electricity to rural areas in Indonesia is the geographical concern which consisted of islands leading to the difficulty of electricity grid development. In some areas, diesel power plant has been being built to supply the electricity. However, it causes another problem to transport the primary energy source to the targeted area which would increase the cost of electricity supply. Besides, it also needs investment cost to build transmission and distribution systems, as well as the maintenance expenses. To resolve this issue, a new scheme of battery-based Tabung Listrik or TaLis (DC-based power bank) and DC house system is proposed to be the solution to provide electricity to rural areas. The potential of local renewable energy sources such as biomass, hydro, wind, and solar could be utilized to be charging sources of batteries (TaLis). This study presents the TaLis prototype, DC house installation, supply chain process and charging scheme of TaLis, and cost comparison between the proposed system with other existing power systems such as communal PV farm and diesel power plant. We found that TaLis and DC house system provided the least cost of electricity production compared to other power systems, i.e. $0.88/kWh for TaLis and DC house system, $1.65/kWh for diesel power plant, and $1.47/kWh for communal PV farm. Implementation of this approach is expected to improve the welfare and quality of life of rural communities immediately.

1 Introduction

Conventional electric grids generate electricity using generators in alternating current (AC) form. In order to deliver electricity from power plants to consumers, transmission and

* Corresponding author: c.hudaya@eng.ui.ac.id

distribution systems are needed at various voltage levels. In AC systems, it is easy to transform the voltage level with transformers [1]. However, this system has disadvantages as well, such as the existence of reactive power and skin effect which leads to power losses in transmitting and distributing electric power. Meanwhile, in direct current (DC) system there are no such issues so the efficiency of power transmission and distribution from power plants to consumers is higher compared to the AC system [2]. Besides, the improvement of renewable energy sources (RES) and DC loads makes DC power systems more feasible to be implemented in the future [3].

A concept of DC microgrid has been established as the evolvement of the renewable energy sources, DC loads, and energy storage systems. In [4], the scheme of power sharing and management for utility connected DC microgrid has been studied. While for isolated DC microgrid, a power control algorithm has been developed in order to attain effective energy management [5]. Another intelligent power control of DC microgrid based on fuzzy algorithm has been proposed as well [6]. For rural electrification, an efficient distribution architecture of solar PV based low-power low-voltage DC microgrid could be achieved with regards to the arrangements of houses [7]. In this study, another scheme to electrify rural areas without constructing transmission or distribution networks or microgrids is proposed, while the renewable energy sources, battery energy storage systems, and DC loads are implemented.

2 The proposed system of TaLis and DC house

This study proposes a simple scheme to deliver power in rural areas, from charging stations to consumers. The centralized charging stations might be supplied from any type of power plants, including conventional power plants and renewable energy sources. Here, we select solar power plant as the power source, in which the potential of solar energy in Indonesia reaches 4.8 kWh/m2/day or equivalent to 112,000 GWp [8].

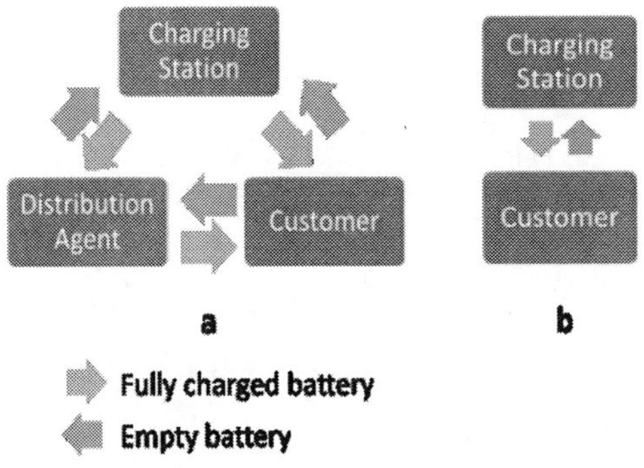

Fig. 1. Flow diagram of TaLis distribution and exchange scheme (a) with a distribution agent, and (b) without distribution agent

All loads are in DC form and this platform is known as DC house systems or DC low voltage house [9]. A new scheme introduced here is the battery bank known as TaLis which is distributed to all households as the power supply. This concept adopts the distribution of gas cylinder as well as mineral water in Indonesia, where the empty gas cylinders or mineral waters are returned to the distribution agents or supplier to be exchanged with the fully charged one. In the same manner, the "empty" TaLis can be exchanged with the fully charged one in the charging station or distribution agent. Distribution agents are then needed if the targeted location is considered too far from the charging station, more than 1 km. Within this scheme, transmission or distribution networks will not be needed, which significantly reduce the investment cost.

2.1 DC house system

In this system, the houses are proposed to utilize DC loads and DC electrical installation. DC house system here is popularized by Dr. Taufik, a Professor of California Polytechnic State University. This system is proposed for rural electrification which is expected to gradually educate people in rural areas about the benefits of electricity [10]. It has been tested as in [11] that 48 V is the most efficient system voltage when the percent loading was varied. For electric equipment or home appliances with different voltage level, it would require a DC/DC converter. Table 1 shows the assumption of daily consumption for each household in rural areas.

Table 1. Assumption of average daily energy consumption for a single household

No	Electric Equipment	Unit	Power (Watt)	Usage (Hour)	Energy (Wh)
1	LED lights	4	7	5	140
2	Television	1	50	4	200
3	Fan	1	30	3	90
4	Others (Hand phone charger, Radio)	1	40	3	120
Total daily energy consumption					450

2.2 TaLis/ battery bank

Figure 2 shows the concept of TaLis main components which consists of 18650 typed lithium-ion batteries arranged in series and parallel, battery management system (BMS), and covered by battery pack from steel plate. In general, a 18650 lithium-ion cell has the capacity around 2.2 – 3 Ah with nominal voltage 3.7 V. There is 13 battery in series and 4 sets of those batteries are connected in parallel. Thus, there are 52 battery cells in total for one battery bank.

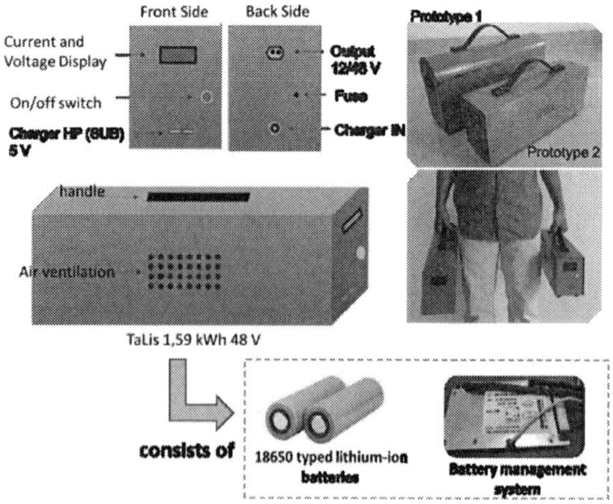

Fig. 2. *TaLis* main components consist of 18650 typed lithium-ion batteries arranged in series and parallel, battery management system (BMS), and covered by battery pack from steel plate.

By using that configuration, each battery bank is able to store 0.58 kWh energy with voltage 48.1 V. The battery bank is connected by nickel plate in its positive and negative polarities using spot welding systems. It is equipped by BMS as well for protecting and monitoring the battery states.

If the daily energy consumption of the households follows the assumption in Table 1, with the designed battery capacity as in 2.2 (0.58 kWh/battery bank), it is estimated that the battery needs to be charged every day. Later for the implementation, the TaLis capacity could be increased up to 1.59 kWh to reduce the charging cycle numbers and prolong the battery lifetime.

2.3 Charging stations

There are several options of power sources for charging stations, such as fossil fuel power stations, nearby power utility grid, and renewable energy sources. In this study, the solar power plant is selected as the power source as it has a large potential in Indonesia. The solar irradiance and the temperature of the target location are also considered enough to generate the electricity by photovoltaic (PV) panels. Besides, it provides the most efficient cost of electricity compared to others.

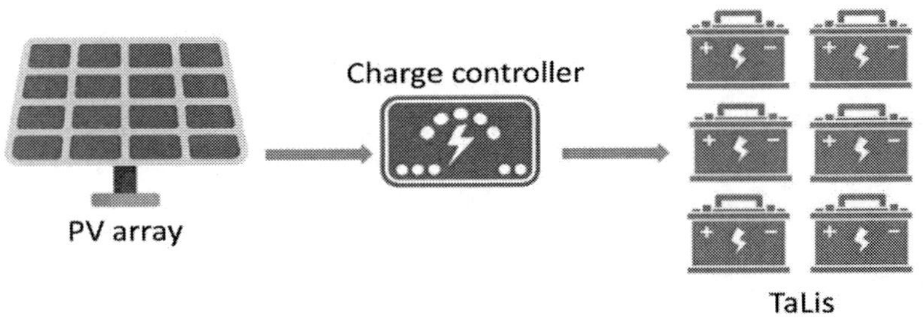

Fig. 3. Power flow during the charging mode of TaLis

Figure 3 shows the load flow from PV array towards TaLis during the charging mode of TaLis. In the charging station, TaLis' are considered as the loads. A solar charge controller is needed to control power flowing from PV to TaLis. Capacity of the power plant depends on the numbers of TaLis distributed in the target location, where numbers of TaLis is depended on the total households in that area. For example, there are 120 houses in one area, meaning 120 packs of TaLis are provided. The total daily energy needed to charge TaLis or E_{ch} in that area is:

$$E_{ch} = n \times C_{TaLis} \tag{1}$$

where n is numbers of the distributed TaLis in that area and C_{TaLis} is TaLis capacity.

Using configuration in Fig. 4, there might be some losses resulting in the efficiency of cables and charge controller is less than 100%. In this study, the system efficiency (η) is assumed 85%. Thus the energy flow from PV (E_{pv}) is:

$$E_{pv} = E_{ch} / \eta \tag{2}$$

The solar power plant capacity (P_{pv}) with h hours charging duration for a set of battery packs is

$$P_{pv} = E_{pv} / h \tag{3}$$

This capacity can be reduced if there is less amount of TaLis to be charged at one charging cycle. It means that the capacity of the solar power plant could be reduced into three fourth or even half of the total amount calculated above.

Using calculations in (1) - (3), it is found that for a region with 120 households and TaLis capacity 1.59 kWh, the total energy demand E_{ch} is 190.8 kWh, while E_{pv} with η = 0.85 is 224.47 kWh. The required P_{pv} for 4 hours charging duration is approximately 60 kW.

3 Cost comparison between the proposed system with existing power systems

In this section, the cost of electricity using the proposed scheme of combined TaLis and DC House are compared with other existing power systems in some rural areas such as diesel power plant and communal solar power plant. This calculation is assumed for 20 years operation and the targeted area has 120 houses.

3.1 Charging stations

In this scheme, the cost of electricity is determined based on the cost of investment, construction, and operational and maintenance. Investment cost refers to some components required for charging station such as PVs, charge controllers, and TaLis chargers, as well as TaLis units. Some components need to be renewed at some particular times, such as TaLis' are replaced in every 7 years and chargers in every 10 years. The total investment cost for the combined TaLis and DC house in 20 years is $162,278.

Construction for buildings in the charging station is renovated in 10 years, with renovation expense takes 50% of the beginning construction cost. The total construction cost for this system is $ 18,371. Meanwhile, operational and maintenance cost is counted annually for 20 years, with a total $159,692.

3.2 Cost analysis for diesel power plant

In case the power system in rural areas generates electricity using diesel power plants, the distribution lines, and towers or poles would be required. It means that the investment cost is used for the diesel generator set and distribution system components. Total investment cost is $25,017. Construction cost is needed for the power plant which is only 38% of the proposed system construction cost, around $7,505.

Operational and maintenance cost is higher than the proposed system, around $408,103. This is due to the distribution system components, which leads to requiring more human resources. Besides, there is an additional cost for fuels of diesel machines which is usually more expensive than the normal cost due to the transportation fee to the rural areas, with a total $198,729 for 20 years.

3.3 Cost analysis for communal solar power plant

In this system, the power source is the same as the proposed scheme, using PV panels with higher capacity even could reach ten times. This scheme also requires a distribution system, so the cost of operation and maintenance is higher than the proposed system, around $266,154. The cost of investment is used for PV panels, batteries, charge controllers, inverters, and distribution system components, around $300,190. Some equipment needs to be renewed as in the proposed system. The construction cost for this system is the highest among all since it requires large plant, around $43,798.

After calculating all elements mentioned above, it is obtained that the cost of electricity for systems in 3.1, 3.2, and 3.3 are $0.88/kWh, $1.65/kWh, and $1.47/kWh respectively. Thus, the cost of electricity for the proposed system using the combined TaLis and DC House system is the most economical solution compared to others. Another way to attain optimum cost of electricity in the isolated area is by utilizing the hybrid system as presented in [12], especially as it was proposed for Papua Province.

4 TaLis and DC house implementation

TaLis and DC house system has been being implemented in a school in Depok since November 2017. Three units of TaLis with the total capacity around 1.9 kWh and 3 kWp rooftop solar panel are used to supply some DC loads (LED lamps) in a class as shown in Figure 4.

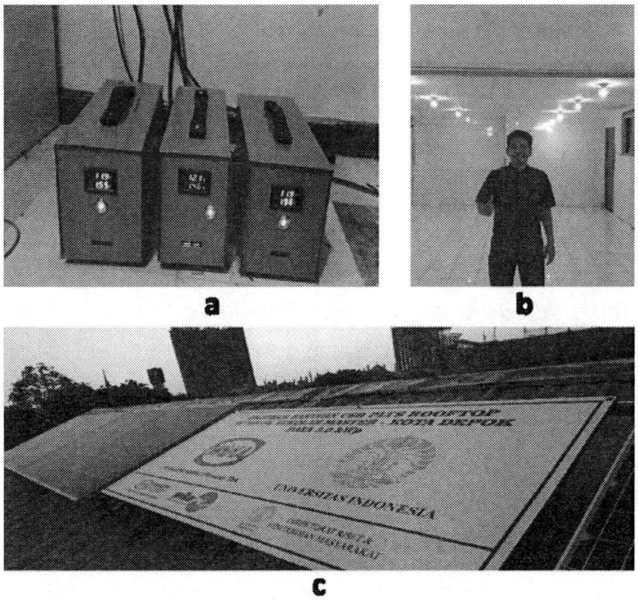

Fig. 4. Implementation of TaLis and DC House in a school; (a) three packs of TaLis as power supply, (b) DC lights supplied by TaLis in a classroom, and(c) rooftop solar panels as another power supply.

5 Conclusion

In this study, we proposed an approach to provide a reliable and economical solution to electrify rural areas in archipelago countries like Indonesia. TaLis and DC house system has a simple scheme to generate and deliver electricity from the power source to the consumer without requiring construction for transmission or distribution grid. The development of renewable energy sources and DC system including energy storages and DC loads also leads this system to be feasibly implemented. Compared to diesel power plant and communal solar power plant with the distribution network, the electricity production cost for TaLis and DC house system is cheaper. Moreover, the DC system is much more efficient and safer than the common AC system.

Acknowledgment

This article's publication is supported by USAID through Sustainable Higher Education Research Alliances (SHERA) Project for Universitas Indonesia's SMART CITY Center for Collaborative Research (No. 0144/UN2.R3.SC/HKP.05.01/2018)

References

1. M. Larruskain, I. Zamora, A. J. Mazon, O. Abarrategui, J. Monasterio, "Transmission and Distribution Networks: AC versus DC" (2018)
2. A. Sannino, G. Postiglione, M. H. J. Bollen, "Feasibility of a DC Network for Commercial Facilities", IEEE Trans. Ind. Appl. **39**, 1499-1507 (2003)
3. S. Backhaus, G. W. Swift, S. Chatzivasileiadis, W. Tschudi, S. Glover, M. Starke, J. Wang, M. Yue, D. Hammerstrom, "DC Microgrids Scoping Study—Estimate of Technical and Economic Benefits",*LA-UR-15-22097* (March 2015)
4. A. J. Datta, A. Ghosh, S. Rajakaruna, "Power Sharing and Management in a Utility Connected DC Microgrid", *Australasian Universities Power Engineering Conference* (Melbourne, Australia, 2017)
5. P. Sanjeev, N. P. Padhy, P. Agarwal, "Effective Control and Energy Management of Isolated DC Microgrid", *IEEE Power & Energy Society General Meeting* (Chicago, IL, 2017)
6. A. Hajizadeh, M. Soltani, L. E. Norum, "Intelligent Power Control of DC Microgrid", *IEEE 17th International Conference on Ubiquitous Wireless Broadband* (Salamanca, 2017)
7. M. Hamza, M. Shehroz, S. Fazal, M. Nasir, H. A. Khan, "Design and Analysis of Solar PV Based Low-Power Low-Voltage DC Microgrid Architectures for Rural Electrification", *IEEE Power & Energy Society General Meeting* (Chicago, IL, 2017)
8. Secretariat General of National Energy Council, *Indonesia Energy Outlook 2016* (Jakarta, 2016)
9. M. Friedeman, A. Timmeren, E. Boelman, J. Schoonman, "Concept for a DC- Low Voltage House", *Smart & Sustainable Built Environments* **8**, 85-94 (Jan 2008)
10. Taufik, R. Muchtar, M. Taufik, "Performance Analysis of Multiple Input Boost Converter with Inherent OR Diode Configuration", *International Seminar on Intelligent Technology and Its Applications* 175-180 (2017)
11. J. E. Chaidez, "DC House Modelling and System Design", *Senior Project of Electrical Engineering Department California Polytechnic State University* (June 2011)
12. F. Ardin, A. Rahardjo and C. Hudaya, "Electricity Price and Subsidy Scenario for Hybrid Power Generations on off-Grid System", *International Conference on Control, Electronics, Renewable Energy and Communications* 132-138 (Yogyakarta, 2017)

Polynomial Load Model Development for Analysing Residential Electric Energy Use Behaviour

Erwin Nashrullah[1] , *Abdul* Halim[1,*]

[1] Department of Electrical Engineering, Faculty of Engineering, Universitas Indonesia

Abstract. Analysing and simulating the dynamic behaviour of home power system as a part of community-based energy system needs load model of either aggregate or dis-aggregate power use. Moreover, in the context of home energy efficiency, development of specific and accurate residential load model can help system designer to develop a tool for reducing energy consumption effectively. In this paper, a new method for developing two types of residential polynomial load model is presented. In the research, computation technique of model parameters is provided based on median filter and least square estimation and implemented by MATLAB. We use AMPDs data set, which have 1-minute data sampling, to show the effectiveness of proposed method. After simulation is carried out, the performance evaluation of model is provided through exploring root mean-squared error between original data and model output. From simulation results, it could be concluded that proposed model is enough for helping system designer to analyse home power energy use.

1 Introduction

Community-based energy system is becoming promising scheme to overcome some fundamental human energy issues such as operational efficiency and local resource utilization. The system allows the participation and ownership of citizens or members of a defined community in operational aspect. The characteristic of the system is formed by at least two of the following elements: 1) Most of a renewable energy project is owned by local stakeholders, 2) Community institution takes the control, 3) Most of social and economic benefits are distributed locally [1]. In this paper, we limit the definition of community-based energy system as electric power system owned and operated by community.

Community-based energy system adopts smart micro grid architecture. Microgrid means a power delivery infrastructure in micro size (<1MW) which has small-scale generating units, energy storage, and loads. Microgrid normally operates connected to a traditional centralised grid and has capability in self-islanding operation. Meanwhile smart grid means that the system accommodates information communication technologies

* Corresponding author: a.halim@ui.ac.id

actively to connect consumers and producer's needs. Smart grid makes the system operation efficient due to sharing information [2,3].

To design the system, the simulation tool is required. The simulation is also needed for performance evaluation. In order to develop a simulator, mathematical model of system must be derived firstly. In this paper, we present mathematical models of home electrical load. The load changes continuously due to the operation of home appliances and other equipment. The models proposed here are called as multiplicative and additive polynomial model. The models derived are static non-linear and derived by using least square method. Furthermore, Almanac of Minutely Power dataset (AMPDs) is used as case study [4]. Median filter is utilized to eliminate the noise. In this research the performance of median filter is also evaluated. By using MATLAB, the calculation of model parameters is conducted, and models are validated.

2 Mathematical Model

2.1 Residential Load Data

Smart meters have become mandatory for smart grid. The use of digital smart meter effects to the volume of information acquired and its quality. By using this tool, the consumption patterns of different appliances inside the house can be grasped in more accurate. The insight into the residential load consumption has advantages for both the energy provider as well as for the consumer such for managing energy efficiently [7].

Residential load data can be classified as low sampling and high sampling. High sampling rate ≥50 Hz (≤ 0.02 s) (up to several kHz). While low sampling rate ≤1 Hz (≥ 1 s). AMPDs has 1-minute sampling rate so that it enters the class of low sampling rate. AMPDs dataset includes voltage, frequency, active and reactive power. In each dataset,
The emergence of spike and noise cause the model derivation more difficult.

2.2 Median Filter

Median filter is a nonlinear mathematical tool for repressing noise while retaining sharp sustained changes (edges) in signal values. The filter is particularly effective in reducing impulsive-type noise. The filter outputs median value of the input data inside the window centered at the point [6]. Median filter can be defined as follows [7]:
For a set of m numbers $\{z_1, z_2, \cdots$, if they are arranged in order such that $z_1 \leq z_2 \leq \cdots$ .The median of m numbers $\{z_1, z_2, \cdots$ is the $(\xi + 1)$ th-order statistic, where $\xi = (m-1)/2$ if m is odd, and if m is even. In this paper, set of m numbers is called n-window.

2.3 Polynomial Load Models

Polynomial load model is commonly used in analysis and design of power system. ZIP model is a model type classified as polynomial model [8,9]. Different to model ZIP model, in this paper we propose two type models that are called as Multiplicative and Additive Polynomial Model. The mathematical models are:

a. n- Order Multiplicative Polynomial Model

$$P_l(V) = P_{lo}\left[a_n\left(\frac{V}{V_o}\right)^n + \cdots \quad \left(\frac{V}{V_o}\right)^2 + a_1\left(\frac{V}{V_o}\right) + a_0\right] \tag{1}$$

$$Q_l(V) = Q_{lo}\left[b_n\left(\frac{V}{V_o}\right)^n + \cdots \quad \left(\frac{V}{V_o}\right)^2 + b_1\left(\frac{V}{V_o}\right) + b_0\right] \tag{2}$$

b. n- Order Additive Polynomial Model

$$P_l(V) = P_l(V_o)\left[\alpha_o + \alpha_1\frac{(V - V_o)}{V_o} + \cdots \quad \frac{(V - V_o)^n}{V_o^n}\right] \tag{3}$$

$$Q_l(V) = Q_l(V_o)\left[\beta_o + \beta_1\frac{(V - V_o)}{V_o} + \cdots \quad \frac{(V - V_o)^n}{V_o^n}\right] \tag{4}$$

where V_o : rated voltage; V : actual operating voltage, P_l, Q_l : active and reactive load powers at operating voltage ; P_{lo}, Q_{lo} : active and reactive powers at rated voltage, a_0, $a_1, \ldots, a_n, b_0, b_1, \ldots, b_n, \alpha_o, \ldots, \alpha_n, \beta_o, \ldots, \beta_n$ are constant.

2.4 Parameter Estimation Method

Least square method is used to estimate the parameters of models (1)-(4). In the next formulas, instead using n order polynomial, for clarity and simplicity, 2-order polynomial is used. The formula to estimate the parameters is

$$\bar{\theta} = \left(\bar{\Omega}^T\bar{\Omega}\right)^{-1}\bar{\Omega}^T\bar{\vartheta} \tag{5}$$

where, for multiplicative polynomial model,

$$;\bar{\theta} = \begin{bmatrix} a_2 & a_1 & a_o & b_2 & b_1 & b_o \end{bmatrix}^T \bar{\vartheta} = \begin{bmatrix} \bar{\vartheta}_P \\ \bar{\vartheta}_Q \end{bmatrix};\; ;\bar{\vartheta}_P = \frac{1}{P_{l_o}}\begin{bmatrix} P_{l_1} & P_{l_2} & \cdots & \end{bmatrix}^T;$$

$$\bar{\vartheta}_Q = \frac{1}{Q_{l_o}}\begin{bmatrix} Q_{l_1} & Q_{l_2} & \cdots & \end{bmatrix}^T;\bar{\Omega} = \begin{bmatrix} \Omega_P \\ \Omega_Q \end{bmatrix};\; \Omega_P = \Omega_Q = \begin{bmatrix} (V_1/V_o)^2 & (V_1/V_o) & 1 \\ (V_2/V_o)^2 & (V_2/V_o) & 1 \\ \vdots & \vdots & \vdots \\ (V_n/V_o)^2 & (V_n/V_o) & 1 \end{bmatrix}$$

and for additive polynomial model,

$$\bar{\theta} = \begin{bmatrix} \alpha_o & \alpha_1 & \alpha_2 & \beta_o & \beta_1 & \beta_2 \end{bmatrix}^T;\bar{\vartheta} = \begin{bmatrix} \bar{\vartheta}_P \\ \bar{\vartheta}_Q \end{bmatrix}\bar{\vartheta}_P = \frac{1}{P_l(V_o)}\begin{bmatrix} P_{l_1}(V) & P_{l_2}(V) & \cdots &) \end{bmatrix}^T$$

$$\bar{\vartheta}_Q = \frac{1}{Q_l(V_o)}\begin{bmatrix} Q_{l_1}(V) & Q_{l_2}(V) & \cdots &) \end{bmatrix}^T;\bar{\Omega} = \begin{bmatrix} \Omega_P \\ \Omega_Q \end{bmatrix}$$

$$\Omega_P = \Omega_Q = \begin{bmatrix} 1 & (V_1 - V_o)/V_o & ((V_1 - V_o)/V_o)^2 \\ 1 & (V_2 - V_o)/V_o & ((V_2 - V_o)/V_o)^2 \\ \vdots & \vdots & \vdots \\ 1 & (V_n - V_o)/V_n & ((V_n - V_o)/V_o)^2 \end{bmatrix}$$

3 Simulation And Analysis

In this section, we will present the performance evaluation of median filters, model parameter computation results and model validation.

3.1 Performance Evaluation of Median Filter

In this reseach, AMPDs data has been studied to evaluate median filter. Before processing by median filter or original data of active power, reactive power and voltage profiles are given in Figure 1. As shown in this figure, the big problem is the existance of noise in each profiles particularly in voltage profile.

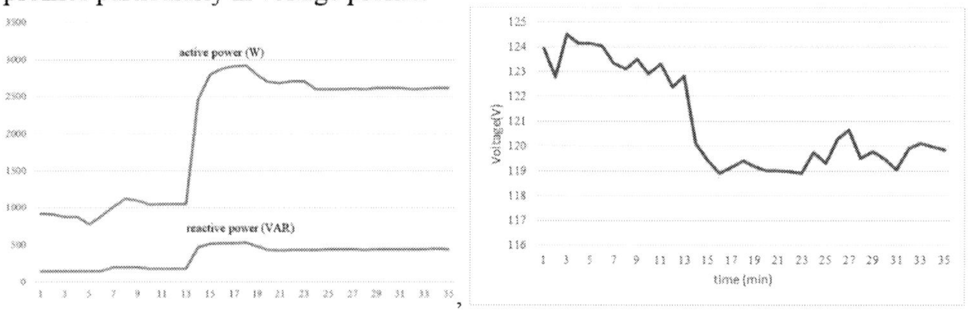

(a) Active and Reactive Power (b) Voltage

Fig. 4. Pre-processing AMPDs Data

In the next, smoothing of active power, reactive power and voltage curve are deployed by using median filter with 3,5,7 and 9-window. The results are shown in Figure 2. From these figures, it could be concluded that:

a. Median filter has ability to smooth the curve and preserve the edge values.
b. The bigger value of *n*-window results smoother curve.
c. However, bigger value of n-window potentially suffers the curve so that some characteristics of data will lost.

Based on the above analysis, in this paper, 5-window median filter is used to proceed raw data and formulate the model.

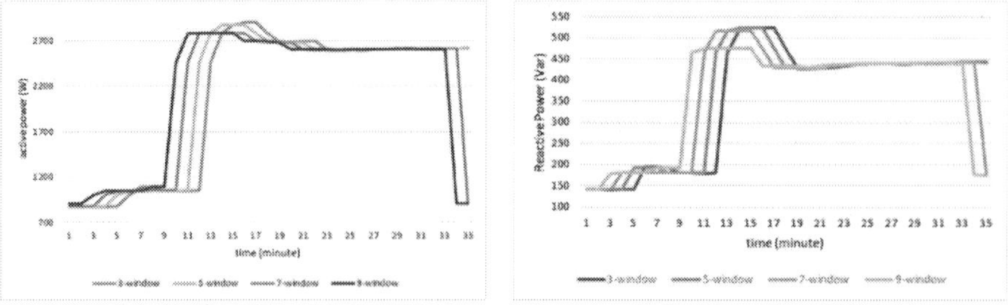

(a) Active Power (b) Reactive Power

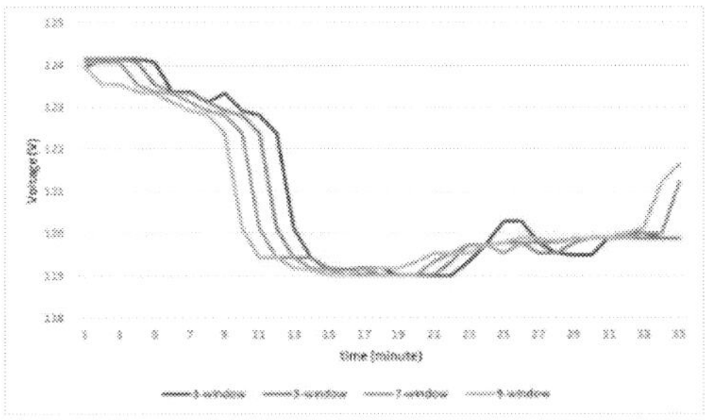

(c) Voltage

Fig. 2. Data Profiles in Four Cases of Windows

3.2 Parameter Estimation and Validation Result

Calculation of parameters is carried out by using MATLAB. The results are as follows. In all case, it is set $V_o = 120V$.

For Multiplicative Model, with $P_{lo} = 3000$ $Q_{lo} = 500$, the model is

$$P_l(V) = P_{lo}\left[58.40\left(\frac{V}{V_o}\right)^2 - 134.95\left(\frac{V}{V_o}\right) + 77.37\right] \tag{6}$$

$$Q_l(V) = Q_{lo}\left[(-25.62)\left(\frac{V}{V_o}\right)^2 + 35.01\left(\frac{V}{V_o}\right) - 8.55\right] \tag{7}$$

RMSE (Root Mean Squared Error) is 0.0438 and 0.0675 resectively. Graphical result is depicted in Figure 3.

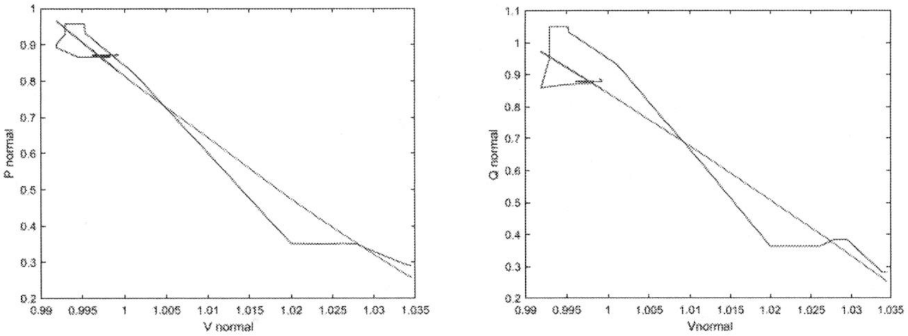

(a) Active Power (b) Reactive Power

Fig.3. Comparison of Real vs Model (Multiplicative Type)

For Additive Model with $P(V_o) = 3000$ and $Q(V_o) = 500$, the model is

$$P_l(V) = P(V_o)\left[0.840 - 18.17\left(\frac{V - V_o}{V_o}\right) + 58.4\left(\frac{V - V_o}{V_o}\right)^2\right] \tag{8}$$

$$Q_l(V) = Q_l(V_o)\left[0.840 - 16.07\left(\frac{V - V_o}{V_o}\right) - 27.84\left(\frac{V - V_o}{V_o}\right)^2\right] \tag{9}$$

RMSE (Root Mean Squared Error) is 0.0438 and 0.0679 resectively. Graphical result is depicted in Figure 4.

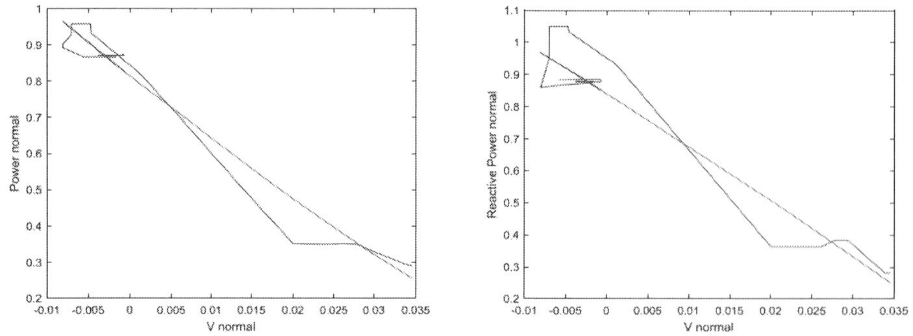

(a) Active Power (b) Reactive Power

Fig.4. Comparison of Real vs Model (Additive Type)

4 Conclusions

The aim of this paper is to provide model for representing dynamic residential load. There are two models presented here which are multiplicative and additive load model. The parameters of models have been estimated by using least squared error method. Before model is derived, median filter with 5-window has been used. Using AMPDs data, it has been verified that two models have almost same performance. Moreover, some improvements can be done to reduce error between model and real data,

We acknowledge the support from Hibah PITTA 2018 funded by DRPM Universitas Indonesia No.2356/UN2.R3.1/ HKP.05.00/2018.

References

1. International Renewable Energy Agency, *Community Energy: Broadening The Ownership Of Renewables*, IRENA Coalition for Action (2018).
2. Adam Hirsch, Yael Parag, Jos ep Guerrero, *Microgrids: A Review Of Technologies, Key Drivers, And Outstanding Issues*, Renewable and Sustainable Energy Reviews, **Vol. 90** (2018).
3. Rehman Zafar et al, *Prosumer based energy management and sharing in smart grid*, Renewable and Sustainable Energy Reviews, **Vol. 82, Part 1** (2018).
4. Stephen Makonin et al., AMPds: **A Public Dataset for Load Disaggregation and Eco-Feedback Research**, 2013 IEEE Electrical Power & Energy Conference(2013).
5. Kaustav Basu, et al., *A Comparative Study Of Low Sampling Non Intrusive Load Dis-Aggregation*, 42nd Annual Conference of IEEE Industrial Electronics Society (2016).
6. Stork M., *Median filters theory and applications*, Third international conference on electrical and electronics engineering papers-chamber of electrical engineering (2003).
7. Yongkui Sun, Wen Tan, Tongwen Chen, *A method to remove chattering alarms using median filters*, ISA Transactions, **Vol.73** (2018).
8. Doron Shmilovitz, *Characteristics Of Modern Nonlinear Loads And Their Influence On Systems With Distributed Generation*, Int. J. Energy Technology and Policy, **Vol. 5, No. 2** (2007)

9. Abdullah Bokhari et al., *Experimental Determination of the ZIP Coefficients for Modern Residential, Commercial, and Industrial Loads*, IEEE Transactions On Power Delivery, **Vol. 29, No. 3** (2014).
10. Adalgiza del Pilar Rios, Kodjo Agbossou, Alben Cardenas, *Taylor series approximation of ZIP model for on-line estimation of residential loads' parameters*, 2017 IEEE International Conference on Industrial Technology (ICIT) (2017).

Leak Current Monitoring System On The Ground Cables Medium Voltage Transformer 150/20 kV

A. Sofwan[*1], *A.Multi*[1], *R.Soleman*[1], *Sugianto*[1] *and Septian* A.Kusuma[2]

[1]Electrical Engineering Program of ISTN, Jl. Moh. Kafli II Jagakarsa, Jakarta Selatan,
[2]PT. PLN Transmisi Jawa Bag.Barat, APP Cawang Technician Substation System Jatirangon

Abstract. On January, 2017 there had been fault trip at PMT 150/20 kV transformer in Jatirangon substation. For fault detection, that rele differential phase T and REF 20 kV have worked to detect this fault. The resulted impact of this fault is Power outage in feeder a 1350 A, 35 MW and 5 Mvar. The cause of this fault is the occurrence of breakdown on ground cables 20kV T-phase-core 1. This results in a short circuit to the ground so that the differential protection relay and REF 20kV work because the relay detects a fault in the ptotection zone. The result of this research is the design of an early detection monitoring tool. This tool is used to determine the amount of leakage current on the ground cable in order to minimize the occurrence of interference that causes the occurrence of electrical power outage. The result of the leakage current monitor on the ground phase cable T obtained a current of 0.6A with temperature 35 °C. With thermal failure calculation method for leakage current obtained result of 0,56180A with temperature 35 °C. Comparison of the calculation with the measurement of leakage current on the ground cable T phase is obtained at 6.36%.

1 Introduction

The distribution of electric power transformer power *step down* voltage of 150 kV System to a 20 kV medium voltage (MV) system in large capacity, is generally done using a medium voltage cable. Distribution of electrical energy step down power transformer from 150 kV voltage system to 20 kV medium voltage system in a large capacity, generally carried out using medium voltage cable network. This is because for the distribution of 20 kV medium voltage power System air conduct is sometimes difficult to implement because it reduces the aesthetics of space. For this reason, an important factor to be considered in using ground cable is the isolation characteristics. One of the obstacles to the use of ground cable is the failure of isolation in carrying out its function as a medium voltage isolation medium. Because in the manufacture of cables is sometimes not perfect so there are cavities in isolation. If the rate of heat at a point in the cable material exceeds the rate of heat dissipation, there will be an unstable state and at some point the material will cause a thermal failure.

---

[1] Corresponding author: asofwan8@gmail.com

If this thermal failure lasts a long time, it will reduce the reliability of the cable so that it will cause isolation failure and damage to the cable. The failure of the isolation of the 20 kV medium voltage ground cable occurred in the 150 kV / 20 kV Transformer #4 at the Jatirangon Substation, which caused a disturbance and the electrical power outages. Before the occurrence of fault, no monitoring of the leakage current of the power cable was made so that it was not detected early that an increase in the leakage current of the medium voltage cable resulted in the Fault. For this reason, it is necessary to have a tool to detect early increase in the MV leakage current so that it can minimize the disturbance that causes the power to be disconnected to the consumer, the productivity stops and causes losses to the Industries due to damaged equipment.

2. Transformer protection

The transformer as the main equipment at the substation that functions to distribute electrical energy to the distribution electricity system must be protected properly. Because when the transformer is damaged will cause the flow of electricity to the distribution network is disconnected. The transformer is protected by using protection relay, the relay is in charge of identifying the fault that occurs in the transformer, fault around the transformer, and fault with the network supplied by a transformer. If a fault condition or abnormal condition is found, the protection relay will give the PMT command to trip so that the transformer is free from system voltage and fault current.

The protection relay is grouped into 2 types of protection, namely mechanical and electrical relay. Beside that, transformer relays are also grouped based on the type of protection that is divided into the main protection relay and backup protection relay. The same mechanical and electrical protection relay function is to protect the transformer from the short circuit current by giving a trip command to the PMT. All mechanical relays are the main protection relay. This relay works without delay time (instant) when there is an internal or external fault of the transformer which causes in an anomaly in the transformer's internal. The following is an explanation of the electrical relay of the transformer.

2.1 Rele *differensial*

Differential relay is relays that work when detecting phasor differences and or instantaneous difference in input current and output current. The principle of this relay work is the comparison of the primary, secondary and or tertiary winding currents (if the tertiary is burdened). The working principle of this relay is based on Kirchoff's Current Law I, namely the amount of incoming current equal to the amount of current coming out at the point of branching of an electrical circuit.

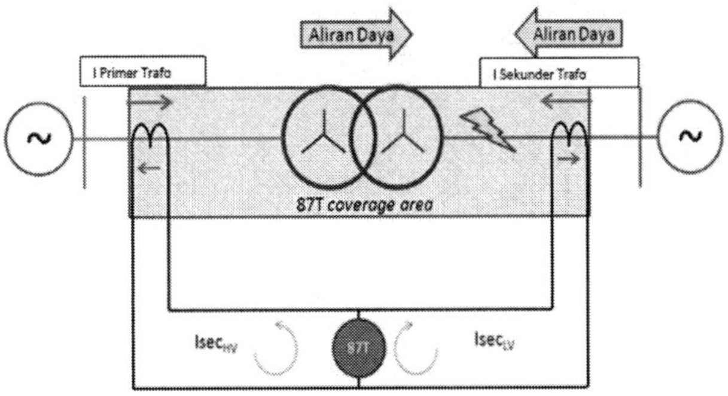

Fig. 1. Scheme of internal fault condition

In Figure 1 explains the workings of the RD when fault conditions. The current will not be detected by rele differensial as fault current when the transformer in case of normal operation because the primary current and the secondary current will cancel each other at point 87T. In this normal condition the differential relay does not detect the difference in the incoming and outgoing currents, so the differential relay does not work. But on the fault condition, *differensial* Relay feels any difference in inflows and outflow. So that RD will operate to give trip commend to PMT.

When the fault is outside the relay protection zone, the current from the primary side and the current from the secondary side will also cancel each other at point 87 T even though the magnitude of the fault current is large. In this condition the differential relay does not detect the difference in the primer current and the secondary current so that the differential relay does not work. Figure 1 decribes the Fault condition in Protection Zone. In this condition *differensial* rele detected any difference in primer current and secondary current, so rele *differensial* will operate.

The different current will be read on the workings of the relay differensial while the transformer is in a State of abnormal in the protection zones rele. Current from the primary side and the secondary side of each other add up at this point 87T. So in this condition RD detect any difference on primer and secondary (primer does not equal secondary). So RD worked for gave orders for the trip because the PMT occur in the protection zone. So in general the workings of RD is when the disruption is outside the protection zone then the sum of the currents is zero at the point of branching (*restraint point*). However, in the event of fault in the area of protection then the incoming current is no longer equal to the secondary current so that it will arise at the point of differensial current in *restraint point*. If the difference in the current *differensial* has already reached the point settings then this relay will *operate*. In the condition of the transformer and the normal protection system the RD must be stable or not work if conditions occur as follows: Inrush current, External Throught Fault Current, Overfluxing on the transformer and Change tap on Load.

2.2 Rele restricted earth fault

REF Relay is used to protect the transformer from phase fault to the ground near the neutral point of the transformer. This relay is installed on the transformer with the YNyn vector group design grounded. This protection zone of REF is an area that is not detected by differential relays, so that the sensitivity of this relay becomes the main point of the setting. The magnitude of the phase fault current to the ground depends on the value of the resistance that is installed on the ground of the neutral point.

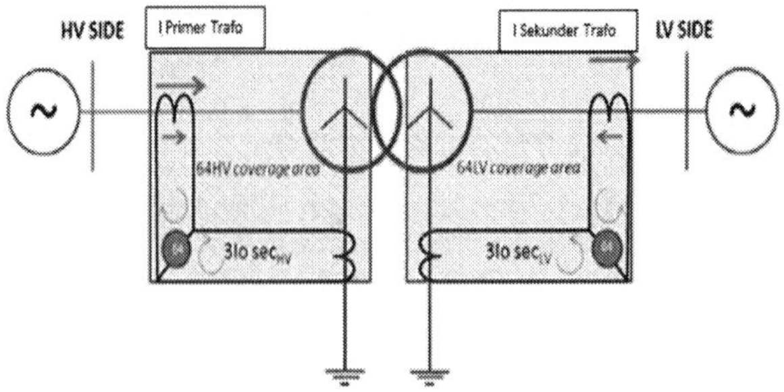

Fig. 2. Scheme of rele REF

Figure 2 describes how to work with the reading of current when the transformer operates normally. The working principle of relay REF is the same as differential relay, using Kirchhoff Current Law I that is the incoming current equal to the current coming out at the point of branching of an electrical circuit or at restrain points. In 1 power transformer with 2 windings there are 2 REF relays each protection the primary and secondary sides of the transformer. The detection of the current on the REF relay in the event of a ground fault on outside the protection zone, the REF relay will not detect because the current passing through the 64 is the same value. On the other hand, if the ground faults occurs inside the protection zone because the current passing through 64 is not the same value, the REF relay will operate.

3 Ground cable construction

Ground cable construction basically consists of 3 (three) important components, namely:

a. Conductor.

The conductor is intended as a good electrical conductor of low voltage, medium voltage or high voltage and called the core of the cable, including in the ground cable section. The electrical conductivity of a conductor material is expressed by conductivity, which is the opposite of resistivity, or the resistance type of conductor, where the resistance type is defined as:

$$\rho = \frac{R.A}{L} \tag{1}$$

Where A= cross-sectional area (m^2), L= Cable length(m), R= conduct resistivity (ohm-m), ρ= Conductivity

Table 1. Properties of aluminum

Properties of Materials	Unit	Alumunium (Ac)
Electrical Conductivity (ρ)	$(\text{ohm meter})^{-1}$	$3{,}8 \times 10^7$
Thermal Conductivity (k)	J / m.s °C	200
	Kkal / m.s °C	$500{,}\ 10^{-4}$
Resistance Type	Ωm	$2{,}65 \times 10^{-8}$

On the research had used medium Voltage ground cable 20kV with a core made of aluminum is then seen table 1. Material aluminum has an electrical conductivity of 3.8×10^7 $(\text{ohm m})^{-1}$, thermal conductivity (k) of 200 J/m s °C and the resistance type of 2.65×10^{-8} Ωm.

b. Isolation

Cable isolation is often called the dielectric, whose function is to limit or to prevent direct contact between the conductors that are voltage with the surrounding objects.

c. Protective

Its Protective functions prevent the isolation from being affected from the outside such as the entry of water into the cable isolation, or humidity mechanical or pressure that could damage the cable isolation.

Figure 3. that is the kind of NA2XSY ground cable 1x150 cm/12/20 25 (24) kV declares a single nucleated, cables for nominal voltage 12/20 kV XLPE insulated, conductor, aluminum with a broad cross-section of 150mm^2 copper layers, on the outside of the core arrangement with a broad cross-section of 25 mm^2, PVC. These cables are used in this research are already installed on a trnasformator 150/20 GI Jatirangon 4 kV.

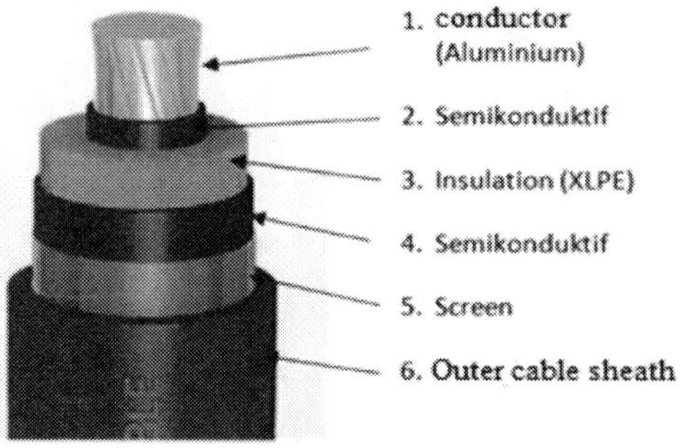

Fig. 3. Nucleated single ground cable

3.1 Process leaked current

3.1.1 Cavities on the cable

Inside the cable there are often cavities that contain gas or air. This gas or air cavity is formed during cable manufacturing or cable usage time.As it is known that a cable consists of several types of layers made of different materials and has different expansion coefficients. In case of heating and cooling, either at the time of manufacture or at the time of loading with the current depreciation expansion and then from each ingredient will differ. As a consequence will thus the cavity-cavity containing the gas or air between the layers that and cavity gas or air has a dielectric strength of materials isolation materials. The cavity on cavity-in isolation material can also arise at the time of manufacture. The air cavity is a hole with pressurized air and low power dielectric isolation, is the weak point of isolation because the permittivity is lower, it will be an increase in the electric field in the air in the cavity exceeds the power to penetrate the air. The cable damage processes is shown in Figure 4.

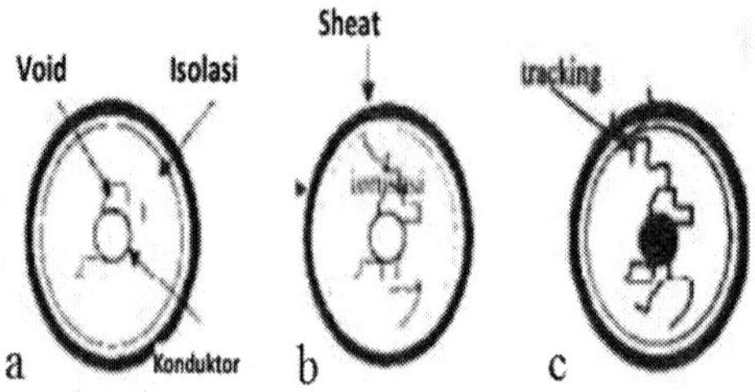

Fig. 4. Process of damage on Cable

3.1.2 Failure isolation

Cable insulation failure is a situation where isolation cannot anticipate an abnormal condition that exceeds the isolation capability. Isolation failure is caused by the type of electrode material, the configuration of the electric field, temperature, pressure, voltage, and the life time (age) of the used isolation material.

3.1.3 Failure of thermal

Thermal failure is a failure that occurs if the speed of heat at a point in the used material exceeds the rate of heat dissipation out. As a result there is an unstable failure so that at some point the material fails. The mechanism of thermal failure follows the laws of energy conversion, i.e., the heat generated is equal to the heat is channeled out through electrodes in the surrounding medium added with the heat is used to increase the material temperature. According to Whitehead, the minimum voltage of thermal failure (Vm) is:

$$V_m = \int_{Tm}^{To} \left| \frac{(8\,k)}{\sigma} \right| \, dt \quad (2)$$

Where : Vm : Voltage fails the minimum thermal [V], k : thermal conductivity [J / m.s ^{o}C]
To : Temperature on material surface (in this case equal to the temperature of the circumference) [^{o}C]
Tm : critical Temperature of faiedl materials, σ : electrical conductivity [ohm meter]$^{-1}$

3.1.4 Flow of leaked power cable

Leakage current is a current that flows through the isolation surface. Isolation serves to electrically separate two or more conductors that are close together, so that no leakage current occurs. Leakage currents also caused by leaky cavities on isolation material, which caused an error in the manufacture of isolation material. Isolation resistance affects the leakage current, the isolation resistance will be greater if the conductor gets longer. The equation of the prisoner's isolation as follows:

$$I_b = \frac{Vm}{R} \quad (3)$$

Where, I_b: Leak current [A], Vm : voltage fails the minimum thermal, R: thermal resistance [Ω]

3.1.5 Leakage isolation

Normally the current will flow through the cable conductor, while unwanted leakage current will flow radially from the conductor through the dielectric to the protective layer. In the cable cross section then it will become larger when it is started from the conductor.

3.2 Monitor concept of leakage current

To be able to monitor the leakage current on ground cables transformer-4 GI Jatirangon, then a device monitoring facilities of leaked cables was designed base on grounding. On operated

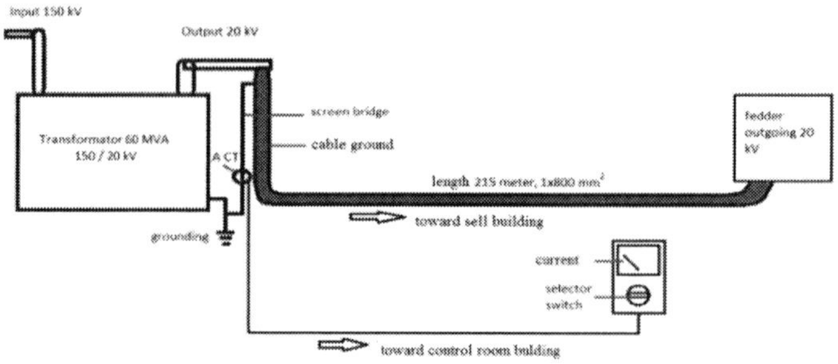

Fig. 5. Monitor of Leaked current on ground Cable

transformer, the only leaked currents of cable ground can be monitored. The leakage current flows from the part of the cable called the screen which is grounded on one side of the cable end. The screen functions as the location of the flow of short circuit current in phase fault to the ground to the nearest ground flow the capacitive current that arises in isolation because of the phase to ground voltage. For this reason, a leakage current monitoring facility can be made by using the screen that is grounded. To be able to monitor the current, a current transformer ring is used for current measurement. After that, the ct ring is connected to a measuring device so that it can monitor the current value. Because in T phase there are 3 cables with a core, then 9 CT rings are needed for each cable. For more details, see Figure 5 below.

3.3 Leaked current measuring

1 Monitoring the leakage current can be done in a calculation based on thermal failure by monitoring the temperature on the cable surface using flir. According to Whitehead the voltage of thermal failure can be calculated using formula 1 and the leakage current can be calculated using the formula 2.
2 Monitoring the leakage current when operating with a facility of leakage current facilities that have been made above by monitoring using ampere pliers, so that it is directly obtained from the leakage of ground cable.

Fig. 6. Monitoring leaked Current on the operation

4. Calculation of leaked current

Specifications techniques for ground cable for transformer 4 the length of the 215 meters with a cross-section area of 1x800 mm^2. Electrical conductivity, thermal conductivity and resistance type of aluminum can be seen in tabel 2.1. Then for conducting Resistance can be calculated by using the formula 1. By substituting the value of the parameter value, then the value obtained for the 20kV ground cable resistance is R = 0.00712 Ω. To get the value of the leakage current, the ground cable must first find the voltage of the thermal cable ground failure then the thermal failure voltage can be calculated by using formula 3.1 as follows:
Ground cable the T phase core 1:

$$V_m = \int_{Tm}^{To} \left|\frac{(8\,k)}{\sigma}\right| dt = \int_{130}^{35} \left|\frac{(8\,.\,200)}{3{,}8 \times 10^7}\right| dt = 0.004 \text{ V} \tag{4}$$

Then obtained for failed cable ground thermal voltage phase T core 1 for $V_m = 0.004003$ V. After a failed voltage rated thermal ground cable, then for the flow of leaked cables land can be calculated using the formula 3.2, with calculation as follows:

$$I_b = \frac{Vm}{R}\ ,\ I_b = 0.56180 \text{ A} \tag{5}$$

After the calculated then obtained for the leaked cable ground T phase of core 1 for I_b = 0.56222A. For other cable core calculation result can be seen in table 4.1. For calculation results are smaller than the results of monitoring the equipment directly. For the calculation of the T core 1 phase ground cable with thermal failure method, the result of leakage current is 0.56180 A with a temperature of 35°C, while the results of monitoring the tool using the ampere pliers as a measuring device obtained the leakage current of 0.6 A with the same temperature 35° C. Based on the table above obtained for the results of manual calculations with monitoring tools have a deviation of results. Obtained for the difference between the calculation with the measurement of 6.36%.

Table 2. Comparison of calculation with the measurement of the leaked current.

Phase	Core Cable	Thermal (°C)	Calculation		Measurement (A)	Calculation : Measurement (%)
			V_m (V)	I_b (A)		
R	1	33	0.00140	0.57163	0.8	28,54
	2	20	0.00463	0.65028	1,1	40,88
	3	30	0.00421	0.59129	0,8	26,08
S	1	20	0.00463	0.65028	1,1	40,88
	2	25	0.00442	0.62079	0,9	31,02
	3	20	0.00463	0,65028	0,8	18,71
T	1	35	0.004	0.56180	0,6	6,36
	2	20	0.00463	0,65028	1,0	34,97
	3	20	0.00463	0.65028	1,2	45,81

The difference between the calculation with the measurement because in the manual calculation only calculates the value of leakage current contained in the ground cable, while the results of the measurement of the value of leakage current monitored on the ground cable and transformer on the secondary side so the results are greater.

5 Summary

To find out the leakage current of ground cable manually, thermal failure method is used by monitoring the cable surface heat results. After that it was analysed by using whtehead theory calculations. The results of leakage current of T phase core 1 ground cable is 0.56180 A with a temperature of 35°C. While using a equipment to monitor grounding-based leakage current for the ground cable results in 0.6 A with the same temperature 35°C. Difference in current 20 kV ground cable leakage of T phase core 1 transformer in Jatirangon substation between calculations with measurements of 6.36%.

With a leakage current monitor equipmen can detect early on the leakage current. If there is a significant increase in leakage currents, there can be a minimization of fault caused by a recurring ground cable breakdown so that electricity distribution to consumers and electrical equipment at the Jatirangon substation becomes more reliable.

References

1. Hasan Basri, Ir., "*Protection System of Electrical Power*", ISTN, Jakarta (2003)
2. Arismunandar, DR, "*Electrical Power Engineering III*" Pradnia Paramita, Jkt, 97.
3. Erhaneli dan Musnadi, " *Leakage Current Effect on Temperatur Change at Ground Cable 20 kV* " , Journal of Electrical Dept Faculty ITP, Padang (2012).
4. Suprame Cable Manufacturing Corp.,P.T., "*Power Cable Catalogue*"
5. Suprame Cable Manufacturing Corp.,PT. NEXANS, "*High Voltage Underground Cables*"
6. Karyono, "*Ptotection System on Transmission and Jawa Bali Substation*", Jkt (2013).
7. PLN P3B JB.*"Guide Book on O & M Secunder Equipment of Electrical Power Distribution (Protection of Transformator) "* Jakarta (2014).
8. Hadi Sadat, "*Power System Analys*", Milwaukee School Of Eng., (1999)
9. PLN. *" Guide Book on Introduction of Ground cable 20 kV",* Jkt (2014).
10. PLN. *" Guide Book on Maintenance of Ground cable MV",* Jkt (2014).
11. PLN. *" Guide Book on Protection of Transformator",* Jakarta (2014).
12. PLN. *" Transformator Guide Book",* Jakarta (2014).
13. S.Rachmat,"*Analysis of Local Electrical Field Effect On Development of Electrical Load at Cable XLPE*", Journal of Electrical Dept. UI, Jkt (2008).
14. S.Aris and Sofwan Agus, "*Design Simulation Modul of ELCB 1-Fasa as Human Protection*", Journal of Electrical Engineering Faculty ISTN, Jakarta (2016).

Design of a Sensor Coil for Electromagnetic Induction Tomography

Dita Ayu Banjarnahor[1], *Rohmadi* Rohmadi[2,*] *Amir Rudin*[2], *Imamul* Muttakin[3], and *Warsito P.* Taruno[2]

[1]Department of Engineering Physics, Institut Teknologi Bandung, Bandung, Indonesia
[2]CTECH Labs Edwar Technology, Tangerang, Indonesia
[3]Engineering Tomography Laboratory, University of Bath, Claverton Down, BA2 7AY, UK

Abstract. Electromagnetic tomography method works by utilizing magnetic field induced by coils that are given an electric current. An object with certain conductivity property interferes the magnetic field which will be sensed by the sensor in the form of voltage difference. Experiment using iron as an object has been conducted. In addition, parameters given are distance between transmitter and receiver coil, and frequency of transmitter signal. The result shows that conductive material gives significant voltage difference, which ranges between 0 – 0.072 V. The optimal transmitter-receiver coil distance is the shortest, while the optimal transmitter signal frequency is at 5MHz and 9MHz. Based on the optimum parameters gained, multi-channel magnetic induction tomography (MIT) sensor is designed. It contains four transmitter coils and four receiver coils. They are arranged in circle, which each transmitter and receiver pairs are in opposite location. The sensor proved to be able to sense voltage difference induced by an object. Samples of imaging are also successfully provided accordingly.

1 Introduction

Magnetic Induction Tomography (MIT) is a relatively new non-contact imaging modality which is used for imaging the internal distribution of the passive electromagnetic properties (conductivity, permittivity, and permeability) of an object by measuring the mutual inductances between pairs of coils placed around its periphery [1].

Since this technique uses magnetic field, it has non-invasive feature and therefore can be applied in many fields, like biomedical application. There has been considerable interest shown over the past decade in the MIT imaging technique and its application to various problems. If there is a contrast or change in dielectric properties of the object of interest, MIT may potentially be used as an imaging tool [2]. MIT can be used for industrial process such as metal production and coating inspection [3].

Maxwell's equation states that the electric voltage applied to a closed path is proportional to the rate of change in the magnetic flux enclosing the closed path [4]. It is also stated that the induced magnetic field around a closed path is proportional to the rate of

* Corresponding author: rohmadi@c-techlabs.com

change in the electric field enclosing the path. In practice, this means that a magnetic field can be created by passing an electric current to a closed path in the form of a wire coil. In addition, an electric field can also be created by exciting the magnetic field to the wire coil. The electromagnetic tomography works by utilizing this principle.

2 Induction Coil Sensor

Sensors will be specifically used for imaging of metal objects. Metal in general has a relatively high conductivity. Based on the need to measure the voltage difference, an MIT sensor will be designed with specific number of coils, number of turns, diameters, shapes, and configurations. The selected design is coils having radius of 21 mm with each 50 number of turns, wire with 0.8 mm diameter, and with four transmitter and receiver coils respectively [5]. This selection is done by considering the optimal diameter of the sensor. The measured voltage will be greater in value when the distance between the transmitter and receiver coils is getting smaller. The object tested on the test itself is a thin beam iron with a width of 10 cm. Thus, the total diameter of the entire sensor of 11 cm is chosen, the maximum number of sensors that meet is eight.

The following Fig. 1 shows a sketch of the sensor configuration.

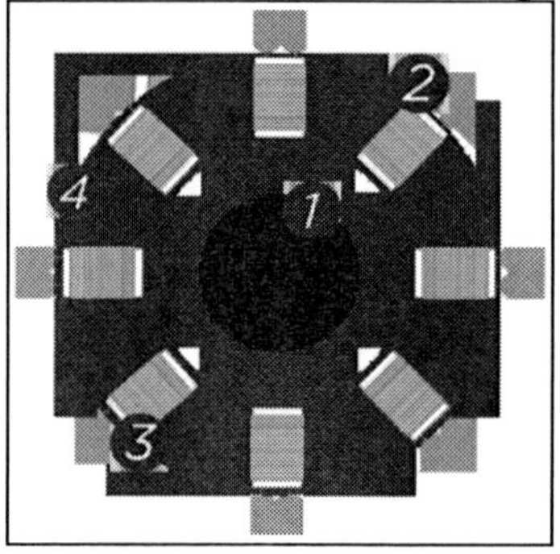

Fig. 1. The Coil Sensor Configuration in Top View

It has specification of 0.8 mm coil wire as long as (8 x 50 x 2 x π x 21) mm = 52.800 mm or approximately 55 m; and cylinder pipe for coil winding with 21 mm radius. The total height is estimated (8 x 6) cm = 48 cm. Transmitter and receiver buffer with area of (π x (5.5 + 6)2) cm^2 = 415.3 cm^2. In **Fig**. 1, the notation 1, 2, 3, 4 each shows the object, transmitter coil, receiver coil, and coil sensor buffer respectively. The transmitter and receiver coil is arranged to be face to face (opposite).

Basically, the magnetic induction tomography (MIT) sensor consists of at least a pair of transmitting and receiving sensors, both of which are coils. The transmitter coil is connected to a power source that provides an electrical signal. In accordance with Maxwell's equations, the receiving coil will detect the magnetic field and convert it into an electric field, which can be measured in voltage and phase quantities. The existence of an object with a certain conductivity between the transmitter and the receiver coil will affect the excited magnetic field so that the measured voltage on the receiver coil will differ in magnitude before and after being given the object. This measured voltage difference is used to then determine the image of the object. The transfer function of the receiver coil sensor is expressed by the following equation which is a derivative of the Maxwell-Faraday equation [6].

$$V = -n \,.\frac{d\Phi}{dt} = -n \,.A \,.\frac{dB}{dt} = -\mu_0 \,.n \,.A \,.\frac{dH}{dt} \quad (1)$$

$V =$ voltage

$n =$ number of coil

$\mu_0 =$ magnetic vacuum conductivity

$A =$ coil surface area

$H =$ magnetic field intensity

To produce high image resolution, a sufficient number of measurements are required to map all output voltages excited by each part of the object. Increasing the number of measurements can be made by increasing the number of coils or by moving the transmitter coil and/ or the receiver and/ or the object [6].

3 Characterization

A characterization is conducted related to the design. Parameters being tested for determining an optimal sensor design are transmitter-receiver distance, and frequency of transmitter signal. Initially, a single pair of transmitting and receiving sensor is used, with variations in distance ranged from 3cm - 11cm, and signal frequency variation in ranges: 10,000Hz – 100,000Hz, 100,000Hz - 1,000,000Hz, and 1,000,000Hz - 10,000,000Hz. The test is performed in accordance with the experimental diagram in Fig. 2 below.

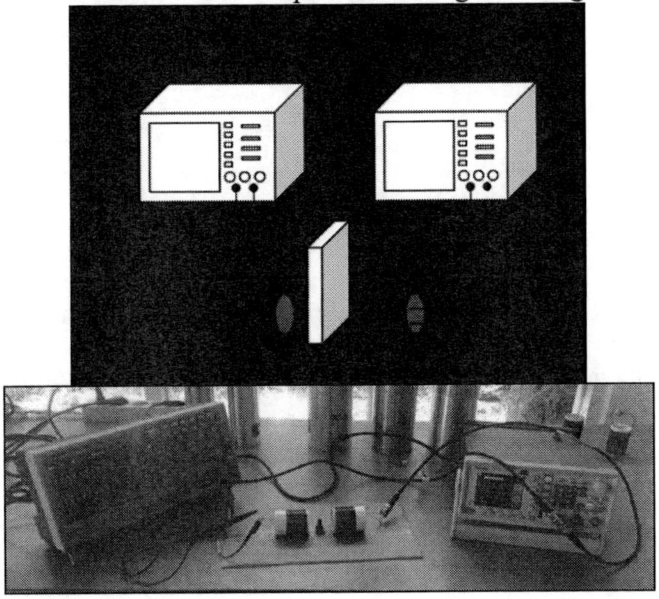

Fig. 2. Characterization Diagram and Experiment Setup

For each variation of these parameters, the optimum condition is indicated by the largest voltage difference measured before and after the magnetic field is disturbed by the object. In graphical representation, the measurement results are shown as follows.

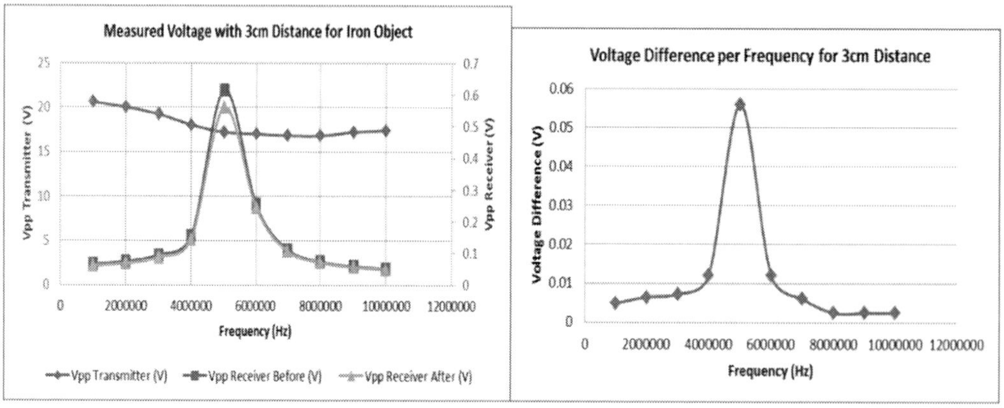

Fig. 3. Measured Voltage and Voltage Difference with 3cm Distance

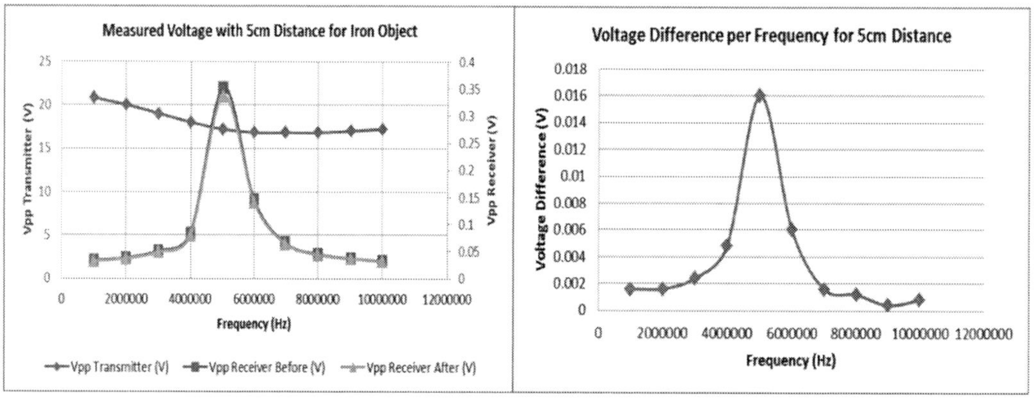

Fig. 4. Measured Voltage and Voltage Difference with 5cm Distance

Based on the tests, it is found that the measured voltage difference will be greater if the distance transmitter - receiver is getting closer. For frequency parameters, 4MHz and 5MHz values show significant voltage differences compared to other frequencies. Thus, the inductance sensor of this coil will be designed to have the closest distance between the transmitter coil and its receiver; while excitation signal is set at frequency range of 4MHz - 10MHz, and is used for imaging metal objects.

4 Experiment Results

Sensor and its testing are made according to the optimum parameters that have been obtained. Figure 5 shows multi-channel sensors measurement setup according to the design.

Fig. 5. Multi-Channel MIT Sensor Experiment Setup

 https://doi.org/10.1051/matecconf/201821802001

There are eight coils with four transmitters and four receivers. Each transmitter coil is arranged in opposite with the receiving coil. The following test results show the measured voltages for the iron object for all four pairs of coil sensors.

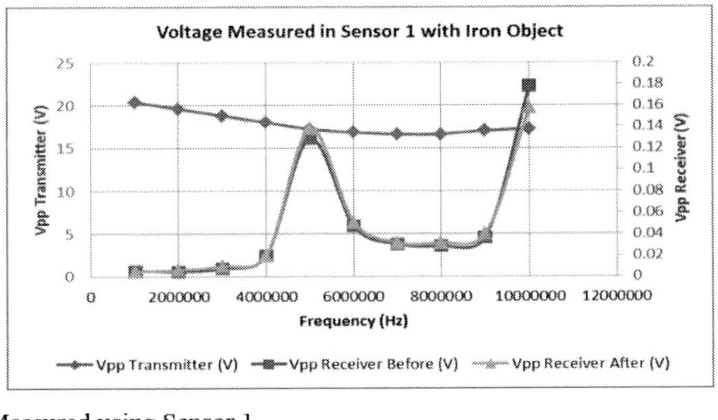

Fig. 6. Voltage Measured using Sensor 1

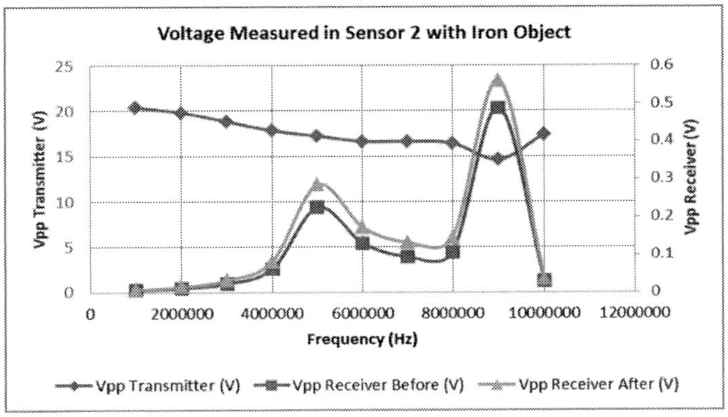

Fig. 7. Voltage Measured for using Sensor 2

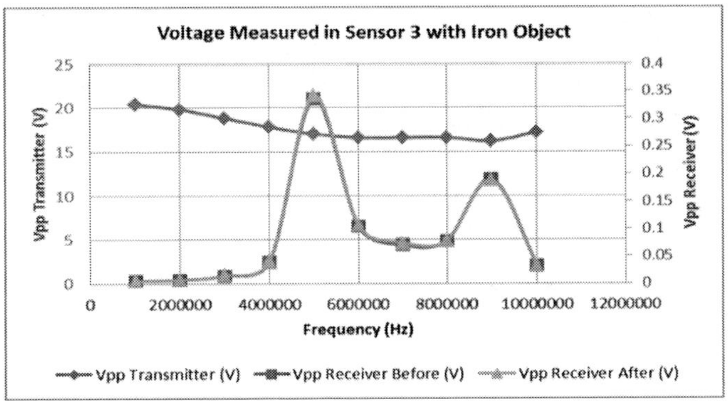

Fig. 8. Voltage Measured using Sensor 3

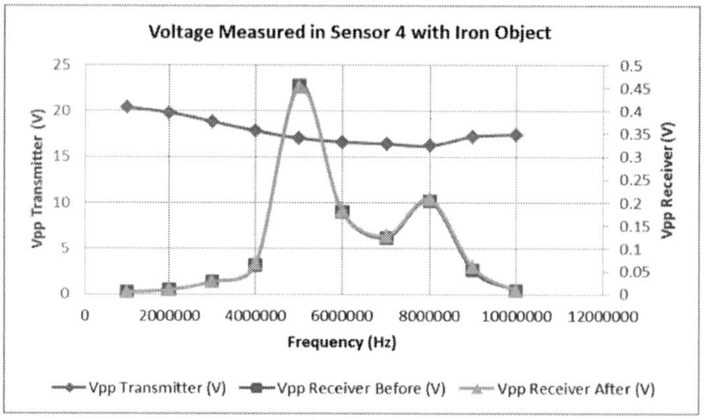

Fig. 9. Voltage Measured using Sensor 4

Based on the measurement results, it can be seen that in general, frequencies that give the most significant voltage difference are 5 MHz and 9 MHz. These two frequencies are then referred to as resonance operating frequencies, where the greatest voltage difference before and after the magnetic field is disturbed by the object occur. Multi-frequency sensor implementation could be beneficial in precision medical use [8].

The date is verified from Bath 8-MIT system used for frequency spectroscopy imaging [9]. Metallic object was placed inside sensing area of similarly constructed coil (with different number of turns) and swept by ac-signal at frequency of interest. Amplitude mode measurements were taken and reconstructed using back-projection algorithm [10] as shown in Fig. 10.

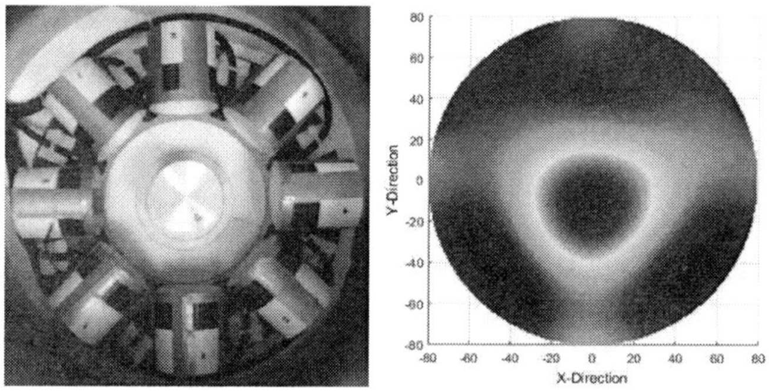

Fig 10. Reconstruction of the Object

Generally, the presence of conductive object can be distinguished, although the image is rough due to simplification in reconstruction. Further analysis of imaging requires more considerations toward physical simulation and algorithm, and is beyond the scope of this work.

5 Discussion and Conclusions

An induction coil sensor with four transmitter and four receiver coils has been constructed. The main goal is to detect the voltage difference in the presence of metal object. Made of 50 turns, 4.2 cm diameter coil, and 0.8 mm wire diameter, arranged in circle and the transmitter and receiver pair is face to face, the sensor is proved to be able to sense the voltage difference, with the most significant value captured at 5MHz and 9MHz frequency. Multi-channel measurement performance has been tested and offers adequate merits for electromagnetic tomography on conductive material.

For higher sensitivity, conductor shield can be attached to separate the transmitter and receiver coils electronically without disturbing the excited magnetic field. The arrangement of coils also influential. By arranging the transmitter and receiver coils asymmetrically, the number of measurements can be increased to improve image resolution. Additionally, the preparation of each coil as an array will allow the excitation of magnetic field of one of the transmitter coils to be captured by the receiver coil as an array as well.

Reference

1. H Griffiths, "Magnetic induction tomography," in *Electrical Impedance Tomography - Methods, History and Applications.*, D S Holder, Ed.: IOP Publishing, 2005, ch. 8, pp. 213-38.
2. M Soleimani, Image and Shape Reconstruction Methods in Magnetic Induction and Electrical Impedance Tomography, PhD Thesis.: The University of Manchester, 2005.
3. Wuliang Yin et al "The Design of a FPGA-based Digital Magnetic Induction Tomography (MIT) System for Metallic Object Imaging," Instrumentation and Measurement Technology Conference, ©2010IEEE doi: 10.1109/IMTC.2010.5488275
4. J. Clerk Maxwell, *A Treatise on Electricity and Magnetism*, Oxford: Clarendon, 1892, pp.68-73.
5. H. Yu Wei, M. Soleimani, "Magnetic Induction Tomography System for Prospective Industrial Processing Applications," Chin. J. Chem. Eng., vol. 20. 2012
6. S. Tumanski, "Induction coil sensors a review," Meas. Sci. Technol., 2007.
7. C. Hannelore Igney, R. Pinter, O. Such, "Magnetic Induction Tomography System and Method," US Patent US8125220B2. 2012.
8. I. Muttakin, S.-Y. Yeap, M. M. Mansor, M. H. M. Fathil, I. Ibrahim, I. Ariffin, C. Omar, and E. Supriyanto, "Low cost design of precision medical ultrasound power measurement system," International Journal of Circuits, Systems and Signal Processing, vol. 5, no. 6, pp. 672-682, 2011.
9. L. Ma, M. Soleimani, "Magnetic Induction Spectroscopy for Permeability Imaging," Sci. Rep., vol. 8, no. 1, p. 7025, 2018.
10. H. Herdian, I. Muttakin, A. Saputra, A. Yusuf, W. Widada, W. P. Taruno, "Hardware implementation of linear back-projection algorithm for capacitance tomography", 2015 4th International Conference in Instrumentation Communications Information Technology and Biomedical Engineering (ICICI-BME), pp. 124-129, 2015.

Image Data Acquisition for NOAA 18 and NOAA 19 Weather Satellites Using QFH Antenna and RTL-SDR

Romi Wiryadinata[1], *Muhammad* Khoirussolih[1], *Neneng* Rohanah[1], *Imamul* Muttakin[1], *Teguh* Firmansyah[1]
[1]*Department of Electrical Engineering, Universitas Sultan Ageng Tirtayasa, Cilegon, Indonesia*

Abstract. Satellite imaging is widely used in monitoring weather patterns and interpreting satellite images. The limited access to weather satellite image data in real-time is still major issues faced by satellite enthusiasts. NOAA 18 and NOAA 19 weather satellites provide free accessible weather data continuously through Automatic Picture Transmission at 137.9125 MHz and 137.1000 MHz frequencies. In this research the data acquisition system is designed with the help of a Quadrifilar Helix antenna and RTL-SDR. In order to receive both frequencies, the Quadrifilar Helix antenna at 137,5 MHz is designed. Recording, decoding, editing, and viewing of Automatic Picture Transmission signals using WXtoImg software. The result showed that the weather satellite image receptions are sufficiently clear and descriptive, although there are 0.0685% frequency offsets of Automatic Picture Transmission signal reception on NOAA 18 and 0.0686% on NOAA 19.

1 Introduction

Access to obtain local weather satellite image data in real-time is still very limited and difficult to decode the hidden information contained in it. The APT (Automatic Picture Transmission) signal which can be accessed by the public makes NOAA weather satellite an alternative choice to get weather satellite image data. The NOAA 18 and NOAA 19 satellite weather image data will be very useful in GIS (Geographic Information System) that can be used for disaster mitigation, scientific investigation, resource management, development planning, cartography and road planning [1].

The NOAA weather satellite carries five types of sensors, one of which is the AVHRR (Advanced Very High-Resolution Radiometer) [2]. The AVHRR sensor has a function to detect the reflection of electromagnetic waves by clouds, objects on the surface of the earth, as well as the wave of surface emissions of clouds and surface waters. The AVHRR sensor has 6 detectors working on a radiometer channel of different wavelengths ranging from visible and thermal infrared. The APT signals transmitted by NOAA satellites are RHCP (Right Handed Circularly Polarized), so it takes an RHCP type antenna to produce a good signal reception [3][4][5].

[1] Corresponding author: wiryadinata@untirta.ac.id

This paper focuses on hardware and software implementation to receive and decode the APT signal on NOAA 18 and NOAA 19 weather satellites. The ideal antenna type for APT receiver at 137 MHz is the QFH antenna (Quadrifilar Helix) [6][7][8]. The QFH antenna we made using copper pipe with the proper dimensions. The presumption that to receive satellite image requires a large parabolic antenna can be broken with a QFH antenna. QFH antennas made with copper pipes are also expected to ease the manufacturing process. SDR is a key area for realizing a variety of software implementations that allow adaptive and re-configurable communication systems [9].

Many SDR devices used as NOAA weather satellite receivers [10][11][12]. In this research, we use RTL-SDR. This SDR uses Realtek® RTL2832U as the controller and Rafael Micro® R820T2 as the tuner.

2 Research Method

The satellite image data acquisition system in this study consists of QFH antenna, LNA, RTL-SDR and a computer that already installed with CubicSDR and WXtoImg. QFH antenna connected at a certain height to receiver and computer shown in Figure 1.

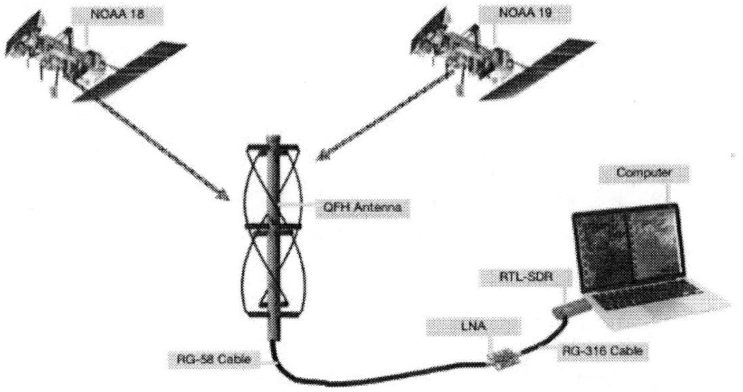

Fig. 1. Simplified Scheme of Satellite Image Data Acquisition System

The diagram block of the satellite image data acquisition system represented by Figure 2. The QFH antenna receives a very weak APT signal from the satellite. We calculated antenna design using QFH calculator [15]. The designed antenna (Figure 4) is a QFH antenna is a ½ rotation, 1 lambda wavelength, and 137,5 MHz working frequency. To connect RTL-SDR and QFH antennas required some adapter connectors. Since the RTL-SDR device used has an MCX connector, the antenna is using the PL-289 UHF connector and connected to the LNA using an SMA connector. LNA acts as a signal amplifier that amplifies the signal from the QFH antenna.

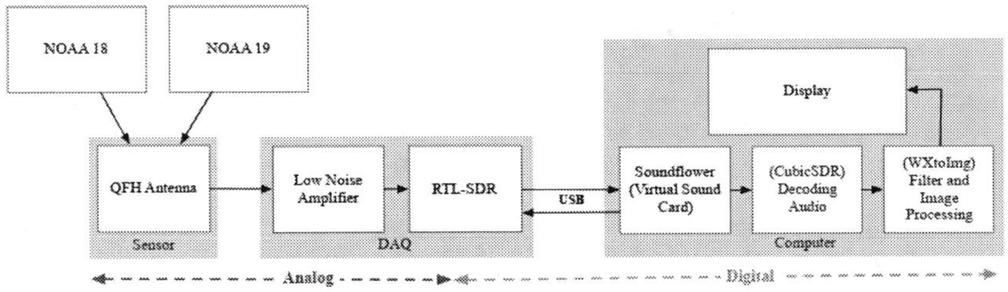

Fig. 2. Block Diagram of Satellite Image Data Acquisition System

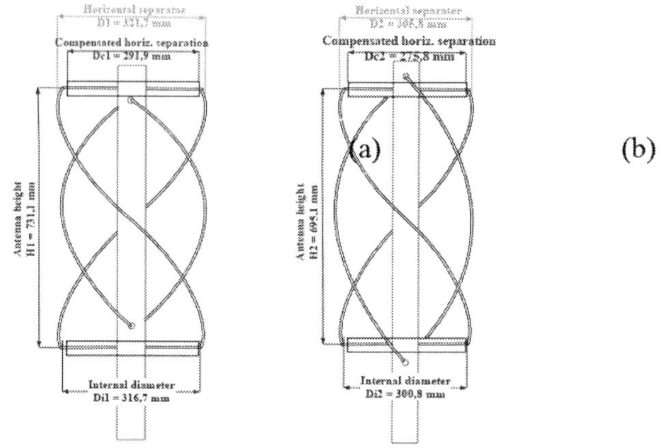

Fig. 3. QFH Antenna Design. a) Larger Loop, b) Smaller loop

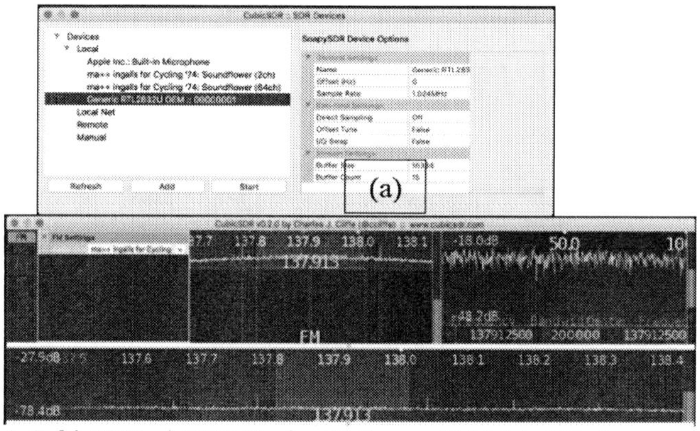

Fig. 4. CubicSDR; a) SDR Device Option and b) Main Window

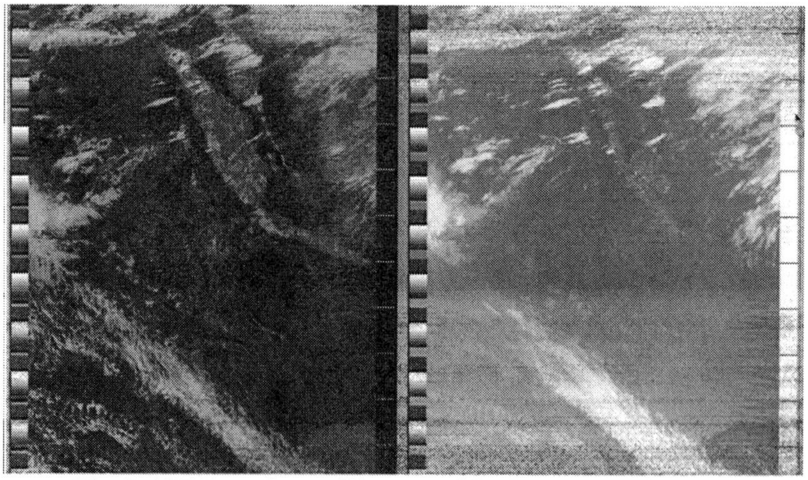

Fig. 5. APT Recording and Decomposition Process on WXtoImg

The information signal received by the QFH antenna is then separated from the carrier signal by RTL-SDR. The RF spectrum was tuned by the tuner on RTL-SDR which is then converted to IF. The demodulator on the RTL-SDR converts from IF to baseband, the ADC digitized the signal at this stage. The CubicSDR software act as the demodulation process control. After opening CubicSDR, SDR RTL2832U device was selected and set sample rate of 1024 MHz (Figure 4 (a)). We adjust center frequency of CubicSDR to the NOAA frequency (Fig. 4 (b)), ie 137,9125 for NOAA 18 MHz and 137,1000 MHz for NOAA 19. The signal was passed to CubicSDR and WXtoImg by the Soundflower kernel extension. CubicSDR sets parameters on RTL-SDR and WXtoImg records the received APT signal from RTL-SDR. APT signal captured by QFH antenna, recorded by WXtoImg to generate wav file. WXtoImg is also enabled for decoding and preprocessing APT signals.

3 Results and Analysis

3.1 QFH Antenna Result

During the measurement, the antenna is placed in an obstacles free area. The MFj-269 antenna analyzer was used for measurements. Based on the measurement result, the working frequency of 137.5 MHz obtained resistance 41 Ω, reactance 10 Ω, impedance 42,202 Ω, and VSWR 1,3. The following Table 1 results from antenna QFH measurement of a frequency from 132 MHz to 144 MHz. The designed QFH antenna meets the specified VSWR specifications, although the best result is shown at 141 MHz.

Tabel 1. QFH Antenna Measurement Result

f (MHz)	Rs (Ω)	Xs (Ω)	Z_{in} (Ω)	VSWR	Γ	RL (dB)
132,000	39	12	40,804	1,400	0,181	14,834
132,500	40	10	41,231	1,300	0,156	16,128
133,000	41	9	41,976	1,300	0,139	17,128
133,500	42	9	42,953	1,200	0,130	17,703
134,000	43	9	43,932	1,200	0,122	18,271
134,500	44	9	44,911	1,200	0,115	18,820
135,000	44	9	44,911	1,200	0,115	18,820
135,500	44	10	45,122	1,200	0,123	18,176
136,000	44	10	45,122	1,200	0,123	18,176
136,500	44	11	45,354	1,300	0,132	17,563
137,000	42	11	43,417	1,300	0,147	16,666
137,500	41	10	42,202	1,300	0,147	16,656
138,000	40	8	40,792	1,300	0,142	16,971
138,500	39	6	39,459	1,300	0,140	17,048
139,000	40	2	40,050	1,200	0,113	18,917
139,500	41	0	41,000	1,100	0,099	20,096
140,000	44	0	44,000	1,100	0,064	23,900
140,500	47	0	47,000	1,064	0,031	30,193
141,000	51	0	51,000	1,020	0,010	40,086
141,500	53	0	53,000	1,060	0,029	30,714
142,000	55	4	55,145	1,100	0,061	24,302
142,500	53	8	53,600	1,100	0,083	21,650
143,000	50	11	51,196	1,200	0,109	19,224
143,500	47	12	48,508	1,200	0,127	17,954
144,000	44	11	45,354	1,300	0,132	17,563

Improper antenna impedance has an impact on reducing the efficiency of the antenna because some of the signals are reflected back to the source. The coefficient reflection of 137,5 MHz was 0,147. Return loss, RL, at 137,5 MHz working frequency obtained value of 16,656 dB. Return loss of 16,656 dB causes more than 97% power absorbed to antenna and less than 3% power reflected. A larger return loss indicates a good impedance matching.

3.2 Signal to Noise Ratio

In this paper, we tested the LNA that connected to QFH antenna with 12 VDC of power, while the RF gain on the RTL-SDR is set at 13 dB. P_{signal} is the average power signal received by the antenna. P_{noise} is the average value of power noise received by the antenna. The average SNR reading is 57,197 dB.

Tabel 2. SNR of VHF Frequencies

Frekuensi (MHz)	P_{signal} (dB)	P_{noise} (dB)	SNR (dB)
137,100	-2,8409	-60,4167	57,5758
137,200	-2,6515	-60,6061	57,9546
137,300	-2,6515	-60,7955	58,144
137,400	-2,6515	-60,0379	57,3864
137,500	-2,8409	-59,2803	56,4394
137,600	-2,8409	-59,4697	56,6288
137,700	-3,2197	-59,4697	56,25
137,800	-3,0303	-60,6061	57,5758
137,900	-3,4091	-60,2273	56,8182
Average			57,197

The internal amplifier in the RTL-SDR has a noise figure of <4,5 dB. The amount of noise received causes the signal to degrade, especially when the gain is increased. SNR is a measure of signal quality relative to noise, the higher the better. SNR can be increased by reducing noise from the receiver. The use of LNA can help to reduce the noise effect that can enter through the coaxial cable and help reduce attenuation on a long cable. Apart from the use of LNA, bandwidth can also affect the SNR because the wider the bandwidth of the receiver, the greater the noise generated.

3.3 Offset Frequncy

The tested RTL-SDR had low-quality crystal oscillator, with ± 150 PPM (parts per million) offset. PPM indicates how often the crystal oscillator frequency can deviate from its original frequency. PPM can be changed into percent, 150 PPM equals 0.015%. Referring to 150 PPM, there should be 20.687 kHz frequencies offset. Cause signal reception at 137,9125 MHz would be 137,892 MHz. The RF spectrum (Figure 6 (a)) shows the signal at 137.818 MHz which means there are 94.5 kHz frequencies offset or equivalent to 0.0685%. Referring to 150 PPM, there should be 20,565 kHz frequencies offset when receiving NOAA 19 satellite signal. Causes signal reception at 137,1000 MHz would be 137.079 MHz. The RF spectrum (Figure 6 (b)) shows the reception of the NOAA 19 satellite signal at 137.006 MHz which means there are of 94 kHz frequencies offset or equivalent to 0.0686%.

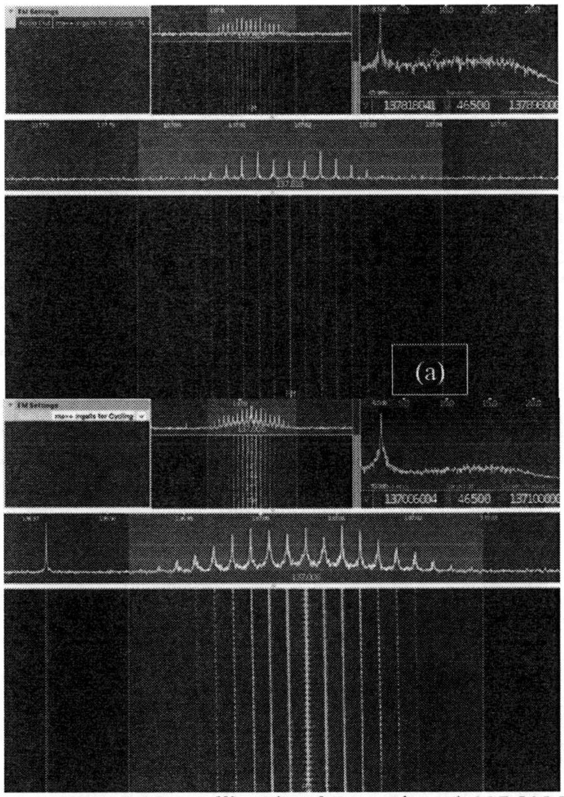

Fig. 6. Offset Frequency on NOAA 18 Satellite Signal Reception, a) 137.818 MHz, b) 137.006 MHz

The frequencies offset on NOAA 18 and NOAA 19 satellite signals reception is not only caused by oscillator crystals but also can depend on environmental factors mainly by temperature. The heating process when a new RTL-SDR is used causes the received frequency signal to deviate over time. The aberrant frequency will be stable after the RTL-SDR reaches thermal equilibrium. The solution that can be done is to calibrate the frequency of the CubicSDR application. The frequencies offset can be adjusted by sett the PPM offset value.

3.4 Acquisition Result

Data acquisition process begins by sett center frequency on CubicSDR with the bandwidth of 34 kHz. The output from CubicSDR needs to be forwarded to WXtoImg so that the recording and decoding of APT signals can be done automatically. Recorded APT signal results are stored automatically via WXtoImg with wav (Waveform Audio File) format. WXtoImg changed the recording of the wav file from the NOAA weather satellite APT signal to image data, which results can be instantly displayed on WXtoImg while still in the recording process. Figure 7. shows the sample acquisition of NOAA 19 satellite image data in real time on August 15, 2017, at 15:28:44 (UTC+7). The satellite passes from south to north with a maximum elevation angle of 69 ° westward. The recording duration lasts for 11 minutes and 45 seconds until 15:40:29 (UTC+7).

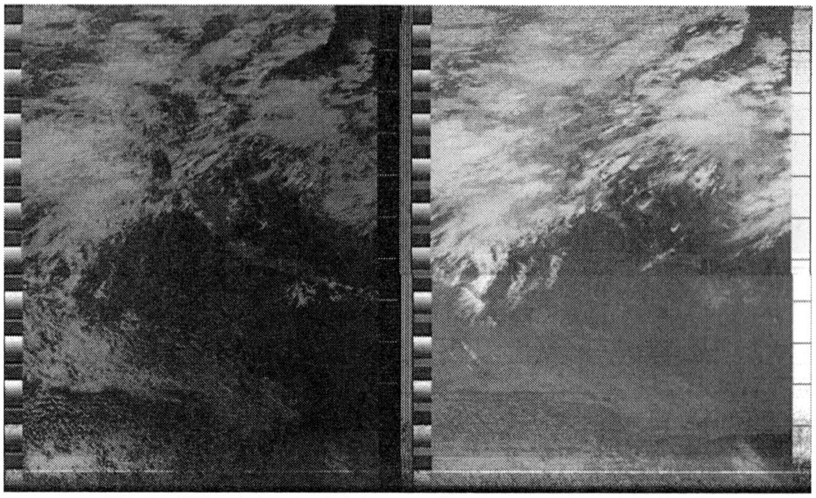

Fig. 7. NOAA 19 On 15 August 2017 at 15:28:44 (UTC+7)

The image data (Figure 7) on the left is the near infrared channel and on the right, is the thermal infrared channel. The image data stored in JPEG format with 2080 × 1405 pixels of image resolution. The AVHRR sensor used in NOAA has a very wide field of view with a spatial resolution of 4 km. Spatial resolution refers to the amount of land represented by each pixel. The higher the spatial resolution means the more pixels per unit area. Suitable mapping scales determined from known spatial resolution [15]. Satellites with a spatial resolution of 4 km produce maps with a 1:8000 of scale.

Samples of image data acquisition when NOAA 18 satellite passes at night on August 15, 2017, at 18:41:32 (UTC +7), shown in Figure 4.11 below. The satellite passes from south to north with a maximum elevation of 52 ° to the east. The recording duration lasts for 11 minutes and 27 seconds until 18:52:59 (UTC+7).

Fig. 8. NOAA 18 On 15 August 2017 at 18:41:32 (UTC+7)

Satellite image data at night shows infrared data on both channels with a different spectrum. Mid-infrared data spectrum on the left is 3.55 - 3.93 μM and thermal infrared on the right is 10.30 - 11.30 μM. The image data stored in JPEG format with 2080 × 1358 pixels of image resolution. Each channel resolutions are 1040 × 1358 pixels. The quality of data reception over analog radio waves affected by unfavorable weather and atmospheres during the trip of APT signals from satellites to ground stations. One of the atmosphere effects is energy absorption by each layer of the atmosphere. This absorption depends on the density and air content at the time of the test. There is also an effect of propagation delay and frequency shift, which affected by the ionosphere layer. Disturbances in the ionosphere layer occur when particles ionized, especially by solar radiation. There is always a possibility for improvement in the image data acquisition system, but so far the one we implemented is the best in APT signal reception.

4 Conclusion

In this paper, we have built an inexpensive NOAA satellite ground station. The designed QFH antenna was able to work well at 137,5 MHz with 1,3 of VSWR and 42,202 Ω of impedance, while the best results in 141 MHz with 1,06 of VSWR and 51 Ω of impedance. RTL-SDR is capable of receiving information from NOAA 18 and NOAA 19 satellites by radio waves quite well, although there are 0.0685% frequency offsets of Automatic Picture Transmission signal reception on NOAA 18 and 0.0686% on NOAA 19.

References

1. Wicaksono A., F. F. Muhsoni, and A. Fahrudin, "Aplikasi Data Citra Satelit NOAA-17 Untuk Mengukur Variasi Suhu Permukaan Laut Jawa," *J. Kelaut.*, **vol. 3**, no. 1, pp. 70–74, (2010)
2. National Oceanic and Atmospheric Administration, *User's Guide for Building and Operating Environmental Satellite Receiving Stations*, no. February. Suitland,: U.S. DEPARTMENT OF COMMERCE, (2009)
3. Gao S.S., Q. Luo, and F. Zhu, *Circularly Polarized Antennas*. Chichester: John Wiley & Sons, (2014)
4. Stutzman W.L. and G. A. Thiele, *Antenna Theory and Design 2nd edition*. Hoboken: John Wiley & Sons, (1998)
5. Balanis C.A. , *Antenna Theory Analysis and Design 3rd Edition*. Hoboken: John Wiley & Sons, (2005)
6. Hollander R.W., *Resonant Quadrafilar Helical Antenna*. Den Haag: Working Group Satellites, (1999)
7. Fusco V.F., R. Cahill, and R. L. Li, "Quadrifilar loop antenna," *IEEE Trans. Antennas Propag.*, vol. 51, no. 1, pp. 115–120, (2003)
8. Govarthanam K.S., M. Abirami, and J. Kaushik, "Economical Antenna Reception Design for Software Defined Radio using RTL-SDR," in *Proceedings of the International Conference on Innovative trends in Electronics Communication and Applications*, 2014, pp. 185–190.
9. Stewart R.W., K. W. Barlee, D. S. W. Atkinson, and L. H. Crockett, *Software Definded Radio using MATLAB & Simulink and the RTL-SDR*. Glasgow: Strathclyde Academic Media, (2015)
10. Patil C., T. Chavan, N. Jhurani, D. Jethani, and M. Chaudhari, "Satellite Image Acquisition Using SDR," *Int. J. Innov. Res. Comput. Commun. Eng.*, **vol. 4**, no. 6, pp. 2257–2263, (2016)
11 Mahmood S., M. T. Mushtaq, and G. Jaffer, "Cost Efficient Design Approach for Receiving the NOAA Weather Satellites Data," in *Proceedings of the IEEE Aerospace Conference*, pp. 1–6, (2016)
12. Hassini A. and A. H. Belbachir, "Hardware and software consideration to use near real time MSG-SEVIRI and NOAA-AVHRR images," *c, ISPS 2013*, pp. 12–16, (2013)
13. Maheshwarappa M.R. and C. P. Bridges, "Software defined radios for small satellites," *Proc. 2014 NASA/ESA Conf. Adapt. Hardw. Syst. AHS 2014*, no. July 2014, pp. 172–179, (2014)
14. Coppens J., "Quadrifilar helicoidal antenna." [Online]. Available: http://jcoppens.com/ant/qfh/calc.en.php. [Accessed: 14-Apr-2017].
15. Arlinghaus S.L. and J. J. Kerski, *Spatial Mathematics: Theory and Practice through Mapping 1st Edition*. Boca Raton: CRC Press, (2014)

Photovoltaic Measurement Using Microprocessor Logger and Current Sensor

Rhezal Agung Ananto[1], and *Rudy* Setiabudy[1,*]

[1] Electrical Engineering Department, Universitas Indonesia, Kampus Baru UI Depok 16424, Indonesia

Abstract. Nowdays, energy demand increase every years. Some obstacles stand in the way of energy supply. Some obstacles are energy limitation, climate change and environmental regulation. The change of conventional energy to renewable energy is the future solution. The development of renewable energy is the main priority, based on Indonesian government regulations. The potential of solar energy in Indonesia is very large because it is located in the equator. The research tries to measure solar energy in Indonesia region, it use atmega328p microprocessor logger and INA219 current sensor module. The measurement result is used to determine the photovoltaic stand-alone system performance. Measurement variables are sun irradiance, voltage, current, power, energy and temperature. The research method is measurement comparison between photovoltaic system with maximum power point tracking (MPPT) and without MPPT. Sun irradiation measurement with solar power meter is compared with sun irradiation data from NASA. Performance Ratio of the photovoltaic system with MPPT is better than the photovoltaic system without MPPT, the best photovoltaic system with MPPT Performance Ratio (PR) result is 0.93.

1 Introduction

Today, energy demand is increasing every year. Some obstacles are energy limitation, climate change and environmental regulation. Currently, most of the primary energy sources are oil and gas. The change of conventional energy to renewable energy is the future solution. One of the biggest potential energy in the future is solar energy. Solar energy is one of renewable energy. The development of renewable energy is the main priority, based on Indonesian government regulations. Indonesia is also located on the equator, which has solar energy every years. Base on the above problems, the research is expected to be useful in the future to develop solar energy in Indonesia.

An effective Photovoltaic power system is needed to convert solar energy into electrical energy. This system is needed because the energy produced by Photovoltaic is not constant, based on weather condition. Some of the Photovoltaic systems are Photovoltaic stand-alone system and microgrid system [1].

* Corresponding author: rudy@eng.ui.ac.id

Performance measurements of Photovoltaic systems are required to obtain effective solar power generation systems. The measurement of solar power performance based on IEC 61724 is the photovoltaic Performance ratio (PR). Example of photovoltaic performance ratio is short-term performance ratio measurements performed in subtropical climates in India. In India, PR (performance ratio) results are 0.74; 0.78; and 0.76 [2].

This Measurement is located in the tropical climates of Indonesia. Malaysia is the closest country, it has the same tropical climates characteristic. The results of performance Ratio (PR) measurements in Malaysia during 4 days of measurements are 0.5940; 0.4481; and 0.280 [3]. The Performance ratio results from other research in the same area are 0.75; 0.77; 0.78; 0.82; 0.84; 0.85; and 0.92. The research locations are located in several city in Indonesia. The locations are Surabaya, Kupang, and Tangerang Selatan[4-6]. In this paper, the photovoltaic measurement is located in kabupaten Malang, East Java Province, Indonesia (tropical climates area).

2 Photovoltaic Stand-alone System and Performance Ratio

Photovoltaic systems require several essential elements to support a Photovoltaic stand-alone system running effectively. One of the necessary essential elements is the battery. Batteries are useful for saving energy during the day, when Photovoltaic produces energy. Charge controller is useful to protect against battery overcharging (protect from burning) and to set power on Maximum Power Point(as MPPT). 12V DC load does not require a DC/DC converter and AC load required DC/AC inverter.

Performance ratio is used to determine the performance of Photovoltaic system. Based on the IEC 61724 standard, performance ratio is the actual energy ratio of Photovoltaic system with reference energy (energy measurement in the region) [7]. Calculating Performance Ratio (PR) takes several parameters, according to IEC 61724 standard, we can see the following parameters; E_{out} (Photovoltaic Output Energy (kWh)), P_0 (Photovoltaic Power rating (kWp)), H_i (In-Plane irradiation (kWh/m^2)), and $G_{i,ref}$ (in-plane irradiance peak reference 1000 (W/m^2)). Formula of Performance ratio calculation (PR) is the formula below,

$$PR = \frac{E_{out}/P_0}{H_i/G_{i,ref}} \tag{1}$$

This formula is commonly called the traditional performance ratio, because this formula does not consider the module temperature.

3 Methodology

This study is comparing performance ratio of two different photovoltaic system. Two different photovoltaic system are installed in the same location. First system is photovoltaic system with MPPT and second system is photovoltaic system without MPPT (photovoltaic direct to load). Power measurement using atmega328p microprocessor logger and INA219 curent sensor. Sun irradiance power measurement using solar power meter. First photovoltaic system has $P_0 = 10\ Wp$ (Polycrystalline PV Panel), $V_{mp} = 17.2$ V, $I_{mp} = 0.58$ A, MPPT = 10 A; and Battery = 12V/60Ah. Second photovoltaic system has $P_0 = 10$ Wp (Polycrystalline PV Panel), $V_{mp} = 17.2$ V, $I_{mp} = 0.58$ A, and load 30 ohm.

Design of logger atmega328p and INA219 current sensor has good error test result. The logger error results are 0.29% - 1.16% (voltage) and 0.19% - 0.57% (current).

First photovoltaic system, current sensor INA219 logger is installed between photovoltaic module and maximum power point tracking (MPPT). Second photovoltaic system, current sensor is installed between photovoltaic module and load. Current sensor

Logger can measure voltage, current and time. The measurement parameter is recorded in micro-sd card using micro-sd module every 0.5 second. The sun irradiance mesurement is recorded every 2 second using solar power meter in micro-sd card. Solar power meter is installed near photovoltaic location. Figure 1 below is the installation of current sensor and research configuration block diagram.

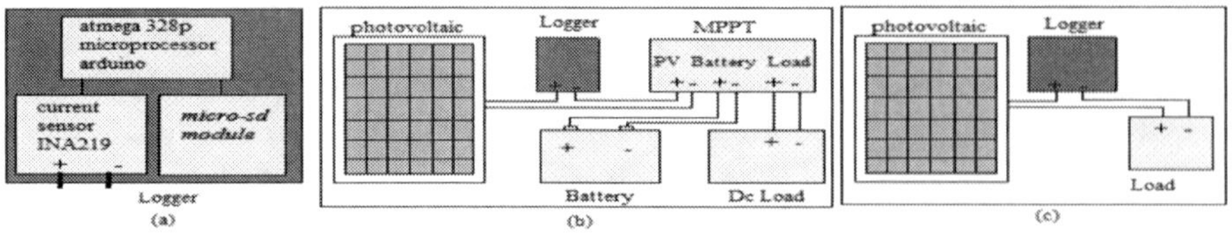

Fig. 1. Block Diagram (a) Logger current sensor (b) PV with MPPT (c) PV without MPPT.

Energy can calculate with equation (2) below,

$$E_{out} = \frac{P}{3600\times(\frac{1000}{\Delta t})} = \frac{V\times I}{3600\times(\frac{1000}{\Delta t})} \quad (2)$$

With V = Voltage (volt), I = Current (ampere), E_{out} = Energy(Wh), P = Power (watt) and Δt = delta sampling time (milliSeconds).

4 Result and Discussion

Photovoltaic performance ratio measurement is located in kabupaten malang, east Java, Indonesia. The comparison photovoltaic system PR and sun irradiation measurement results can be seen in Table 1 below,

Table 1. Sun Irradiance Measurement Using Solar Power Meter and Performance Ratio.

Date	H_i (Wh/m^2/day)	Without MPPT		With MPPT		T(°C)
		E_0 (mWh)	PR	E_0 (mWh)	PR	
28 April	7054.54	55513.13	0.79	32384.03	0.46	25.64
29 April	7255.42	54970.67	0.76	32384.03	0.45	25.33
4 May	7138.51	58007.39	0.81	37758.06	0.53	24.42
5 May	5976.52	46036.15	0.77	53859.38	0.90	24.74
6 May	6592.79	52417.99	0.80	58723.57	0.89	26.32
7 May	6881.85	55351.16	0.80	63495.24	0.92	25.79
8 May	6951.75	56031.09	0.81	52645.93	0.76	25.75
9 May	6938.77	55917.11	0.81	32716.02	0.47	25.20
10 May	6659.63	53716.27	0.81	60810.84	0.91	26.32
11 May	6736.32	54648.34	0.81	61970.78	0.92	25.01
12 May	6643.46	55334.03	0.83	61985.51	0.93	24.76

H_i, and E_{out} data in Table 1 are obtained with solar power meter measurement output data and microprosessor logger measurement output data. Performance Ratio (PR) photovoltaic system in Table 1 can calculate with equation (1). The calculation below is Example of PR calculation 12 may 2018 (with MPPT). PR = [61985.51 mWh/10000mWp]/[(6643.46 Wh/m^2/day)/(1000 W/m^2)] = 0.93.

The lowest Performance ratio (PR) result of the two systems is 0.45. The highest PR result of the two system is 0.93. The performance ratio result is affected by the solar irradiance. The solar iradiance is directly proportional to the performance ratio of the system. The performance ratio of the system with MPPT is affected by the battery condition when charging process occurs. When the battery condition is full performance ratio will decrease by less than 0,5. The power measurement 28 april 2018(full battery conditon) and 12 may 2018 (battery condition is not full) results can be seen in Figure 2 below. The Figure 2 below can be used to understand the battery condition effect on Performance ratio result.

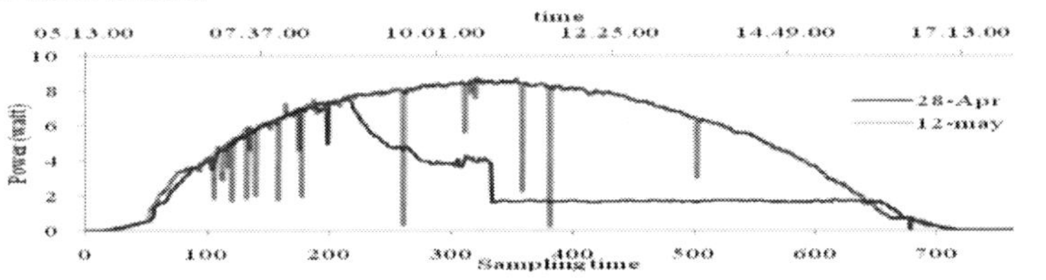

Fig. 2. Power Chart with MPPT 28 April 2018 and 12 may 2018 (comparison of battery condition).

Figure 2 above is photovoltaic system with MPPT power chart with different battery condition. The chart start at 05.13 o'clock in the morning and finish at 18.10 o'clock in the evening. The highest power on 28 april 2018 (battery condition is full) is 7.473 watts at 08.45 o'clock, and the highest power on 12 may 2018 (battery condition is not full) is 8.703 watts at 10.34 o'clock.

The performance ratio decreases due to the battery condition. The current limiting is active, when battery condition is full. This system works because MPPT has current limiting system, so the battery is not overcharged (burned). Performance ratio will be high when battery condition is not full. The condition can be seen in Figure 2 (12 may). The chart indicates that the current limiting system is not active.

The photovoltaic system without MPPT produces voltage based on the characteristics of the photovoltaic loadline I/V characteristic curve. The photovoltaic can not set voltage to the maximum power point (MPP).

The photovoltaic system with MPPT produces voltage base on maximum power point (MPP) voltage. DC/DC converter is setted by maximum power point tracking (MPPT) to set maximum power point (MPP) voltage. The comparison voltage of two photovoltaic system and sun irradiance can be seen in Figure 3 below.

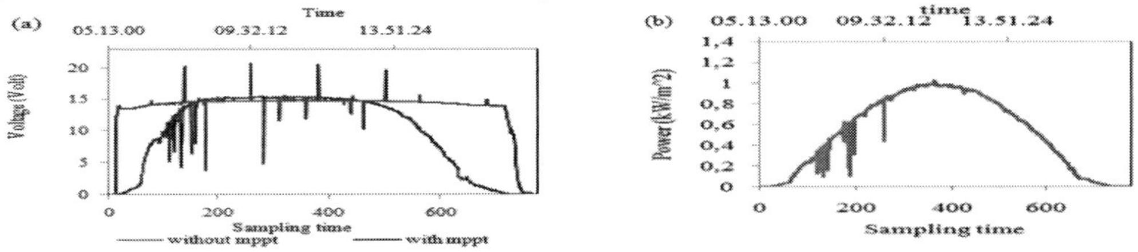

Fig. 3. Chart 12 May 2018 (a) Comparison voltage with MPPT and without MPPT (b) sun Irradiance.

Figure 3. (a) above is voltage chart of photovoltaic system. The chart start at 05.13 o'clock and finish at 18.10 o'clock in the evening. Photovoltaic system with MPPT, the voltage start at 05.27 o'clock in the morning (11.58 volts) and finish at 17.57 o'clock in the eveing (0.2 volts). Photovoltaic system without MPPT, the voltage chart start at 05.27 o'clock in the morning (0.01 volt) and finish at 17.21 o'clock in the evening (0.02 volts).

Figure 3. (b) above is sun irradiance power chart. The chart start at 05.13 o'clock and finish at 18.10 o'clock in the evening. The Power start at 05.27 o'clock in the morning (0.8 watts/m^2) finish at 17.45 o'clock in the evening (0.1 watt/m^2). The highest Sun irradiance power is 1021 watts/m^2 at 11.15 o'clock.

The two figure above show that the voltage of the photovoltaic system with MPPT does not follow the solar radiation power, which means it does not follow the sun irradiance and I/V characteristics curve.

The average performance ratio of the photovoltaic system without MPPT is 0.79 and the average performance ratio of the photovoltaic system with MPPT is 0.74 (measurement solar radiation using solar power meter). The PR result of photovoltaic system without MPPT is higher than photovoltaic system with MPPT, but photovoltaic system can not be used to charge the battery 12 volts, because some voltages are below 12 volts. The PR result of the photovoltaic system with MPPT is better than the photovoltaic system without MPPT, when the battery is not full.

Performance ratio calculation based on average solar radiation data from NASA (6.899167 kWh /m^2/ day) [8] can be seen in Table 2 below;

Table 2. NASA Irradiance Data and Performance Ratio.

Date		H_i (Wh/m^2/day)	Without MPPT		With MPPT	
			E_0 (mWh)	PR	E_0 (mWh)	PR
28	April	6899.16	55513.13	0.80	32384.03	0.47
29	April	6899.16	54970.67	0.80	32384.03	0.47
4	May	6899.16	58007.39	0.84	37758.06	0.55
5	May	6899.16	46036.15	0.67	53859.38	0.78
6	May	6899.16	52417.99	0.76	58723.57	0.85
7	May	6899.16	55351.16	0.80	63495.24	0.92
8	May	6899.16	56031.09	0.81	52645.93	0.76
9	May	6899.16	55917.11	0.81	32716.02	0.47
10	May	6899.16	53716.27	0.78	60810.84	0.88
11	May	6899.16	54648.34	0.79	61970.78	0.90
12	May	6899.16	55334.03	0.80	61985.51	0.90

Table 2 above is the results of performance ratio using NASA data. Photovoltaic system without MPPT, the lowest performance ratio result is 0.67 and the highest performance ratio result is 0.84. Photovoltaic system with MPPT, the lowest performance ratio result is 0.47 and the highest performance ratio result is 0.90. The measurement start on 28 april 2018 and finish on 12 may 2018.

The photovoltaic system without MPPT, The average result difference between sun irradiance measurement and NASA sun irradiance is 0.028561. The photovoltaic system with MPPT, the average result difference between sun irradiance measurement and NASA sun irradiance is 0.028236. The result difference occur because the calculation of PR is only influenced by Photovoltaic power.

Comparison different photovoltaic system performance ratio with other reasearch in same area (at least on the same grid) can be seen in tabel.3. below;

Table 3. the research PR results and another research PR results

Measurement area	Performance Ratio	Measurement area	Performance Ratio
Malang	0.74; 0.79	Kupang [5]	0.82
Malang (NASA data)	0.72; 0.79	Surabaya [6]	0.75; 0.77; 0.84
Tangerang Selatan[4]	0.78; 0.85; 0.92	Malaysia [3]	0.28 0.45; 0.59

The lowest Performance Ratio (PR) result from other research in same area is 0.28. The highest PR result from other reasearch in same area is 0.92. Performance Ratio (PR) result in tangerang selatan is the highest result (0.92). Malaysia performance ratio (PR) is the lowest result (0.28). The Photovoltaic system with MPPT, the result difference between malang area and another area can be calculated with Table 3 data.

The lowest Performance Ratio (PR) difference results is 0.01 and the highest PR difference result is 0.51. PR diference results between malang (PV system with MPPT) and Tangerang selatan are 0.01; 0.06; and 0.13. PR diference result between malang (PV system with MPPT) and Kupang is 0.03. PR diference results between malang (PV system with MPPT) and Surabaya are 0.02; 0.04; and 0.05. PR diference results between malang (PV system with MPPT) and Malaysia are 0.2; 0.34; and 0.51.

The PR result betwen malang and surabaya has almost the same value. This result indicated that the measurement is accurate, because surabaya is the closest area from malang. The PR result difference in indonesia area is not significant. the PR results difference are 0.01; 0.02; 0.03; 0.04; 0.05; 0.06 and 0.13. The PR result difference is not significant because the area is located in same tropical climate. The PR result difference is affected by weather condition. For example, average sun irradiance between malang and kupang is different.

Average sun irradince in Malang 6899.16 kWh /m^2/ day and average sun iradiance in Kupang 6362.50 Wh/m^2 /day [8].

5 Conclusions

Photovoltaic system performance ratio result is affected by sun irradiance and the DC/DC converter system of MPPT. DC/DC converter set voltage to the MPP voltage. The photovoltaic power will be optimum and photovoltaic performance ratio will be optimum. The average PR of the photovoltaic system without MPPT is 0.79 and the average PR photovoltaic system with MPPT is 0.74.

Performance ratio of Photovoltaic stand-alone systems is affected by battery condition (battery condition is full or battery conditions is not full). Performance Ratio (PR) will decrease, when the battery condition is full. The lowest PR result system with MPPT is 0.46, when the battery condition is full.

Performance Ratio of photovoltaic system with MPPT is better than the photovoltaic system without MPPT, when the battery condition is not full. The best PR result of the system with MPPT is 0.93, when the battery condition is not full.

The design of logger microprocessor and current sensor has good error test result. The logger error results are 0.29% - 1.16%(voltage) and 0.19% - 0.57%(current).

The PR result difference in Indonesia area is not significant. The PR results difference are 0.01; 0.02; 0.03; 0.04; 0.05; 0.06 and 0.13. The PR result difference is not significant because the area is located in same tropical climate. The PR result difference is affected by weather condition.

Acknowledgement

This research financed by DRPM UI, PITTA grand scheme No: 2524/UN2.R3.1/HKP.05.00/2018.

References

1. B.S. Hartono, Y. Budiyanto and R. Setiabudy, *Quality in Research*, 127-132 (2013)
2. T. Georgitsioti, G. Pillai, N. Pearsall, G. Putrus, I. Forbes and R. Anand, IET Renewable Power Generation, **9**(5), 438-445 (2015)
3. Ghazali. A M., and Rahman, AMA, Energy and Environment Research, **2**(1), 235-243 (2012)
4. O.A. Rosyid, *Smart Cities, Automation & Intelligent Computing System*, 81-86 (2017)
5. I.N.S. Kumara, M. Ashari, A.S. Sampeallo, and A.A.G.A. Pawitra, International Journal of Engineering and Technology innovation, **7**(2), 117-129 (2018)
6. Elieser Tarigan, Djuwari, and Lasman Purba, Energy Procedia, **47**, 85-93 (2014)
7. IEC 61724, *Photovoltai system performance monitoring. Guidelines for measurement, data exchange and analysis,* (1999)
8. NASA Surface Meteorology and Solar Energy, https://eosweb.larc.nasa.gov/cgi-bin/sse/grid.cgi?email=skip@larc.nasa.gov (Accesed 1 May 2018)

Study and Analysis of Shading Effects on Photovoltaic Application System

Handoko Rusiana Iskandar[1,a], *Yuda Bakti* Zainal[1] and *Susanto* Sambasri[1]

[1]*Electrical Engineering, University of Jenderal Achmad Yani, Cimahi, Indonesia*

Abstract. This study investigates the shading on PV systems. Shading has considerable influence on the solar cells characteristics, temperature and radiation on site need to be considered as the basis for designing PV systems. Modeling conducted on a 1 kWp PV system at a latitude of 6°53'2.69"S and a longitude of 107°32'28.69", to find the magnitude of solar radiation, surface temperature, and tilt angle, and partial shadow simulation using PVsyst software. Testing result shows the characteristic PV 1 kWp is obtained with the angle of solar cell shade at 18° and azimuth 0°, the shading per year generates 4.71 kWh/m^2 in a solar active area at 6.9m^2, the power losses of 20.8% of its nominal power.

1 Introduction

1.1. Background

The global issue of reduced fuel oil resources encourages human beings to play a role in the expansion of the utilization of renewable energy sources such as Photovoltaic Systems [1-2]. The PV system has an important role as a source of one of environmentally friendly renewable energy. The PV system has been widely used in various public facilities. To design a PV system some testing must be done, i.e the characteristics of solar panels. These characteristics will affect the maximum working of the PV. Optimum characteristics of solar cells in terms of cell material and panel. They are closely related to the surrounding conditions, both of the solar irradiance, surface temperature, tilt angle, and shade. The shading effect will be modeled to find the loss of the PV output and the tilt angle match.

The characteristics of solar cells are seen from the parameters in the (I_{SC}) curve with (V_{OC}) and several other factors affecting the characteristics of solar cells are the rise and fall of temperature and solar irradiance, as well as the effect of shading, were adopted from [3,4]. Several related studies linked with percentage efficiency of solar panels [1], combining the effect of temperature and solar intensity on the power of solar panels used simulated [2]. Another influence is the deterioration of power caused by the placement of panels in the shade, the results of these experiments show an accuracy increase of 5 - 10% which is modeled by [3], another approach to the analysis of shading effects or losses caused by[4], partial shading was done by the investigation conducted using Matlab to show an increase. Other methods ANN used to detect errors occurring on or the PV module

[a] Corresponding authors: handoko.rusiana@lecture.unjani.ac.id

is tested under normal partial shadow conditions. these inputs are randomly tested performed by [5][1]. Decreased power due to shading effect can be optimized using fuzzy logic by [6], this approach can increase the Maximum Power Point Tracker although the power characteristics of PV influenced shading effect [7-8]. It can be concluded that there are some losses to the solar cell power rating such as the decrease caused by the influence of irradiance and temperature, the effect of shading on the solar panel due to the environment condition. It is, therefore, necessary to have an optimal installation or slope study of solar panels, by reducing the effect of shading, irradiance and temperature accordingly.

1.2 Location and Geographical

In this research, 1 kWp PV system will be applied in the Electrical Engineering Laboratory, University of Jenderal Achmad Yani, Cimahi, Indonesia. Based on NASA Surface Meteorology & Solar Energy the location of this solar insolation is 4.81 kWh /m^2/day, wind speed 3.81 m/s and average air temperature 27.0°C and humidity 81.9% per year at a latitude of 6°53'2.69"S and a longitude of 107°32'28.69"E or -6.884082 and 107.541304.

Table 1. Monthly meteo data

Month	At tilt angle 18° and azimuth 0°			Temperature (°C)	Wind Velocity (m/s)
	Global Irradiance (kWh/m^2.day)	Tilt Global (kWh/m^2.day)	Shed Global (kWh/m^2.day)		
Jan	4.18	4.59	4.55	26.0	3.31
Feb	4.25	4.46	4.42	25.9	3.333
Mar	4.76	4.75	4.71	26.5	2.66
Apr	4.82	4.53	4.49	27.2	3.02
May	4.74	4.24	4.20	27.8	3.97
Jun	4.58	4.00	3.96	27.8	4.34
Jul	4.86	4.26	4.22	27.5	5.29
Aug	5.24	4.78	4.74	27.5	5.38
Sep	5.56	5.41	5.36	27.4	4.93
Oct	5.32	5.52	5.48	26.9	3.71
Nov	4.80	5.26	5.21	26.6	3.09
Des	4.63	5.22	5.18	26.3	2.69
Average/year	4.81	4.75	4.71	2.69	3.8

2 Modeling of photovoltaic systems

2.1 Photovoltaic characteristics

According to [9]. solar cells consisting of a row of panels, the shadow effect will reduce the output power is generally there are two things, the first is during the winter season in northern latitudes and the second if the available area of the system is limited. Leading to a very small spacing between the rows. Other parameters are characteristic of the solar cell at current-voltage, among other, I_{SC}, V_{OC}, maximum power also fill factor. Solar cells can be modeled as current sources paralleled with diodes based on Schockley diodes. By using the source of solar radiation it is obtained the characteristics when compared with the modeling results then the difference and similarity will be any further analysis, the factors that affect these characteristics is the level of solar radiation and temperature in solar cells[1-5,-10-11].

$$i_{pv} = i_{ph} - i_o\left[exp\left(\frac{v_{pv}+i_{pv}R_s}{nk_BT/q}\right) - 1\right] - \frac{v_{pv}+i_{pv}R_s}{R_p} \quad (1)$$

This junction absorbs photon energy from sunlight and generates electron-hole pairs that propagate throughout the junction so as to produce an electric field passing through the junction and causing an electric current flow. Equation (1) shows I_{PV} and V_{Ph} is the current and output voltage of the PV cell, I_{Ph} is the current generated from the photon energy absorption process.

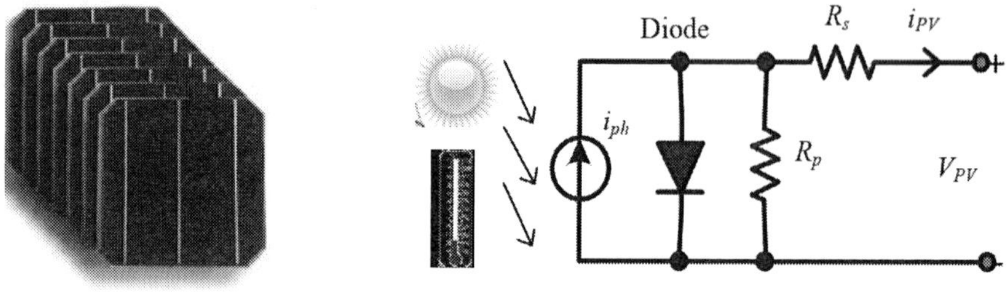

(a) (b)

Fig. 1. Solar cell; a) physical model, b) equivalent circuit of a solar cell

Figure 1 shows the standard equivalent circuit for solar cells is a circuit model consisting of 1 diode, a constant source for photo-generated current (i_o), series resistance (R_s) and parallel constraints (R_{sh}) [12]. The calculation of the voltage and current of the PV system depends on the configuration. The value V_{mp_n} and I_{mp_n} each solar module can be obtained from the data provided by the manufacturer.

Table 2. Mechanical data of solar cell

Module	Cells
Length 1.23 m	Si-mono
Width 0.56 m	1 in parallel
Thickness 350 mm	36 in series
Weight 6.50 kg	Cell area 191 cm^2

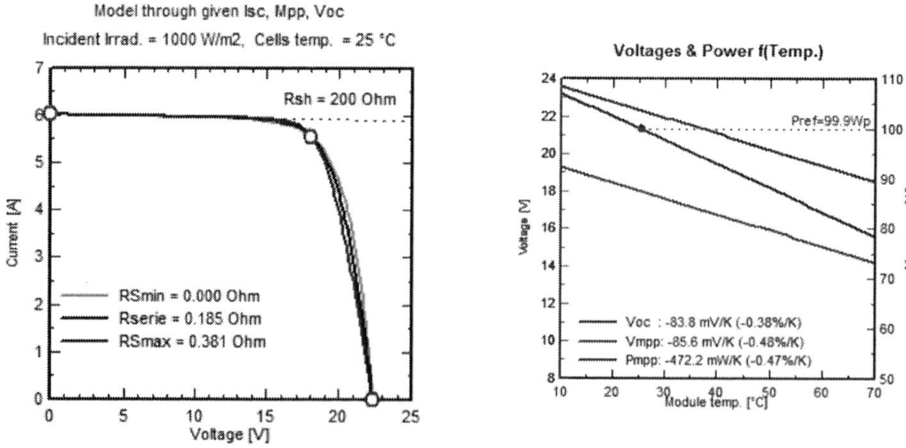

(a) (b)

Fig. 2. Simulation of cell characteristics; a) I-V curve, b) temperature coefficient setting

This photon current is proportional to the solar radiation level of the sun. While i_0 is the diode saturation current, which shows the diode characteristics. q is the charge of the electron, and n is an ideal factor, k. B is Boltzman's constant, then T is the temperature of the PV cell in Kelvin, R_p is the shunt resistance associated with the leakage current of the cell through the junction to the inside of the cell. Furthermore, R_S is a series resistance showing the losses of the circuit interconnection, the other two important parameters in the characteristic PV cell are I_{SC} and V_{OC}. Figure 2 shows the I-V curve and power and voltages under module temperature (°C).

2.2 Configuration of PV systems

Total operating voltage and current calculations depend on the configuration of installed solar modules in series, parallel, or series-parallel [13]. Series have given by equation (3), (4).

$$Vmp_{total-series} = \sum_{n}^{i} = 1Vmp\_n \quad (3)$$

$$Imp_{total-series} = Imp\_n \quad (4)$$

The type of solar cell used is silicone mono-crystalline, with open land installation mode. The value V_{mp_n} and I_{mp_n} each solar module can be obtained from the data provided by the manufacturer. Specification and calculation result gave by table 3. Optimum operating voltage (V_{mp}) 18 Volts, an open circuit voltage (V_{OC}) is 223.6 V and the sum of voltage for 10 solar panels of 180 V. The temperature relationship take effect on the temperature drop in the solar panel, determination of the temperature dropping factor given by equation (5),

$$f_{temp.} = 1 - \gamma\left(Tcell_{eff} - T_{STC}\right) \quad (5)$$

Definition of γ is air temperature coefficient in (%/°C), $Tcell_{eff}$PV cell temperature in average and T_{STC} is the standard test condition temperature in 1 kW/m^2 at 25°C. Figure 3 (a) and (b) similarly shows the I-V curve with variant irradiance to give the Maximum Power Point at the PV system. This simulation using table 3 for calculation.

Table 3. PV module manufacturer specification

Parameters	**Information**
Brands and Type	SUMICO TN-100M
Maximum power (P_{max})	1008 Watt peak
Max. operating current ($I_{mp_{max}}$.)	5.56 Ampere
Short Circuit Current (I_{SC})	6.02 Volt
Temp. The coefficient of P_{max} (%)	-0.47/°C

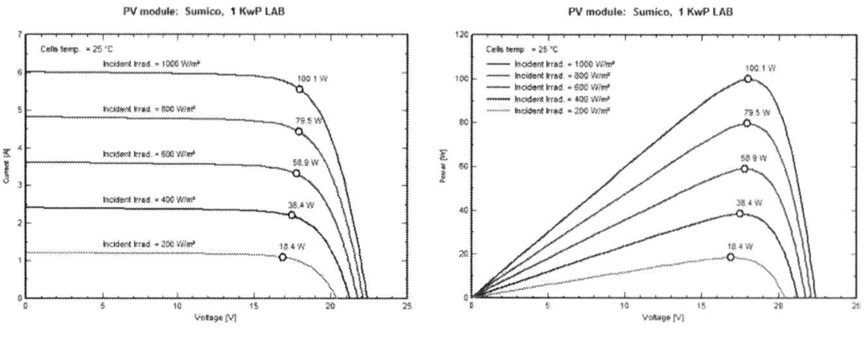

(a) (b)

Fig. 3. Simulation result; a) I-V curve according to irradiance (W/m^2), b) P-V curve according to temperature (°C)

2.3 Shading calculation

Depending on the position of the Sun, a single line of panels may be partial or all will be shaded. The shading effect is a change in the current-voltage characteristics (I-V) of the system, and consequently a reduction in the power output. For certain shadows in the system, the optimum tilt is generally not the same and will be different.

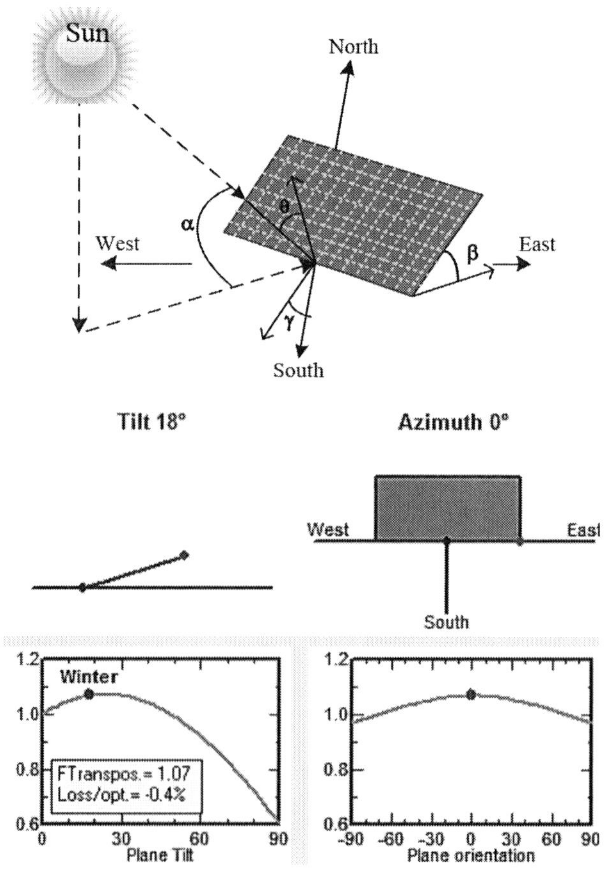

(a) (b)

Fig. 4. Definition; a) angle on PV panel, b) simulation

The construction of PV system the result of shaded areas per total area involving other electrical losses[8]. We have to define the sun position given by time function, latitude and longitude and then used to calculate the size and position of the shading area. The altitude α of the Sun as a function of latitude ϕ, solar declination δ and angle ω subtended by the sun at a particular hour is given by[9].

$$\sin\alpha = \sin\phi\sin\delta + \cos\phi\cos\delta\cos\omega \tag{5}$$

The angle θ from the direction of the solar radiation beam on the sloping surface of PV cell is following equation (6),

$$\cos\theta = cos\delta\cos\omega\,(\cos\gamma\sin\beta\sin\phi + cos\phi\cos\beta) + \sin\delta(\sin\phi\cos\beta - \cos\gamma\, cos\phi\, cos\beta) + \sin\gamma\sin\beta\, cos\delta\, sin\omega \tag{6}$$

For a south facing surface, γ = 0 and equation (12) so that it becomes and shows by Figure 4, Where γ is the azimuth angle and β is the tilt of the surface azimuth.

$$\cos\theta = \sin(\phi - \beta)\, sin\delta + \cos(\phi - \beta)\cos\delta\cos\omega \tag{7}$$

PVSyst. Software used for model and simulation. Figure 4 used to set the angle tilt solar panel, this a supporting application in analysis shading and dynamically simulating photovoltaic systems on the site [14]. Figure 5 is the simulation for a movement from the angle of the sun as it passes through the sky, we must know its height and azimuth. The altitude is the angular distance above the horizon measured perpendicular to the horizon. It has a maximum value of 90° on the zenith, which is the dot above our head. Azimuth distance of the measured angle is seen in a clockwise direction. Other things to consider is Albedo, albedo is a reflection effect on the surface. This effect affects the amount of solar energy reflected back into space. Albedo on earth shows a decrease in temperature. Albedo values in some places will be different where dry albedo value will be less than cold or wet areas.

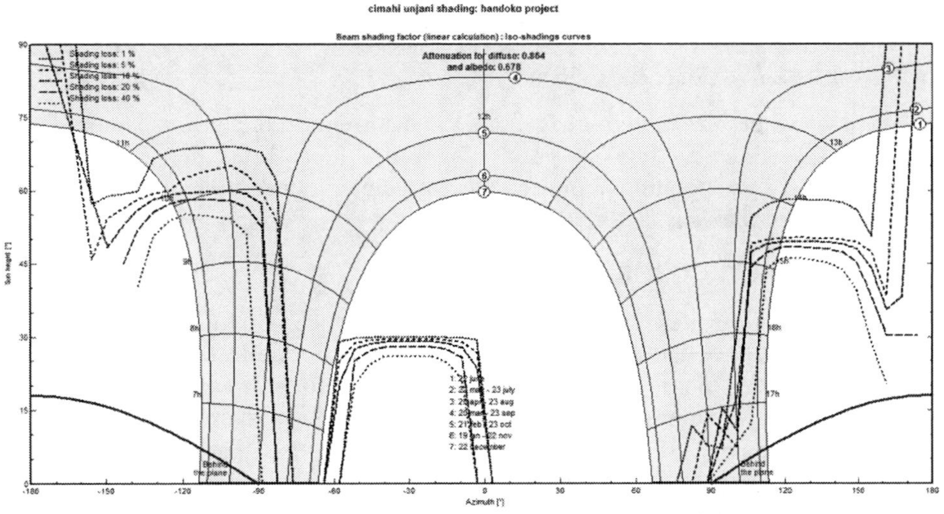

Fig. 5. Beam shading factor on linear condition

Table 4. Shading factor table (linear), for the beam component.

Azimuth / Height	-180°	-160°	-140°	-120°	-100°	-80°	-60°	-40°	-20°	0°	20°	40°	60°	80°	100°	120°	140°	160°	180°
90°	0.800	1.000	1.000	1.000	1.000	1.000	1.000	1.000	1.000	1.000	1.000	1.000	1.000	1.000	1.000	1.000	1.000	1.000	0.800
80°	0.791	1.000	1.000	1.000	1.000	1.000	1.000	1.000	1.000	1.000	1.000	1.000	1.000	1.000	1.000	1.000	1.000	1.000	0.800
70°	0.785	1.000	1.000	1.000	1.000	1.000	1.000	1.000	1.000	1.000	1.000	1.000	1.000	1.000	1.000	1.000	1.000	1.000	0.800
60°	0.776	1.000	1.000	0.934	0.898	1.000	1.000	1.000	1.000	1.000	1.000	1.000	1.000	1.000	1.000	1.000	1.000	1.000	0.800
50°	0.764	1.000	0.833	0.237	0.403	1.000	1.000	1.000	1.000	1.000	1.000	1.000	1.000	1.000	1.000	0.947	0.952	1.000	0.800
40°	0.741	1.000	0.686	0.000	0.226	1.000	1.000	1.000	1.000	1.000	1.000	1.000	1.000	1.000	1.000	0.000	0.269	1.000	0.800
30°	0.800	1.000	0.686	0.000	0.224	1.000	1.000	1.000	1.000	1.000	1.000	1.000	1.000	1.000	1.000	0.000	0.000	0.795	0.800
20°	0.800	1.000	0.686	0.000	0.224	1.000	1.000	0.000	0.000	1.000	1.000	1.000	1.000	1.000	0.793	0.000	0.000	0.576	0.800
10°	0.800	1.000	0.686	0.000	0.224	1.000	1.000	0.000	0.000	1.000	1.000	1.000	1.000	1.000	0.943	0.000	0.000	0.358	0.800
2°	Behind	Behind	Behind	Behind	0.224	1.000	1.000	0.000	0.000	1.000	1.000	1.000	1.000	1.000	1.092	Behind	Behind	Behind	Behind

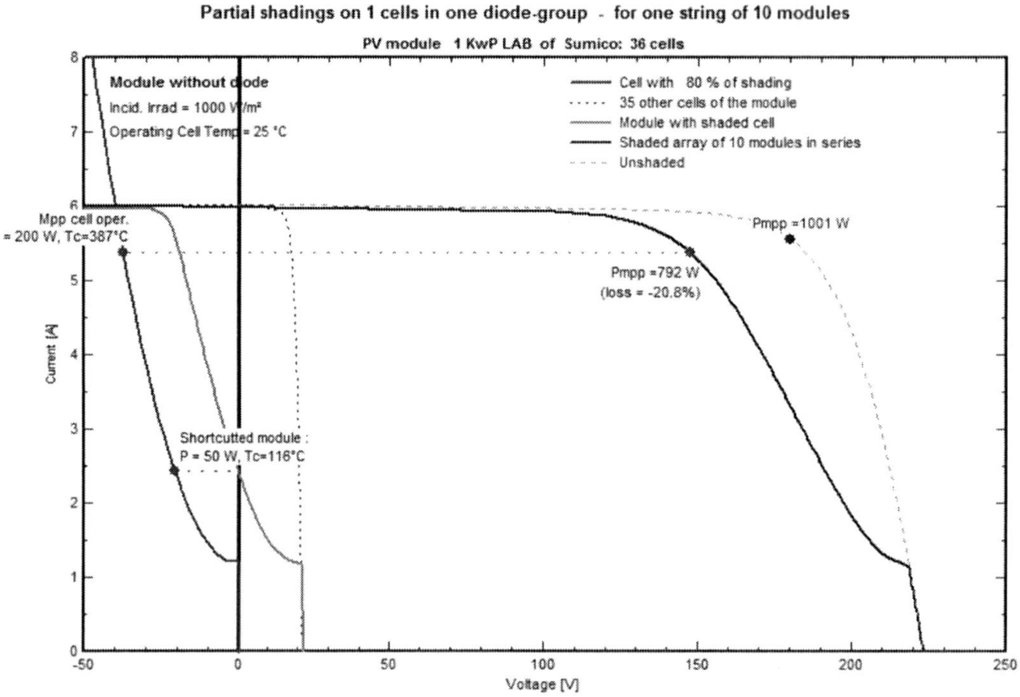

Fig. 6. P_{loss} at 80% of shading on one module

Figure 6 visualizes the behavior of voltages and currents at the terminals of a shaded (or bad) cell, placed in a module or a group of modules. It evaluates the energy dissipation of this cell for various operating points, and its resultant temperature by making reasonable analysis for the thermal loss factor.

3. Result and Discussion

When the solar cell operates, one of the most important measurable values is the power. In contrast, the behavior of the current source is observed on the left side of the I-V curve. There is an operating point where the output power will be maximized. This point of operation is called Maximum Power Point (MPP). The PV system is required to always operate close to the point of MPP it is intended to obtain the energy gain in the maximum state. The linear current of the linear I_{SC}depends on the degree of solar radiation whereas the V_{OC} open-circuit voltage shows the dependence of the temperature present in the solar cell. I_{SC}, and V_{OC} are measured under T_{reff} are 25°C, and irradiance λ_{reff} in 1000 W/m^2, T_{reff} and λ_{reff}. The value of muP_{max} and muV_{OC} temperature coefficient are output one diode model which are dependent on the Gamma, in this simulation Gamma we put 1.30

with Gamma coefficient temperature -0.0007°C. Diode saturation current, quality factor and voltage temperature coefficient determined by this requirement. According to the nameplate module, the solar panel temperature coefficient has -0.47%/°C at maximum power. Generally, silicone crystalline type modules have -0.4 to -0.5%/°C (see table 3) result of the simulation of the efficiency per cell area is 14.56 % (6.880 m^2) and per module area is 0.15% (0.68 m^2).

Shading analysis performed on each cell, these specifications in one panel consisting of 36 cells tested with Irradiance 1kW/m^2 and T_c 25°C. The other parameters, the losses are about 20.8%, the P_{mpp} value will decrease to 792 W of nominal power. This shows the phenomenological study of the resultant I-V characteristic of a module or PV array, composed of non-identical cells or modules shows the partial shading curve between currents and voltages on one module. Near shading simulation result from the showing a shading factor from the simulation of Electricity losses in PV systems are generated from shaded areas per total area. Table 4 describe the shading factor is a complex calculation which spends too much time to be performed during a simulation at each time step. Therefore the program builds a table in sun's height (10° steps) and azimuth (20° steps). The simulation can then interpolate in this table. During the elaboration of the shading table, the points (sun's positions) situated behind the plane of the PV field appear in lower height and azimuth ± 180° until 120°. The perspective of PV system is placed in surrounded by a variety of building blocks that result in the shadow around the PV area.

4. Conclusion

Simulation result the existing data has power maximum at S_{TC} 1008 Watt with operating voltage (V_{mp}) 18 Volts, an open circuit voltage (V_{OC}) is 223.6 V with voltage for 10 solar panels of 180 V in series. Generally, silicone crystalline type modules have -0.4 to -0.5%/°C so the result of the simulation of the efficiency per cell area is 14.56 % (6.880 m^2) and per module area is 0.15% (0.68 m^2) under partial shade. Shading analysis performed on each cell, these specifications in one panel consisting of 36 cells tested with Irradiance 1kW/m^2 and T_c 25°C. The other parameters, the losses is about 20.8%, the P_{mpp} value will decrease to 792 Watt of nominal power. The maximum performance of solar panels at an angle of 18°is 1.001 kilo Watt peak as desired.

References

1. A. Sangwongwanich, *Mas. Thes.* 11, (2014)
2. H. R. Iskandar, A. Purwadi, A. Rizqiawan, N. Heryana, *ICPERE.* **3**, 171 (2016)
3. M. F. Nayan, S. M. S. Ulah, *ICGET.* **3**. 6 (2015)
4. H. M. Al-Masri, A. Abu-Errub, W. R. Ayyad, M. Ehsani, *SPEEDAM.* 902 (2016)
5. A. Hadj. Dida, M. Bhekti, *ICRERA,* **6,** 2 (2017)
6. C. S. Chin, P. Neelakantan, H. P. Young, S. S. Yang, K. T. K. Teo *CICSyN,* 7 (2011)
7. Aravinda, Padmavathi, *ISPACE,* 1 (2017)
8. P. Bharadwaj, A. John, 4407 (2017)
9. D. Passias, B. Kalbak. Sol. Cell. **11,** 282 (1984)
10. P. Srinivasa Rao, G. S. Ilango, C. Nagamani, IEEE J. of Photovoltaic. **4,** 680 (2014)
11. S. Kumar, A. Selvakumar, *Int. Conf. Innov. Power. Adv. Comput. Tech.* 5 (2017)
12. A. Karina, S. Satwiko, *Pertemuan Ilmiah XXV HFI Jateng & DIY,* 163 (2012)
13. H. R. Iskandar, A. Purwadi, A. Rizqiawan, N. Heryana, *SENKA.* 1 (2015)
14. A. Mermoud, Mod. Syst. Loses. In PVsyst. 15 (2010)

The Power Usage Allocating System to Optimize Based on the Priority Load

Hartina N. Aidah[1], *Sekina* Y. Trieksa[11], *Mohamad* Ramdhani[1], and *Ahmad* Sugiana[1]

[1]Electrical Engineering, Electrical Department, 40257 Telkom University, Indonesia

Abstract. Monitoring energy consumption in a household is an essential thing for optimization of electrical power usage. Unnecessary or excessive power usage can lead to power outages and damage the electronic devices, so the end-user needs a control device to overcome it. In this paper, we propose an easy-access device that can provide real-time information about energy consumption and optimization of power consumption. The method for current measuring is used ZMCT103c whose works on some transformer and for voltage measuring is used ZMPT101b whose works on step down transformer. This system is focused on loads management priority in study case boarding house with total active power 900 watts which are created to scaled loads allocation based on electronic devices susceptible to damage that caused by power outages with assumption 2 (two) rooms and 3 (three) loads each room. The system has priority mechanism for load usage that could not be shut down such as PC (personal computer) and refrigerator. The device works by distributing maximum 55% of active power from excess power room to needed power room so that can avoid power outages and increase the efficiency of power usage.

1 Introduction

Recently, The conservation of energy consumption is become very important [1] because the power source is limited, likewise the conservation of electrical energy consumption. Electrical energy usage increases on a building every year, so that increases the tariff also.

An excess of the power usage causes over capacity. Therefore it cuts off the current suddenly. In consequence, the optimization is required so that the power consumption more maximal. The optimization depends on the device identification which consumes more power and the development energy management technique [2, 3] which made on hardware. Every of these hardware has to communicating and to synergy for power usage accumulation, consumption cost of power usage, and allocating to efficiency electrical power [4].

The system implementation made on boarding-house with two rooms that have three loads in every room. In this system implementation, many researchers want to replenish the features of calculating and power consumption monitoring as the main feature on an

[1] Corresponding author: ysekina@gmail.com

electrical system, so that used the current transformer as the current calculating which is the key for data acquisition [5]. Either wise, control device is needed for power consumption monitoring and allocating compatible with user's necessity. The process of electrical power allocate happened to compare the room's necessity with other room if the room needs more electrical power while the other room does not use it then the power on that other room would be allocated to the room whose needed. In other condition is when every room needed more electrical power so some loads would be shut off with the relay based on the load's priority scale [1].

The concept in this allocating system based on priority load research needs so that can maximal the available power without the user replenishing the power's capacity, with the result that more efficiency and resolving the problem that the electrical power source are limited. The system design is presented in section II, in section III is presented in the implementation design, the next section has presented the conclusion, and the last section is presented the future work.

2 Design System

Fig. 1 represents a general block diagram of the proposed system. This system made connected between electrical source voltage 220 V and MCB as a protector each room. The relay which midst MCB uses to do the power allocated. The main of a system is consist of the loads wiring, current and voltage sensor as measures and the relay located between the load and the Arduino as cuts the current to load.

Hardware planning consist of the component selection would be used, design planning and wiring. The point in this system is Arduino-Mega2560 usage. This component is a microcontroller that could be saved as the result of data processing to EEPROM memory [6]. Furthermore, the Arduino has a lot of port I/O so that could be implemented in a various room.

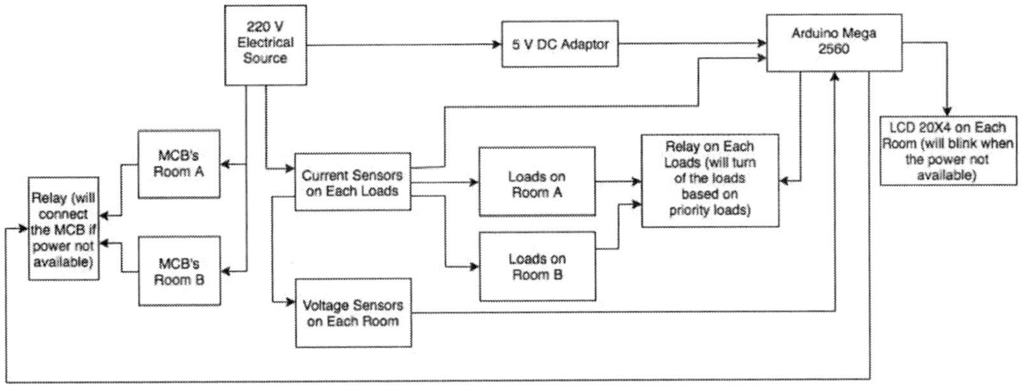

Fig. 1 General Block Diagram

3 Implementation of the System

In this implemented, there have been three processes to be used. The first is for current and voltage measurement [4], in part two is the processing of current and voltage data become the electrical power and long-time that the user used the loads with the assist by Real Time Clock DS3231. The next process is power management and allocating of power based on several conditions. This research is used in six loads with three loads in each room.

To manage the loads, there are two systems [1]. The systems are the priority loads management and the allocating power.

3.1 Allocating Power System

The allocating power system happened when the room needs more power, and the other room only used less than the maximum total power in one room. This allocating coupled by connecting the relay inter each MCB in room 1 and the other room. For taking the resolve by the relay regarding the power that could be allocated or could not be allocated needs 1 minute and for taking back the power that would be observed by the current spring that happened on the room whose gave the power.

LCD installed on a panel for each room and displayed the total power consumption, tariff, and information that become the notification if the power is allocating approval or not. This LCD is useful for monitoring the power consumption.

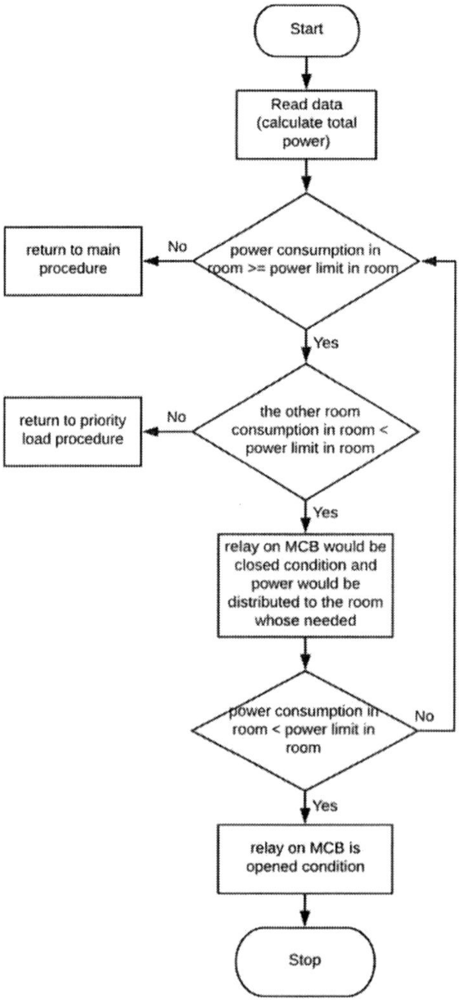

Fig. 2 Flowchart Allocating Power System [2]

3.2 Priority Loads Management

The sequence of this priority needs to preserve the devices continuously turn on so that would not have happened the damage. The system is made by connecting the relay on every load's socket. The relay would be cut the current based on a specific condition and the sequence of loads priority x, y, and z those presented in **Table 1**

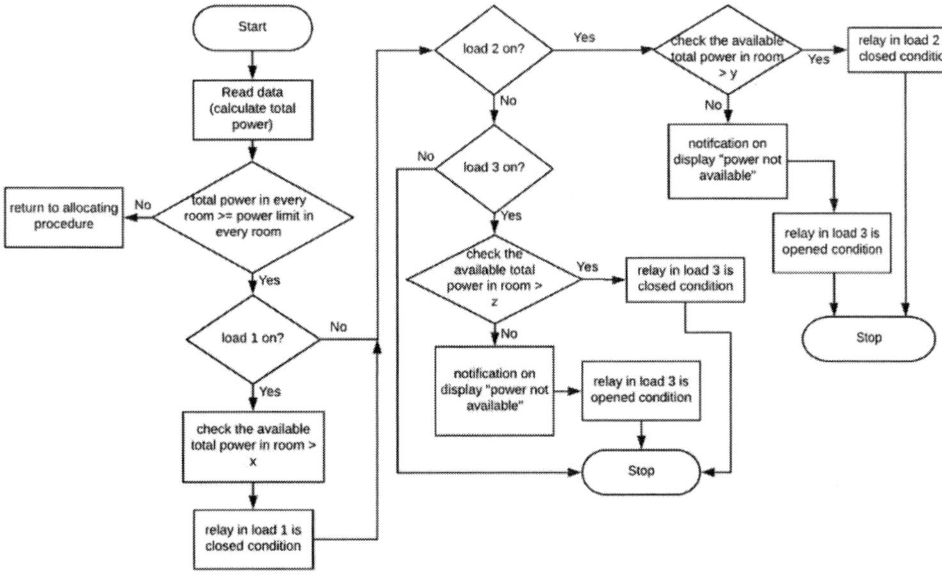

Fig. 3 Flowchart Priority Loads Management [2]

The allocating power and priority loads management, those two ways to manage power consumption that created the efficiency of power consumption. **Table 1** showed the rules based on an amount the total power consumption for each room.

Table. 1 The rules for allocating power use system or priority load management

Conditions	Load Priority in Room 1			Load Priority in Room 2			Notes
	x	y	z	x	y	z	
Power 1<450 and power 2<450	On	On	On	On	On	On	Normal Condition
power 1>450 and power 2<450	On	On	On	On	Off	Off	Allocating power by room 2
Power 1<450 and power 2>450	On	Off	Off	On	On	On	Allocating power by room 1
power 1>450 and power 2>450	On	Off	Off	On	Off	Off	Priority Load Management

3.3 Result

Table. 2 Loads Priority and these Power Consumption [1]

Measurement Electricity Scale Variable			
	Measurement Results		Calculation Results
Loads	Voltage RMS (V)	Current RMS (A)	Real Power (Watt)
Lamp 1	232.2	0.0387	8.98
Lamp 2	232.2	0.0216	5.01
Laptop	232.2	0.3445	79.99
Personal Computer	232.2	0.8613	199.99

Laptop and Personal Computer are devices which have tendency easier to be broken if the overpower happened so that these devices placed on top priority to maintained in on condition permanently when the current disconnected suddenly. These devices are created as a reference for determining power amount which maintained on each room. Based on measurement and real power calculation results in **Table 2**, then accepted the most significant total power is $P \approx 200\ W$. Accordingly a percentage of the power could be allocated to other room :

$$P_{share} = P_{total} - P$$

$$\frac{P_{share}}{P_{total}} \times 100\% = \frac{250}{450} \times 100\% \cong 55\% \qquad (1)$$

4 Conclusion

The power optimization can do with allocating the power from 1 room to another room in the amount of 55% by concern about the type of devices that tendency easier to be broken if the overpower happened. The first thing to do for optimization is set the devices that the user want to maintain in on condition even though the overpower happened. Furthermore, if the condition in each room uses the power >450 Watt, then automatically the relay will turn the device off based on the priority's scale.

5 Future Work

To develop this research in the future will be more useful if this system could be worked on a larger scale such as office complex, industry and others with many loads then this research.

References

1. R. Dubey, S. Nath, K. Harsha, D. R. S. Vinay, M. Bayya, Dr. BVSSN. P. Rao, "Smart Home Management with Online Power Measurement," IEEE: Humanitarian Technology Conference (R10-HTC), Dec (2016)
2. T. R. Melo, S. O. D. Luiz, "Experimental Platform for Power Measurements of Computer System," IEEE: Consumer Electronics – Berlin, Sept (2014)

3. Y. S. Son, T. Pulkkenin, K. D. Moon, C. Kim, “Home Energy Management System based on Power Line Communication," IEEE: Transactions On Consumer Electronics, Vol: 56, No. 6, (2010)
4. E.Orsi, S. Nesmachnow, "Smart Home Energy Planning Using IoT and the Cloud," IEEE: URUCON, (2017)
5. Y. Thongkhao, W. Pora, “A Low-cost Wi-Fi Smart Plug with On-off and Energy Metering Functions," IEEE, (2016)
6. Arduino Mega 2560 Rev3, May 2018. [Online]. Available on: https://store.arduino.cc/usa/arduino-mega-2560-rev3

Implementation of Naive Bayes for Classification and Potentially MSMEs Analysis

Meta Amalya Dewi[1,1], *Tri Wahyu Widyaningsih*[2]

[1]Tanri Abeng University, Information System Department, Jakarta Selatan, Indonesia
[2]Tanri Abeng University, Informatic Engineering Department, Jakarta Selatan, Indonesia

Abstract. Micro Small and Medium Enterprises (MSMEs) have an important role for a country in significantly increasing its economic growth, its high absorptive capacity to labor can reduce unemployment, and has become the biggest contributor of gross domestic product value. Therefore, the government should give more attention to MSMEs. However, the government does not have any information on the results of clustering analysis and prediction of potential business types from existing MSMEs data. This study aims to assist the government by presenting the results of potential MSMEs processing analysis in Tangerang region based on business characteristics in each region, using Naive Bayes. From the data of the number of MSMEs in Tangerang region, it has been successfully classified and the result of its analysis has become recommendation for the government in establishing the grow up as well as the provision of business assistance for the potential business field.

1 Introduction

MSMEs (Micro, Small and Medium Enterprises) is an easy-to-use term for business segmentation and other organizations that are between the sizes of "small office-home office" (SOHO) and larger companies [1]. In its development perspective, MSMEs are classified into 4 groups, namely [2]:

1. *Livelihood Activities*, is the MSMEs used to make a living, which is more popular as an informal sector. Examples are street vendors.
2. *Micro Enterprise*, is the MSMEs that has the nature of craftsmen but does not have the entrepreneurial nature
3. *Small Dynamic Enterprise*, is the MSMEs that has an entrepreneurial spirit and is able to accept subcontracting job and export
4. *Fast Moving Enterprise*, is the MSMEs that has an entrepreneurial spirit and will transform into a Big Business

MSMEs are going to be one of the fields that contribute significantly to spurring economic growth in Indonesia. MSMEs present have a direct impact on the economy, reduce unemployment and become the largest contributor to the value of gross domestic product [3]. In Europe, MSMEs are an important part of the economy [4] which is the

[1] Corresponding author: meta.amalya@tau.ac.id

driving force for industrial growth and economic development. The growth of MSMEs in Indonesia is increasing rapidly from year to year, data from the Ministry of Cooperatives and MSMEs in 2013 were 57,900,787 MSMEs increasing by 1,361,227 from 2012 [5].

Tangerang has 1850 officially registered MSMEs spread across 29 region, 28 sub-districts and 246 villages [6] and will continue to grow like the vision of the Department of Cooperatives and MSMEs to move the populist economy in Tangerang, but there has been no mapping of potential MSME analysis and business results. which can help provide information and knowledge to the government regarding potential types of businesses in certain areas to be upgraded.

Data mapping can be done by various methods, one of which is by using data mining for the process of extracting information from a very large collection of data through the use of algorithms and withdrawal techniques in the statistics, machine learning and database management systems [7].

Naive Bayes is used as probability and statistical method to classify data and predict future opportunities based on previous experience [8], easily implemented and proven to provide good predictions [9]. The advantage of using Naive Bayes is that it only requires small training data to determine the estimated parameters needed in the classification process. Naive Bayes often works much better in many complex real-world situations than expected [10]. Extracting information on a data that has a large number of records and fields cannot be done easily. Data mining techniques are one of the tools for extracting data in large databases and with complexity level specifications that are widely used in many application domains such as banking and telecommunications [11].

Data mining tools can help hidden knowledge in large amounts of data [12]. Data mining is a process that uses statistical, mathematical, artificial intelligence, and machine learning techniques to extract useful information and then get knowledge from large databases [13]

2 Research Methodology

This research was carried out in the following steps:

1. Designed Measurement Research
   a. Determined Research Variables
   b. Determined Performance Parameters
2. Data Retrieval
   This is the stage of gathering the required data. Data obtained from secondary sources, through the official website of the cooperative service and the Tangerang government MSME.
3. Data Mining Mapping
   As a series of processes, data mining has several stages that are interactive, where users can interact directly or through an intermediary knowledge base. Among the stages of the process are [14]:
   a. Data cleaning and data integration
   b. Data selection and data transformation
   c. Mining process
   d. Pattern evaluation
   e. Knowledge presentation
4. Analysis
5. Documentation

The steps of this research are illustrated through the diagram in Figure 1 below :

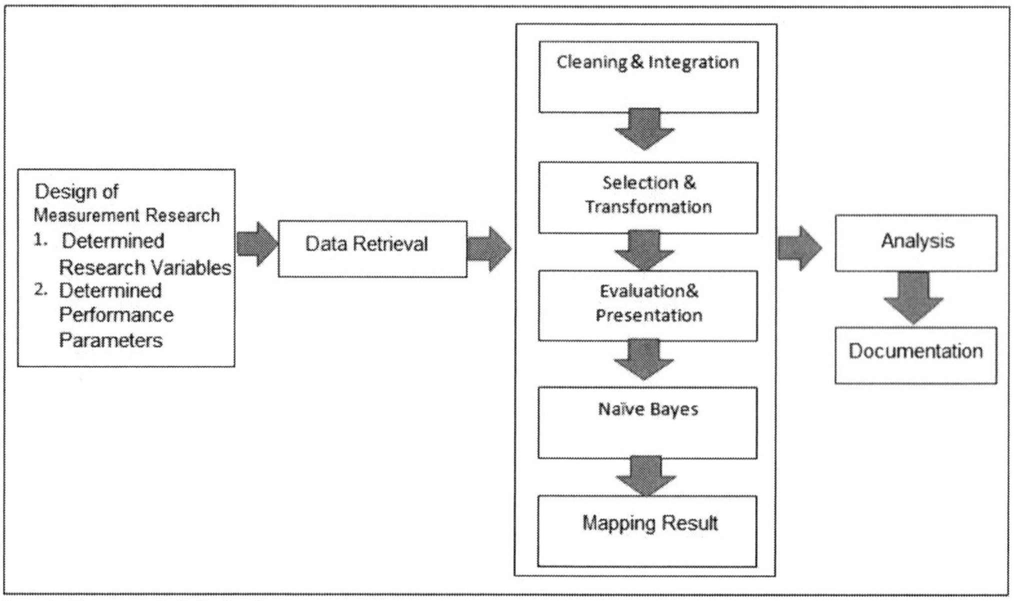

Fig. 1. Research Methods

3 Results And Discussion

The MSMEs data in Tangerang has 7 fields including id, business name, business name, business address, telephone number, business form, and business fields as in table 1 below.

Table 1. Tangerang MSMEs Field Data

No	Field Name	Information
1	Id	Serial number
2	Name of Entrepreneur	Name of registered business owner
3	Business Name	Business name registered
4	Business Address	Complete address of the place of business including the name of the road, RT / RW, region, sub-distric
5	Phone Number	Phone number or cell number (HP)
6	Form of Business	Individual, cooperative, or CV / FA
7	Business fields	Types of businesses that are run

3.1 Cleaning and Integration

This activity is a process of cleaning up data from imperfect entries, inconsistent writing, invalid data, typographical errors, and removing items with empty values. Data adjustments are made to ensure that all data has a complete address to facilitate the mapping process based on sub-district and business fields. Correction techniques can be seen in table 2 below:

Table 2. Inconsistent Correcting Technique Data For Each Table Attribute

Atributte	Information
Address	Eliminate data records with incomplete addresses and addresses with sub-districts outside Tangerang

Business fields	Eliminate data records that are business fields with incorrect data, for example: selling, trading, traveling, and so on.

From this cleaning process, all data records are generated with complete contents by adding sub-district attributes which are the results of the fraction of the address attribute, and adding business classification attributes which are the result of business data grouping. All attributes can be seen in table 3 below

Table 3. Data Integration

No	Field Nama	Information
1	Id	Serial number
2	Name of Entrepreneur	Name of registered business owner
3	Business Name	Business name registered
4	Business Address	Complete address of the business place
5	Sub- districts	The area where the business is run
6	Phone Number	Phone number or cell number (HP)
7	Form of Business	Individual, cooperative, or CV / FA
8	Business fields	Types of businesses that are run
9	Business Classification	Specifications of business groupings

3.2 Evaluation and Presentation

After the data integration process, information is obtained on the percentage of business fields per sub-district as shown in Figure 2 below :

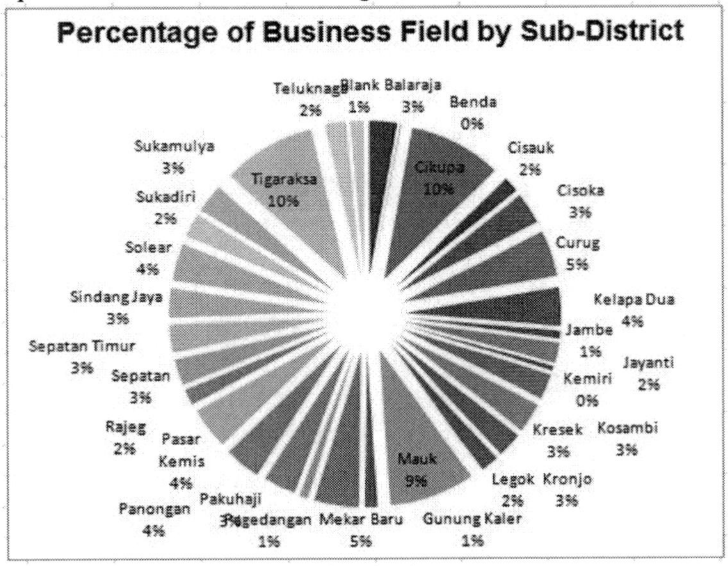

Fig. 2. Percentage of Business Field by Sub-District

And the percentage of business information based on business classification can be seen in Figure 3 below :

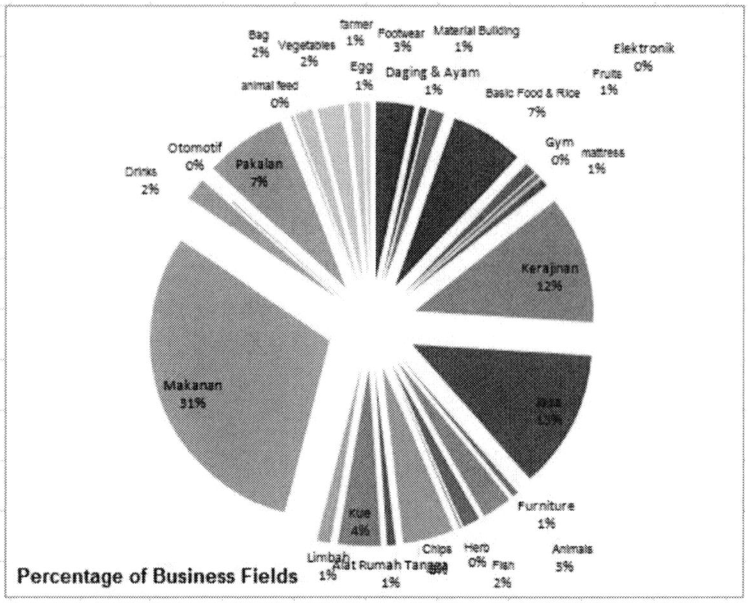

Fig. 3. Percentage of Business Fields

3.3 Naive Bayes

In this process the file and select column are in the data menu to select the file to be analyzed, select column functions to select the variable target. Naive bayes is on the Model menu as a statistical classification model, test and score is in the Evaluate menu to see the results of the analysis with data input and output evaluation result. The process can be seen in Figure 4 below :

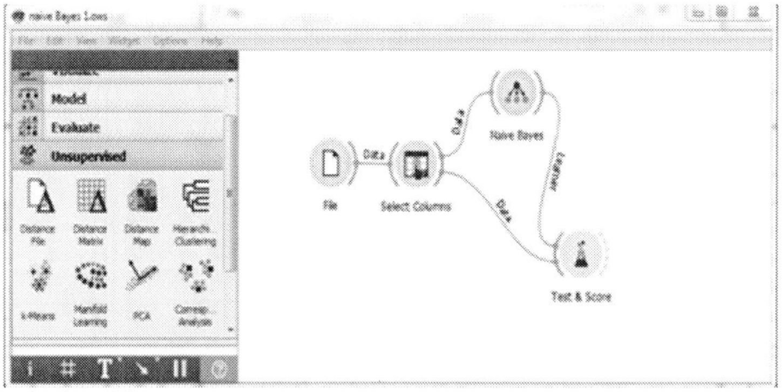

Fig.4. Data Analysis with Orange

The next step is select column as a variable target. In select column there are features, target variables and meta attribus. The target variable must be determined first to determine the results of a better analysis. In this process, a business form field is selected as shown in Figure 5 below :

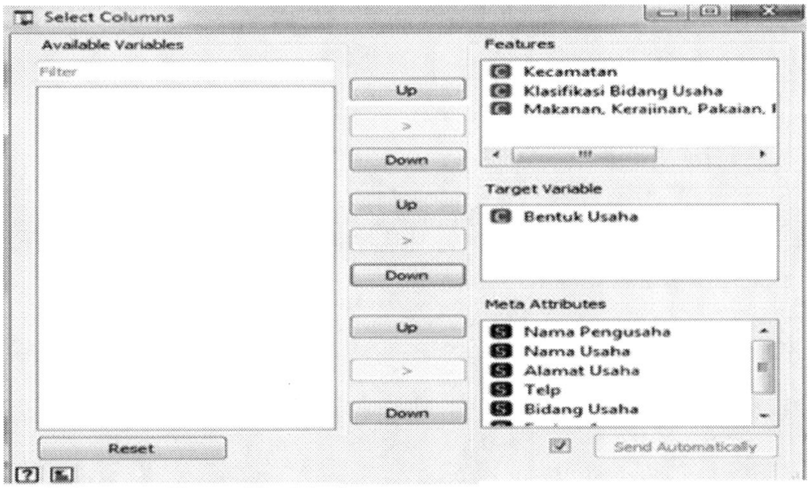

Fig.5. Select Column as Target Variabel

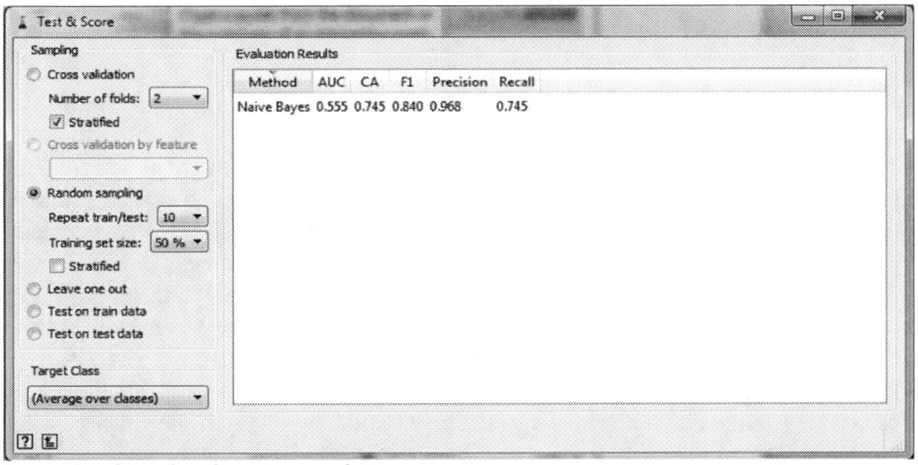

Fig.6. Result of Evaluation Test and Score

Based on Figure 6 above, the level of data accuracy is measured from 5 criteria [15] namely AUC, CA, F1, Precision, and Recall. F1 score results are valid based on the following formula [16] :

$$2\,x\,\frac{precision * recall}{precision + recall}$$

AUC (Area Under Curve) : 0.55 shows the accuracy of the diagnostic test in general (getting closer to 1 is getting better). CA (Capability Analytics) : 0.74, F1 (matrix that combines recall and precision using harmonic mean) : 0.84, precision (the value of a classification model to return only the relevant instances) : 0.968 and recall (the value of a classification model to identify all relevant instances) : 745

3.4 Classification Analysis Results

Based on all the processes that have been completed, the results of ranking 9th District in Tangerang which are considered the most productive are:

1. Tigaraksa with a productivity level of 10%
2. Cikupa with a productivity level of 10%
3. Mauk with a productivity level of 9%
4. Curug with a productivity level of 5%
5. Mekar Baru with a productivity level of 5%
6. Solear with a productivity level of 4%
7. Pasar with a productivity level of 4%
8. Panongan with a productivity level of 4%
9. Kelapa with a productivity level of 4%

Based on the results of the classification analysis of business fields that are in demand by the public, shows the ratings:

1. Food is 31%, with the most productive area is Cikupa
2. Services is 13%, with the most productive area Mekar Baru
3. Crafts is 12%, with the most productive area Tigaraksa
4. Basic food and rice are 7%, with the most productive area Tigaraksa and Cikupa
5. Clothing is 7%, with the most productive area Curug

4 Conclusion

After completing research on potential MSMEs classification and analysis, some conclusions can be drawn as follows :

a. The Naive Bayes method has been successfully carried out in generating the most productive MSMEs classification and analysis based on sub-districts and regions based on business specifications from 1850 Tangerang MSMEs data containing 7 attributes (id, name of enterpreneur, business name, business address, telephone number, business form, and business fields) which results in 9 data integrity (id, name of enterpreneur, business name, business address, telephone number, sub-seconds, business forms, business fields, and business classifications)
b. From the analysis results can be used as recommendations in establishing a grow up policy and business assistance based on potential business fields in areas that have high business productivity.

References

1. M. Rouse (2011) small to medium enterprise (SME), Business intelligence - business analytics retrieved from http://whatis.techtarget.com/definition/small-to-medium-enterprise-SME Accessed Maret, 17th 2018
2. Sudaryanto, Regimun, and R.R. Wijayanti,"Srategi Pemberdayaan UMKM Menghadapi Pasar Bebas ASEAN", 2014.www.kemenkeu.go.id/sites/default/files/strategi%20pemberdayaan%20umkm.pdf
3. D. Sena, M. Ozturkb, and O. Vayvay, "An Overview of Big Data for Growth in SMEs", Elsavier 12th International Strategic Management Conference, pp. 159-167 (2016)
4. B.O. Ogbuokiri, C.N. Udanor, and M.N. Agu, "Implementing bigdata analytics for small and medium enterprise (SME) regional growth", IOSR Journal of Computer Engineering, Vol. 17 (6), pp.35-43 (2015)
5. Departemen Koperasi, http://www.depkop.go.id/pdf-viewer/?p=uploads/tx_rtgfiles/sandingan_data_umkm_2012-2013.pdf diakses tanggal 18 Maret 2018

6. website kab tangerang https://tangerangkab.go.id/
7. Taruna R., S., Hiranwal, S., "Enhanced Naive Bayes Algorithm for Intrusion Detection in Data Mining", *International Journal of Computer Science and Information Technologies*, Vol.6 (4), pp. 960-962 (2013)
8. Bustami., "Penerapan Algoritma Naive Bayes Untuk Mengklasifikasi Data Nasabah Asuransi", Jurnal Penelitian Teknik Informatika, Vol. 3 (2), pp. 127-146 (2013)
9. S. Sarkar, R.S. Sriram, "Bayesian models for early warning of bank failures. Management Science", Management Science,Vol. 47 (11), pp. 1457–1475 (2001)
10. S. A. Pattekari, A. Parveen, Prediction System for Heart Disease Using Naive Bayes, *International Journal of Advanced Computer and Mathematical Sciences,* ISSN 2230-9624, Vol. 3 (3), pp. 290-294. (2012)
11. Jananto, Arief. *Memprediksi Kinerja Mahasiswa Menggunakan Teknik Data Mining (Studi kasus data akademik mahasiswa UNISBANK.* Tesis Tidak Terpublikasi. Yogyakarta: Universitas Gajah Mada (2010)
12. E.W.T. Ngai a,*, Li Xiu b , D.C.K. Chau a, Application of data mining techniques in customer relationship management: A literature review and classification, Expert Systems with Applications 36. 2592–2602 (2009)
13. Turban, E., Aronson, J. E., Liang, T. P., & Sharda, R. Decision support and business intelligence systems (Eighth ed.). Pearson Education (2007)
14. J. Han and M. Kamber, Data Mining: Concepts and Techniques. Morgan Kaufmann Publishers Inc, San Francisco (2006)
15. https://www.scribd.com/doc/15123416/Kurva-Receiver-Operating-Character diakses Agustus 2018
16. https://mragungsetiaji.github.io "Machine learning: matrics recall and precision" diakses Agustus 2018

Design of Auto-tuning Relay Feedback controller for Shell Heavy Oil Fractionator

Wahyudi[1,*], *Sela* Martasia[1], *Budi* Setiyono[1], and *Iwan* Setiawan[1]

[1]Control Laboratory, Department of Electrical Engineering, Diponegoro University, Semarang 50271, Indonesia

Abstract. Auto-tuning relay feedback is one of the control techniques, which is used to solve the non-linear, long delay time, and disturbance's problems. This control technique is the development of Ziegler-Nichols that can be done automatically without doing system modeling. In this paper, auto-tuning relay feedback is used in the control system response to optimization of Shell Heavy Oil Fractionator (SHOF) system so the output of product composition as expected. SHOF is a distillation column type used to separate crude oil into desired products based on the difference in the boiling point of each product. PI regulators of relay feedback are used to control the valves on the SHOF with three inputs and three outputs that has been decoupled. Based on the tests, the average values of IAE at top end point composition (Y_1) obtained with disturbance and no disturbance are 83.17 and 10.933, respectively. At the side end point composition (Y_2), the average values of IAE with disturbance and no disturbance are obtained respectively, 336.38 and 42.3467. The average values of IAE at bottom reflux temperature (Y_3) with disturbance and no disturbance are obtained 0.15 and 0.13, respectively.

1 Introduction

Shell Heavy Oil Fractionator (SHOF) is a of distillation column type used to separate crude oil into desired products based on the difference in the boiling point of each product[1, 2]. The design of control on SHOF has several constraints, it caused by non-linearity of the process, multivariable interactions, long dead time, and disturbances. It needs a control method that is able to the control system response to optimization of SHOF system so the output of product composition as expected.

Previous research that relating to the control of SHOF is suggested that the combination of Subspace Identification Method (SIM). The Multivariable Output Error State Space (MOESP) method is adopted in SIM. This SIM-based double-layered is illustrated through an example of the heavy oil fractionator model[3]. Chunqing Huang and Shifu Chen used performance assessment of multi input multi output (MIMO) systems in the presence of plant uncertainty. It shows that only delay matrix is required, and the benchmark is obtained without any knowledge of interact matrix or Markov parameter matrices when condition regarding the delay matrix is satisfied. Effectiveness of this technique is

*Corresponding author: wahyuditinom@elektro.undip.ac.id

demonstrated by the SHOF gain is uncertain[4]. Relay feedback that illustrated using the Ogunnaike Ray (OR) distillation column model is carried out on MIMO system. The model parameters thus obtained can be used to develop model based control schemes[5]. Auto-tuning relay feedback based design of fractional order PI controller for 2-by-2 MIMO system is performed on a Wood and Berry (WB) distillation column. The performance was measured by using Integral of Time and Absolute Error (ITAE)[6].

This paper proposed design of auto-tuning PI controller using Ziegler-Nichols method with signal input in fixed frequency and variable amplitude. The outputs which controlled are top end point composition (Y_1) and side end point composition (Y_2) of SHOF. The performance from the system was measured by using Integral Absolute Error (IAE).

2 Methodology

2.1 Auto-tuning relay feedback controller

The relay feedback auto-tuning method is the development of the second tuning method of Ziegler-Nichols, which uses the proportional gain value as the ultimate gain to obtain the oscillation graph output constantly replaced by ultimate gain value K_u using relay. The block diagram of auto-tuning relay feedback as a control system is shown in Figure 1[7].

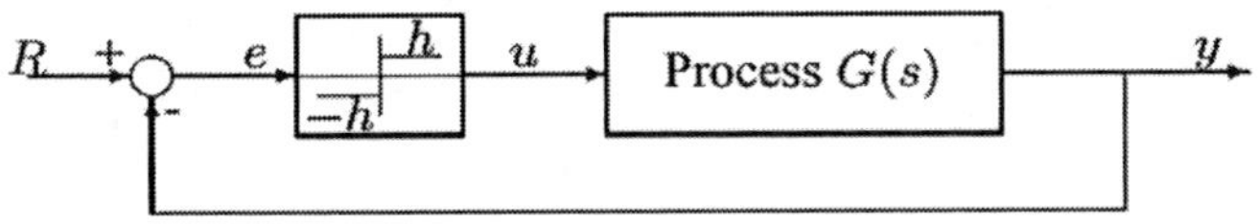

Fig. 1. Block diagram of auto-tuning relay feedback.

The basic principle of this method is the limitation period value of the oscillation process. It is controlled using a closed-loop method with a relay as a control system. The input and output signal obtained with the control signal (u) is shown in Figure 2[7].

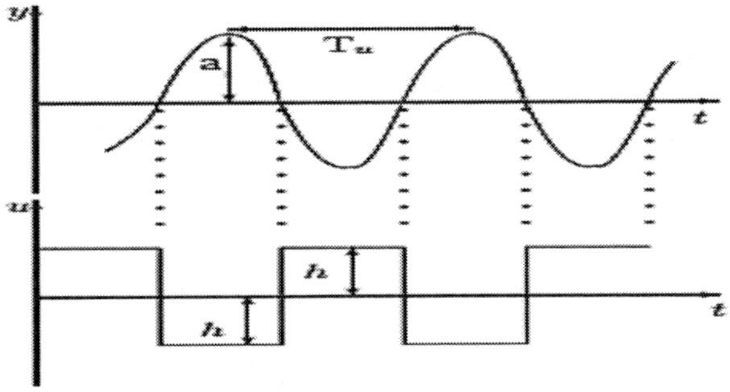

Fig. 2. Input and output of relay feedback system.

The input to the process G(s) is a square wave with the amplitude h and frequency ω_u. The output is a sinusoidal with the amplitude $4d/\pi$. Relation of the input and output of the process is shown in equation 1.

$$a = \frac{4h}{\pi}|G(j\omega_u)| \tag{1}$$

In order to have an oscillation, the fundamental component of the input and the output must be have the opposite phase. The conditions for oscillation are here,

$$arg\ G(j\omega_u) = -\pi \tag{2}$$

$$|G(j\omega_u)| = \frac{a\pi}{4h} = \frac{1}{Ku} \tag{3}$$

where Ku is called the ultimate gain and can be regarded as the equivalent gain of the relay for transmission of sinusoidal signals with amplitude a and Tu is similarly called the ultimate period. The calculation of the PID control parameter with relay feedback is determined by the ultimate period and the ultimate gain as shown in Table 1[8].

Table 1. Ziegler-Nichols PID parameter according to the ultimate point.

Controller	Kp	Ti	Td
P	0.5 Ku	-	-
PI	0.4 Ku	0.8 Tu	-
PID	0.6 Ku	0.5 Tu	0.12 Tu

2.2 Shell heavy oil fractionator

The heavy oil fractionator is characterized by three outcomes draws and three sides circulating loops as shown in Figure 3[2].

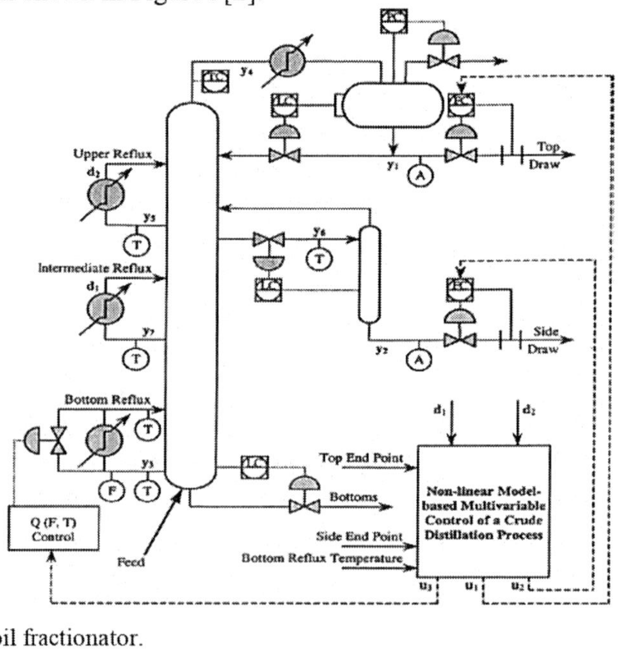

Fig.3. Shell heavy oil fractionator.

The transfer function of SHOF plant is shown in equation (4)[1,2]. SHOF is a multivariable system with three inputs and three outputs (MIMO). Interaction between subsystems is reduced by applying the method of Relative Gain Array (RGA) and decoupling, so the system can be transformed into three Single Input Single Output (SISO).

$$\begin{bmatrix} Y_1 \\ Y_2 \\ Y_3 \end{bmatrix} = \begin{bmatrix} \frac{4.05e^{-27s}}{50s+1} & \frac{1.77e^{-28s}}{60s+1} & \frac{5.88^{-27s}}{50s+1} \\ \frac{5.39^{-18s}}{50s+1} & \frac{5.72^{-14s}}{60s+1} & \frac{6.90^{-15s}}{40s+1} \\ \frac{4.38^{-20s}}{33s+1} & \frac{4.42^{-22s}}{44s+1} & \frac{7.20}{19s+1} \end{bmatrix} \begin{bmatrix} U_1 \\ U_2 \\ U_3 \end{bmatrix} + \begin{bmatrix} \frac{1.20\,e^{-27s}}{45s+1} & \frac{1.44\,e^{-27s}}{40s+1} \\ \frac{1.52\,e^{-15s}}{25s+1} & \frac{1.83\,e^{-15s}}{20s+1} \\ \frac{1.14}{27s+1} & \frac{1.26}{32s+1} \end{bmatrix} \begin{bmatrix} d_1 \\ d_2 \end{bmatrix} \tag{4}$$

The RGA was used to determine the best input manipulative–output controller pairing and can be calculated from the expression here [9]

$$\Lambda = K_{ij} \otimes H_{ij} = K \otimes (K^{-1})^T \tag{5}$$

where ⊗ denotes the Schur product (element by element multiplication):

$$K_{ij} \otimes H_{ij} = K_{ij} H_{ij} \quad (6)$$

The value of RGA that obtained with equation (5) and determinate pairing rule, with the paired elements along the diagonal and avoid pairings that correspond to negative steady-state RGA-elements[10], so can be determined the controlled output pair of the best manipulative input for multivariable control leads to a so-called dominant interaction strategy[11].

2.3 Design of auto-tuning relay feedback Ziegler Nichols

Design of auto-tuning relay feedback method has two steps. Step one, make the SHOF system on top end point composition (Y_1) and side end point composition (Y_2) in close loop circuit using relay. The magnitude of the relay is added from 0 until a critical magnitude relay is obtained to get a graph of the constant of amplitude and the period oscillation ouput process. The graph of each output can be seen in Figure 4.

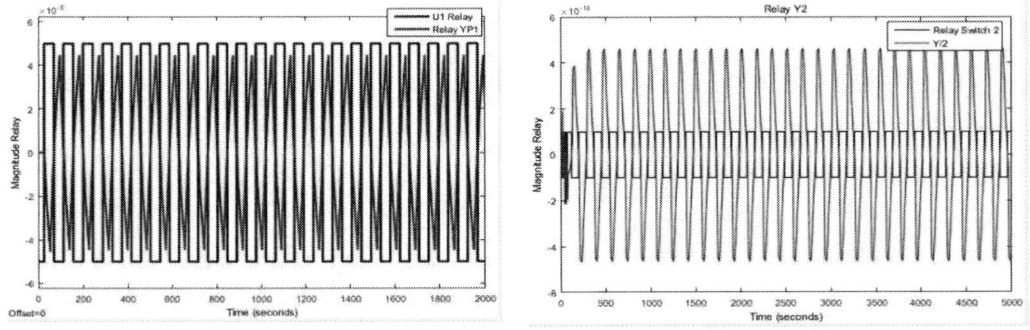

(a). output Y1 (b). output Y2

Fig. 4. Magnitude and period oscillation relay feedback.

Based on Figure 4, the values of h, a and T_u at top end point composition (Y_1) are respectively $6x10^{-15}$, $7.11x10^{-14}$, and 80.85305 also the values of side end point composition (Y_2) the values of h, a, and T_u is $4x10^{-5}$, $4.3306x10^{-5}$, and 50.3283 respectively. Step two, looking for the values of K_u, K_p, and K_i. K_u is obtained by using equation (3), while K_p and K_i using the equation PI Ziegler Nichols in Table 1.

3 Result

3.1 The response without disturbance

The set point of top end point composition (Y_1) increase test from 0.0 to +0.5; response system is shown in Figure 5. Figure 5a shows that auto-tuning relay feedback Ziegler-Nichols controller is capable to reach set point value and steady state of top end point composition (Y_1) and capable to maintain the initial set point value of side end point composition (Y_2), also PI controller is capable to maintain the initial set point value of bottom reflux temperature (Y_3). Based on Figure 5b, flow rate on top draw (u_1), side draw (u_2), and bottom reflux duty (u_3) gave a graph result are within high and low limits of 0.5 and -0.5. The IAE value in this test is obtained by auto-tuning relay feedback Ziegler Nichols controller at top end point composition (Y_1), and side end point composition (Y_2) with 31.06 and 6.13, respectively and by PI controller at bottom reflux temperature (Y_3) with 0.04.

The set point of side end point composition (Y_2) is the increase test from 0.0 to +0.5; response system is viewed in Figure 6. Auto-tuning relay feedback Ziegler-Nichols controller is capable to reach set point value and steady state of side end point composition (Y_2) and capableto maintain the initial set point value of top end point composition (Y_1). Flow rate on top draw (u_1), side draw (u_2), and bottom reflux duty (u_3) gave a graph result are within high and low limits of 0.5 and -0.5. The IAE value in this test is obtained by auto-tuning relay feedback Ziegler Nichols controller at top end point composition (Y_1), and side end point composition (Y_2) with 1.22 and 100.47, and by PI controller at bottom reflux temperature (Y_3)with 0.06.

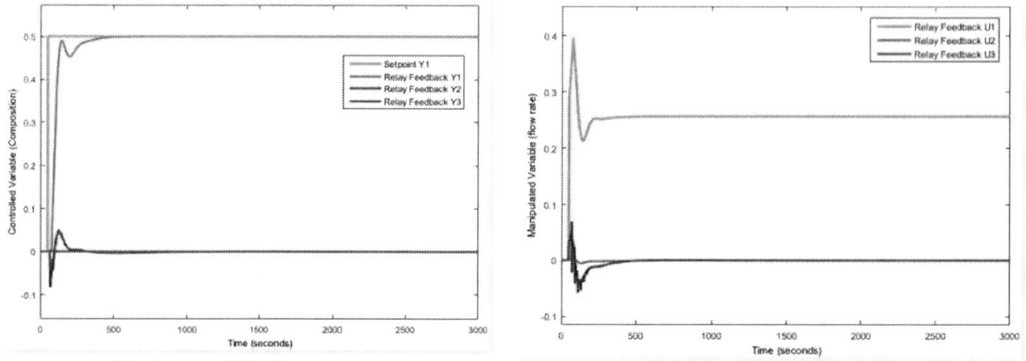

(a). Output SHOF response (b). Input SHOF response

Fig. 5. SHOF response to the increase of top end point composition (Y_1) +0.5

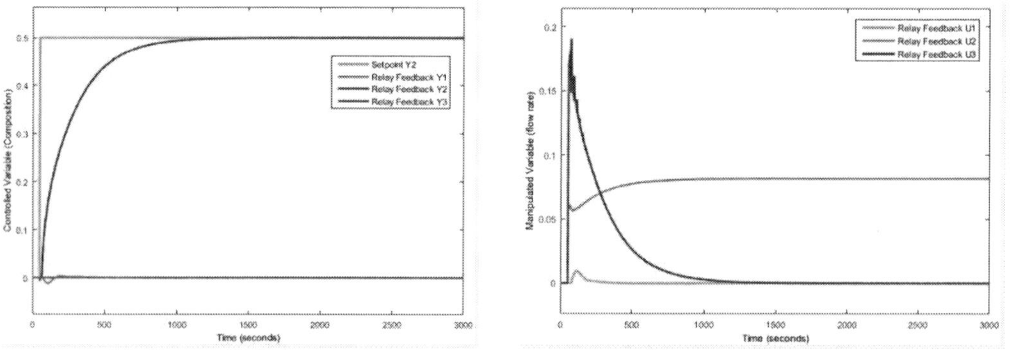

(a). Output SHOF response. (b). Input SHOF response

Fig. 6. SHOF response to the increase of side end point composition (Y_2) +0.5

3.2 The response with disturbance

The step input and output of intermediate reflux duty (d_1) and upper reflux duty (d_2) by +0.5 are shown in Figure 7.

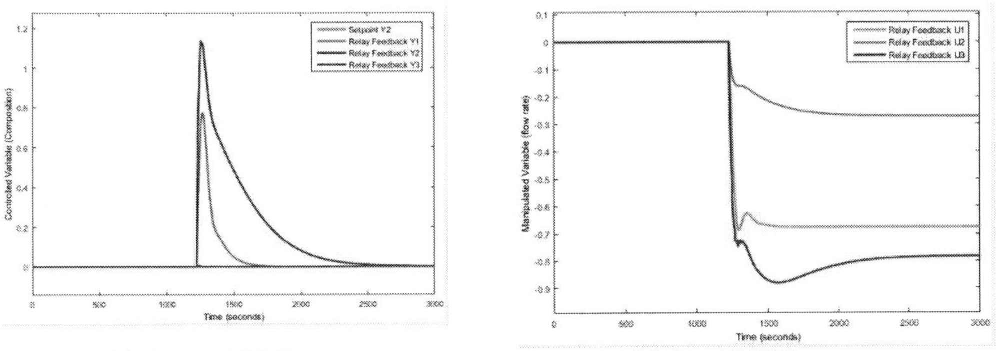

(a). Output SHOF response (b). Input SHOF response

Fig. 7. SHOF response with input disturbance +0.5.

PI controller is capable to eliminate the disturbance which is the step input of intermediate reflux duty (d_1) and upper reflux duty (d_2) by +0.5 and capable to maintain the initial set point value and the initial steady-state condition of each product composition (Y_1,Y_2, and Y_3). The IAE value in this test is obtained by auto-tuning relay feedback Ziegler-Nichols controller at top end point composition (Y_1), and side end point composition (Y_2) with 83.17 and 336.38 respectively, and by PI controller at bottom reflux temperature (Y_3) with 0.15.

4 Conclusion

Based on all test results, the auto-tuning relay feedback controller using Ziegler-Nichols tuning method was able to eliminate the disturbance in the Shell heavy oil fractionator and able to control the system to maintain initial set point value and initial steady-state condition as well as achieving new set point value and new steady-state condition. The average IAE value with auto-tuning relay feedback controller at top end point composition (Y_1) obtained with disturbance and no disturbance of 83.17 and 10.933, at side end point composition (Y_2) obtained with disturbance and no disturbance of 336.38 and 42.3467, and with PI controller at bottom reflux temperature (Y_3) obtained with disturbance and no disturbance of 0.15 and 0.13.

References

1. C. E. García, D. M. Prett, and B. L. Ramaker, *Fundamental Process Control*, 1st ed. Boston: Butterworths, (1988).
2. R. M. Ansari and M. O. Tadé, *Nonlinear Model-based Process Control Applications in Petroleum Refining*, 1st ed. London: Springer, (2000).
3. H. Pan and H. Gao, "A combinatory study on MPC with subspace identification, steady-state target calculation and dynamic quadratic control," *Proc. World Congr. Intell. Control Autom.*, vol. 2015–March, no. March, pp. 3164–3169, (2015).
4. C. Huang and S. Chen, "Performance assessment of process with uncertainties based on minimum variance *," no. Iccse, pp. 848–854, (2016).
5. K. Bijeesh, D. Kalpana, and T. Thyagarajan, "Identification of 3-by-3 MIMO system using relay feedback," *Proc. 2014 IEEE Int. Conf. Adv. Commun. Control Comput. Technol. ICACCCT 2014*, no. 978, pp. 157–162, (2015).
6. D. Kalpana, T. Thyagarajan, and N. Venkatachalam, "Design of fractional order PI

controller for MIMO system using relay feedback," *Proc. - TIMA 2017 9th Int. Conf. Trends Ind. Meas. Autom.*, (2017).

7. C.-C. Yu, *Auto-tuning of PID controllers - A Relay Feedback Approach*, 2nd ed., vol. **11**, no. 1. Taiwan: Springer, (2001).
8. K. J. Åström and B. Wittenmark, *Adaptive Control*. New York: Dover Publications, Inc., (2008).
9. D. E. Seborg, T. F. Edgar, and D. A. Mellichamp, *Process Dynamics and Control*, 2nd ed. United States of America: John Wiley & Sons, Inc, (2004).
10. S. Skogestad and I. Postlethwaite, *Multivariable feedback control: analysis and design*, 2nd ed., vol. **8**, no. 14. New York: John Wiley & Sons, Inc, (2005).
11. M. T. Tham, "Multivariable control: An Introduction to Decoupling Control," *Univ. Newcastle upon Tyne, Newcastle upon Tyne*, vol. 1999, no. May, (1999).

Phytoplankton Concentration Measurements Base On Arduino Microcontroller

Gunady Haryanto [1,*], *Vector* Anggit Pratomo [1] *Agung* Saputra [1]
[1]Electrical Engineering University of Pancasila, Jakarta, Indonesia

Abstract. In this research is designing and manufacture Phytoplankton concentration measurements in liquid medium based on arduino microcontroller. The design works with by exploiting fluorescence phenomenon, by using Purple Light Emitting Diode (λ = 405 nm, P = 10 mW, frequency modulation 625 Hz) as a light source absorbed by phytoplankton, mesuring vessel, optical filters, and photodiodes. Data obtained of the photodiode sensor is then conducted data acquisition by arduino, the quantitiy of phytoplankton concentration will be displayed on the lcd. From testing to Phytopathton for the concentration range it was found that for the concentration range 102 - 106 cells / ml obtained a consistent relationship the intensity of fluorescence with the increase in cell concentration. Gradients obtained for high range concentration with R^2 = 0.9632 lower than the low range concentration with R^2 = 0.9642.

1. Introduction

Indonesia is an archipelago country where the geographical area of 75% is the ocean, one of nature potential with various marine fauna such as fish, terumbu karang, zooplankton, phytoplankton and others. Phytoplankton or so-called microalgae has a major role for the balance and sustainability of marine and terrestrial biota ecosystems [1]. Fish life and a variety of marine living things are highly dependent on phytoplankton which is the beginning of the food chain [2]. Phytoplankton function as a carbon cycle globally that can regulate the earth's temperature and oxygen-producing. So we need an observation to know the amount of phytoplankton concentration in a water area. The size of the phytoplankton in micrometer units and having unique characteristics can only be seen using a microscope to identify and calculate the amount of concentration [3] so it takes expertise in the field of biology. For the purposes of the field on fish hatchery and phytoplankton culture required a measuring instrument that is easy to operate to measure the concentration of phytoplankton. In this study the method used to detect the presence of phytoplankton using optical methods as light media. Optical method has the effect of phenomenon of light interaction on phytoplakton particles causing phenomenon of reflection, absorption, scattering, refraction, diffraction, interference and disperse. The use of optical method is aimed at this research is to see the existence of phytoplankton by way of physical process of light. Usually to identify the presence of phytoplankton requires a high-resolution microscope and a UV-Vis current spectrophotometer. This tool can only be operated in laboratory labs and can not operate in real time, this tool also has an economical value is expensive. By using optical

* Corresponding author: gunady.haryanto@univpancasila.ac.id

method it can be designed a measuring instrument that can be used in real time, economical and easy to operate.

The basis of the optical measurement method utilizes the effects of light absorption on a particular region by phytoplankton. With this method the light source will be generated by laser diode with light dectektor (PIN photodiode). Interferometry method is used for optical set-up configuration in design in this research. Interferometry method is used for optical set-up configuration in design in this research. Fluorescence or phosphorescence absorption and emission is the natural and dominant process of light physics when phytoplankton responds to light. To measure the proposional level that occurs in the absorption of light technically will compare the presence of phytoplankton and its inability.

The main basis for selecting optical methods in the measurement of phytopathton concentrations is:

a. The optical measurement method is a method of measurement whose resolution is within the wavelength dimension of the micrometer-nanometer scale so that expected by the optical method used in this study can yield accurate measurement data in identifying the presence / types and quantities of phytoplankton species.
b. Optical method is a method that can be integrated with electronic systems so that the measurement in real time can be done without having to do chemical testing in the laboratory and signal processing can be done electronically and data processing can be integrated with computer in the form of software.
c. The optical method is an appropriate aplicative technology for designing phtyoplankton concentration measuring devices economically so that the success of the device design can be widely used by Indonesian fishermen and fish farmers.
d. The optical measurement method is an in-situ and non-destructive measurement method so that measurements can be made on-site and do not endanger the life cycle of phytoplankton.

The main issues that will be solved and analyzed related to the achievement of the objectives of this research are:

- The size of phytoplankton in the micrometer scale will be an obstacle for measuring instruments to be designed in terms of precision and accuracy so that it should be a primary consideration in the detection system.
- Different types and species of phytoplankton are serious problems to consider in this study because each phytoplankton has unique biological characteristics and varies according to its ecosystem.

The optical phenomenon used as a reference analysis is the phenomenon of absorbance and fluorescence of phytoplankton occurring in response to the light beam falling on the microbial creature, fluorescence emission characteristic of the phytoplankton as the determinant of the laser wavelength to be used as the primary detecting medium. So that the accuracy of the fluorescence fluorescence characteristic analysis will have a big effect on the accuracy of the measurement result.

The main objectives to be achieved in this research are:

- Designing sensors capable of accurately identifying the presence of phytoplankton according to the type of species desired.
- Designing precise and accurate quantity of phytoplankton devices in measurement resolution of 50 cells / ml.
- Designing phytoplankton detectors that have an in-situ, compact, non-destructive, user friendly and economical advantages.

- Designing microcontroller based phytoplankton measuring instrument as a data processing center.

2. Theory

2.1 Fluorescence Spectroscopy Study

Fluorescence is a photon emission process that occurs when the molecule is in a state of relaxation so that the conditions result in photon emission at a duration of approximately 10-9 s duration. The excitation process that occurs involves a single state process so that the lifetime of the upper state increases and ultimately leads to a transition to a metastable state. Quantum emissions in fluorescence depend on the degree of balance between radiative and non-radiative relaxation. While the intensity of light emitted from the process is to follow Lambert's law;

$$I = I_0 \exp(-\varepsilon cl) \tag{1}$$

where ε : absorption coefficient;c = medium concentration; l is the time of photon emission

3. Measurement Device Component

Measurements of the concentration of fitolankton dissolved in a liquid medium comprise a probe, signal processor, and driver circuit. The probe consists of a light source and detector an adjacent signal amplifier for maximum fluorescence intensity. The desired device is portable and easy to use in the field, therefore the probe is placed separately from the processor driver circuit. These three sections are connected with cables as shown in Figure 1 Probes can be immersed into a liquid medium and confronted in different directions or moved

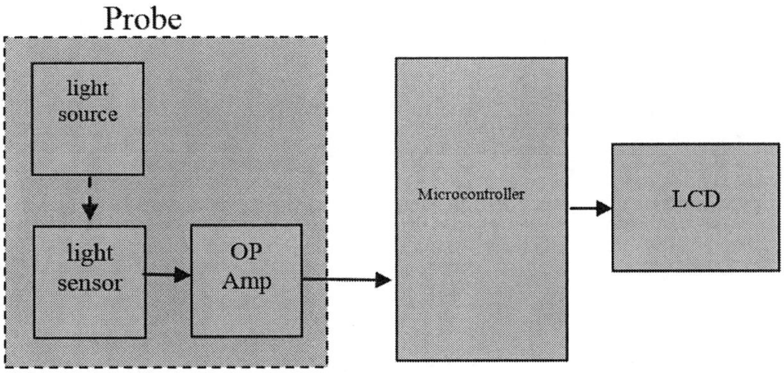

Fig. 1. Fitoplankton concentration gauge block diagram.

3.1 Light source

Based on the absorbance spectrum obtained from Phytoplankton test results. To be able to produce light that is able to excite the phytoplankton used a light source LED diode 405E (Thorlabs) can be seen in Figure 2.

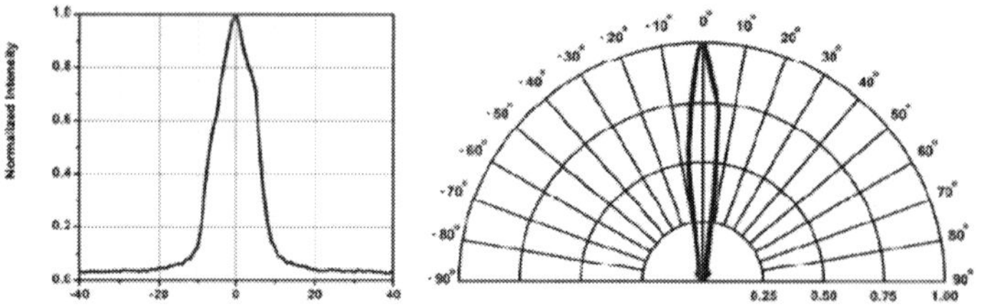

Fig. 2 The direction of the intensity distribution of LED405E Thorlabs

3.2 Photodioda Sensor

Based on the fluorescence characteristic of Phytoplankton obtained the dominant spectrum at 685 nm wavelength. To be able to detect the fluorescence produced by phytoplankton is used a photodiode. In this research the photodiode is used PIN type from Si type FDS100 made by Thorlabs. This photodiode has a spectrum response of 350 - 1100 nm and is used as a light detector worked in a conductive photo mode and coupled with an operational amplifier (Op - Amp).

3.3 Driver LED

In the measurement process, the photodiode functions to receive the fluorescent light due to the light coming from the LED. But in reality, there is light coming from other sources such as scattering. To solve this problem in the process of measurement, the light emitted by the LED must be differentiated and detected separately. Therefore a method is sed by laying the amplitude modulation signal on the light emitted by the LED

3.4 Analog Signal Boosters and Processors

Photodiode output signal is very small, ie in the order μV. Therefore it takes a signal amplifier circuit (pre-amplifier) to obtain a signal that can be easily processed further. An amplifier Operational Amplifier IC LM358 is selected for this purpose with the consideration that its characteristics meet the requirements.

3.5 Optical Filters

To avoid detecting light other than fluorescence light, optical filters are used only through the fluorescence light. The filter used is made of plastic with a spectral transmission response of 600 - 730 nm [17]. The installed filter r5 attaches to the photodiode surface and is protected by glass to avoid direct contact with the culture water.

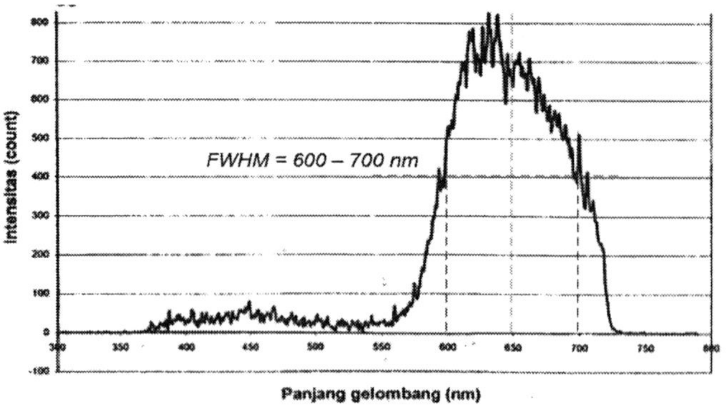

Fig. 3. Spectrum of optical filter used

3.6 Design and Manufacture of Optical Probes

By involving the previously described parts, the arrangement of the optical set up (probe). Before going to the LED light beam measuring container is focused by the lens d = 3 mm, f = 10 mm in the sleeve. The sleeve is made of plastic material and has a size adjusted to the diameter of the LED to excite the phytoplankton, while the photodiode is used to detect fluorescence intensity. In this voltage range, the light energy that can be used to excite is 0 - 20 mJ, while the modulation frequency generated by the LED Driver is 625 Hz. The fluorescence intensity detection result is then fed to the pre-amplifier and measured by Voltmeter.

3.7 Microcontroller

The microcontroller used as instrument control center is arduino uno which operates at 5V DC voltage. In the measurement instrument of phytoplankton concentration there are several ports used, including: digital port pin 2 as input data from light sensor of A5 digital port port. The main series of arduino uno can be seen in Figure 3.4

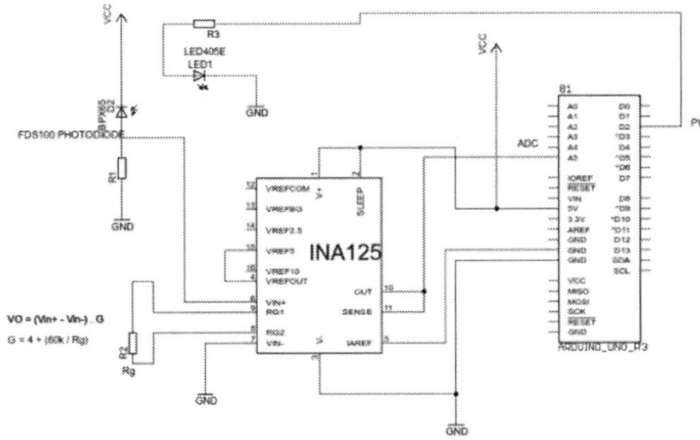

Fig. 4. microcontroller system design circuit

4. Measurement Results Concentration Scenedesmus sp.

Furthermore, the measurement results representing the intensity of the fluorescence and the intensity of light before penetrating the culture for each concentration (102 to 106 cells / ml) at room temperature 26 ° C is processed and presented in graphical form Figure 5.2 a and b. In the graph a results of Scenedesmus sp measurements of 102 up to 1 x 104 cells / ml and on the graph b of Scenedesmus sp measurements from 5x104 to 1 x 106 cells / ml.

From the measurement result of Figure 4.1 (a) the concentration of 102-104 cells / ml, it is known that the normalized intensity increases linearly along with the increase of Scenedesmus sp. The gradient in the range 102 to 104 cells / ml is 2 x 10-8 and the standard deviation value (R^2) is 0.9642. As for Fig. 5.2 (b) the gradient in the 5 x 104 - 106 cell / ml range is 1 x 10-9 and the standard deviation (R^2) is 0.9632.

These results show that for the high concentration range (5x104-106 cells / ml) the resulting curve gradient is relatively smaller compared to the low concentration range (102-199 cells / ml) due to the influence of reabsorption by phytoplankton.

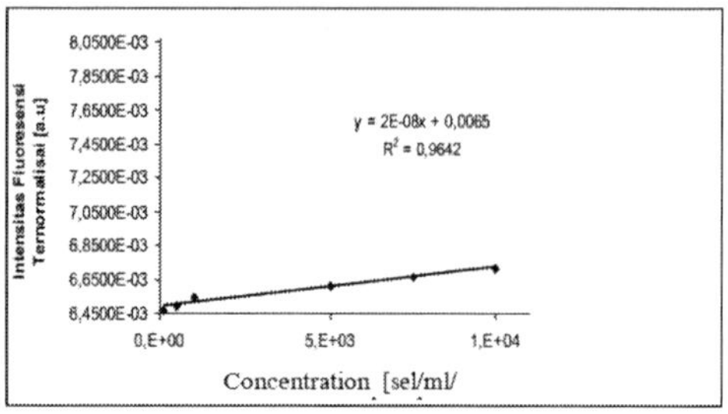

(a)

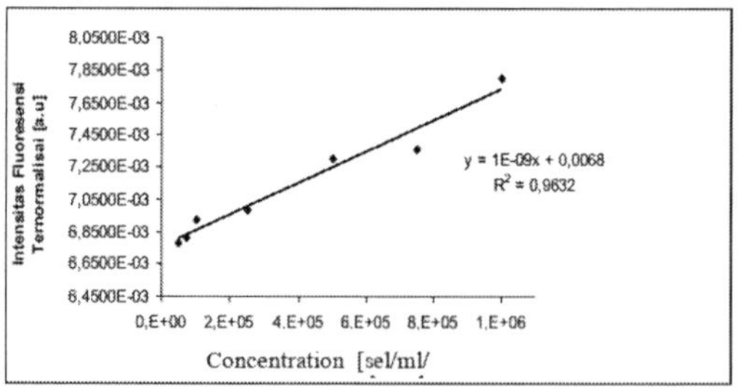

(b)

Fig. 5. Effect of Scenedesmus sp concentration on fluorescence intensity.
(a.) concentration 0 - 104 cells / ml
(b.) concentrations 5x 104 - 106 cells / ml

Can be seen from Figure 5. that the intensity of fluorescence increases with the increase in the consentration of Scenedesmus sp cells.

5. Conclusion

From the design results, the Phytoplankton test and measurement are:

1. Absorption of light by Phytoplaknton occurs at 400 - 500 nm spectral.
2. Phytoplankton has a typical fluorescence spectrum in which there is a large intensity fluctuation in the range 652 - 789 nm and its intensity peaks occur at 681nm, 752nm, and 777nm.
3. In order to produce accurate phytoplankton concentration value, an analog 10 to bit digital converter is required.

References

1. Maria Saburova, Faiza Al-Yamani, *Diversity and distribution of winter phytoplankton in the Arabian Gulf and the Sea of Oman*, Elsevier, Continental Shelf Research 119 (2016).
2. YuZhaoa,c, SanlingYuanb, TonghuaZhangd, *The Stationary distribution and ergodicity of stochastic phytoplankton allelophaty model under regime switching*, Commun Nonlinear Sci Numer Simulat **37,** 131–142 (2016)
3. Thomas Leeuw Emmanuels and Dana L.W, *in situ measurments of phytoplakton sensor* , (2013)
4. Julia Uitz , Dariusz Stramski , Rick A. Reynolds, *Assessing phytoplankton community composition from hyperspectralmeasurements of phytoplankton absorption coefficient andremote-sensing reflectance in open-ocean environments*, Elsevier, (2016)

Body Design Concept of Remotely Operated Vehicle (ROV) of Observation Class with the Method of Concept Screening and Concept Scoring

Dhimas Satria[1,*], *Dovan* Pujangga Asmara Lanank Esiswitoyo [1], *Ni Ketut* Caturwati [1], *Erny* Listijorini [1], *Rina* Lusiani[1]

[1] Mechanical Engineering Department, Universitas Sultan Ageng Tirtayasa, Indonesia

Abstract. The Remotely Operated Vehicle (ROV) underwater observation equipment has previously been developed, but it had restriction for not being hydrodynamic still due to the unique conditions of the Sunda Strait waters. This study aims to further develop the previously developed ROV by improving the design structure of the ROV body. The method of design concept selection used was concept screening and concept scoring. In concept screening, the initial concept was evaluated relatively to the same preferences using screening matrices. On the other hand the Concept Scoring was conducted by determining the relative performance value which was the assessment used to evaluate how the comparison between concepts and number as the comparison. The result of the study obtained the selected design concept, which was small ROV with streamlined / half-streamlined body geometry which had rear-middle rotor configuration as motion component of hydrodynamic 3 DOF (surge, heave, and yaw).

1 Introduction

ROV design which is previously developed has limitation in design structure and hydrodynamic response [1]. The solution of those problems is the development of correct design structure suitable for hydrodynamic [2][3], and also has current design. Geometry and the body shape of moving object in flowing fluid affect hydrodynamic respond design, which is ROV body design structure [4][5][6]. The geometry of equipment working in moving fluid is determined by the parametric geometry [7]. Meanwhile the shape of body is determined by design optimization with form follow function method which refer to parametric geometry [8].

Design optimization is started with the concept formulation. Two concept selection needs to be through to obtain the concept; those are concept screening and concept scoring. Therefore this study aims to obtain ROV body design concept structure is in accordance with the criterion of hydrodynamic used as the basis the development of design ROV of

* Corresponding author: dhimas@untirta.ac.id

observation class [9]. The second part gives method of concept selection, in which product specification developed into solution and concept alternatives. Meanwhile the third part is the result obtained after concept selection step.

2 Method

The method of concept selection used is a method that has been widely used in design planning that is Pahl & Beitz method [10]. Pahl & Beitz stated that design is a process of creativity but is limited by the provisions that have been established to produce design through creativity. Fig. 1 illustrate Pahl & Beitz method is a conceptualization method that is composed from the process of translating the specifications and needs of prospective users into the technical language, to form the concept of design of the results of elimination and assessment concept against the criteria.

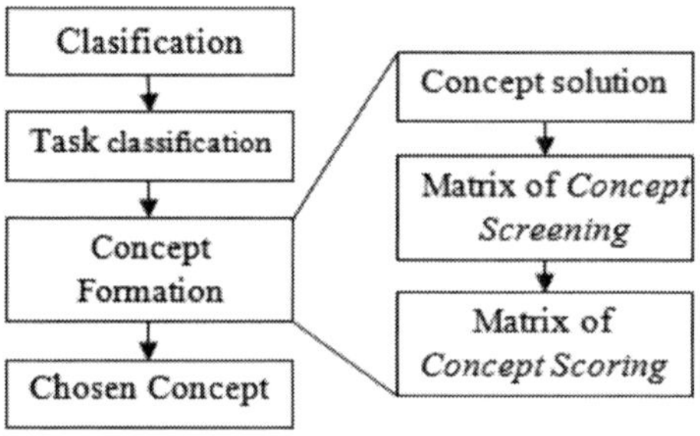

Fig. 1. Conceptualization Step of Pahl & Beitz Method. [10]

Specifications and the needs of potential users are obtained by collecting qualitative data expected by prospective users. Qualitative data were obtained by interviewing prospective users and document studies of the literature collected from various sources. Qualitative data are collected in the product specification sheet and reviewed together with quantitative data from the previous design structure specifications. Quantitative data processing is classified according to the task to be qualified by the ROV, which contains; structure, mobility, operating conditions, operating time, effectiveness, and interaction.

Data specifications and user requirements compiled in the specification sheet are reviewed by identifying the subject matter. Identification is conducted by abstraction and problem formulation through five steps of the abstraction process [10]. The five steps of the abstraction process produce the formulation of the problem; geometry and/or vehicle design with operating conditions determined by hydrodynamic response, with variations in speed and acceleration, and degrees of freedom.

3 Result

The product specifications which have been identified in the permeate formulation determine the main functions that describe the relationship between incoming and outgoing data values (outcomes). The problem formulation forms the overall design function in the block diagram.

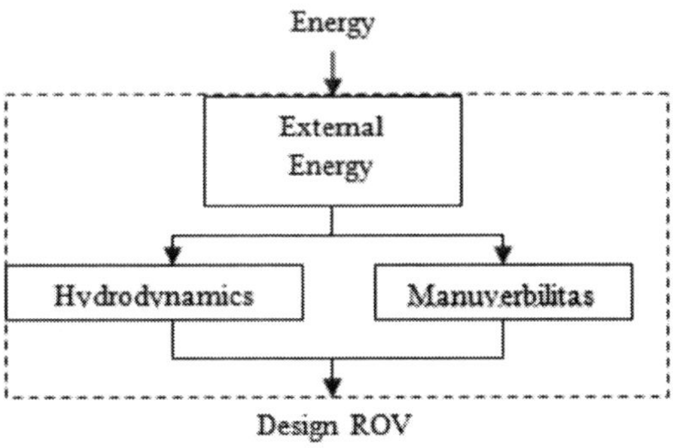

Fig. 2. Block Diagram Sub function.

Fig. 2 illustrates block diagrams composed of sub-functions that influence concept formation. Fig. 3 illustrates each sub-function that is the definition of the problem requires a solution. Possible solutions for Hydrodynamics sub-functions are; ROV size and ROV geometry, and solutions for manuverbility sub-functions are; driver configuration and motion system model.

Based on literature [10][11], the selection of concepts consists of two stages. The first stage is the concept screening and the second stage is the concept of scoring. In concept screening, solutions are evaluated relative to references using screening matrices. At this stage the quantitative comparison evaluation uses values relative to the criteria. Once the solution is evaluated, each solution is formed in an alternative concept that may be formed. Alternative concepts are evaluated by concept scoring using scoring matrix.

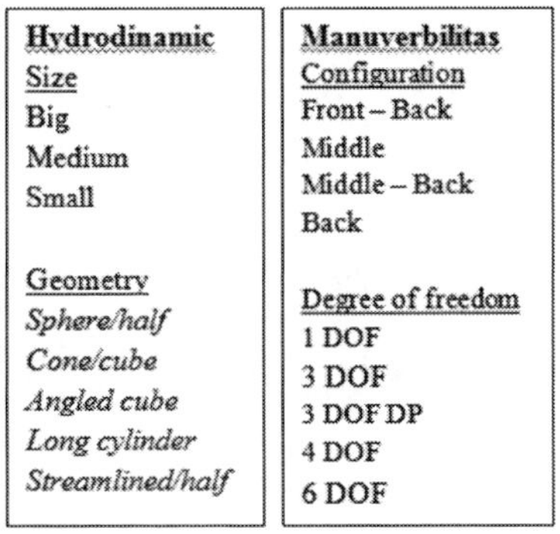

Fig. 3. Solution Sub-function.

3.1 Concept Screening

In this study, the concept screening method used is a method developed by Stuart Pugh in 1980 known as Pugh Concept Selection. Table 1 is the concept screening matrix. Selection criteria are criteria formed from problem formulations and specification sheets. Each solution in its sub-functions is given an assessment of the selection criteria by taking one of each solution in the sub-function as an assessment reference. Sub-functional solutions that most likely meet the selection criteria are given a positive (+) value, while those with a small chance of meeting selection criteria against reference are assigned as a negative (-). The assessment has a weight of 1 for a positive value (+), 0 for an assessment that has the same criteria as the reference, and -1 for the negative (-). The weights obtained from the solution are summed to form a ranking.

Table 1. Concept screening matrix.

	Concept																			
	Shape											Manuverbility								
	Size			Geometry								Thruster Arrangement				Model				
Selection Criteria	Large ROV	Medium ROV (ref.)	Small ROV	Sphere	Half Sphere (ref.)	Cone	Cube	Angled Cube	Cylinder	Stremalined Body	Streamlined Half Body	Front – Rear	Colleteral at Middle	Rear – Middle (ref.)	Rear	1 DOF	3 DOF	3 DOF DP	4 DOF (ref.)	6 DOF
Better ext. load resistance	0	0	0	0	0	0	-	-	0	+	+									
Low drag	-	0	+									0	0	0	0					
Dive ability	0	0	0	+	0	0	-	0	-	+	0	-	0	0	-	-	0	0	0	0
Easy to manufacture	0	0	0	0	0	0	0	0	0	0	0	0	0	0	0					
Manuverbility	-	0	+	0	0	0	-	-	0	+	+	-	-	0	-	-	0	0	0	+
Portability	-	0	+	0	0	0	0	0	0	0	0									
Realibility	0	0	0	+	0	0	-	0	0	+	+	0	0	0	-	0	0	0	0	0
Total +	0	0	3	2	0	0	0	0	0	4	3	0	0	0	0	0	0	0	0	1
Total 0	4	7	4	4	6	6	2	4	5	2	3	3	4	5	2	1	3	3	3	2
Total -	3	0	0	0	0	0	4	2	1	0	0	2	1	0	3	2	0	0	0	0
Sum	-3	0	3	2	0	0	-4	-2	-1	4	3	-2	-1	0	-3	-2	0	0	0	1
Rank	3	2	1	3	4	4	7	6	5	1	2	3	2	1	4	3	2	2	2	1
Continue?	no	no	yes	yes	com	com	no	no	no	com	com	no	yes	yes	no	no	com	com	yes	yes

The concept screening process gets the solution that has the greatest possibility of being a concept. The solution used for the process of forming alternative concepts is indicated by the words yes and comb, which means the solution needs to be combined to meet the selection criteria, in the section that indicates "continue?". However, one solution is that the model has been determined by potential users.

3.2 Concept Scoring

An alternative concept variant is formed from a solution that has been evaluated. The concept variant connects each solution into a unified concept of alternative sub-functions. The concept variant comprising one connected solution of each sub-functions, given the symbol of Variant x, x = A, B, C, ..., N. Table 2 is score relative performance.

Table 2. Score relative performance.

Relative Performance	score
Worse than reference	1
Slightly worse than reference	2
Same as reference	3
Slightly better than reference	4
Worse than reference	5

Table 3. Solution sub-function.

R	Q	P	O	N	M	L	K	J	I	H	G	F	E	D	C	B	A	Concept	
Small ROV																		Size	Solution
Streamlined Body/Half Streamlined Body						Half Sphere/Cone						Sphere						Geometry	
Rear – Middle			Colleteral at Middle			Rear – Middle			Colleteral at Middle			Rear – Middle			Colleteral at Middle			Thruster Arrangement	
6	4	3	6	4	3	6	4	3	6	4	3	6	4	3	6	4	3	Model (DOF)	

The concept screening stage uses the number of value distribution of each concept. Assessment of concept scoring uses the Karl method [11], where the karl method uses a relative performance value of 1 to 5 (Table 3). Relative performance values are assessments used to evaluate how the comparison of concepts to references by using the score significance as a comparison.

Concept Variants formed is evaluate by comparing the selection criteria. Selection criteria are judged by the weight of each criterion from the highest to the lowest in scale of 100. As in the prior conception process, concept screening, in the conception process of concept scoring it is necessary to define conceptual reference, where one of the references has been specified in the table. In the concept scoring table each concept is assessed against the selection criteria of relative performance value, which is the value of the ratio (rating) of the concept. For the final assessment, the value of the ratio and the weight of the criteria are multiplied and summed in one concept resulting in the total value that is the rank of each concept. The concept that has the greatest total value of reference is the chosen concept that will be further developed for analysis.

$$S_j = \sum_{i=1}^{n} r_{ij} w_i \qquad (1)$$

whence

r_{ij} = score of raw concept j for i criterion

w_i = score for i criterion

n = number of criterion

The concept has been selected through the concept screening stage and concept scoring. Both stages eliminate alternative concepts that have no score better than the chosen concept. Selected concept to be tested (concept testing) with simulation method, that is:

The concept of P small ROV with streamlined / half-streamlined body geometry which has a rear-middle rotor configuration as a motion component of hydrodynamic 3 DOF (surge, heave, and yaw) models.

4 Conclusion

The result of the research got the selected design concept, that is small ROV with streamlined / half-streamlined body geometry which has rear-middle rotor configuration as motion component of hydrodynamic 3 DOF (surge, heave, and yaw).

Acknowledgments

The authors would like to send the gratitude to the Directorate of Research and Community Service (DRPM) of DIKTI who has funded this research in the 2018.

References

1. R. Wiryadinata, A. S. Nurliany, I. Muttakin, and T. Firmansyah, Design of a Low Cost Remotely Operated Vehicle with 3 DoF Navigation, *Bull. Electr. Eng. Informatics*, vol. **6**, no. 1, pp. 13–23 (2017)
2. K. Amjith, V. Arundas, M. Harikrishnan, J. Sebastian, and V. Vineeth, Underwater Search and Rescue Device, *Int. J. Innov. Res. Sci. Technol.*, vol. **1**, no. 11, pp. 481–488 (2015)
3. D. L. Paster, Importance of Hydrodynamic Considerations for Underwater Vehicle Design, *IEEE*, pp. 1413–1422 (1986)
4. F. Aguirre, State of the Art of Parameters for Mechanical Design of an Autonomous Underwater Vehicle, *Int. J. Ocean. Oceanogr.*, vol. **11**, pp. 89–113 (2017)
5. H. G. Park, Hydrodynamic Design of an Underwater Hull Cleaning Robot and Its Evaluation, *Int. J. Nav. Archit. Ocean Eng.*, vol. **4**, pp. 332–352 (2012)
6. R. D. Christ, *The ROV Manual*. Oxford: Elsevier (2007)
7. P. S. Granville, *Geometrical Characteristic of Streamlined Shapes*. Washington D.C: Department of the Navy Naval Ship Research and Development Center (1969)
8. H. . Nguyen, Design, Modelling, and Simulation of a Remotely Operated Vehicle, *J. Comput. Sci. Cybern.*, vol. **29**, pp. 299–312 (2013)
9. W. Chuanfeng, Z. Fumin, and D. Schaefer, Dynamic modeling of an autonomous underwater vehicle, *J. Mar. Sci. Technol.*, no. August (2015)
10. G. Pahl and W. Beitz, *Engineering Design: A Systematic Approach*, Second Edi. London: Springer-Verlag (1996)
11. K. T. Ulrich and S. D. Eppinger, *Product Design and Development*, 6th Editio. New York: McGraw-Hill, Inc (2016)

Development of the Groundwater Level Changes Detector for Earthquake Prediction at Yogyakarta Region – Indonesia

Herlambang Laksamana Firdaus [1*], *Sunarno* [2], *Memory* Motivanisman Waruwu [2], and *Rony* Wijaya [3]

[1] Student of Magister Engineering Physics, Department of Nuclear Engineering and Engineering Physics, Universitas Gadjah Mada, Jl. Grafika 2, Sleman, D.I.Yogyakarta, Indonesia
[2] Sensor and Tele-control Laboratory, Department of Nuclear Engineering and Engineering Physics, Universitas Gadjah Mada, Jl. Grafika 2, Sleman, D.I.Yogyakarta, Indonesia
[3] Production House Unit, Faculty of Medicine, Universitas Gadjah Mada, Jl. Farmako no 2 Sekip Utara, Sleman, D.I.Yogyakarta, Indonesia

Abstract. Groundwater level (GWL) change is one of the earthquake precursors that used for earthquake prediction. The groundwater level change system detector for earthquake prediction at Yogyakarta region - Indonesia contains the GWL detector, signal conditioner, controller, data storage and electrical power supply. The GWL changes detectors which are developed in this report are expected to be used for the earthquake precursors prediction in Yogyakarta region - Indonesia. The detector system shows the detection of groundwater level changes with a sensitivity of 0.01 centimeters. The data from the detector is stored on the storage media, then the data which are collected in weekly will be confirmed with the earthquake data from the BMKG (Indonesian Meteorology, Climatology, and Geophysics Agency). On June 11, 12, 15, and 21 the GWL have a change of 4 centimeters to 19 centimeters respectively, the earthquake occurs within the next 2 days. It shows that the system could be used as an earthquake precursors monitoring system based on GWL changes. This report in order to give the valuable information of the development system that can be used for earthquake prediction, even though it needs further study and development of the instrumentation system from other earthquake precursors anomaly.

1. Introduction

A reliable earthquake precursor can save thousands of lives each year all over the world. Several instrumentation systems have been developed related to earthquake precursor anomaly. Many engineers and scientist are trying to establish this fact as a reliable precursor of the earthquake. Study of earthquake precursors may hold promise in advancing the science of short-term earthquake prediction. As a short-term earthquake

* Corresponding author: edhalaksamana@gmail.com

prediction, earthquake precursors are usually based on observations of various physical phenomena that are reported at least some earthquakes before. The precursor phenomena include groundwater level changes, induced electric and magnetic fields, surface deformations, gas emissions, temperature changes, radon concentration anomalies, and many more [1].

Earthquake prediction is poorly understood, how the future earthquakes occurred can be determined from the behavior before the earthquake occurred. The poor understanding is reflected in the inability to predict major earthquakes in seismically active regions within a short period of time. The observable signal before earthquakes that indicates with high probability the location, time, and magnitude of an impending event is the most proposed prediction methods rely on the concept of a diagnostic precursor. The research for diagnostic precursors has not yet got a successful short-term prediction scheme [2].

The previous research has found that the earthquake occurred when the detector gets a groundwater level change at 4 to 50 days before the earthquake [3]. This paper discusses the earthquake prediction state of the art and the main issues which are deduced from fluctuations in groundwater levels in the Yogyakarta-Indonesia region, and it concluded by discussing the challenges of the future of the domain.

2. Literature

2.1. Deterministic Earthquake Prediction deduced from Changes in Groundwater Level

216 wells were observed in China mainland to figure out the relationship of groundwater level changes induced 2011 Mw = 9.0 Tohoku earthquake as an earthquake prediction at epicenter distances between 1300 and 5400 km. Out of 216 wells operating during the time of the event, 85 wells showed the groundwater level oscillations during the passage of the seismic waves. The sustained changes that are evenly distributed between those experiencing groundwater level rise and falls were showed on 73 wells. At the remaining 58 wells, no response to the earthquake was observed. Observed response in terms of distance or in terms of azimuth with respect to the epicenter showed was no spatial pattern in the different well response types is evident. Statistical analysis showed that wells that responded to the Tohoku earthquake were characterized by the changes above 0.5m in water level per microstrain in accordance with 5 kPa changes in water pressure per microstrain or 5 GPa, while most wells those who respond are characterized by lower tidal factors. The sensitivity of the well-aquifer response is about 80 percent. However, co-seismic oscillations/fluctuations showed has no systematic difference in tidal admittance factors or phases can be seen between wells that displayed sustained changes. To shows the changes in the aquifer permeability, post-seismic phase shifts were observed at 31 sites, positive in 22 cases and negative in nine cases. Accordingly, 43 percent of those wells that displayed sustained water level changes as earthquake-induced temporal variations in the permeability might have occurred, but less than 15 percent of all observed wells. Statistical methods also used to analyze the relationships between the different types of groundwater level changes and the earthquake system [4].

2.2. Identification the groundwater level as an earthquake signals

The coseismic signals of groundwater levels are generally caused by Earth tides, the diminished of atmospheric pressure response and precipitation from the observed data. However, the coseismic signal is often difficult to be extracted if observations are made far from a barometer, pluviometers or tidal data for correction. In this case, To obtain the instantaneous frequency and amplitude for each point of the decomposed intrinsic mode function (IMF) of groundwater level data to distinguish the response that depends on the associated frequency without further additional input, the Hilbert–Huang transform (HTT) is usually used. The extracted coseismic signal shows a strong amplitude pulse seen clearly in the third IMF. Furthermore, two types of coseismic signals can be easily distinguished in the results of the HHT transformation. One of them is the instantaneous short time signal caused by the passage of seismic waves. Another peak of the earthquake and the magnitude of the earthquake at Hualien Station, Taiwan and showed a positive correlation between the distance of the earthquake and its magnitude [5].

3. Methodology

3.1. The System Detectors

The system detectors were placed at remote area consisting of the GWL transducers, and signal conditioning. The system detector contains the GWL detector, signal conditioning, controller, data storage and electrical power supply. It has specification such as Photovoltaic 50WP, PV controller MPPT 5A 12V, accumulator battery power 12V, data storage 16 GB, and system sensitivity 0.01 centimeter. The block diagram of the monitoring station is shown in Fig. 1.

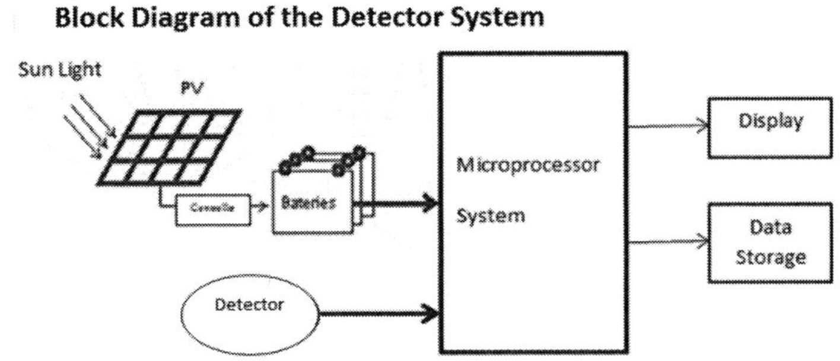

Fig. 1 Block Diagram of The Monitoring Station System

3.2. Tests and Evaluations

The stages of development are:

1. Create and compiled the instrumentation system on the basis of the developed model.
2. Create the program for the purposes of control and collection of data as well as developing a data management system.
3. Test in simulation stages to obtain the level of accuracy and speed of data access from data sources.
4. Revise the model developed on the basis of the data obtained and re-testing the system to obtain the results as expected.
5. Test in laboratory scale to make sure the result of the system is calibrated according to standard.
6. Placed the system into the remote area (chosen water wells) to validate the system are related to the anomaly before the earthquake occurred.

4. Result and Discussion

4.1. Data System Test

In this section, the detector output shown in the system display is the centimeter unit and will be compared to the standard gauge. The detector will be drowned every centimeter and then the output yield that shown on the system display will be recorded. The test was performed as many as 25 variations of the depth and replicated as much as 5 times test. After the data obtained, the next step is to do the analysis using the statistical method i.e comparison method. Fig. 2 presents the results of the comparison method analysis with an accuracy of 85% which is displayed in the box plot.

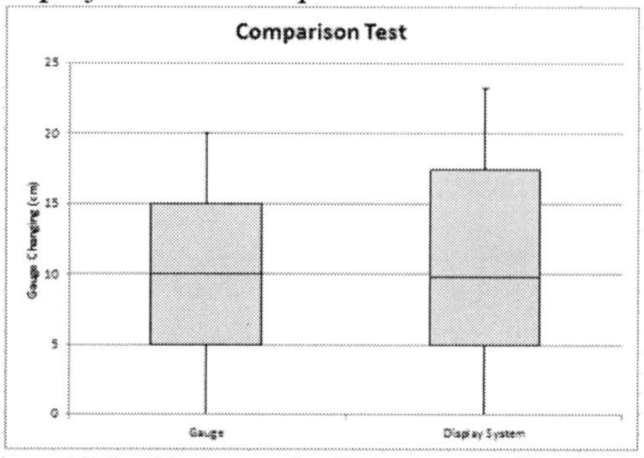

Fig. 2 Box Plot from Comparison Method between the Detector and Standard Gauge

Fig. 2 shows that the output of the system is not much different from the standard gauge.

4.2. System Validation

The groundwater level changes detector placed on the chosen water wells with the requirement the wells is rarely to use on a human activity. The system supported with a power system that contains photovoltaic, controller, and accumulator battery. Data that been recorded in the data storage, will be validated to the earthquake data from BMKG (Indonesian Meteorology, Climatology, and Geophysics Agency). It will be shown the relationship between groundwater level changes and the occurred earthquake.

Fig. 3 Implementation of Groundwater Level Detector

The data has been recorded by the detector system stored on the SD card. The data is analyzed on a weekly interval test, and the reliability of the system against earthquake will be recorded outside the system. The data shows that the detector system can detect groundwater level changes with a sensitivity of 0.01 centimeters.

The Results of Groundwater Level Monitoring for 14 days, as follow:

- Location: Detector 1 at 07°81'90.13"S, 110°52'07.14" E (Wukirharjo water well) and Detector 2 at 07°76'20.43"S, 110°53'12.82" E (Muruh water well)
- Range: June 8, - June 21, 2018
- The earthquake was occurred in Mentawai Indonesia (June 13, 2018), Mentawai Indonesia (June 14, 2018), Bantul Indonesia (June 17, 2018), and Indramayu Indonesia (June 23, 2018).
- The groundwater level changes that detected by the system several days before the earthquake is ± 4 centimeters per day for Wukirharjo water well and ± 19 centimeters per day for Muruh water well.

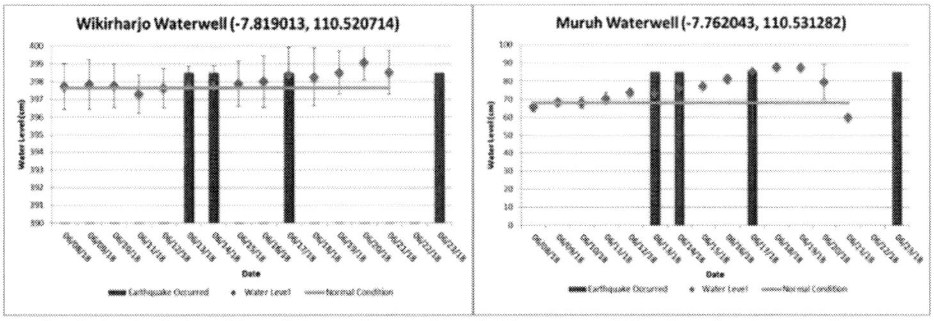

Fig. 4 Graph of The Relationship of Groundwater Level Changes and Earthquake Occurred

Fig. 5 shows the example of an earthquake that occurred in the research range time, i.e an earthquake event that occurred in Bantul Indonesia.

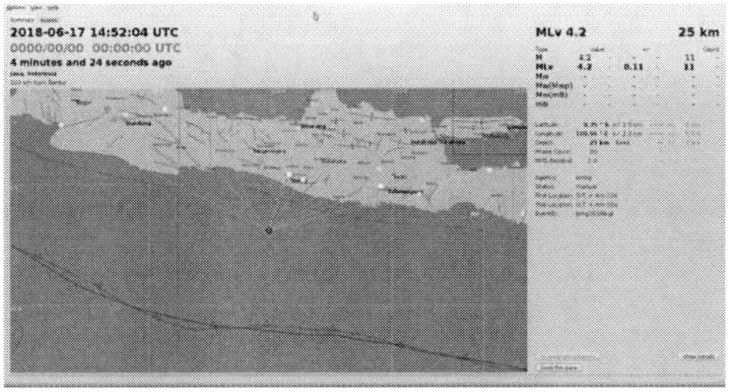

Fig. 5 Picture of Bantul Indonesia Earthquake on June 17, 2018 [6]

5. Conclusion

This research informs the detector system which is used for detecting the fluctuation of groundwater level as an earthquake prediction that placed on Yogyakarta Region - Indonesia. The system which developed in this report is expected to be used in detector system for earthquake prediction deduced from fluctuations at Yogyakarta Region - Indonesia.

The results of the detection system test at Yogyakarta region indicate that the groundwater level changes that recorded on the system have a relationship with the occurred earthquake. On June 11, 12, 15, and 21 the GWL have a change of 4 centimeters to 19 centimeters respectively, the earthquake occurs within the next 2 days.

It shows that the system can be used as an earthquake monitoring system based on the groundwater levels changes. It can be used for earthquake prediction even though needs further study and development of the database system. This report is expected to give valuable information from developing the system that can be used for earthquake prediction.

Acknowledgment
I would like to thank the Sensor and Tele-control Laboratory UGM and Production House UGM who supported my work in this way and helped me get the result of better quality.

References:

1. R. D. Cicerone, J. E. Ebel, and J. Britton, "A systematic compilation of earthquake precursors," *Tectonophysics*, vol. 476, no. 3–4, pp. 371–396, Oct. 2009.
2. T. Jordan *et al.*, "Operational Earthquake Forecasting. State of Knowledge and Guidelines for Utilization," *Ann. Geophys.*, vol. 54, no. 4, Aug. 2011.
3. Sunarno, M. M. Waruwu, and R. Wijaya, "Development of the real-time telemonitoring system for earthquake prediction deduced from fluctuations in groundwater levels at Yogyakarta region-Indonesia," *J. Theor. Appl. Inf. Technol.*, vol. 83, no. 1, pp. 95–99, 2016.
4. R. Yan, H. Woith, and R. Wang, "Groundwater level changes induced by the 2011 Tohoku earthquake in China mainland," *Geophys. J. Int.*, vol. 199, no. 1, pp. 533–548, 2014.
5. C.-H. Chen *et al.*, "Identification of earthquake signals from groundwater level records using the HHT method," *Geophys. J. Int.*, vol. 180, no. 3, pp. 1231–1241, Mar. 2010.
6. BMKG, "Gempabumi-Terkini." [Online]. Available: http://www.bmkg.go.id/gempabumi/gempabumi-terkini.bmkg. [Accessed: 11-Jul-2018].

Feasibility study of photovoltaic water pump for rice paddy irrigation

Ri Munarto[1a], *Arif* Faishal[2]

[1,2] Department of Electrical Engineering, Universitas of Sultan Ageng Tirtayasa, Cilegon- Banten

Abstract. Diesel pumps are commonly used for irrigation purposes of paddy fields at locations far from the grid. Diesel pumps have low reliability because they require high maintenance costs. The use of photovoltaic is a valuable option because the selection decision is not only based on direct biya modal but also includes environmental costs. These solutions reduce the impact of global warming, mitigate negative effects on air quality and contribute to sustainable economic development. Daily sun conditions in Indonesia are average in the horizontal plane within the range of 5-7 kWh/m^2/day and the average daily sun shines within the 7-9 hour. The availability of solar radiation is quite abundant and due to the high cost of grid connections, the pumping of irrigated rice paddy water is beneficial to the water needs for remote communities in Indonesia. It is known from some literature that the photovoltaic irrigation pump system is technically and economically feasible for very small systems in hectares orders. Technical and economic feasibility is determined as a function of the plant and geographical location, including climatic factors, soil quality, ground water depth and recharge rate. The objective of the research is to design a system of pumping irrigation of paddy field of photovolatics to replace previous diesel pumps that meet the criteria of feasibility standard. From simulation using PVSYST 5.0 software, it is known need of solar panel as many as 5 series module and 2 parallel module which fulfill the required eligibility criteria.

1 Introduction

Technical feasibility is determined as a function of plant type and geographical location which includes climate factor, soil quality, ground water depth and recharge rate of water. The maximum daily volumetric flow rate required for crop irrigation is determined from the maximum evatranspiration rate of the plant during its growing season [1]. Maximum evapotranspiration rate is a function of plant type, soil conditions and climatic conditions. The maximum daily flow rate required is used to determine the required pumping power. The solar panel areas needed to provide the required power use a monthly average solar insolation and are compared with the total field area [2]. Solar insolation represents the amount of solar radiation received over a single day on the Earth's surface measured in kWh/m^2/day. On cloudy days, light is scattered over sunny days, so the amount of light

[a] Corresponding author: rim.moenarto@gmail.com

reaches the surface is reduced. Areas with the same latitude that have a cloudy day, will have a very low insolation rate{1]. Solar photovoltaic water pumping has been recognized as suitable for grid-isolation rural locations in poor countries where there are hight levels of solar radiation. The power output for the photovolatic array depend upon insolator and availability of sun per daya i.e sunshine hours available on particular location per day. The insolation varies from one location to another, mounth to mounth because of seasional and climate chnge. [3]. Photovoltaic irrigation systems are technically feasible if there is enough land available to install solar panels. Technical feasibility is determined from the maximum power required for irrigation, which depends on the type of plant and geographic location.

The high cost of photovoltaic irrigation requires prior to further study, it is necessary to measure the dimensions of the photovoltaic installation as accurately as possible. The method used consists of three stages: (1) Determining irrigation needs according to soil type characteristics and climate, (2) Hydraulic analysis of pumping system made according to the aquifer depth and altitude required to stabilize pressure on the water distribution network, (3) After determines the peak photovoltaic power required for irrigation by taking into account the overall yield of the photovoltaic pump irrigation system. Use of photovoltaic irrigation pumps to increase socially significant plant output by optimizing the use of water from solar energy sources and at the same time preserving the environment. The main objective of providing a simple procedure for estimating the design requirements of a photovoltaic system to provide photographic power is based on a separate photovoltaic pumping system from the water requirements of the plant. In order for connections between irrigation systems and pumping equipment with solar energy to be energy efficient and economical, some things must be taken into account: (1) efficient use should be made to water resources, ie only the amount of water needed for crops to be raised, taking into account the rainfall prevention capacity of the soil during the rainy season, (2) the amount of water should be raised to the minimum height above the soil required to stabilize the pressure on the irrigation head and (3) the most efficient irrigation system for a particular plant that should be used.

These guidelines, written in the style of a submission to *MATEC Web of Conferences*, explain how to prepare your paper using Microsoft Word. Please submit sources files directly to the conference organizers. If they chose to provide print-ready PDF documents to the publisher, you have to submit high-resolution PDF file with all fonts embedded (see PDF guidelines) instead of the sources files. In this case, please remember that no final corrections will be made by the publisher.

2 Methodology

2.1 Rainfall data

Rainfall data that have been obtained from Meteorology and Geophysics Agency Serang is the average rainfall data period of two weekly or half monthly. Then do the sorting of data from the largest value to the smallest value. From the literature, then can be obtained the amount of rainfall 80% of R80 is from the fourth order data. The amount of effective rainfall for rice plants is 70% of the rainfall of 80% R80, to obtain effective rainfall for rice crops in January the first period of 158.83, or the January rice re 10.58 mm/day, and 70% of R80 paddy in January second period was 65,87 or re of January second period that is 4,39 mm /day.

2.2. Need of water

2.2.1 The total need of water in the field

The total need of water in the field is the water required from the start of the land preparation, the land preparation, so that it is ready for planting, until the harvest time. In other words, water is needed from the beginning until the completion of planting. The total water requirement in the rice field can be calculated by the formula:

$$GFR = Etc + P + WLR \tag{1}$$

GFR : the need for total water in the rice field (mm/day) or (Liter/day hectare)
Etc : Evapotranspirasi costant (mm/day)
WLR : Water Layer Replacement (mm/day)
P : Perlocation.

2.2.2 Needs water retrieval (DR)

$$DR = \frac{NFR}{8.64 * ef} \tag{2}$$

DR : the need for water in the field (Liter/sec/hectare)
NFR : Net water need in paddy fields (mm/day)
ef : Efficieny of irrigation (65%)

2.2.3 The need for clean water in the fields (NFR)

The need for clean water in the fields is the need for total water in the fields reduced by effective rainfall.

$$NFR = Etc + P + WLR - \mathrm{Re} \tag{3}$$

Re : Effectice rainfall (mm/day)
After the net water needs are known, the water requirement for irrigation of rice crops can be obtained, so that the following values are obtained:
IR in the first January period -2.69 mm/day
DR in January the first period -0.311 liter/second/ hectare
IR in January second period 6.83 mm/day
DR in January second period 0.79 liter/second/hectare,
and then the calculation can be continued covering the next months in a year. Taking water through a photovoltaic pump to irrigate a rice plant area of 2101.44 m2 or 0.2101 hectares, then the monthly water requirement for flowing rice cultivation can be shown in Table 3.1. From table 3.1, it can be seen that the maximum water requirement for rice cultivation process is August at 31.74 m^3/day.

2.3 Batterey

The required battery capacity can be calculated at 77.90Ah / day. With a voltage rating of 48 V, it takes 4 batteries that strung the series. Design calcultation for solar pumping system[4]:

- Head of water level from the pump to the vertical lift (6.42 m)
- Pipe length 34,42 m

When installed a water storage tank with a height of 2.82 m and a diameter of 4.37 m, then can be calculated the following :

- Flow rate of 190 liters/minute
- Friction loss 1.37 m (2 inch diameter pipe)
- Total Dynamic Head 7.84 m
- Hydraulic pump energy 0.67740 kWh/day
- Hydraulic power 135.48 W (assuming 5 hours of solar irradiance)
- Mechanical power, P_m = 301.06 W
- Power, P_e = 327.24 W
- System efficiency 41.09%
- Duration of operation of pumps 2,983 hours/day (rounded 3 hours/day) .

2.4 The size of photovoltaic array

The size of photovoltaic array is determined based on the required electrical power, so that the photovoltaic area, PV area = 2,875 m^2. From the photovoltaic area, Peak Solar Insolation (100 W / m^2) and solar panel efficiency, obtained power generated by solar panels, $P_{watt\ peak}$ = 421.18 Watt peak. For a 72 V voltage rating, it takes a series of solar panel series on one string of 4.11 and rounded to 5. The number of string strings is 1.02 (rounded 2), $V_{mpp\ array}$ = 87.5 V, $I_{mpp\ array}$ 11.42 A.

2.5 Solar panels

The laying of the elevation angle of the solar panel is determined by the location of the laying area of the solar panel. Geographical location of Banjarsari village can be seen in Table 1. Due to the geographical position of the Banjarsari village area in the southern hemisphere, the solar altitude reference is based on the north pole of the earth and the tilt angle is oriented towards the North Latitude (LU).

Table 1. Geographical location of banjarsari village

	Winter solstice	Equinox	Summer solstice
Declination	+23,5 °	0	-23,5 °
Solar Altitude	107,43 °	83,93 °	60,43 °
Tilt angle	-17,43 ° LU	6,07 ° LU	29,57 ° LU

The determination of the elevation angle of the solar panel is determined based on the average tilt angle value during winter solstice and summer solstice, then the angle of elevation of Banjarsari is 23° LU. Determination of fixed tilt angle is recommended at least 15° [8]. This is because a low tilt angle can make solar panels covered with dust and stagnant water when exposed to rainfall that will result in reduced output energy. The azimuth angle of solar panel used using azimuth angle obtained from research result using solar tracker from case study of Serang area (Hardiansyah, 2012).

3 Results and discussion

Table 2 shows the balance and main results of the paddy irrigation pump rice irrigation system. The global horizontal irradiation per year is 1381 kWh/m^2. The virtual energy of photovoltaic arrays at Maximum Power Point, E_{ArrMPP} is 1139.5 kWh. The required energy operates an average annual pump of 35.36. The energy required to fill the water tank to the full averaged annual 17.43 kWh. The height of the pump head is 7.092 m. The effort required for pumping an annual average of 23.61 m^3/day. While the business is actually used for pumping 23.55 m^3/day. Miss occurs in several months of the year, ie the months of August, September, October and December amounted to 4.109 m^3/day, 2.735 m^3/day, 1.764 m^3/day and 0.445 m^3/day. Average miss 0.865 m^3/day

Table 2. Balance and Main Result

Months	**GlobEff (kWh/m^2)**	**E_{arrMPP} (kWh)**	**E_{pumpOp} (kWh)**	**E_{tkFull} (kWh)**	**H_{pump} (meter)**	**W_{pump} (m^3/day)**	**W_{used} (m^3/day)**	**W_{miss} (m^3/day)**
January	145,5	120,3	29,59	50,09	7,001	16,23	14,35	0,000
February	112,4	92,6	28,39	34,25	6,970	17,00	17,29	0,000
March	131,4	108,5	39,32	31,35	7,095	21,52	21,55	0,000
April	114,0	93,9	44,24	16,20	7,107	25,07	25,08	0,000
Mei	109,4	90,8	46,22	10,35	7,084	25,40	25,53	0,000
June	97,8	81,1	46,34	4,22	7,115	26,58	26,47	0,000
July	105,9	88,2	38,84	15,79	7,095	21,32	21,39	0,000
August	92,0	75,8	46,50	1,86	7,215	25,73	26,63	4,109
September	96,4	79,5	47,96	0	7,205	27,51	28,32	2,735
October	143,8	117,4	54,76	21,63	7,090	30,26	27,46	1,764
November	119,2	98,5	47,35	17,43	7,092	26,81	26,73	0,000
Desember	113,6	93,0	35,56	24,42	7,065	19,52	19,45	0,445
Years	1381,4	1139,5	501,17	227,59	7,107	23,61	23,55	0,865

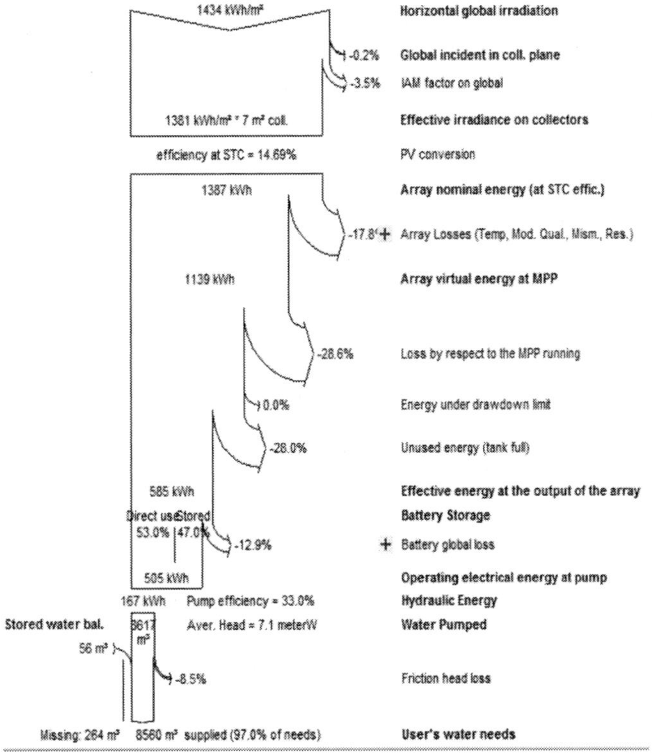

Fig. 1. Overall loss diagram

Figure 1 shows the overall system disadvantage diagram of irrigation pump irrigation system of agricultural land in Banjarsari village. Horizontal global irradiation 1434

kWh/m^2. Irradiation is effective on the collector field 1381 kWh/m^2. The magnitude of irradiation is effectively influenced by two environmental factors, namely the incident global factor of 0.2% and the Incidence Angel Modifier on global resulting in a decrease in irradiation that actually reaches the 3.5% photovoltaic cell surface associated with radiation in the normal event. Further photovoltaic cells convert solar energy into electrical energy. After photovoltaic conversion, the nominal array energy is 1387 kWh/m^2. Efficiency of photovoltaic arrays 17.8% in Standard Test Condition. Virtual energy array obtained 1139 kWh. Effective energy at 585 kWh array output. Battery efficiency 12.9%. The electrical energy used operates a 505 kWh pump. Pump efficiency 33%. Hydraulic energy 167 kWh. Water pumped 8617 m^3. Friction head loss of 8.5%. Water required user 8560 m^3. Water lost 264 m^3.

4 Conclusion

In this research, the design result of irrigation pumping system of rice field photovoltaic in Banjarsari village, Serang regency with the software used. Maximum solar irradiation at tilt angle 23° for Banjarsari village is close to the latitude of its location (23°28') and no shadding effect should be considered. In the irrigation pumping system of paddy fields of photovoltaic rice, the energy used injects pumps is 505 kWh (pump efficiency 33%). A Performance Ratio of 0.76 and some power losses can be calculated. Known miss occurs on several months of the year, with the largest miss in August of 4.109 m^3/day.

References

1. Leah C. Kelley, Eric Gilbersertsin, Anwar Sheikh, Steven D. Espinger, Steven Duboorsky, *On the Feasibility of Solar-Power Irrigation* , Renewable and Sustainable Energy Review.
2. E. PDNarale, "Study of Solar PV Water Pumping System for Irrigation of Horticulture Crops," *Int. J. Eng. Sci. Invent. ISSN (Online*, vol. **2**, no. 12, pp. 2319–6734 (2013)
3. Er. P.D. Narale, Dr. N.S. Rathore, Dr. S. Kothari, *Study of Solar PV Water Pumping System for Irrigation of HolticultureCrops*, International Journal of Engineering Science Invention, ISSN (Online), 2919-6734.
4. Naveen Kumar Lodha, K. Sudhakan, *Theoritical Design and Simulation Analysis of PV Based Pumping System for Domestic Application in Bhopal, M.P., India,* International Journal of Science and Research, ISSN (Online): 2319-2064..

Proposed statistical-based approach for detecting distribute denial of service against the controller of software defined network (SADDCS)

Mohammad A. AL-Adaileh*, *Mohammed* Anbar*, *Yung-Wey* Chong 1, and *Ahmed* Al-Ani1

[1]National Advanced IPv6 Centre, Universiti Sains Malaysia, 11800 Gelugor, Penang, Malaysia

Abstract Software-defined networkings (SDNs) have grown rapidly in recent years be-cause of SDNs are widely used in managing large area networks and securing networks from Distributed Denial of Services (DDoS) attacks. SDNs allow net-works to be monitored and managed through centralized controller. Therefore, SDN controllers are considered as the brain of networks and are considerably vulnerable to DDoS attacks. Thus, SDN controller suffer from several challenges that exhaust network resources. For SDN controller, the main target of DDoS attacks is to prevent legitimate users from using a network resource or receiving their services. Nevertheless, some approaches have been proposed to detect DDoS attacks through the examination of the traffic behavior of networks. How-ever, these approaches take too long to process all incoming packets, thereby leading to high bandwidth consumption and delays in the detection of DDoS at-tacks. In addition, most existing approaches for the detection of DDoS attacks suffer from high positive/negative false rates and low detection accuracy. This study proposes a new approach to detecting DDoS attacks. The approach is called the statistical-based approach for detecting DDoS against the controllers of software-defined networks. The proposed approach is designed to detect the presence of DDoS attacks accurately, reduce false positive/negative flow rates, and minimize the complexity of targeting SDN controllers according to a statistical analysis of packet features. The proposed approach passively captures net-work traffic, filters traffic, and selects the most significant features that contribute to DDoS attack detection. The general stages of the proposed approach are (i) da-ta preprocessing, (ii) statistical analysis, (iii) correlation identification between two vectors, and (iv) rule-based DDoS detection.

1 Introduction

The Internet has grown rapidly on heterogeneous network structures with complicated networking protocols. Thus, conventional networks are now complex, network traffic is difficult to manage, and maintaining network security is significantly critical [1].In

* Corresponding author: m_aladaileh2003@nav6.usm.my and anbar@nav6.usm.my

addition, the rapid development of information and communication technologies has facilitated the exchange of enormous amounts of data and the sharing of network assets; this change has exposed these networks to attacks Moreover, network threats are on the rise [2].

Software-defined network (SDN) architecture [3], [4] has emerged as a new and novel method that meets network requirements in terms of network management and flexibility. SDNs are designed to establish high-level abstraction on top of hardware and software infrastructure. A fundamental characteristic of SDN architecture is the ability to isolate the control plane from the data plane. Therefore, the advantage of SDNs is centralized network control.

The present study illustrates the architecture of the proposed approach, which is called the Statistical-Based Approach For Detecting DDoS Against Controllers Of Software-Defined Networks (SADDCS) that aims to detect the presence of DDoS attacks against SDN controllers based on the statistical analysis of TCP and UDP features. This paper is organized as follows. Section 2 presents the existing studies pf detection techniques and its limitations. Section 3 describes the proposed SADDCS approach. Section 4 provides the conclusion and the future works.

2 Related Work

SDNs provide many advantages over traditional networks, but SDN also has some of drawbacks such as security threats due to depending the SDN on a centralized controller to monitor over the entire network. Hence, any threat leads to the creation of a single point of failure. One of the most widely known threats is the Distributed Denial of Service (DDoS) attack. This attack aims to exhaust network resources by sending a huge number of packets to the controller, which in turn causes network congestion. That is, a DDoS attack is considered as a one of the most dangerous threats security on SDN networks[5]. SDNs can be extremely affected by DDoS attacks because SDN controller is main part, which controls and manages all network actions. Thus, a breakdown of a controller might lead to a network collapse. Many techniques have been proposed to secure SDNs against DDoS attacks by detecting and mitigating this type of attacks. Entropy is known as a measure of randomness variable [6]. The detection based entropy works based on calculating the distribution of packets in a network according to different variables in the network, such as source IP address, source port or some other values of packet characteristics [7]. Therefore, entropy-based metrics are used in traffic analysis and anomaly detection. Mousavi and St-Hilaire [8] proposed a lightweight detection method to identify DDoS attacks to controllers in early stages. The method collects network statistics from the flow table and calculates entropy, which compares entropy value with a specific threshold in a fixed window size. However, time characteristic of this method is not considered during anomaly traffic detection, because it causes an excessive overhead in SDN controller. A hybrid mechanism based on security analysis is proposed in [9] to defend against DDoS attacks. This mechanism involves the use of a trust value and an entropy concept for detection purposes. The trust value is assigned by a trust value server, the trust value is higher than the attacker and lower than the legitimate users based on client access behavior. This mechanism allows a SDN controller to identify malicious users by monitoring their access behavior.

IDS is a system for detecting intrusions attempting to misuse data or to consume network resources. IDS collects network information from all sources, analyzes the information to identify network intrusion, and uses the information to detect DDoS attacks [10]. Thus, the use of IDS could help identify any abnormal traffic. Existing systems could be merged with different approaches to achieve network behavior with dynamic access control, such as rate bounds. A lightweight flow that is based on IDS is proposed in [11];

this method collects network statistics from OpenFlow (OF) switches periodically and employs statistical analysis for the collected information thus, the controller will process the information obtained from switches and then it sends the new rules to the switches. This proposed method can extract and collect proper information to classify flows as legitimate or abnormal traffics.

Attackers might generate low traffic flow that is similar to the normal traffic in order to trigger malicious flooding requests by connecting to different switches. A suggested an effective detection method is proposed in [12] that is designed to detect the low traffic flows by using the sequential probability ratio test (STRT). A mathematical model for estimating the impact of low-rate DDoS attacks by analyzing the vulnerability of the retransmission timeout mechanism [13]. The results showed that increasing the bottleneck buffer size in a network effectively protects against DDoS attacks. However, vulnerabilities may emerge when an attack is of different types, in which case it becomes difficult to identify.

Attackers may use a duration of time to achieve target destinations, that is, they send out attacks with a time period. Several DDoS attack detection methods do not consider time characteristics in detecting abnormal traffic. A methodology to detect SYN flooding attacks according to specific packet features is presented in [14]. Although this method is capable of large traffic flow detection, it does not consider the time characteristics of DDoS attacks. Table 1 summarizes the limitations of related works.

Table 1. Limitations of related works

Author/ Year	Limitations
Mousavi and StHilaire (2015)	-Time characteristic of this method is not considered. -Attack method handle a single host and then makes it more vulnerable.
Jantilaet al (2016)	-High false positives. -Does not distinguish DDoS attack traffic in the flash crowd traffic.
Georgi et al(2017)	- Takes long time to the attack detection - Cannot detect low traffic rate.
Dong et al (2016)	-Resources consumption in the controller.
Luo et al (2014)	- Unable to distinguish several kinds of attacks. -Consumes time during detecting low rate DDoS attacks.
Muhammad Nugraha et al (2014)	-Increase detection time.

3 Proposed Approach (SADDCS)

Detecting DDoS attacks that target SDN controller is still a challenging problem that has caught the attention of researchers and has prompted the proposal of efficient approaches to detect DDoS attacks. To achieve the study objectives, we present a set of stages for the proposed approach. The proposed approach aims to detect UDP and TCP DDoS attacks by monitoring the behavior of UDP and TCP traffic packets. The proposed approach consists of four stages, as shown in Fig 1. The first stage, i.e., data collection and preprocessing, aims to receive network traffic and filter TCP and UDP traffic. This stage consists of two steps: (a) network traffic capture and (b) network traffic filter. The second stage is the traffic statistical analysis that is generated from the switch table. This stage consists of three steps: (a) aggregated values related to entropy values, (b) aggregated values related to TCP and UDP packet header features, and (c) dynamic threshold calculation. The third stage, i.e., correlation identification between two vectors, aims to come up with the second group of statistical features by checking the existence of a correlation between the two vectors. The fourth stage, i.e., rule-based DDoS detection, aims to detect the presence of DDoS attacks on the basis of the randomness of network traffic using the proposed rule 1 and the presence of a correlation between two vectors of statistical features using the proposed rule 2.

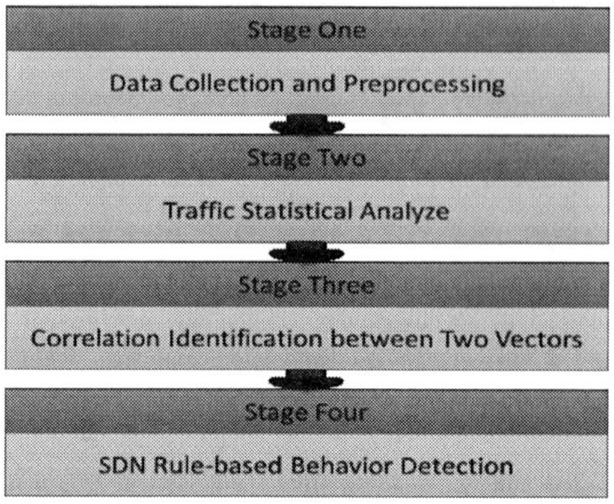

Fig.1. General Stages of Proposed Approach

3.1 Data Collection and Preprocessing

This stage aims to collect network traffic packets from switches after they are received from hosts. This stage consists of two steps: (1) network capture of all incoming packets targeting the switch and (2) network filtering that selects the significant features that contribute to the detection of TCP and UDP DDoS attacks.

3.1.1 Network traffic capture

In this step, all incoming traffic packets from hosts are received by the switches without losing any packet feature. The switch device monitors all incoming packets that match the packet with rules in the switch flow table that received rules from the controller regardless of their resource and destination IP. However, this step might include redundant features that do not contribute to the detection of DDoS attacks against SDN controllers.

3.1.2 Network traffic filtering

This step involves filtering the incoming network traffics. It consists of two main layers. The first layer aims to filter all the incoming network traffics and holds TCP and UDP packets to be used to detect DDoS attacks that relate to it. The second layer aims to extract TCP and UDP features that could be used to detect DDoS attacks.

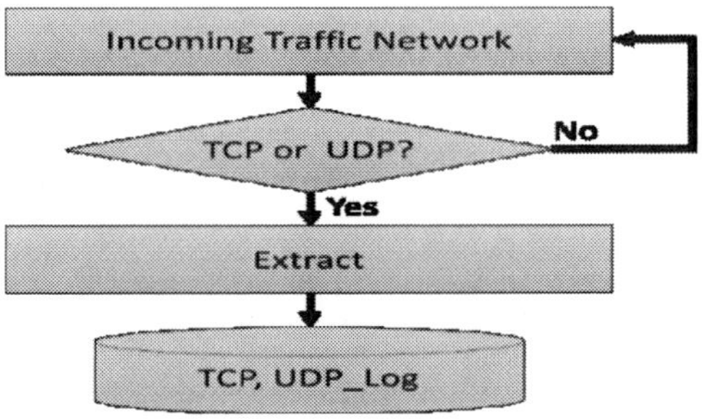

Fig. 2. Incoming network traffic filtering

Therefore, only the significant features that contribute to detect the DDoS attacks are filtered. The filtered TCP and UDP features will be stored into TCP_Log and UDP_Log files. Fig 2 shows the flow chart of network traffic filtering steps. The log files will be used as input to the next stage which is traffic statistical analyses (refer to section 0).

3.2 Traffic statistical analysis

Statistical analysis is used in SADDCS to detect different types of attacks. This stage is responsible for providing details about the selected features by calculating the aggregated values according to an analysis of TCP and UDP logs received from the previous stage. The aggregated values are divided into two groups: (1) aggregated values based on TCP and UDP packet header features, (2) aggregated values based on the entropy approach and (3) dynamic threshold. .

3.2.1 Aggregated values related to entropy values

Entropy approaches show impressive results in terms of detecting different types of attacks. Entropy is used for DDoS detection mainly because of its ability to measure randomness associated with a source IP and destination IP in the packets that are entering a network using the equation below.

$$H(X) = -\sum_{i=1}^{n} P(x_i)\, log_b\, P(x_i) \tag{1}$$

Therefore, in the case of a DDoS attack, the entropy value of the source and destination IP address increases because the attacker always spoofs its source IP address on each packet transmission or changes the destination IP address. Traffic randomness can occur if the entropy value exceeds a threshold (Th), the value of which is calculated dynamically on the basis of the adaptive threshold algorithm.

3.2.2 Aggregated values related to packet features

This step extracts the features of TCP and UDP from TCP_log and UDP_Log and performs the following statistical analysis for each log table. (i) The number of packets that have sent from one source to specific destination (Outbound). (ii) The number of packets that have received by specific destination (Inbound). Thus, the output of this step comprises two log files: (1) statistical_TCP_Log and (2) statistical_UDP_Log. Moreover, it facilitates the process of attack detection by an alert that is triggered if the number of packets exceeds a predefined threshold (t).

3.2.3 Dynamic threshold

The threshold value, which is a critical component for detecting DDoS attacks, is determined on the basis of the aggregation of the data for the adaptive threshold algorithm parameter. In increasing the accuracy of detection, the threshold value must be updated according to the incoming network traffic condition. The adaptive threshold algorithm relies on testing whether the traffic measurement over a given interval exceeds a particular threshold; it can be defined as follows:

$$X_n \geq (\alpha + 1) \cdot \mu_{n-1} \tag{2}$$

where: X_n: number of packets in the nth time interval

α: a positive parameter (α>0) that indicates the percentage above the mean value that is considered evidence of anomalous behavior.

$\mu\_(n-1)$: mean rate estimated from measurements prior to n . The mean value (μ_n) can be computed using the Exponential Weighted Moving Average (EWMA) which represented as the following formula:

$$\mu_n = \lambda \cdot \mu_{n-1} + (1-\lambda) \cdot X_n \tag{3}$$

3.3 Correlation Identification between Two Vectors

This step aims to calculate (r), which represents the existence of a correlation between two vectors A and B. Vector A represents the data of statistical TCP_log and statistical_UDP_log while B represents the probability_TCP_log and probability_UDP_log. Pearson's correlation coefficient is adopted to calculate the value of r on the basis of the equation below.

$$r_{AB} = \frac{\left(n\sum AB - (\sum A)(\sum B)\right)}{\sqrt{((n_A \sum A^2 - (\sum A)^2\ )(n_B \sum B^2 - (\sum B)^2\ ))}} \tag{4}$$

The correlation coefficient(r) ranges from −1 to 1. A value of 1 implies that a linear equation describes the relationship between (A) and (B) perfectly, with all data points lying on a line for which (B) increases as (A) increases. A value of −1 implies that all data points lie on a line for which (B) decreases as (A) decreases. Finally, a value of 0 implies that there is no linear correlation between the variables.

3.4 SDN Rule-based Behavior Detection

This stage is a core stage in the proposed SADDCS approach. It aims to propose rules to detect DDoS attacks against SDN controllers according to the statistical analysis. The assumption is that DDoS attacks can be detected if one of the statistical analysis results exceeds the threshold. The rules are as follows.

3.4.1 Rule 1: Entropy-based rule

The first proposed rule will be applied to probability_TCP_log or probability_UDP_log. The entropy value that matches the proposed rule will be identified as DDoS attack. For probability_TCP_log and probability_UDP_log the following rules will be applied:

If Entropy (probability_TCP_log)) >Th or Entropy (probability_UDP_log.)) >Th, then they have accessed suspicious traffic behavior.

3.4.2 Rule 2: Correlation-based rule

The second proposed rule will be applied to probability_TCP_log and statistical_TCP_log or probability_UDP_log and statistical_UDP_log. The correlation coefficient (r) value that match the proposed rule will be identified as DDoS/DoS attack. For probability_TCP_log and probability_UDP_log the following rules will be applied:

If correlation (statistical_TCP_log, probability_TCP_log)) >Th or If correlation (statistical_UDP_log, probability_UDP_log)) > Th, then they have accessed suspicious traffic behavior.

4 Conclusion and Future Work

SDN controllers are important in SDNs, and several approaches have been proposed to secure SDNs controllers against DDoS attacks. However, these approaches suffer from various limitations, such as additional overhead for SDN controllers, high false positive rates, and the need for external resources to detect DDoS attacks. Therefore, the proposed approach intends to capture network traffic, to filter traffic, and to select the most significant features that contribute to the detection of DDoS attacks. Moreover, the SADDCS approach is designed on the basis of a statistical approach, entropy variation based on statistical TCP and UDP packet header feature, a correlation between statistical_TCP, UDP_Log and probability of packets, and rule-based behavior DDoS detection. The main stages of SADDCS are (i) data preprocessing, (ii) statistical analysis, (iii) correlation identification between two vectors, and (iv) rule-based DDoS detection. In the future, an implementation of the proposed approach will be presented and an evaluation of it against existing approaches will be done.

References

1. J. Chen, X. Zheng, and C. Rong, "Survey on software-defined networking," Lect. Notes Comput. Sci. (including Subser. Lect. Notes Artif. Intell. Lect. Notes Bioinformatics), vol. 9106, no. 1, pp. 115–124, (2015).
2. D. Kreutz and F. Ramos, "Software-Defined Networking: A Comprehensive Survey," arXiv Prepr. arXiv …, p. 49, (2014).
3. A. Samson and N. P. Gopalan, "Software Defined Networking," Proc. Int. Conf. Informatics Anal. - ICIA-16, pp. 1–6, (2016).
4. S. . A. Scott-Hayward, S.a , Natarajan, S.b , Sezer, "Survey of Security in Software Defined Networks," Surv. Tutorials, vol. 18, no. 1, pp. 623–654, (2016).
5. H. D. Zubaydi and M. Anbar, "Review on Detection Techniques against DDoS Attacks on a Software-Defined Networking Controller," (2017).
6. A. X. Liu, "l An Advanced Entropy-Based DDOS Detection Scheme Jie Zhang, Zheng Qin, Lu Ou, Pei Jiang , JianRong Liu," pp. 67–71, (2010).
7. M. Jackson, D. Nelson, and S. Stirk, Lecture Notes in Computer Science. (2005).
8. S. M. Mousavi and M. St-Hilaire, "Early detection of DDoS attacks against SDN controllers," 2015 Int. Conf. Comput. Netw. Commun. ICNC 2015, pp. 77–81, (2015).
9. Y. Jiang, X. Zhang, Q. Zhou, and Z. Cheng, "An Entropy-Based DDoS Defense Mechanism in Software Defined Networks," in International Conference on Communicatins and Networking in China, pp. 169–178, (2016)

10. A. S. Syed Navaz, V. Sangeetha, and C. Prabhadevi, "Entropy based Anomaly Detection System to Prevent DDoS Attacks in Cloud," Int. J. Comput. Appl., vol. 62, no. 15, pp. 975–8887, (2013).
11. G. A. Ajaeiya and A. F. B. Ids, "Flow-Based Intrusion Detection System for SDN," pp. 787--793, (2017).
12. P. Dong, X. Du, H. Zhang, and T. Xu, "A detection method for a novel DDoS attack against SDN controllers by vast new low-traffic flows," 2016 IEEE Int. Conf. Commun. ICC 2016, (2016).
13. J. Luo, X. Yang, J. Wang, J. Xu, J. Sun, and K. Long, "On a mathematical model for low-rate shrew DDoS," IEEE Trans. Inf. Forensics Secur., vol. 9, no. 7, pp. 1069–1083, (2014).
14. M. Nugraha, I. Paramita, A. Musa, D. Choi, and B. Cho, "Utilizing OpenFlow and sFlow to Detect and Mitigate SYN Flooding Attack TT -," J. Korea Multimed. Soc., vol. 17, no. 8, pp. 988–994, (2014).

Steady State Modification Method Based On Backpropagation Neural Network For Non-Intrusive Load Monitoring (NILM)

Sigit Tri Atmaja[1,*], *Abdul* Halim[1]

[1] Department of Electrical Engineering, Faculty of Engineering, Universitas Indonesia, Depok, Indonesia

Abstract. Household electric power sector is highlighted as one of significant contributors to national energy consumption. To reduce electric energy usage in this sector, a technique called Non-Intrusive Load Monitoring (NILM) has been developed recently. NILM is a load disaggregating and monitoring tool that can be used to identify the daily usage behavior of individual electric appliance. Different to conventional method, NILM promises the reduction of sensor deployment significantly. NILM commonly uses either transient or steady state signal. Based on load/appliance signal condition, many NILM's research results have been published. In this paper, steady state modification method of backpropagation neural network (NN) is applied for developing NILM. We use steady state signal to disaggregate the sum of load power signal. In the proposed method, NN is explored for feature extraction of electric power consumption of individual appliance. The presented method is powerful for load power signal which has almost same value. To verify the effectiveness of proposed method, data provided by tracebase.org has been used. The presented method can be applied for local data. It is obvious from simulation results that the proposed method could improve the recognition rate of appliances until 100 %.

1 Introduction

Mankind is facing energy problem seriously. Energy consumption is growing rapidly; however, the depletion of fossil energy resources is unstoppable. Nowadays, the average growth of electrical energy consumption is about 2.3% - 2.5%, and by the end of 2030 it is estimated that the increasing will be twice or around 16,000 TWh per year [1]. The growth of energy consumption will give a negative impact on the environment such as increasing of Carbon dioxide (CO2), Nitrogen oxide (NOx), and Sulfur dioxide (SOx) emission. Many researchers are developing technologies in order to address this problem like renewable energy technologies, environmentally friendly material technologies, energy efficient appliances etc. [2].

* Corresponding author: sigit.tri@ui.ac.id

Excessive consumption of energy takes place in various sectors. One of them is household sector. Electric power use in this sector has been underlined as significant contributor to national energy consumption. Efforts to reduce energy consumed in residential have been exercised by introducing a tool to monitor electric power consumption and give this information back to consumers [3,4]. Generally, conventional load monitoring system uses a lot of sensors mounted on each appliance. Thus, it is very high cost both for installation and maintenance and not practice for household application [5].

Dissimilar to conventional one, nowadays, development of energy monitoring system has been focused to develop a technique that be able to minimize sensor number. And Non-Intrusive load monitoring (NILM) has become iconic research result. NILM is a method that can disaggregate electrical load changes and then identify the daily usage behavior of individual electric appliance [5,6]. The advantage and disadvantage of NILM method can be summarized as follows [7]:

- Advantages: Low cost and ease of installation and maintenance, No needs large space for mounting, and Able to cover large number of appliances to be monitored,
- Disadvantages: Continuously variable appliance cannot be detected, Electrically identical appliances cannot be distinguished, Greater potential for undetected error, and difficult to recognize unusual appliances.

Various NILM techniques have been proposed. Some mathematical models have been explored such as Hidden Markov Model (HMM), Fuzzy Model, Graph Signal Processing, Linear Programming Model etc. One of the powerful models is Neural network (NN). Many research publications have discussed the advantages of artificial neural networks to improve the performance of NILM system. Some improved neural network methods have been presented to identify the load signal on transient [8] and steady state conditions [9]. In our research, an improved type of backpropagation NN is exercised for developing more efficient and effective NILM method. NN in the proposed method is utilized for feature extraction of individual electric power consumption.

In our research, an input layer of NN is added to overcome the difficulty when there is a steady state signal which has almost same value. That additional input layer is set for maximum power value of appliance during switch on. By using simulation, the proposed method has proved that recognition rate could be improved even when the active power of each appliances has almost same value.

2 Methodology

2.1. Problem Definition

The main idea of NILM system is to calculate the power consumption of individual appliances in a house based on the analysis of the aggregated data measured from a single meter. The total aggregated measurement of the household power consumption of the individual appliance from the main meter can be estimated according to equation [7]

$$p_t = \sum_{i=1}^{n} p_i(t) + e_t \tag{1}$$

where p_t is aggregated household power consumption, $p_i(t)$ is individual appliance power consumption at a given period of time and is measurement errors and line loss. Specific characteristic of power signal appears in each electrical appliance during the normal working cycle. This characteristic contains information about values of electrical parameters during steady-state or transient states. The aggregate power signal from the main meter is shown in Figure 1. This graphic represents steady state condition. This curve is derived from database of tracebase [11].

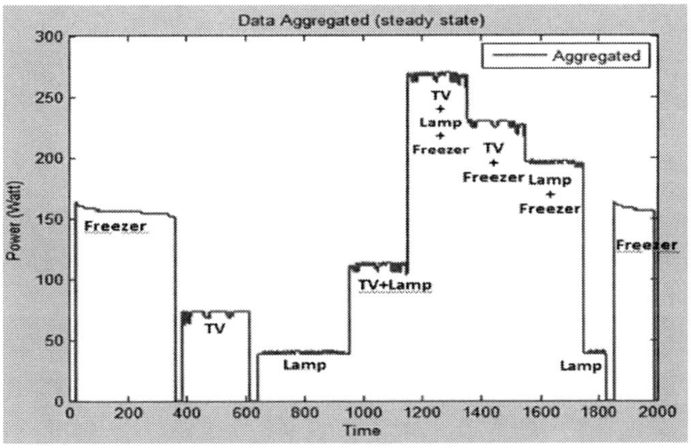

Fig. 1. Aggregated Power Signal

As shown in Figure 1, power signal level at each period of time has significant different. In this case, NILM system using neural network can work well. However, when power signal level has almost same value at different period as depicted in Figure 2, the problem of NN application has occurred. In this paper, a modified method for overcoming this challenge is provided.

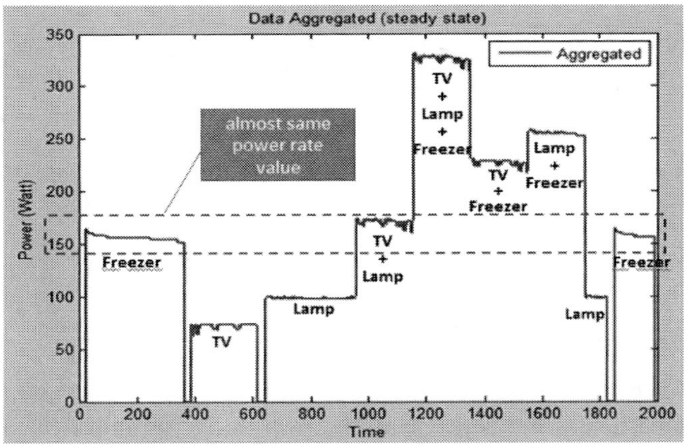

Fig. 2. Aggregated Signal With Almost Same Power Level

2.2. Steady state method Modification

Steady state method can be classified as conventional and modification type. Each type is described in this section.

2.2.1 Conventional Type

In this section, we present two types of NN for disaggregating the load. Firstly, the conventional type is urged. This method uses one neuron on input layer of neural network architecture. Active power is set as the input layer. The architecture of the backpropagation neural network can be seen in detail in Figure 3 below. In general, the backpropagation neural network structure consists of three layers: the input layer, the hidden layer and the output layer including the neuron[10].

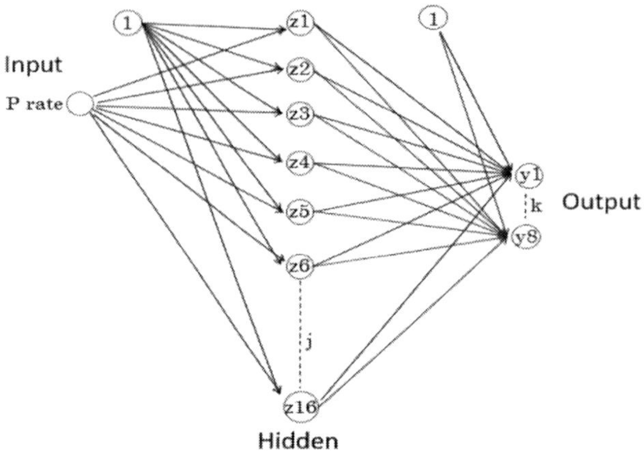

Fig. 3. Architecture NN with steady state method

The hidden layer has 16 neuron and the output layer has 8 neuron. This output layer gives information about the status of each appliance. Output layer contain 1 if the appliance switch on and vice versa. Detail of each output layer can be seen in the Table 1 below.

Table 1. Output Layer

Appliances ON	**Neuron of Neural Network Output Layer**							
	y1	***y2***	***y3***	***y4***	***y5***	***y6***	***y7***	***y8***
Freezer	1	0	0	0	0	0	0	0
TV	0	1	0	0	0	0	0	0
Lamp	0	0	1	0	0	0	0	0
Freezer+TV	0	0	0	1	0	0	0	0
Freezer+ Lamp	0	0	0	0	1	0	0	0
TV+Lamp	0	0	0	0	0	1	0	0
Freezer+TV+Lamp	0	0	0	0	0	0	1	0
OFF	0	0	0	0	0	0	0	1

The simulation of backpropagation neural network is designed using MATLAB. The specification of this neural network are as follows : maximum epoch 2000, target MSE value < 1e-06, the amount of learning data per probability of appliance when turned on is 1000, and the value for (alpha) is set to 0.3. The result of the learning is the value of neural network weights. Iteration learning will stop when it has achieved the target of mean square error (MSE) or maximum epoch value. The MSE is the performance index function commonly used to learn feed forward neural networks. The error is occurred between the output neural network and the target output. MSE can be computed by

$$MSE = \frac{1}{n}\sum_{i=1}^{n}\left(\hat{Y}_i - Y_i\right)^2 \tag{2}$$

where variable n is the number of learning samples, variable $\hat{Y}_i$ is the target output, and variable Y_i is neural network output. After the value of the weight has get from learning neural network, the testing of NN is carried out. The testing of NN results Recognition Rate (RR) value.

2.2.2 Modification Type

The difference between the previous type is an additional one input layer set on an input layer. That additional input layer is the maximum power of appliance. Hence, this type uses two neuron on input layers. Detailed this architecture of the backpropagation neural network can be seen in Figure 4 below.

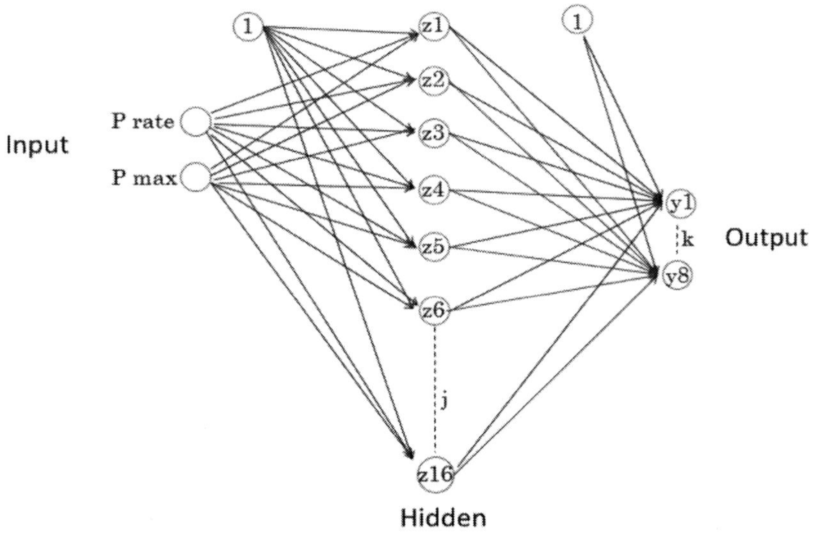

Fig. 4. Architecture NN with steady state modification method

3 Simulation and Analysis

Before starting learning of backpropagation NN, data for input layer should be normalized firstly. In this study, data for learning normalized between minimum 0 and maximum 1 using min max formula.

3.1. Simulation using steady state method

In this simulation, we use aggregated data with significant differences of power rate value. The results of learning backpropagation NN with steady state method on this data using same specification parameter of learning. The MSE value of learning backpropagation NN is 0.0015 and the epoch value is 2000. For testing the results of RR is 100%. Other experiment has done by using data which has the almost same power rate value. The MSE value of learning backpropagation NN is 0.051 and the epoch value is 2000. Moreover, RR value of this testing is 81.9%.

3.2. Steady State Modification Method

In this study, modification type of NN is tried for performance verification. Simulating is performed using aggregated data with almost the same power rate value. We use the same NN learning specifications. The results the MSE value of learning NN is 4.0426e-06 and the epoch value is 2000. RR for testing is 100%.

Table 2. Result Of Simulation

	Data : significant difference power rate			**Data : almost the same power rate**		
	MSE Learning	***Epoch***	***RR (%)***	***MSE Learning***	***Epoch***	***RR (%)***
Steady State	0.0015	2000	100	0.051	2000	81.9
Modify Steady State	-	-	-	4.0426e-06	2000	100

The above simulation result is summarized in Table 2. It can be stated that the proposed modification method proves the significant improvement compared with conventional one. In this paper, the data used is provided by foreign institution which is tracebase. However, the above method can be applied for domestic data. Now we are preparing for providing power signal of local appliance.

4 Conclusion

Based on the result of simulation, it shows that steady state modification method has significant performance for identifying or disaggregating each appliances. The Recognition Rate of proposed method is higher than conventional one. Even though foreign data has been used, the presented method can be applied for data acquired locally. In the next future, this method will be developed to identify appliances that has certain pattern of signal power and can be applied to the smart meter in real time.

We acknowledge the support from Hibah PITTA 2018 funded by DRPM Universitas Indonesia No.2356/UN2.R3.1/ HKP.05.00/2018.

References

1. P. A. Dahono, konversi ITB, *URL: https://konversi.wordpress.com/2011/03/07/menghemat-energi-dengan-menggunakan-listrik,* (2011) [Accessed 1 Jan. 2018].
2. A. Zoha, A. Gluhak, M.A Imram, and S. Rajasegarar, *Non-Intrusive Load Monitoring Approaches for Disaggregated Energy Sensing: A Survey*, Sensors, **Vol.12, Issue 12** (2012).
3. K.E Martinez, K.A Donelly, J.A Laitner, *Advanced Metering Initiatives and Residential Feedback Programs: A Meta-Review for Households Electricity-Saving Opportunities*, Technical Report of American Council for an Energy-Efficient Economy (2010).
4. Department of Energy & Climate Change, *Energy Consumption in United Kingdom, Technical Report* (2010).
5. G. W. Hart, *Nonintrusive Appliance Load Monitoring*, Proceedings IEEE, **Vol. 80**, **No. 12**, (1992).
6. J. G. Roos, I. E. Lane, E. C. Lane,G. P. Hanche, *Using Neural Networks For Non-Intrusive Monitoring Of Industrial Electrical Loads*, Conference Proceedings of 10th Anniversary. IMTC (1994).
7. E.J. Aladesanmi, KA Folly, *Overview of non-intrusive load monitoring and identification technique*, IFAC-PapersOnLine **Vol. 48**, **Issue 30, p. 415-420** (2015).
8. H.H Chang, K.L Chen, Y.P Tsai, W.J Lee, *A New Measurement Method for Power Signatures of Nonintrusive Demand Monitoring and Load Identification*, IEEE Transactions on Industry Applications, **Vol. 48** (2012).
9. D. Srinivasan, W. S. Ng, A. C. Liew, *Neural Network-Based Signature Recognition For Harmonic Source Identification*, IEEE Transactions Power Delivery, **Vol 21** (2006).
10. Yi Du, Liang Du, Bin Lu, Ronald Harley, Thomas Habetler, *A Review of Identification and Monitoring Methods for Electric Loads in Commercial and Residential Buildings*, Proceeding of 2010 IEEE Energy Conversion Congress and Exposition (2010).
11. GitHub, *areinhardt/tracebase. [online] Available at :https://github.com/areinhardt/tracebase,* (2018) [Accessed 1 Jan. 2018].

Development of Hand Gesture Based Electronic Key Using Microsoft Kinect

Arief Ramadhani [1*],, *Achmad* Rizal[1], and *Erwin* Susanto[1]

[1]School of Electrical Engineering, Telkom University, 40257 Bandung, Indonesia

Abstract. Computer vision is one of the fields of research that can be applied in a various subject. One application of computer vision is the hand gesture recognition system. The hand gesture is one of the ways to interact with computers or machines. In this study, hand gesture recognition was used as a password for electronic key systems. The hand gesture recognition in this study utilized the depth sensor in Microsoft Kinect Xbox 360. Depth sensor captured the hand image and segmented using a threshold. By scanning each pixel, we detected the thumb and the number of other fingers that open. The hand gesture recognition result was used as a password to unlock the electronic key. This system could recognize nine types of hand gesture represent number 1, 2, 3, 4, 5, 6, 7, 8, and 9. The average accuracy of the hand gesture recognition system was 97.78% for one single hand sign and 86.5% as password of three hand signs.

1 Introduction

Security access using conventional keys is at risk of being lost or duplicated. Applying technology to the door lock system can make the door open only through the authentication process. This authentication process can be in the form of passwords, electronic key, and even biometrics. On of modality that can be used as authentication is hand gesture.

The hand gesture is commonly used in everyday life to communicate or help someone who has a lack of verbal language to communicate [1]. Hand gesture can be understood by looking at the signals that are delivered and then sent to the brain and processed by the brain to mean the meaning of the signal. With the same logic, a machine can also define a signal when it has 'eyes' and 'brains' as in humans. In this context, the use of cameras and processors or controllers can replace the role of the eyes and brain in humans.

Several studies have been conducted related to electronic key based on this hand gesture. An automatic door lock was developed using hand gesture using a webcam [1]. The image segmentation process was carried out with skin detection. The hand image and background were separated based on the color similarity with the skin color. This method could distinguish five different sign namely signals A, B, C, D, and E. The authors in [2] performed hand gesture recognition based on an accelerometer embedded in a user's smart watch. Data from the accelerometer was then processed using the Multi-Layer Perceptron artificial neural network so that the sign could be recognized.

* Corresponding author: arieframadhani48@gmail.com

2 Related work

Several studies have been conducted regarding hand gesture recognition through different methods. In [3], an RGB camera was used to capture the hand image. The image segmentation process was carried out by determining the limits of red, green, and blue colors. In [4], the authors used Kinect sensor depth in the gesture recognition system. Their system could recognize ten different gestures. By using the support vector machines (SVM), the accuracy could reach more than 95%. The author in [5] also uses Kinect sensor depth in the system. Nine difference gestures can be classified using K-Mean Clustering with the accuracy was from 84% to 90%. By using a different approach, the author in [6] transformed geometrical properties of the hand including area, centroid, Euclidian distance into a feature vector. Then, neural network is used for recognition and classification. The average accuracy of this gesture recognition was 85%.

In this study, we developed a hand gesture based electronic key based on the depth sensor of Microsoft Kinect. By utilizing the Kinect sensor depth, lighting conditions did not affect the performance of this system [7]. The difference with previous research in [3-6], we used Raspberry Pi as the processor which makes this system was portable and more efficient in space considering this hand gesture recognition will be applied to the electronic key system. In this study we propose a simpler method of hand gesture recognition by scanning each pixel to detect the presence of the thumb and the number of open fingers. Using this method our system has an accuracy of 97.78 percent for single hand gesture.

3 Proposed method

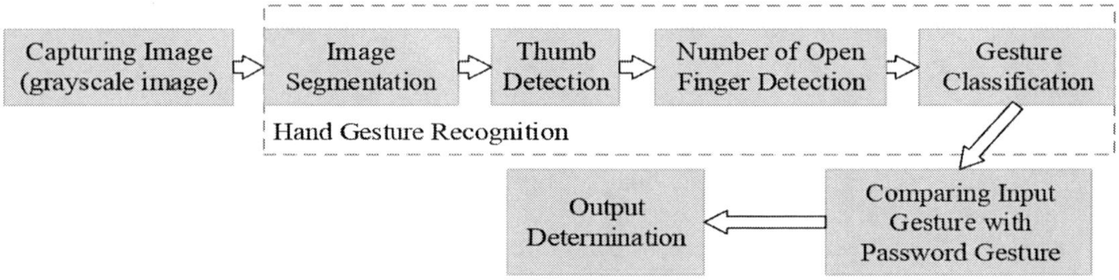

Fig. 1. Proposed method

Fig. 1 displays the proposed method. The hand gesture recognition method proposed in the paper consists of four main components: image segmentation, identification of thumb presence, detection of the number of open fingers and classification of hand signals. The rest subsection explains each process in detail.

3.1 Image Segmentation

Image segmentation is the process of dividing images into meaningful parts and not overlap based on similar criteria [8]. The purpose of image segmentation is to separate the desired object from the background. The segmentation of the grayscale image was done by determining the intensity limit (brightness) of pixels. In the depth image of the Kinect, the closest object to the sensor had the smallest intensity. In this case, the closest object to the sensor was the hand. So we simply took the smallest intensity in the grayscale image in the segmentation process.

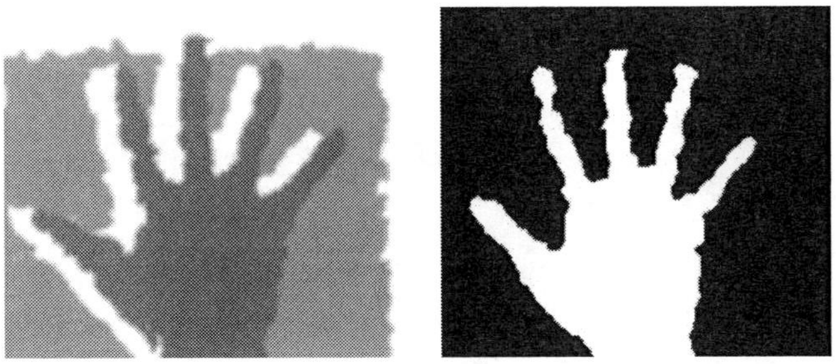

Fig. 2. Image segmentation result

3.2 Thumb detection

Fig. 3 displays the thumb detection process. The thumb detection process was done by scanning the pixels of each column from the leftmost column. The scanning process stopped when it finds the first 255 value and records the pixel row coordinates. The line coordinate data was then compared to the row coordinate limit. If the row coordinate of the leftmost white pixel was smaller than the row coordinate limit (Fig. 3. (a)), it means that there was no thumb on the image. If the row coordinate of the leftmost white pixel was greater than the row coordinate limit (Fig. 3. (b)), it could be assumed that there was a thumb. In some instances, the row coordinates of the leftmost white pixel were greater than the row coordinate limit, but in reality, there is no thumb on the image (Fig. 3. (c)).

To complete this process, we needed to calculate the area of the white area (pixel value of 255) from the column where we found the first 255 value for 20 columns to the right. Figure 4. (a) shows the area (the red line). This area was used to distinguish that the area is part of the thumb or not. If the area obtained did not exceed the limit, then it can be said that there was a thumb, and vice versa. Based on the experiment on ten subjects, we determined the line coordinate limit at 55 and area limit of 400 pixels.

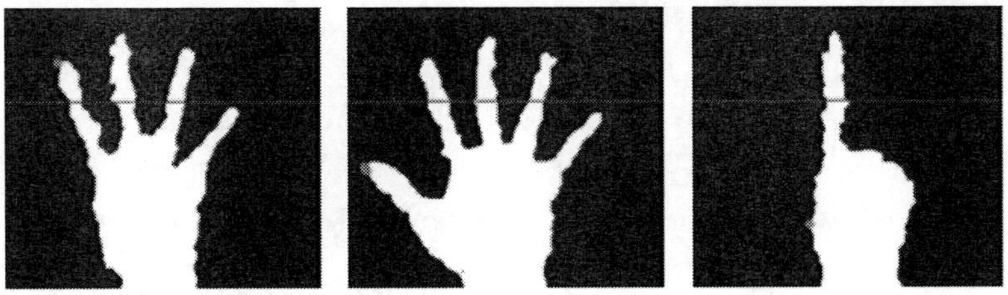

(a) (b) (c)

Fig. 3. (a) The coordinate of the leftmost white pixel is smaller than row limit, (b)(c) The coordinate of the left most white pixel is higher than row limit

(a) (b)

Fig. 4. (a) Area of white pixel on the thumb, (b) Area of white pixel when thumb was not detected.

3.3 Number of open fingers detection

We scanned each pixel from the top of each row. An open finger was obtained if the white pixels with 5 pixels before are black, and the next 4 pixels are white. Each time this condition was fulfilled in the same row, the count would increase by 1. Each time the scan changed row, the count would be reset to 0. The end value of each row would be saved. The end value of each row would be compared, and the maximum value was taken, this value indicated the number of open finger.

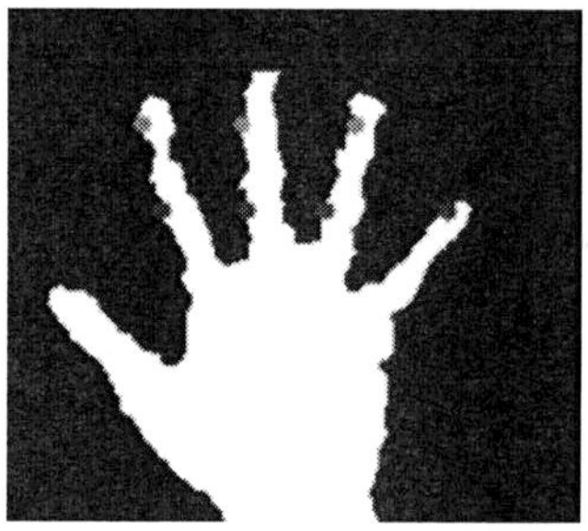

Fig. 5. Condition when open finger detected

3.4 Hand gesture classification

We classified the hand gesture base on thumb and number of open fingers. The hand gesture was classified into two conditions, namely when the thumb was detected and when the thumb was not detected. Table 1 presents the signs formed by the hand gesture.

Table 1. Combination between Thumb and number of open finger

Thumb	Number of open finger	Classification target
Not detected	1	Sign 1
Not detected	2	Sign 2
Not detected	3	Sign 3
Not detected	4	Sign 4
detected	4	Sign 5
detected	0	Sign 6
detected	1	Sign 7
detected	2	Sign 8
detected	3	Sign 9

4 Results and discussion

Tests were carried out on 10 subjects with two test scenarios. The first test is performed on the hand gesture classification, the second test is performed on the hand gesture sequence as a password.

4.1 Hand gesture classification testing

Tests were carried out on ten subjects who demonstrated signals 1 to 9 with each signal was taken five times. The distance between hand and sensor was 60-65 cm that produced the best hand image. Table 2 summarizes the classification result of 450 experiment.

Table 2. Hand gesture classification result

	Detected sign	Number of errors	Cause of error		Accuracy
			Thumb detection	Number of open finger detection	
Sign 1	1	-	-	-	100%
Sign 2	2	-	-	-	100%
Sign 3	3	-	-	-	100%
Sign 4	4	-	-	-	100%
Sign 5	5 and 9	1	-	1	98%
Sign 6	6	-	-	-	100%
Sign 7	7 and 8	9	-	9	82%
Sign 8	8	-	-	-	100%
Sign 9	9	-	-	-	100%

Based on Table 2, misclassification occurred at most when performing a classification of sign '7' which was classified as sign '8'. There were ten errors from 450 experiment data, produced an accuracy of 97.78%.

4.2 Hand gesture sequence as password testing

At this stage, hand signals were used as passwords with three hand movement sequences. The test was performed on ten subjects who demonstrated three hand signs as a password; signal '3', sign '4' and sign '5'. Each subject demonstrated a password sign 20 times. Table 3 shows the experiment results. There were 27 errors from 200 trials. Therefore, the accuracy of the password recognition was 86.5%.

Table 3. Result of password testing

Hand sign	Target	Detected sign	Number of trial	Number of error detection
First sign	3	3 and 9	200	1
Second sign	4	4 and 3		8
Third sign	5	5, 4 and 9		18

Table 4. Average time required by the system to identify passwords

Subjects	Time to recognize gesture (s)	Total time to unlock the door (s)
Subject 1	1,41	8,85
Subject 2	1,59	9,04
Subject 3	1,49	9,03
Subject 4	1,45	8,91
Subject 5	1,29	8,06

Subject 6	1,66	9,21
Subject 7	1,25	8,73
Subject 8	1,49	8,25
Subject 9	1,54	9,06
Subject 10	1,37	8,80
Average	**1.45**	**8,79**

From the test, the average processing time for the recognition of hand signals is 1.45 seconds, while the total time taken by user perform the three hand sign until the key was unlocked is 8.79 seconds. The system built depended on the distance between the hand and the sensor. The distances of hand and sensors that were too close will make the image degrade while if it was too far away, it was difficult to separate hands and background.

The advantage of the system that it did not require machine learning for the classification process but only used pixel analysis. Hand sign recognition was not determined by who the person was but the sequence of hand signs. Meanwhile, the weaknesses of this system are some parameters determined from empirically trained data. In the next study, more data will be tested and longer hand signal sequence as passwords.

5 Conclusion

This paper describes the design of electronic key systems using hand gesture with Kinect sensors as data acquisition devices. Based on the experiment, the system could recognize nine types of hand sign that represent numbers 1 through 9 using the right hand. Optimum distance for this system was 60 cm - 65 cm from the sensor. The hand gesture recognition system would be implemented into an electronic key system that was able to recognize three different hand signs as a password. The average accuracy of the hand gesture recognition system was 97.78% for one single hand sign and the recognition accuracy of password reached 86.5% with the average time processing of 1.45 seconds.

References

1. R. Passarella and M. Fadli, "Hand Gesture Recognition as Password to Open The Door With Camera and Convexity Defect Method," *Proceeding of The 1st International Conference on Computer Science and Engineering*, pp. 69–73 (2014).
2. H. J. Ahn, J. S. Kim, J.-Y. Shim, and J. S. Kim, "Hand Gesture Recognition for Doors with Neural Network," *Proc. Int. Conf. Res. Adapt. Converg. Syst.*, pp. 15–18 (2017).
3. A. Chaudhary, K. Vatwani, T. Agrawal, and J. L. Raheja, "A vision-based method to find fingertips in a closed hand," *J. Inf. Process. Syst.*, vol. 8, no. 3, pp. 399–408 (2012).
4. T. Q. Vinh and N. T. Tri, "Hand gesture recognition based on depth image using kinect sensor," in *2015 2nd National Foundation for Science and Technology Development Conference on Information and Computer Science (NICS)*, pp. 34–39 (2015).
5. Yi Li, "Hand gesture recognition using Kinect," *2012 IEEE Int. Conf. Comput. Sci. Autom. Eng.*, pp. 196–199 (2012).
6. N. B. Bhoyar and P. M. Bartere, "Neural Network based Static Hand Gesture Recognition," *Int. J. Innov. Res. Comput. Commun. Eng.*, vol. 3, no. 2, pp. 112–116 (2014).

7. N. Febriana, A. Rizal, E. Susanto, "Sleep Monitoring System Based on Body Posture Movement using Microsoft Kinect Sensor," *Int. Symp. Biomed. Eng*, vol. 3 (2018).
8. J. Chen, B. Guan, H. Wang, X. Zhang, Y. Tang, and W. Hu, "Image Thresholding Segmentation Based on Two Dimensional Histogram Using Gray Level and Local Entropy Information," *IEEE Access*, vol. 6, no. c, pp. 5269–5275 (2017).

Expert System To Detect Human's Skin Diseases Using Forward Chaining Method Based On Web Mobile

Jenie Sundari[1,]* *Hamimah*[2], *Popon* Handayani[2], *Yunita*[1], *Ibnu* Dwi Lesmono[3], *Mulia* Rahmayu[2], *Paramita* Kusumawardhani[4], *Yusnia* Budiarti[1], *Fadilah*[4], *Nurhayati*[4]

[1]STMIK Nusa Mandiri, Information Technology Program Jakarta, Indonesia
[2]STMIK Nusa Mandiri, Information System Program Jakarta, Indonesia
[3]AMIK BSI Purwokerto Manajemen Informatic Program Purwokerto, Indonesia
[4]ABA BSI Jakarta, English Program Jakarta, Indonesia

Abstract. Skin is a sense of touch in humans' body which one of its functions is to feel the touch. Another function of the skin is to release the residual substances such as sweat. This part of the human body is very sensitive, easy to get hurt, and feels a sense of sensitivity. Human skin consists of epidermis and dermis. Skin diseases can be caused by a decrease in the immune system, allergies, viruses, or other causes. Skin disease is generally caused by a less guarded hygiene, bacteria, viral, allergic reactions, and low body resistance. If the cause of skin disease is only due to a lack of maintaining cleanliness, it can be prevented by changing lifestyles to be cleaner and healthier. Especially in Indonesia which is a tropical country, where the humidity is very high and can increase the development of bacteria on the skin. In another situation, not all people understand skin diseases for treatment or prevention. By the reason, The Forward Chaining Method is made to provide information about diagnoses of several skin diseases and produces conclusions. The existence of this expert system application makes it easy for people to get any information about skin diseases

1 Introduction

Skin is a human-body part which is the widest area that arranges body and system that covers all body. Skin is the part which can accept the stimulus such as touching, feeling hurt and others influence from outside. Based on the most important thing of covering the organs inside and under it, it is important to protect the skin-health from the very beginning. Many skin diseases can be factored by many aspects such as climate changing, environment

[1] Corresponding author: jenie.jni@nusamandiri.ac.id

surrounding and a bad-self health which is caused by bacteria, viruses, self immune, allergic reaction, and others. Skin defining cannot be done carelessly because it can endanger the skin care and handling. Based on the reasons, it is important to consult about skin problems to the doctors or the experts.

Skin diseases in Indonesia can be high categorized and make a serious problem. It can be happened because there are still many Indonesian citizens who do not really care with the skin diseases. The citizens' ignorance can cause skin diseases spread out quickly. Bad environment and habits, such as climate changing, viruses, bacteria, allergic, self-immune, etc. are some factors which can cause skin diseases. [1]

Indonesian health profile in 2008 showed that the distributions of outpatients patients based on International Classification of Diseases-X (ICD-X) in Indonesia's hospitals, based on skin diseases, are about 115.100 visitors with 64.557 new cases. Skin diseases increased up to the third grade from the ten-most diseases in outpatients' patients around Indonesia, 192.414 visitors with 48.576 new cases in 2011. [2]

The idea behind creating an expert system is that it can enable many people to benefit from the knowledge of one person – the expert. It is done in order to simulate the judgment and behavior of a human that has expert knowledge and experience in a particular field is expert system. [3]

To apply forward chaining inference techniques which are applied, the data and facts has been obtained so it can be made to conduct the research process that will provide a conclusion or solution based on a set of data and facts. By using inference techniques are also opportunities to get a more specific conclusion can be easily obtained. [4]

2 Literature review

The expert system for memory losing disease is should be appropriate to the rules and facts. It can also be used based on case to save the cases to compare between the new cases with the previous case. It is done by discussing the different approach by focusing to all information about losing memory. Different symptoms and caused of losing memory in different ages group can make different way to avoid all losing memory diseases. The patients can be helped to get the advices needed about the symptoms based on their nervous system by using the expert system. [5]

In this study, an automated diagnosis system to detect breast cancer based on data preprocessing and Bayesian networks is presented. This study shows that Bayesian networks in the diagnosis of the disease according to the nature of the algorithms work very well. Relief algorithm is also used to reduce the number of the database features that shows how it can produce useful results in combination with Bayesian network classifier. Also, we tried to show that the classifiers can reach to a certain extent of accuracy for this database. Thus for obtaining more accuracy we should do enough preprocessing on the data, resolve missing data and data unbalancing phenomenon for the data. Finally, the significant accuracy for diagnosing breast cancer disease by using 3-fold cross validation test method can be obtained. [6]

This paper has presented an expert system for rickets diagnosis, which provides the patients with the diagnosis, recommendation and treatment; based on the expert system knowledge base and data collected from the patients. The aim of the proposed expert system was to recognize the symptoms of the disease by the user to enable him/her to identify diseases and causes of rickets. This expert system saves the patient's time and effort by allowing the patient to diagnose rickets seamlessly and easily through a simple user interface, so often is dispensed with the traditional diagnosis. [7]

The web based expert system (WES) is created to diagnose the skin disease in Pakistan especially in flood prone areas. The system has the newest knowledge to tract the previous disease based on the skin expert, medical students and pharmacists. The system is developed by using Java application and forward chaining inference system. The result of this research is to diagnose more than 13 skin diseases. [8]

Online system to diagnose skin disease in children is created and developed. The system is created by using forward chaining inference system so parents can get the information about the skin disease happened to their children then suggest the suitable medicine related to the skin disease itself. [9]

The lungs are one of the vulnerable respiratory organs. The purpose of this research is to implement decision tree and dempster shafer method on lung disease diagnosis and measure the accuracy of the system. The symptom was searched using forward chaining decision tree and the diagnosis was calculated using dempster shafer method. [10]

The development cycle of an expert system is shown in Figure 5. Referring to Figure 5, after selecting the orthopedic domain, knowledge was acquired by interviewing an orthopedic surgeon. The knowledge is all about the different patient's problems, such as shoulders, knees, hips, feet, thumbs, and wrists. The knowledge base is incomplete, and has six different problem areas and 370 rules with certain orthopedic diseases, but it could be updated with new symptoms and diseases. [11]

Expert System on the organization was aimed at adding value, increasing productivity as well as the area of managerial can make decisions quickly and accurately. The benefits of the expert system was able to diagnose quickly and accurately to the symptoms of the disease caused and it was expected to help the farmers in anticipating many losses caused by disease. Required accuracy and the accuracy of the counting in diagnosing the symptoms of the disease is in order to summarize the results by using forward chaining method. [12]

This paper describes Expert System (ES) for online diagnosis and prescription of red-eye diseases. The types of eye diseases which can be diagnosed with this system are called Red-eye diseases. It is rule based web-supported expert system, assisting ophthalmologists, medical students doing specialization in ophthalmology. Researchers as well as eye patients have computer know-how. System was designed and programmed with Java Technology. The expert rules were developed on the symptoms of each type of Red-eye disease, and they were presented using tree-graph and inferred using forward-chaining with depth-first search method. [13]

In this research, the expert system is created. It is created to decide about kinds of eyeglass lens using forward chaining method. Forward chaining method is started by using preliminary information (preliminary symptom) and it moves forward to adjust to the latest information. It is done in order to give the solutions based on eyeglass lens. [14]

3 Research method

3.1 Inference method

The Inference Engine forms the heart of the expert system; the knowledge base serves as the brain of the expert system. The inference engine chums through countless potential paths and possibilities based on some combinations of rules, cases, models or theories. Some rules such as predicate logic mimic human reasoning and offer various mathematical arguments to any query. Two methods of inference often are used, forward and backward chaining; forward chaining is a top down method which takes facts from satisfied conditions and backward chaining is a top down method which takes facts from satisfied conditions in rules. [15]

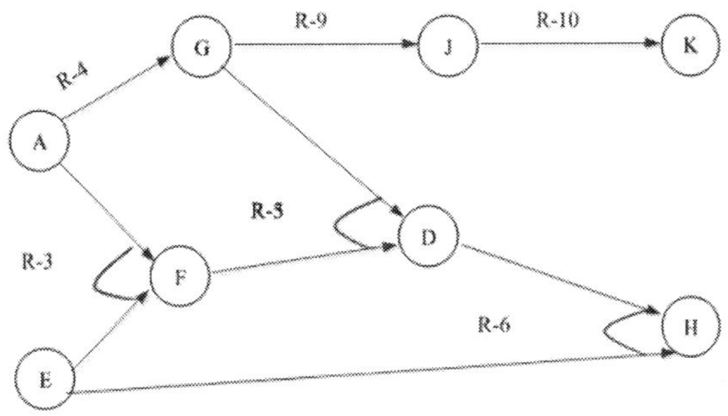

Fig. 1. How the forward chaining inference machine works

3.2 Decision tree

Decision Tree is a tree structure like a flow-chart, in which the rectangular boxes are called the node. Each node represents a set of records from the original data set. Internal node is a node that has a child and leaf (terminal) node is nodes that don't have children. Root node is a topmost node. For finding the best way to distinguish a class from another class, the decision tree is used. [16]

4 Result & Discussion

4.1 The expert

The table of expert consists of facts obtained from an expert, science, research and experiences in diagnosing skin diseases. The following are the data obtained from the experts:

1. Table of skin disease symptoms

Table 1. Table of skin disease symptoms

Code	Symptoms
G01	an itchy rash
G02	open sores or lesions
G03	burning skin
G04	exfoliation or dandruff on the mustache, beard and eyebrow area
G05	bubbles on the skin with yellowish liquid
G06	0.5 centi-meter bubble which quickly swells and pops easily
G07	cracked skin such as blistered skin exposed to cigarette burns
G08	a circular reddish or silvery rash
G09	scaly skin
G10	skin inflammation/dermatitis
G11	easily irritated if exposed to rubbing
G12	raised bumps or wrinkles

2. Table of skin disease types

Table 2. Table of Skin Disease Types

Code (s)	Type of Skin Disease
P01	Dermatitis Seboroik
P02	Vitiligo
P03	Keloid
P04	Tinea Korporis
P05	Impetigo Bulosa
P06	Psoriasis
P07	Seborcheic Keratosis (Keratosis Seboroik)
P08	Dyshidrotic Eczema
P09	Angioedema
P10	Kista Sebaceous

3. Decision tree

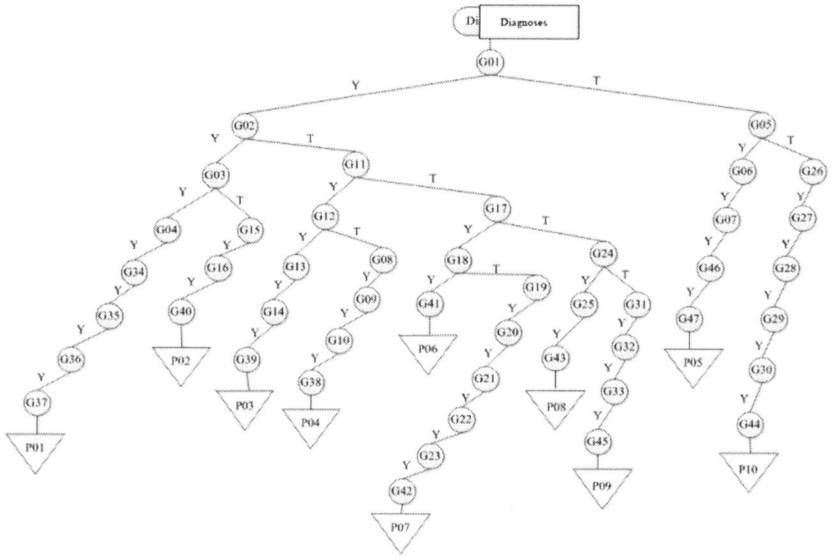

Fig. 2. How the forward chaining inference machine works

4. Implementation of Expert System Web

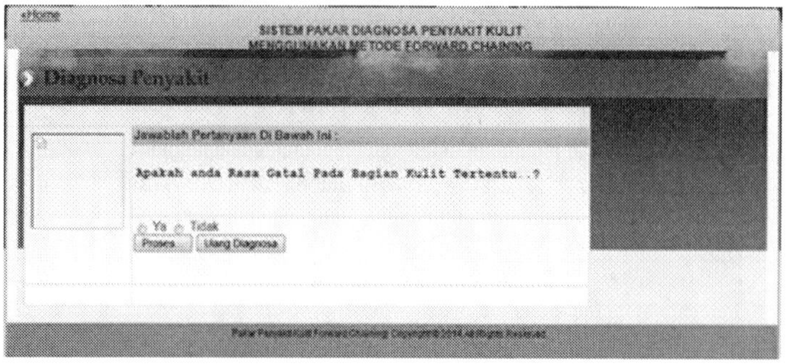

Fig. 3. Consultation page

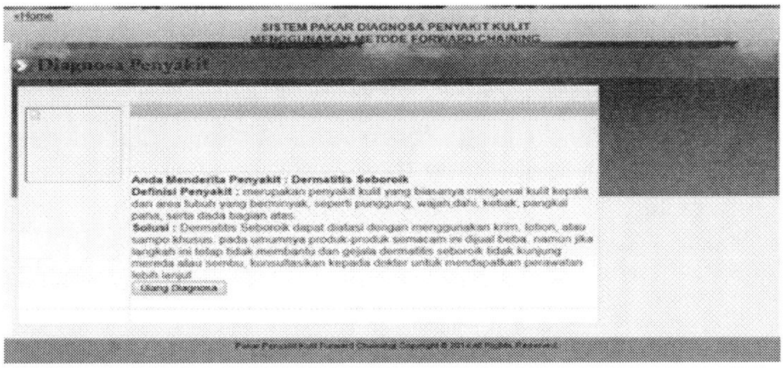

Fig. 4. Diagnosis result

5 Conclusion

From the analytical descriptions above, the following are the conclusions of this research:

The expert system could help the users (patients) to diagnose the suffered disease based on the selected symptoms by those who did a direct consultation.

The process of designing the expert system application program to diagnose skin diseases in humans includes several steps that must be considered such as knowledge acquisition, knowledge representation, and compilation of databases, inference machines, UML diagrams, interface design, implementation and testing.

The use of the forward-chaining method with the search process can be used to create an expert system for detecting human skin diseases.

References

1. Pardiansyah, R., *Association Between Personal Protective Equipment With the Irritant Contact Dermatitis in Scavengers Faculty of Medicine*, (Lampung University 2015).
2. Kementerian Kesehatan Indonesia, 2012. Profil Kesehatan Indonesia Tahun 2009. Jakarta: Kementerian Kesehatan RI
3. Shikhar Kr. Sarma, Kh. Robindro Singh & Abhijeet Singh, *An Expert System for diagnosis of diseases in Rice Plant International Journal of Artificial Intelligence*, **Volume(1): Issue(1) 26.** (2010)
4. Baur Gregory R. & Pigford D.V. 1990. *Expert System for Business: Concept and Implementations*, Boyd &Fraser Publishing Company, Boston-USA (1990).
5. Sasmito WG, Bayu Surarso, Aris Sugiharo. Application Expert System of Forward Chaining and The Rule Based Reasoning For Simulation Diagnose Pest and Disease Red Onion and Chili Plant. ICISBC. (2011)
6. Fallahi Amir, Shahram Jafari. An Expert System for Detection of Breast Cancer using data Processing and Bayesian Network. International Journal of Advanced Science and Technology. **Vol 34 September.** (2011)
7. Al Rekhawi, Abdullah A Ayyad, Abu Naser. Rickets Expert System diagnoses and treatment. International Journal of Engineering and Information Systems. **Vol 1 issue 4.** (2017)
8. Amarathunga, EPWC Ellawala, GN Abeysekara, CRJ Amalraj. Expert System for Diagnosis of Skin Diseases. International Journal of Scientific & Technology Research. **Vol 4 issue 01. January** (2015).

9. Yusof M Munirah, Ruhaya A Aziz, Chew S Fei. The Development of Online Children Skin Disease Diagnosis System. International Journal of Information and Education Technology. **Vol 3 no 2. April** (2013)
10. Alfatah Muis Abdul, Riza Arifudin, Much Aziz Muslim. Implementation of Decision Tree and Dempster Shafer on Expert System for Lung Disease Diagnosis. Scientific Journal of Informatics. **Vol 5 no 1. May** (2018)
11. Zuhra Tul Fatima, Adnan Ahmed Arain, Mohsin Ali Tunio. An Implementation of Expert System for Orthopedic Patient Diagnosis. Quaid E Awam University Research Journal of Engineering Science&Technology. **Vol 15 no 1. Jan** (2016).
12. Hayadi Herawan B, Kasman Rukun, Rizky Ema Wulansari, etc. Expert System of Quail Disease Diagnosis Using Forward Chaining Method. Indonesian Journal of Electrical Engineering and Computer Science. **Vol 5 no 1. January** (2017).
13. Dr Asghar Muhammad Zubair, Dera Ismail Khan, Muhammad Junaid Asghar. Expert System for Online Diagnosis of Red-Eye Disease. International Journal of Computer Science&Emerging Technologies. **Vol 1 issue 2. August** (2010).
14. Pramesti Ari Atikah, Riza Arifudin, Endang Sugiharti. Expert System for Determination of Type Lenses Glasses Using Forward Chaining Method. Scientific Journal of Informatics. **Vol 3 no 2.** (2016)
15. Folorunso I O, Abikoye OC, Jimoh RG, Raji KS. A-Rule Based Expert System for Mineral Identification. Journal of Emerging Trends in Computing and Information Sciences. **Vol 3 no 2. February** (2012)
16. Saini Priyanka, Sweta Rai, Ajit Kumar Jain. Decision Tree Algorithm Implementation Using Educational Data. International Journal of Computer-Aided Technologies. **Vol 1 no 1. April** (2014).

Performance Analysis of Internet Key Exchange Algorithms on IPSec Security Association Initiation

Supriyanto Praptodiyono*[1], *Moh.* Furqon[1], *Alief* Maulana[1], *Iznan* H. Hasbullah[2], *Shafiq* Ul Rehman[3]

[1]Universitas Sultan Ageng Tirtayasa, Indonesia
[2]Universiti Sains Malaysia, Malaysia
[3]Singapore University of Technology and Design, Singapore

Abstract. Naturally, the Internet as an open network allows millions of users to interact each other. Furthermore, the giant network vulnerable to various malicious activities that potentially causes many (losses) for both resources and humanity. In order to reduce the potential attacking activities, Internet Engineering Task Force has standardized a network level security so-called IP Security. The key process behind the IP Security is a Security Association that is identified by Security Parameter Index. The initiation of Security Association performance depends on the encryption algorithm used. Hence, the performance of the algorithm candidate is an important consideration. This paper aims to evaluate some encryption algorithm to be used in IPv6 network. Results of the experiments shows that ECP 384 based on the elliptic curve algorithms outperforms than the traditional algorithm such as MODP 1024 and MODP 3072.

1 Introduction

Nowadays, the Internet as the giant computer network has connected millions of people around the world. The network is able to share information to all users every time and everywhere. The emerging of the Internet gives the human many benefits and advantages such as ease of accessing information, remote communication, entertainment and so forth. However, but this technology also opened many vulnerabilities and risks that can make losses for both resources and humanity. Security holes on the Internet network include data theft, identity fraud, privacy theft and denials of service [1]. Since the Internet network includes a variety of sensitive data that is very important, both for individuals and organizations, it needs a high security system. Even though, it is almost impossible to get very high security without any vulnerabilities [2], the security system used is at least able to handle one of the security threats that exist in the Internet. Hence, the users can use the security system based on the vulnerability.

[1] Corresponding author: supriyanto@untirta.ac.id

There are three aspects on the network security including confidentiality, data integrity and availability. Currently there are several methods in network security to serve the three aspects. The method is usually developed based on the security hole that will be faced. One of the methods used in network security is the IP-Based security system [3]. This protocol can co-exist with IP datagrams to support secure data transmission over unprotected public networks. IPSec is a set of protocols that provide security for Internet communication on the network layer [4]. The most common use of IPSec is to provide a virtual network or VPN (Virtual Private Network) that is used between gateway-to-gateway, host-to-host, or end-to-end [5]. IPSec consists of several subprotocols, they are Authentication Header (AH), Encapsulating Security Payload (ESP), and Internet Key exchange (IKE). AH and ESP aims to provide cryptographic services, while IKE handles the Security Association parameter management.

One of the considerations on the implementation of IPSec with IKEv2 is the selection of key exchange cryptography algorithms. This is very important because the key exchange is the process of connecting two or more hosts that want to connect in a secret communications network but through the public network without having to enter into an agreement between the parties involved. In addition, choosing the right key exchange algorithm will affect the performance of the exchange speed, the performance of the encryption strength and the performance of the network itself. This research will evaluate some candidates to obtain the best cryptographic algorithm to be implemented on IPSec and IKEv2.

2 Overview of IPSec

IPSec is a security standard for the use of Internet-based communication protocols by encryption or authentication of all passing IP packets [6]. IPSec is designed as a cryptographic protocol that works for data security and key exchange. The IPSec protocol is implemented into the network layer. IPSec has two modes in operation, which are the transport method and tunnel method. Transport Mode is commonly used for end-to-end communication. Transport Mode provides data security or is called IP Payload which consists of TCP/UDP headers and data, via AH or ESP protocol. This payload is encapsulated by headers and trailers from IPSec. IP Header originated intact, the IP protocol field changed to ESP or AH and the contents of the original protocol will be stored in the IPSec trailer and will be returned when the packet is decrypted.

Tunnel mode is the default mode used in IPSec that is used to secure all IP packets, the packet is encapsulated and encrypted and then added a new IP header and sends it to the other side of the VPN tunnel. The tunnel mode is commonly used between the gateway or on the end-station to the gateway, this gateway will serve as the proxy of the host using this mode. In tunnel mode an IPSec Header (AH or ESP Header) is inserted between IP headers and the upper layer protocols.

2.1 Security Association

IPsec requires some values on some of its security parameters in order to operate properly. Hosts that communicate within a virtual network will share the same situation and parameters previously negotiated before building a virtual network session. The creation of security parameters in IPsec is done through a mechanism called security association (SA) [7]. SA controls the data traffic of security protocols in one direction. As an illustration, one SA can control the ESP protocol data traffic from host A to host B, but it takes another SA to control traffic from the opposite direction. Hence, a SA pair is required for each protocol used, one for input and one for output. If a VPN session uses the ESP protocol, then each endpoint will use two SAs. After the negotiation, then SA will be applied to the protocol concerned. IPsec stores SA data in a database called SAD (Security Association Database).

2.2 Internet Key Exchange Version 2 (IKEv2)

IKEv2 (Internet Key exchange version 2) is an updated of the old IKE protocol and a standard defined in RFC 7296. IKEv2 is a protocol used to perform SA management tasks such as creation, renewal, and deletion [4]. IKEv2 uses the Diffie-Hellman exchange technique to create a shared secret, which will be used to generate derived cryptographic keys. The IKEv2 header format is shown in Figure1. In addition, IKEv2 is responsible for performing authentication functions on each party that communicates. IKE messages on IKEv2 protocol implementation using UDP port 500 and port 4500. Each UDP packet contains one IKE message. If the message is sent using the UDP port 500, then the IKE message is exactly after the UDP header. Whereas if sent with UDP port 4500, then IKE message preceded by four octets (32 bits) of zero. These four octets of zero are not part of the IKE message and are not included in the length field or checksum. Each IKE message starts with an IKE header followed by IKE payload in accordance with the identification of the Next Payload field in the IKE header or the previous payload.

```
                     1                   2                   3
 0 1 2 3 4 5 6 7 8 9 0 1 2 3 4 5 6 7 8 9 0 1 2 3 4 5 6 7 8 9 0 1
+-+-+-+-+-+-+-+-+-+-+-+-+-+-+-+-+-+-+-+-+-+-+-+-+-+-+-+-+-+-+-+-+
|                       IKE SA Initiator's SPI                  |
|                                                               |
+-+-+-+-+-+-+-+-+-+-+-+-+-+-+-+-+-+-+-+-+-+-+-+-+-+-+-+-+-+-+-+-+
|                       IKE SA Responder's SPI                  |
|                                                               |
+-+-+-+-+-+-+-+-+-+-+-+-+-+-+-+-+-+-+-+-+-+-+-+-+-+-+-+-+-+-+-+-+
|  Next Payload | MjVer | MnVer | Exchange Type |     Flags     |
+-+-+-+-+-+-+-+-+-+-+-+-+-+-+-+-+-+-+-+-+-+-+-+-+-+-+-+-+-+-+-+-+
|                          Message ID                           |
+-+-+-+-+-+-+-+-+-+-+-+-+-+-+-+-+-+-+-+-+-+-+-+-+-+-+-+-+-+-+-+-+
|                            Length                             |
+-+-+-+-+-+-+-+-+-+-+-+-+-+-+-+-+-+-+-+-+-+-+-+-+-+-+-+-+-+-+-+-+
```

Fig 1. IKEv2 Header Format

The key exchange algorithm used for internet key exchange applications (IKEv1) originally used a collection of Diffie-Hellman algorithms called the Oakley Group [8]. There are 4 main groups used, two of which are MODP group and the other are ECN group. This key group is estimated to have the power equivalent of a symmetry key algorithm with 70-80 bits, while for the Internet Key Exchange Version 2 (IKEv2) application, based on the RFC7296 standard document defines two Diffie-Hellman groups, 768-bit MODP Group and 1024-bit MODP Group.

However, based on the update contained in RFC 8268 documents [9] the United States Information Assurance Directorate (IAD) at the National Security Agency (NSA) dealing with National Security Systems (NSS) published "Commercial National Security Algorithm Suite and Quantum Computing Frequently Asked Questions ". The document states that based on the security interests of using MODP Group of less than 2048-bit size is no longer recommended. In the document also, the user must select the DH group that has been formed strongly and has met the minimum needs and have been validated. The suggested DH Group examples are:

- Elliptic Curve with P-384 group size
- RSA modulo with minimum size of 3072 bits
- MODP Group with a size of at least 3072 bits

3 Network Topology

In order to evaluate the cryptography algorithm on IPSec with IKEv2, a limited network topology was setup as shown in Figure 2. The network system was built by IPSec writer based on IKEv2 protocol using strongSwan software. After the IPSec system has been established, the next step is to conduct experimental testing, at this stage the test is done by replacing the key exchange cryptography algorithm used in IKEv2, this is done to find the best algorithm in terms of performance and security strength that can be applied to the system. In this experiment will determine the best key exchange algorithm that can be implemented on IPSec by considering the performance of algorithm and network performance.

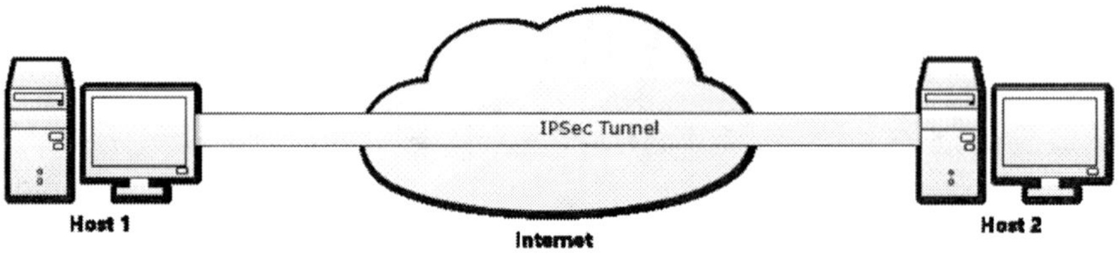

Fig 2. Network Topology

Experiments were conducted to test key exchange cryptographic algorithms running on an IPSec system. There three candidates of algorithms as follows:

1. 1024-bit MODP, an algorithm that has been widely used and no longer recommended
2. 3072-bit MODP, the size increase algorithm from the 1024-bit MODP and it is one of the recommended algorithms
3. ECP 384 bit, a lightweight elliptic curve algorithm and recommended for use as a key exchange algorithm.

4 Result and Discussion

This section provides results of the experiments that was discussed in Section 3. The candidates were implemented at IPSec system to get some parameters such as key exchange speed and network throughput. Experiments have been performed on the network topology for the 1024-bit MODP, 3072-bit MODP and ECP 384-bit respectively.

4.1 Key Exchange Speed

In this first experiment, the test was conducted to do one pair key exchange generation, by measuring the time required to generate the key. This test is purely to measure the performance of the algorithm that has not been implemented into the IPSec system. Table 1 shows the result of key formation experiments on modp1024, modp3072 and ecp384 algorithms for 10 times. In the table, the results with the shortest time, achieved by ecp384 with time less than 0.01 seconds, followed by modp1024 and the last position modp3072.

Table 1. Time Statistics for Key Exchange Generation

Parameter	modp1024	modp3072	ecp384
Mean (s)	0,0172	0,0718	0,007
Min (s)	0,017	0,07	0,007
Max (s)	0,019	0,072	0,007
Standard Deviation	0,001	0,001	0

It can be spoken that the ECP Group algorithm is a relatively more lightweight algorithm compared to the MODP Group. The table also shows that MODP1024 is faster than MODP3072. This because processing of larger bits size requires more time to form a key. It was found that ecp384 had the most superior results compared to other tested algorithms. The ecp384 has the smallest time for the average of key generation time (mean), min key generation as well as max key generation time. The standard deviation of ecp384 shows the value 0 that indicates the data of key generation time for ecp384 has a high stability. Based on this experiment result and analysis on the key generation speed can be concluded that the ecp384 algorithm is the most lightweight from the three candidates.

4.2 Throughput

As discussed in the previous subsection, the key exchange speed is measured to select one of the most lightweight algorithms to be implemented in IPSec. Further, this subsection provides results of IPSec implementation using the three candidates. This is done to know the impact of the key exchange algorithms implementation on the network performance especially on the network throughput. This is to know the actual bandwidth consumed by the network when implementing IPSec using the three key exchange algorithms.

This second experiment includes the plain scenario that means data transmission without IPSec as the reference. This is to know the impact of IPSec implementation using the three candidates on the network throughput. The experimentation results are shown in Table 2. The table consists of four columns: plain, modp1024, modp3072 and ecp384.

Table 2. Network Throughput of IPSec Implementation

Parameter	plain	modp1024	modp3072	ecp384
Mean (Mbit/s)	31,67	19,81	17,3	23,39
Min (Mbit/s)	29	15,6	14,8	19,8
Max (Mbit/s)	37,5	23,5	20,7	27,6
Standard Deviation	2,32	2,52	1,49	2,24

The plain scenario as the reference has the highest throughput because there is no IPSec implementation meaning no overhead on the transmission. Compared to the plain throughput, the nearest value is obtained by the ecp384. The average value of network throughput shown by the ecp384 is 23,39 Mbps or 26% lower than the plain. This is considered as the nearest one compared to the modp1024 that has 37% lower and the modp3072 that has 45% lower. Hence, the ecp384 has the smallest difference compared to the plain scenario. That can be seen on the table the minimum and maximum value of the ecp384 also shows the nearest value to the plain scenario.

From the two experiments, there are found that the ecp384 algorithm is superior to the other candidates. The first experiment shows this algorithm has the smallest generation time that means this is the most lightweight algorithm. The second experiment also shows the

IPSec implementation using ecp384 has the highest network throughput compared to other candidates. At the same time, this algorithm shows the nearest value to the plain scenario. In other word, the IPSec implementation using ecp384 only adds 26% overhead to the plain scenario. This can be concluded that the ecp384 algorithm is the most suitable algorithm to be implemented on IPSec security association. Further, the IPSec implementation using ecp384 can be faster than the current implementation that still using modp1024 algorithm.

5 Conclusion

IPSec is a network layer security protocol that has implemented on the Internet. The protocol consists of some other protocol such as AH, ESP and IKEv2. The implementation of IPSec requires a security association that needs a key exchange. The current IPSec implementation uses modp1024 algorithm on its key exchange. However, the algorithm is considered as heavy if implemented on mobile devices that has limited energy. Furthermore, there is required to find out another algorithm that more lightweight.

The selection of key exchange cryptography algorithm applied to IKEv2 has found that the ecp384 is the most lightweight of the three. The 384-bit cryptography of Elliptic Curve Group Modulo a Prime (ECP) is able to offer the best performance in terms of key generation performance, cryptographic performance applied to IPSec and network performance in general, compared to other cryptographic algorithms. However, there needs to do a next research on the implementation of the selected algorithm on mobile technology.

References

1. A. Marwa, B. Malika, and G. Nacira, "Contribution to enhance IPSec security by a safe and efficient internet key exchange protocol," 2013 World Congr. Comput. Inf. Technol. WCCIT 2013, pp. 1–5, (2013).
2. P. Lubomski and H. Krawczyk, "Practical Evaluation of Internet Systems' Security Mechanisms," IEEE Secur. Priv., vol. 15, no. 1, pp. 32–40, (2017).
3. J. L. Shah and J. Parvez, "Impact of IPSec on Real Time applications in IPv6 and 6to4 Tunneled Migration Network," ICIIECS 2015 - 2015 IEEE Int. Conf. Innov. Information, Embed. Commun. Syst., pp. 0–5, (2015).
4. C. Kaufman, "Internet Key Exchange Protocol Version 2 (IKEv2)," RFC 7296, pp. 1–142, (2014).
5. S. Frankel, "IP Security (IPSec) and Internet Key Exchange (IKE) Document Roadmap," RFC 6071, pp. 1–63, (2011).
6. R. S. Pavel, Network Security Advance, vol. 3. Mumbay: University of Mumbay, (2015).
7. B. A. Forouzan, Data Communications and Networking Third Edition. New York: McGraw-Hill, (2003).
8. D. Harkins, "The Internet Key Exchange (IKE)," RFC 2409, (1998).
9. M. Baushke, "More Modular Exponentiation (MODP) Diffie-Hellman (DH) Key Exchange (KEX) Groups for Secure Shell (SSH)," RFC 8268, (2017).

KWh Meter Smart Card Model Token For Electrical Energy Monitoring

Riki Ruli A. Siregar[1*] , *Hengki* Sikumbang[1], *Iriansyah* BM. Sangadji[1] , *Indrianto*[1]

[1] Informatics Engineering, Sekolah Tinggi Teknik PLN Cengkareng West Jakarta. Indonesia

Abstract. This research Producer Electric meter Appliance model base on smart card kWh (kilo Watt hour). The card power meter Is use to calculate the prepaid electricity meter using the 20-digit voucher number as a charging medium. Utilizing a smart card, the token number in the prepaid electricity meter can help reduce the error rate when entering the token number, the model destination is created. Hardware design and with apps to detect how much energy is used to store basic data to help view history of usage. Using smart cards and microcontrollers in apps that come from a usage history in the Pulse usage category will help users know their usage in efficient, normal, and wasteful categories. The results of this research is a meter model with credit card for charging pulses, and data processing applications as devices that can be used for various electricity consumption. The information system is used as a model validation with pulse input that provides information about the use of electrical energy and as a determinant to save the time and time required to make optimal energy

1. Introduction

Prepaid Electricity is one form of technological development of PLN which is used to regulate the use of household electrical energy through electricity meter that has been integrated directly with the central database in PLN [1]. This innovation is expected to make people feel comfortable in electric energy payment transactions, because people no longer need to come to the counter-office of PLN area to make electricity payment but by using electric pulses people can easily fill their own amount of electrical energy they want to use. Although this innovation is very helpful to the people in the process of purchasing electricity transactions, but some people are still difficult in using pre-paid electricity meter. Of the 20 digit digit of electric vouchers, people sometimes find it difficult to include digit vouchers into the prepaid meter. In addition, if there is an error in input token number then must re-enter the token number, also from the purchase of token numbers that must be purchased to the counter sales of electric pulses or via ATM each will charge the electric pulse [1]. The model of monitoring the use and saving of electrical energy with wireless sensor network technology [2] [3] [4], in this study aims to monitor the use of electrical energy and austerity solutions. Model made in the form of the design and operation of automation of power tools by placing the spread of several sensors that act as

* Corresponding author: riki.ruli@sttpln.ac.id

sensors and monitor the use of electrical appliances operate and can be controlled remotely by implementing wireless sensor and actuator network (WSAN) technology with scheduling method node [5] [2] [6]. This technology serves to monitor the electricity consumption and electrical energy savings [7] [8] [9]. Smart meter design for monitoring and identification of electrical energy consumption in the household sector using backpropagation neural network [10]. Analog and digital kWH meter models do not have features that can monitor load currents, identify electronic equipment equipment and record actual electrical energy usage covering the time of use of electronics equipment and display costs to be paid by consumers [11]. Develop an IoT-based utility system (Internet of things) that is useful for efficiently powering electricity and gradually reducing electricity bills using the fuzzy method [12] [13]. The fuzzy method is used to control the device being used so that the user is always alert to the power consumption and reduces the current charge [4] [12]. Several studies have done by charging pulses through cellular transmission. The kWh meter model used is equipped with SMS-based communication module, the system can communicate with the users via SMS, to check and fill quota kWh meter, and provide warning (reminder) if the remaining quota on kWh meter is near the level of discharged [14].

2. System Design

The Proposed System designed to replace the old system in this design is as follows:

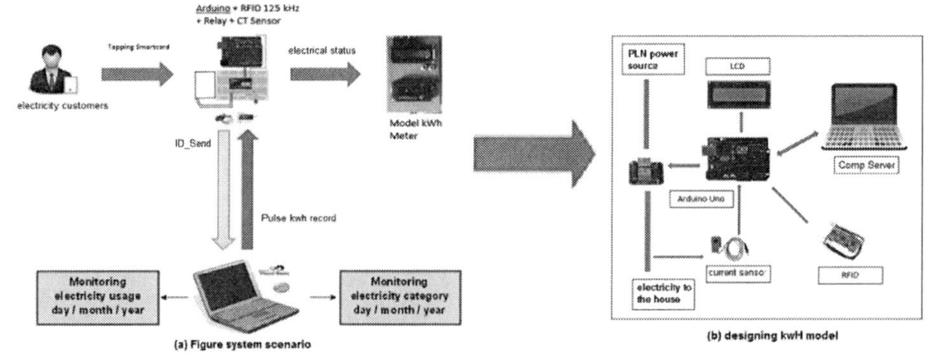

Fig 1. Design of Electric Kwh meter model

The method used for system validation as a way of categorizing step of use by testing on the 2200 VA kWh meter model, with the following test scenarios:

a. Electrical Power : 2200 VA
b. Average usage per month ± : 380 kWh
c. 380 kWh / month = 380,000 Wh / 2.592.000 sec = 0.14667
d. To simplify the calculation rounded to 0.15 Wh / sec,

Membership Function Electricity Usage per **second (X)**

$$\text{Low category} \begin{cases} 1 & , X \leq 0.15 \\ \frac{0.45-X}{0.3} & , 0.15 \leq X \leq 0.45 \\ 0 & , X \geq 0.45 \end{cases}$$

$$\text{Normal} = \begin{cases} 1 & , 0.3 \le X \text{ OR } X \ge 0.6 \\ \frac{X-0.3}{0.3} & , 0.3 \le X \le 0.3 \\ \frac{0.75-X}{0.15} & , 0.6 \le X \le 0.75 \end{cases}$$

$$\text{Large Category} = \begin{cases} 0 & , X \le 0.45 \\ \frac{X-0.45}{0.3} & , 0.3 \le X \le 0.75 \\ 1 & , X \ge 0.75 \end{cases}$$

Scenario Membership Function Total Electricity Usage (Y)

$$\text{Less} = \begin{cases} 1 & , Y \le 200 \\ \frac{380-X}{290} & , 200 \le Y \le 380 \\ 0 & , Y \ge 380 \end{cases}$$

$$\text{Normal} = \begin{cases} 1 & , 290 \le Y \text{ OR } X \ge 470 \\ \frac{X-290}{290} & , 290 \le Y \le 290 \\ \frac{560-X}{200} & , 470 \le Y \le 290 \end{cases}$$

$$\text{Wasteful} = \begin{cases} 0 & , Y \le 380 \\ \frac{X-380}{290} & , 380 \le Y \le 560 \\ 1 & , Y \ge 560 \end{cases}$$

Rule used:
R1, if X = low dan Y = Less then Z = Less
R2, if X = low dan Y = Normal then Z = Normal
R3, if X = low dan Y = Large then Z = Wasteful
R4, if X = normal dan Y = Less then Z = Normal
R5, if X = normal dan Y = normal then Z = Normal
R6, if X = normal dan Y = Large then Z = Wasteful
R7, if X = large dan Y = Less then Z = Wasteful
R8, if X = large dan Y = normal then Z = Wasteful
R9, if X = large dan Y = Large then Z = Wasteful

3. Results

The kwH model uses hardware in a microcontroller that moves as a brain or controller in activating each component of the model. The process of tapping with a smart card then automatically the information contained in the smart card will be read by the application of the number of pulses, after the pulse is read in the application then the customer can make electrical energy. Once the electricity in the instrument hardware is active then the application can automatically read how much current is used by the instrument and converted into electrical energy.

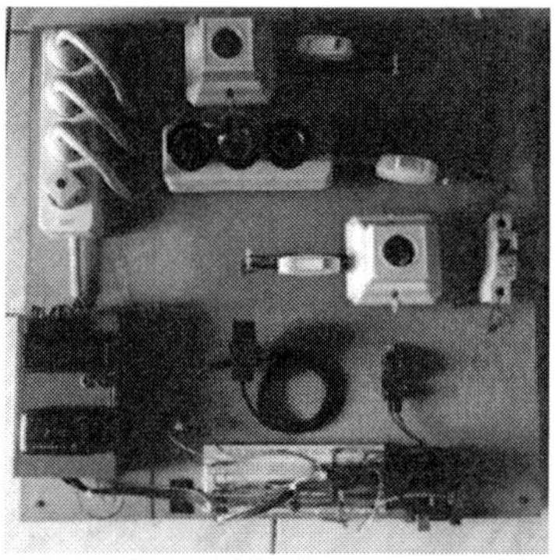

Fig 2. Prepaid kwh meter model

Users have more permissions that allow the electrical energy shown in Figure 3, viewing electricity consumption per day, month and year can also see efficient, normal or wasteful category of electricity usage every day, month and year.

Fig 3. Main Menu

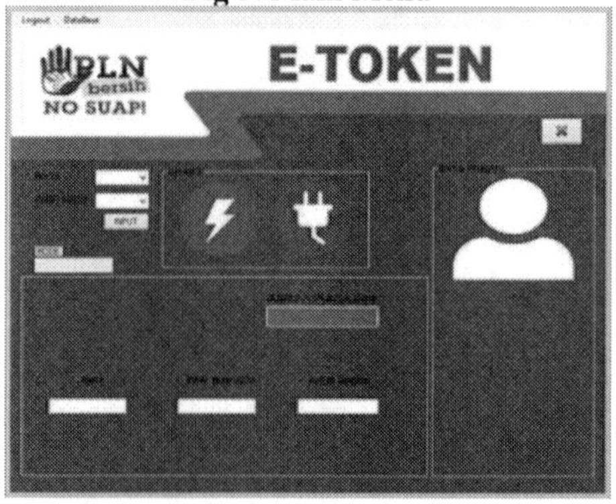

Fig 4. Controlled Application kWh Meter

Figure 4, is the menu that will set the ON / OFF power set on the hardware. In this menu can see a large amount of energy and see the amount of energy found from the application. This menu is accessible to a power user.

Fig 5. History of electricity usage

Figure 5, is a menu to see the amount of electricity consumption every day, month and year. Automatically each time shows 24.00 hours then the application will issue a command that automatically saves energy consumption into the database on the server.

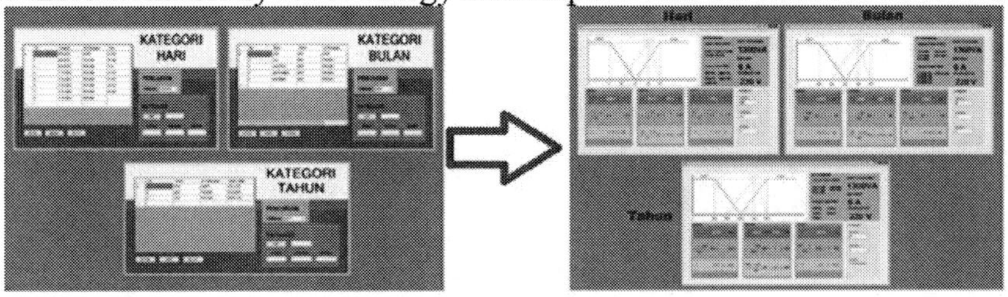

Fig 6. The category of electricity used

Picture. 6 is an application menu that shows detailed calculations used for the determination of electricity usage categories every day, month and year. It will also display graphs as a constraint to determine the efficient, normal or wasteful use of electricity.

4. Conclucion

The process of inserting electrical pulses into the power meter by using a smart card can be done by connecting from the hardware and software design so that when the user conducts the lead the smartcard pulse will automatically read into the application. So the process of determining the electricity consumption in a household can be done using a CT sensor is a sensor that will read the flow. So from the resulting current can be calculated how much Energy is used. And the determination of the category of electricity consumption can be done by utilizing the choice of categories in the application. The results obtained can show the values obtained for testing on decision support systems

Acknowledgements

We would like to thank the Ministry of Research and Technology of Higher Education who has fully supported the research activities in the financing of the Young Research scheme in 2018.

References

1. D. R. Tobergte and S. Curtis, "*Penerapan Sistem Listrik PLN Prabayar Dengan*

Penggunaan Dan Pengoprasian KWH Meter Prabayar Secara IT Dalam E-payment Sistem Pulsa Listrik," *J. Chem. Inf. Model.*, **vol. 53, no. 9**, pp. 1689–1699, 2013.
2. B. Santoso, I. W. Mustika, and S. S. Kusumawardani, "*Pemodelan Monitoring Pethenian Dan Penghematan Energi Listrik Dengan Teknologi Jaringan Sensor Nirkabel,*" *Semin. Nas. Teknol. Inf. dan Komun.*, **vol. 2014, no**. Sentika, pp. 2089–9813, 2014.
3. I. F. Muhammad, M. Abdurohman, and A. Herutomo, "*Implementasi Smart Metering Menggunakan Internet Of Things Dengan Transport Protocol Web Socket Berbasis OpenMTC di Universitas Telkom,*" *e-Proceeding Eng.*, **vol. 4, no. 1**, pp. 1075–1082, 2017.
4. M. M. Rahman, Noor-E-Jannat, M. O. Islam, and M. S. Salakin, "*Arduino and GSM based smart energy meter for advanced metering and billing system,*" ***2nd Int. Conf. Electr. Eng. Inf. Commun. Technol. iCEEiCT 2015***, no. May, pp. 21–23, 2015.
5. R. R. A. Siregar, N. Wardana, and Luqman, "*Sistem Monitoring Kinerja Panel Listrik Tenaga Surya Menggunakan Arduino Uno,*" *JETri J. Ilm. Tek. Elektro*, **vol. 14, no. 2**, pp. 81–100, 2017.
6. P. Studi and T. Informatika, "*Perancangan Alat Monitoring Arus KWH (Kilo Watt Hours) Meter Tiga Phasa Dengan Memanfaatkan Mikrokontroler Arduino dan Sms Gateway Berbasis Web,*" pp. 1–8.
7. D. Risqiwati, "*Rancang Bangun Sistem Monitoring Listrik Prabayar dengan Menggunakan Arduino Uno,*" *Kinetik*, **vol. 1, no. 2**, pp. 47–54, 2016.
8. R. R. A. Siregar, R. Raymond, and R. Arianto, "*Model Sistem Telemetri Untuk Monitoring Tandon Air Pada Daerah Kekurangan Air Bersih,*" **pp. 8–15, 2016**.
9. D. E. Paul, "*Smart Energy Meter Using Android Application And Gsm Network,*" *Int. J. Eng. Comput. Sci.*, **vol. 5, no. 16058,** pp. 16058–16063, 2016.
10. K. Hutoro, A. Soeprijanto, O. Penangsang, J. T. Elektro, and F. T. Industri, "*Desain Smart Meter Untuk Memantau Dan Identifikasi Pethenian Energi Listrik Pada Sektor Rumah Tangga Menggunakan Backpropagation Neural Network,*" pp. 1–6.
11. D. D. Prihartomo, R. D. Nyoto, and A. S. Sukamto, "*Rancang Bangun Aplikasi Pencatatan Pethenian KWH (Kilo Watt Hour) Listrik Digital Menggunakan NFC (Near Field Communication) Sebagai Komunikasi Perangkat Pada Platform Android,*" pp. 1–5.
12. L. Ciabattoni, M. Grisostomi, G. Ippoliti, D. Proietti Pagnotta, G. Foresi, and S. Longhi, "*Residential energy monitoring and management based on fuzzy logic,*" *2015 IEEE Int. Conf. Consum. Electron.*, **no. 2, pp. 536–539, 2015**.
13. J. Surya and R. Perumalraja, "*a Real-Time Iot Implementation for Efficient Energy Consumption,*" *Int. Res. J. Eng. Technol.*, **vol. 4, no. 6**, pp. 1525–1528, 2017.
14. A. Mulyana, T. A. Riza, D. N. Ramadan, M. D. Falih, U. Telkom, and T. Buahbatu, "*Sistem Pengisian Pulsa Pada Kwh Meter Prabayar Menggunakan Ponsel Pulse Filling System on Prepaid Electrical Energi,*" pp. 560–569, 2017.

Usability Evaluation Using Multi-Method for Improvement Interaction in M-Commerce

Daniel Hadrian Yohandy[1,] and **DjokoBudiyanto* Setyohadi[1,1,]
[1]Magister Teknik Informatika, Universitas Atma Jaya Yogyakarta, 55281 Jl Babarsari No. 44, Yogyakarta, Indonesia

Abstract. Today, M-Commerce has become one of the common choices as a way of shopping for society. Therefore, the transaction in m-commerce grows significantly. Moreover, the competition of M-Commerce increases and the performance should be improved primarily for the satisfaction of buyers. These conditions are seen from an improvement of the features and service options that can be used by users, even outside the process of buying and selling goods. Increasing amount of features may reduce the usability of the application which may impact on the performance of M-Commerce. Therefore, this study aims to evaluate the usability of M-Commerce service. Our study addressed to evaluate the usability of one famous M-Commerce in Indonesia. The evaluation process was performed by using cognitive walkthrough and heuristic usability method. Data collection is done by using depth interview and questionnaire to collect quantitative and qualitative data. Our findings show some new insight for designers and developers about M-Commerce especially about usability and the potential of usability to improving the satisfaction of users in terms of using the M-Commerce application overall.

1. Introduction

Usability is one of the important factors in a mobile application. Usability is used as a guide to improve the ease of use of using an application, so users can conveniently and easily understand how the application works. Usability assessment is usually done by user-evaluation process from the prototype stage to the final product before it is launched in order to guarantee ease of use for everyone [1]. Online shopping through smartphones or m-commerce is the most widely used and continues to grow today due to the characteristics of smartphones which are simple, efficient and high mobility [2].

In addition to the positive impact, smartphones also provide users to easily switch to various other m-commerce. The two aspects that affecting are customer retention and customer attraction owned by a m-commerce [3]. Therefore m-commerce does not far from the effort to increase usability [4] in order to improve customer retention through increased comfort and user satisfaction [5]. Some aspects related to user satisfaction are interaction quality, environment, and output [6].

Good usability will result in user comfort and satisfaction and improve customer retention of m-commerce. Also increased customer retention will have a positive impact on increasing user consumption and directly increase the amount of trading activity and profit

[1] Corresponding author: djokobdy@gmail.com

on the m-commerce [7]. Therefore, this research will deepen the experience and opinion from the respondent about usability of the application that support customer retention toward m-commerce T with an interview about interaction, efficiency, effectiveness, learning, and satisfaction toward m-commerce evaluated [8].

2. Literature Review

There have been several previous research related to Cognitive Walkthrough[9]–[12]and Cognitive Walkthrough was originally developed as an additional tool for evaluation processes in engineering areas, such as giving a preliminary assessment whether users will be in favor to accessing and exploring the system [13]. The goal of the cognitive walkthrough is to assist designers in getting user-side assessments and potentially predicting problems that will arise in the process of interaction with the system [10]. Cognitive Walkthrough also focused on understanding the application learnability for new users by evaluator work through given tasks and answer set of questions [11]. Cognitive walkthroughs can produce deep knowledge from the user's point of view and also help in categorizing and positioning the reasons for the usability problems encountered by respondents.

There have been several previous studies related to usability M-Commerce and various methods applied [13]. One of them is a research on m-commerce in general with a background that m-commerce is more personal than e-commerce. Problems faced are the appearance of issues in the m-commerce application design process are not only from the technical side but also from the end-user point of view which then sought the solution with the evaluation of the respondents [14]. Evaluation of m-commerce usability continues to be profound because m-commerce continues to emerge and develop with its own characteristics to compete in attracting people's attention and trust.

3. Research Methodology

This research tries to evaluate the usability of one m-commerce in Indonesia with Cognitive Walkthrough method and assisted by the adoption of the Heuristic method and UPT for data retrieval and processing. The outline of this research (refer to Figure 1) is (1) Scenario Making, 2) Data collection 3) Analysis of evaluation data and analysis of usability problem which then proceed with the conclusion and possible solution to solve usability problems [15].

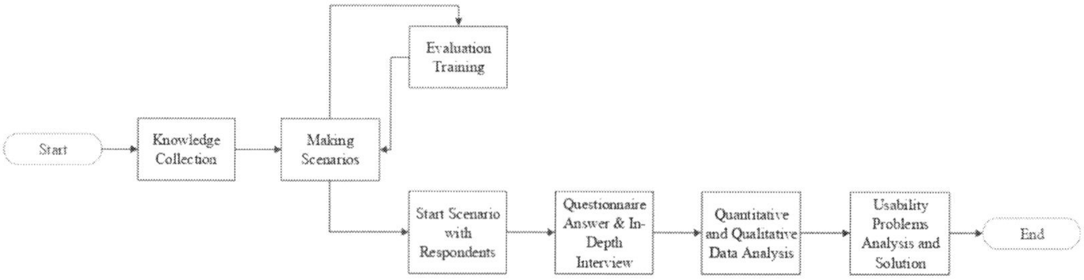

Fig 1. Flow of research

3.1 Scenarios Development

Table 1. Heuristic compotation on usability evaluation scenario (*S = Scenario)

Heuristic Usability Elements	S1	S2	S3	S4	S5
Aesthetic and minimalist design	x				x
Consistency and Standards	x		x		
Error Prevention				x	
Help users recognize, diagnose, and recover from errors				x	
Help and Documentation		x			
Match between system and the real world		x			

Recognition rather than recall			x		
User Control and Freedom			x	x	x
Visibility of System Status			x		x

The scenario development used 9 Heuristic Usability points to be included in five scenarios with the composition shown in Table 1. The heuristic points used are adopted from heuristic points and adapted to the conditions m-commerce selected and nine heuristic usability points [16], [17]. These points cover the interaction that can be done with m-commerce. Scenarios initiated from observation of application interfaces, service trials, evaluation of assistance facilities, product purchasing process, assessment of errors, charging of buyer data, and payment processing [18], [19].

3.2 Data Collections

Data collected after the respondent conducted the scenario and performed in 2 ways for 2 data types, the first was the questionnaire for the quantitative data and the second was the interactive interview for the qualitative data. The given questionnaire will ask the respondent's assessment of some elements of the M-Commerce application that attempted when doing tasks in the scenario. In each question in the questionnaire will be provided Likert Scale [20]rating that divided into 5 level consisting of Strongly Agree (SA), Agree (A), Neutral (N), Disagree (D), Strongly Disagree (SD). In the qualitative part, the interview process is done by following the flow of respondents or according to the flow scenario to facilitate the respondent. During the interactive interview process, respondents were invited to issue all opinions about m-commerce usability and accompanied by exploring the reason of the respondents to the problems of reusability encountered.

3.3 Data Analysis

Quantitative data from the questionnaire will be processed and calculated mean and standard deviation on 5 scenarios from 10 respondents. For qualitative data will be done several steps to analyze and decode qualitative data into structured data. The following steps will be taken to approach the qualitative data.

3.3.1. Framework Development

The initial phase of framework identification is to understand the data collected. Qualitative data obtained from the in-depth interview will be processed with user action data from the video as well as data from the user questionnaire to be aligned and will be processed using the specified framework. This research, leaning toward analysis of qualitative data, therefore, the result will be framed with a table that contains the description of a usability problem, place of occurrence of the problem, and percentage of problem occurrence. Description of the problem of usability is a decoding of raw qualitative data obtained from in-depth interviews with respondents.

3.3.2. Classification of Usability Problems

After determining the framework, the next step is to classify the usability problem using the adoption of a predetermined framework which is also followed by giving severity to each reusability problem. To determine the severity, this study will use the categorization of the [21]Travis skeleton. The framework (referred to in Figure 3) uses three main questions, and then the assessment will be given between 1) Low, 2) Medium, 3) Serious and 4) Critical.

3.3.3. Charting Data

This step is done by making charts based on the usability problem. The chart will be made based on 3 aspects, namely the classification of the problem of reusability, the occurrence of the problem, and the severity (severity)[22].

4. Result

All selected are undergraduate and postgraduate students and know about online shopping on the smartphone. Age range of respondents are in 18-29 years and the majority over 20 years. Of the 10 respondents are included 1 who have never done online shopping through m-commerce and 9 of them already doing online shopping through m-commerce. Among those who have been shopping through m-commerce, 5 of them on average purchase 1 item per month, and 4 others on average purchase 2 to 3 items per month. For the type of goods purchased the majority are clothing and electronics. Respondents mostly choose a good transaction process as an important characteristic in a m-commerce.

4.1 Mean and Standard Deviation Questionnaire Data

The obtained questionnaire data will be used by calculating the mean and standard deviation (refer to Table 2) of the total of all 10 respondents indicating the UP (Usability Problem). The questions order will be sorted by scenario and assessment of scenario projected by 6 questions for each S (scenario).

***Table** 2. Mean and Standard Deviation*

Scenario	UP
S 1	13
S 2	10
S 3	9
S 4	23
S 5	6
Mean	12.2
Std. Dev.	6.534524

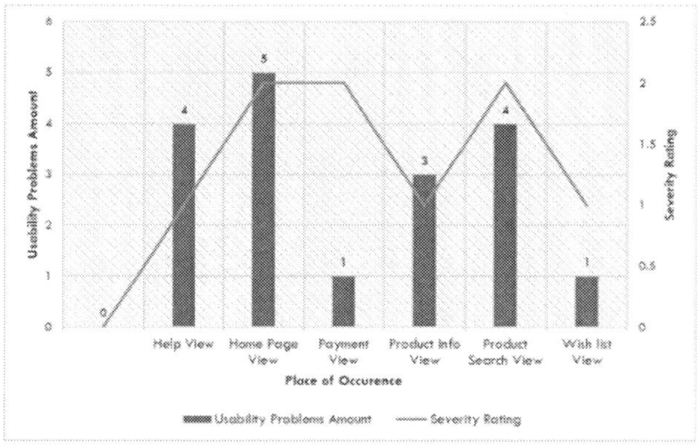

Fig 2. Chart of Usability Problem & Severity

Chart of usability problem (refer to Figure 2) is describing the location of the incident, the number of usability problems and the severity. This is expected to give a clear picture of the comparison of reusability problems at each location/page in m-commerce applications.

4.2 Discussion

There were 90 reusability problems from 10 respondents obtained from the whole method, of which 37 were obtained from the questionnaire and 53 were obtained from the interview. The overall problem of usability is composed into 13 unique usability problems with the percentage requirement in interview method above 30% that means the problem often encountered by the respondent. Table 3 is list of usability problems based on the percentage of occurrence of the highest problem (preferred method interview), along with the data on the occurrence of the usability problem and severity. Usability issues with the highest percentage are in the help view and product search results view. The help view is not included in the red route in m-commerce applications and only has an average of severity one so that despite having the most complaints from respondents this problem is not critical or critical but needs to be fixed. For product search result page which is the main feature (red route) and the average severity is 2 or medium; therefore, a solution is needed to overcome the problem of found usability.

Problem with the highest percentage and severity level 2 is about the product suggestion as an alternative in case of problematic search. In interviews, 60% of respondents felt that they did not get a suitable or desirable product suggestion. Suggested products have different brands and even sometimes different types of products with similarity on the category. In addition to that, the suggested products have little amount therefore indirectly make the user must search again using different keywords to find the desired outcome. This usability problem refers to heuristic points of "*Help users recognize, diagnose, and recover from errors*".

The issue of usability problems with moderate severity (or 2) with a 40% percentage occurring on the home page view with a problem of difficulty reaching or finding the desired application function or service on the main page. The home page location is the reason causes a moderate severity because the home page is the main interface in the mobile app to greet users and should be comfortable and not confusing. Then another reusability is that it is difficult to understand the documentation of goods return assistance. From some help to m-commerce main functions, the item return section has a 50% or high enough reusability problem level even if the severity is only 1 or low.

Table 3. List of Usability Problems found (I=Interview, Q=Questionnaire, S=Severity)

Usability Problems	Place of Occurrence	Pct. (%)		S
		I	Q	
Difficulty to get suggestions in problematic search from systems that match the search being performed	Product Search View	60	40	2
Difficulty to understand the help regarding the return of the goods provided by the application.	Help View	50	30	1
Difficulty to understand the contents of the FAQ	Help View	40	20	1
Difficulty in knowing the wishlist button to add the product to the wishlist	Product Info View	40	20	1
Difficulty reaching or finding the desired application function or service	Home Page View	40	20	2
Difficulty finding documentation and help pages	Help View	30	10	1
Difficulty finding help for goods return problems	Help View	30	-	1
Difficulty in finding the desired product in search view	Product Search View	30	10	2
Difficulty understanding the error message on the problematic search provided by the application	Product Search View	30	20	2
Difficulty in knowing the status of the wishlist button on the selected product	Help View	30	20	1
Difficulty to choose the preferred payment method.	Payment View	30	-	2
Difficulty to manage the wishlist status of the product	Wishlist View	30	20	1
Difficulty in finding the desired services such as the contents of the credit or paying the bills	Home Page View	30	40	2

5. Discussion& Conclusion

Overall, respondents prefer m-commerce applications that provide comfort to users when searching, purchasing goods, and satisfaction after payment because online shopping process ends when the goods are received. Satisfaction resulted by interaction with application can assured the buyer that the order will arrive at the destination without any problems. According to qualitative and quantitative data as support, the problem of reusability problems with the highest percentage occurring on the search results page is the problem of product proposals as a problematic search solution of a system that has a 60% percentage (moderately high) with moderate severity (or 2) since some respondents may overcome searches with other keywords. Another problem with reusable problems with moderate (or 2) severity with a large percentage is the difficulty of reaching or searching for the desired function on the main page.

Based on the results of data analysis, there are three solutions as ideas or suggestions to improve customer retention and customer satisfaction for both new users or old users to remain loyal. The solution includes 1) Create a simple tutorial for the main task process from the main menu to payment. 2) Improve search suggestions or search solutions by adding suggestions generated from each keyword entered so that users can choose. 3) And the addition of structured help documentation with points as a pointer so that users avoid reading all the documentation to find the right solution. For further research, evaluation is expected to be differentiated based on respondent's experience on m-commerce (novice or experienced) so that the usability problem can be categorized to know the priority problem to find the solution in order to improve the usability, convenience, and user satisfaction in online shopping.

References

1. N. Ahmad, M. W. Boota, A. H. Massom, ***J. Softw. Eng. Appl.* 7,** 1045 (2014) https://doi.org/10.4236/jsea.2014.712092
2. R. Agarwal, V. Venkatesh, ***Inf. Syst. Res*** **13**, 168-186 (2002) https://doi.org/10.1287/isre.13.2.168.84
3. C. Xu, D. Peak, V. Prybutok, ***Decis. Support Syst.*** **79** , 171 (2015) http://dx.doi.org/10.1016/j.dss.2015.08.008
4. Q. Min, S. Li, Q. Zhong, An Empirical Study of M-Commerce Adoption from Usability Perspective (ICMB, 2009) https://doi.org/10.1109/ICMB.2009.44
5. A. B. Hussain, A. T. Mahmood, R. K. Naser, Investigating the effect of m-commerce design usability on customers' trust (AIP,2017) https://doi.org/10.1063/1.5005410
6. L. Zhao, Y. Lu, L. Zhang, and P. Y. K. Chau, ***Decis. Support Syst.*** **52**, 645 (2012) http://dx.doi.org/10.1016/j.dss.2011.10.022
7. J. Peng, J. Quan, and S. Zhang, ***Electron. Commer. Res. Appl.*** **12**, 321 (2013) http://dx.doi.org/10.1016/j.elerap.2013.05.002
8. D. B. Setyohadi, S. Kusrohmaniah, E. Christian, L. T. Dewi, B. P. Sukci, M-Learning Interface Design Based On Emotional Aspect Analysis (IHCI, 2016) https://doi.org/10.1007/978-3-319-52503-7
9. D. Jadhav, G. Bhutkar, V. Mehta, Usability Evaluation of Messenger Applications for Android Phones using Cognitive Walkthrough (APCHI, 2013) https://doi.org/10.1145/2525194.2525202
10. J. Rieman, M. Franzke, D. Redmiles, Usability Evaluation with the Cognitive Walkthrough (CHI, 1995) https://doi.org/10.1145/223355.223735
11. A. Solano, C. A. Collazos, C. Rusu, H. M. Fardoun, ***Adv. In H.C.I,*** 2016 (2016) http://dx.doi.org/10.1155/2016/4089520S.
12. S. S. Chan, X. Fang, Y. Zhou, and S. Xu, ***J. Electron. Commer. Res.*** **3**, 187 (2002) http://citeseerx.ist.psu.edu/viewdoc/summary?doi=10.1.1.85.5067
13. A. Mourouzis, M. Antona, and C. Stephandis, ***Univers. Access Inf. Soc.*** **10**, 337 (2011) https://link.springer.com/article/10.1007/s10209-010-0211-y
14. A. Hussain, E. O. C. Mkpojiogu, N. H. Jamaludin, S. T. L. Moh, A usability evaluation of Lazada mobile application (AIP, 2017) https://doi.org/10.1063/1.5005392
15. S. L. Keenan, H. R. Hartson, D. G. Kafura, R. S. Schulman, ***Emp. Soft. Eng*** 4, 71-104 (1999) https://doi.org/10.1023/A:1009855231530
16. Duggirala, S. (2016, Agustus 17). https://blog.prototypr.io/10-usability-heuristics-with-examples-4a81ada920c

17. J. Nielsen, R. Molich, Heuristic Evaluation of User Interfaces (CHI, 1990) https://doi.org/10.1145/97243.97281
18. S. Chan, X. Fang, Usability Issues in Mobile Commerce (AMCIS, 2001) http://aisel.aisnet.org/amcis2001/88
19. D. Pinelle, N. Wong, T. Stach, Heuristic Evaluation for Games: Usability Principles for Video Game Design (CHI, 2008) https://doi.org/10.1145/1357054.1357282
20. Likert Scales and Data Analyses. (2007, 7). Retrieved 5 7, 2018, from Quality Progress: http://rube.asq.org/quality-progress/2007/07/statistics/likert-scales-and-data-analyses.html
21. Travis, D. (2009, 10 5). How to prioritise usability problems. Retrieved 5 7, 2018, from USERFOCUS: https://www.userfocus.co.uk/articles/prioritise.html
22. M. Georgsson, N. Staggers, ***J. Biomed. Inform.*** **59**, 115 (2016) https://doi.org/10.1016/j.jbi.2015.11.008

Designing Automatic Guided Device of Books

Yulius Advent Putra[1,1,], *Porman* Pangaribuan[1], and *Rizki* Ardianto Priramadhi[1]
[1]Assistant Lecturer, School of Electrical Engineering (Telkom University, Indonesia)

Abstract. There are several methods returning scattered books in library, such as asking visitors to put them back to the bookshelf, recruiting more employees, or creating system that can return the books automatically. But, by using the system, visitors no need to walk to certain bookshelf. The system itself called Automatic Guided Device included of Automatic Guided Vehicle (AGV), RFID tag, scissor-lift, and book thruster. Every books have RFID tags which contains the book database. There are name of books and shelf location inside of the database. Visitors just need scan the book(s) and laid on the device, then the system will begin to operate. After that, this Automatic Guided Device sensor equipped will start its journey and go through to the location of bookshelf by following tracked line on the floor. To show them, it uses ASCII Table in the C Source Code Program. This table is to represent information substitutes which contains certain number, then it will be declared as the bookshelf location spot. After reaching certain distance, scissor-lift lifts as high as the table. Then scissor-lift stops, while the book thruster removing the book from device into bookshelf. So, this system not only returning the books without any manual operation but also can do more things such as lifting and putting the books automatically at once. Based on several tests, AGD reaches at certain spot within an average 60% of successfulness by using ASCII Table numbering format.

1. Introduction

Most visitors are not only read a kind of book, but also several books, and usually left the books when they have finished reading, so it will make scattered books on the table. There are several methods to return all of the scattered books, first, asking visitors to put them back to the bookshelf, then second, recruiting new employees, and the third one is creating system that can return the books automatically. The first method shows that the notice from librarians will not give any attention to the visitors, and sometimes it is ignored by them. The second one is not suitable because the library is already has enough staffs. Third, creating a system that can return the books automatically which is called Automatic Guided Device (AGD). Using this device, this system not only returning the books without any manual operation but also can do more things such as lifting and putting the books at once. So, by comparing all of methods, we consider that the third one should be more effective. Also, this method applying automatic guided of books without any manual operation, so it will keep the book tidiness. It confusing another visitor if they didn't find any book they wanted. In other words, the book circulation is restricted. So, we hope that this system will

[1] Corresponding author: yuliusadventp@gmail.com

make visitors are easier looking for their books. Besides, the librarians don't need to worry of all visitors forgetting to return their books because it has been done by automatic guided device.

2. Direct Observation

Direct observation is a directly contact method to get any valid information based on true range, in case, inside Open Library Telkom University. The data is obtained from librarian's database center which contain the number of visitors a day and book status whether it's returned to shelf or read only by visitor.

2.1 Existing Method and Their Limitation

The amount of visitors and borrowed books each day are relative. But there's a big difference between the sum of books list. Based on obtained data, returned-books is always more than read-only books every day. There were 852 borrowed books per day and 11 staffs, which means that a staff must returned it approximately 85 books per day. Now, there are 79,506 books, 1031 borrowed books per day, and 11 staffs, so there are 102 of books for a staff must to do the same thing.
Below is **their limitation of existing system:**

- Book circulation (returned or borrowed books) is restricted.
- Most visitors can't return the book to proper shelf
- Shelving is only done when the library is closed or before it opens

2.2 Proposed System

This **Designing Automatic Guided Device of Books** help both librarian staff and visitors in putting back the books. **Advantages of using this device:**

- Keep the bookshelves tidiness
- Decreasing the restricted book circulation
- Visitors just scan the books then the device automatically operates, so visitors no need to walk to proper shelf.

3. Design Of System

3.1 Block Diagram

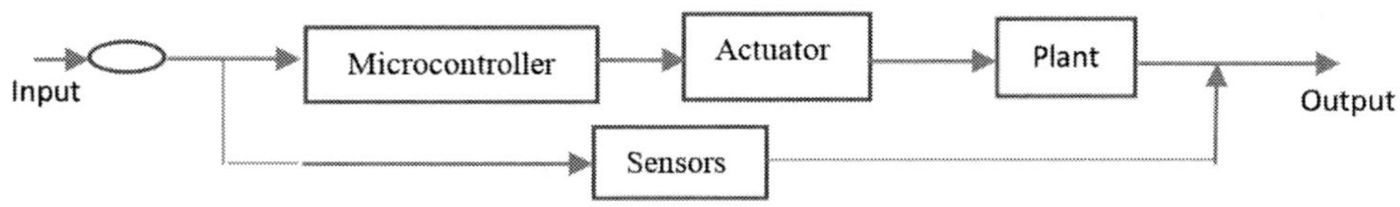

Fig 1. AGD Control System Block Diagram

- Input : Scanned books
- Microcontroller : Raspberry Pi 3 and Arduino Nano
- Actuators : DC Motors
- Plants : AGV, scissor-lift, book thruster
- Sensor : IR sensor
- Output : Moving AGD through the tracked lines

The book is scanned, then it will be processed by Raspberry Pi 3 microcontroller for database reading. Microcontroller which have found the database will transmit its data to Arduino Nano by serial communication to operate DC motor and activate the sensor. The DC motors drive the AGV to proper shelf while the sensor reading through the tracked line that indicate where should the device moved. After it reaches the proper shelf, AGV stops and then scissor-lift motor start to operate, changing rotation-to-vertically linear motion until it reaches certain height of shelf. After that, scissor-lift stops and then book thruster begins to push the lowest book until the book itself are removed from device to proper shelf. Next, the book thruster and scissor-lift are consecutively going back to their initial position to continue the journey.

3.2 Architectural Design

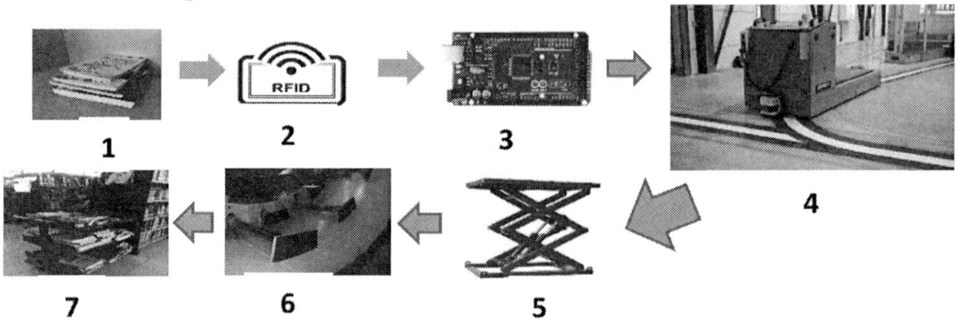

Fig.2. Architectural Diagram of AGD System

3.2.1 Scattered Books

This is the main problem from some visitors who don't return their books after they read them. The books are laid on table only, and need to be returned.

3.2.2 RFID (Radio Frequency Identification)

There are 2 main components in RFID system, Tag and Reader. There are 2 main components in RFID Tag, IC (Integrated Circuit) as a transponder, and Antenna as a receiver also sending RF signal [4]. Each book is inserted by a tag and need to be scanned. The system, scanned books will be identified by the reader. For books RFID tag is used as identifier which reduces manual interference for returning and issuing books [2]. But in some libraries barcode is used for book identification [3].

3.2.3 Arduino Nano

Arduino Nano is one of microcontroller board which is based on the ATmega 328. It is a microcontroller platform that can easily connected into many computers, so users can program any ATmega chip [1]. The system needs more inputs, and this microcontroller can fulfill them. Also, the microcontroller contains the database which shows the bookshelf location and type of book.

3.2.4 Automatic Guided Vehicle (AGV)

Some examples of AGV device products are being optimized, for example intelligent navigation research in warehousing [5], and otherwise, there is also an AGV design that is made to have spring absorption with a double spring solution [6]. But the writer's project is about optimizing AGV into returning books automatically by tracked line.

3.2.5 *Scissor-Lift*

[3] Scissor lift mechanism is one kind of lifter using hydraulic system which can help lifting up vertical easily. To lift up, there's a motor which are connected to metal sheets, changing rotation into vertically linear motion. Using this method as lifting book is because can be set its height.

3.2.6 *Book Thruster*

To remove books from device into bookshelf is using thruster which made of long basic bolt connected to a motor.

3.2.7 *Proper Shelf*

The device moving through its point which is proper shelf. This is where the book should be put, based on the database.

4. Result

The data below present about AGD's depart and arrived time, also the result about how many books, their weight, and how long the scissor lift to do up-lifting and down-lifting experiment. The scissor-lift load is negligible.

Table 1. Test Result of AGD's Depart Time and Arrived Time

Test	Distance (cm)	Depart Time (s)	Arrived Time (s)	Additional Information
1	200	6.19	6.73	SUCCESS
2	200	6.24	-	NOT SUCCESS
3	200	6.16	-	NOT SUCCESS
4	200	6.07	6.63	SUCCESS
5	200	6.11	6.68	SUCCESS
6	200	6.04	-	NOT SUCCESS
7	200	6.13	6.75	SUCCESS
8	200	6.06	6.69	SUCCESS
9	200	6.14	-	NOT SUCCESS
10	200	6.18	6.78	SUCCESS
Average		**6.132**	**6.71222**	

Based on 10 tests that have been carried out, not all have been successful, as in tests 2, 3, 6, and 9. It is because the failure of sensor readings. In addition, a great friction factor between the freewheel and the floor due to the influence of the load, resulting in AGD sometimes are not successfully moving backwards to follow tracked line. This is the reason why arrived time longer than when the device moving forward. The human error factor, like error reading, and the movement of the device system also involves the time value are various. Six of the ten tests were successfully carried out, so in other words, the success percentage is 60%.

Table 2. Relationship between Number of Books, Weight of Book/s (kg), and Time Needed for Up and Down Lifting

No	Number of Books	Weight (kg)	Time (s)		Additional Information
			Up-lift	Down-lift	
1	-	-	29.9	28.75	Scissor-lift operates smoothly.
2	1	0.5	32.93	28.66	When the motor operate at t = 5s it always indicates that the motor is hardly lifting the scissor lift in rotary-to-linear motion because of the book's weight and scissor-lift load itself.
3	2	0.825	33.17	29.84	
4	3	1.275	34.22	28.74	
5	4	3.175	46.84	32.21	
6	5	4.35	53.41	29.73	
7	6	5.25	-	-	Scissor-lift won't lift up because of excessive load.

According to **Table 4.1,** the bigger the load, the longer it takes the time lifting the scissor-lift. We take the book randomly, because not all the books have the same weight. They're relative. Scissor-lift maximum load capacity is under 5 kilograms, so it's not lifted if we put more than its maximum. Meanwhile, time needed for down-lifting is nearly close. They are different because watching-time human error. In addition, there're several other factors, such as scissor-lift high friction, bolt's thread, lifting movement. The friction makes hard moving of scissor-lift, the higher its friction, the longer time it will takes. High friction will make the device quickly damaged and it slowly leaves the rust. Bolt's thread also determines the vertical distance, the longer thread's distant, the higher scissor-lift lifting vertically in a spin. Besides, lifting movement gives an effect to scissor-lift. The device which is not steady while lifting takes more time to reach certain position.

5. Conclusion

1. Many cases of scattered books in the library need to be solved in several methods, and we prefer to AGD considering to their limitation.
2. To find out the position of the bookshelf, use the ASCII table to help align or AGD stepping.
3. The average percentage of success for AGD to run on the line on the floor is 60%, this is due to the influence of several external factors.
4. On scissor lifts, the heavier the load of the book, the more uplifting it takes, but there's no affection in down-lifting.

References

1. Wolfram Donat "Learn Raspberry Pi Programming with Python". 2014 ed.: Apress, c2014

2. Tao Sun, Shuming Tang, Jinqiao Wang, and Weibin Zhang, "A Robust Lane Detection Method for Autonomous Car-Like Robot", 2013 Fourth International Conference on Intelligent Control and Information Processing (ICICIP,) Beijing, China June 9 – 11, 2013.
3. J.Y. Tigli and M.C. Thomas, "Use of Multi Agent Systems for Mobile Robotics Control", IEEE 1994.
4. Susy d'Hont, "The Cutting Edge of RFID Technology and Applications for Manufacturing and Distribution", Texas Instrument TIRIS, 2002.
5. *https://ieeexplore.ieee.org/document/8242766/*
6. *https://ieeexplore.ieee.org/document/8078899/*

Design of Android Application for Telemedicine System to Improve Public Health Services

Sumiati[1, 1], *Haris Triono Sigit*[1],
[1] Informatic Department, Universitas Serang Raya, Banten, Indonesia

Abstract. One of the services provided by health facilities in Indonesia is the heart condition's examination using Electrocardiogram machine. This service is important to provide because heart disease is one of the 10 most common causes of death in Indonesia. The electrocardiogram machine produces a print out that contains a graph of patient's heart condition status. A cardiologist reads an ECG Scan to diagnose whether the patient's heart condition is normal or abnormal. Health facilities that has a cardiologist is a secondary health facility located in the city but in the primary health facility in the village, although the examination by electrocardiogram machine can be done, the patient still have to wait for the diagnosis result because the print out result is sent first to the secondary health facility which has a cardiologist. Therefore it is necessary to design an Android Application for supporting telemedicine system as an effort to accelerate diagnostic service so that patient in primary health facility can also get immediate medical treatment.

1 Introduction

The rapid development of information technology has influenced various aspects of life. One of them in terms of health services for the community. The existing health facilities seek to improve the quality of services by utilizing information technology to process various types of medical information to assist medical procedures. The existence of health facilities such as government hospitals and private hospitals and community health centers is needed by the community to overcome various health problems.

Health problems that need to be addressed at this time is the disease that can cause the highest mortality rate. One of them is coronary heart disease where this disease in Indonesia mostly occur in age group which is 65-74 years old (3,6%), 75 years up (3.2%), 55-64 years (2,1 %) and 35-44 years (1.3%). Coronary heart disease is the leading cause of death in Indonesia [1].

People who lives in the city can check the condition of heart health in health facilities that belong to the category of secondary healthcare facilities ie hospitals and villagers can check the condition of heart health in health facilities that belong to the category of primary health care facilities namely Puskesmas. In primary health facilities, patients who have been examined for cardiovascular health with ECG scan machines should wait for the diagnosis of ECG scans by a cardiologist present in secondary health facilities. Efforts to improve health services in Puskesmas can be done with a technological innovation that is able to accelerate the delivery of medical information from Puskesmas to a cardiologist in

[1] Corresponding author: sumiati82@yahoo.com

the hospital.

Innovation that needs to be done is to build applications that follow the concept of telemedicine systems. With this application, the quality of health services become better because people who lives in villages will still get optimal medical care from cardiologists. Applications built in addition to running on computer devices also need to be attempted to run on mobile devices because today mostly people in Indonesia use smartphone devices in their daily activities.

2 Telemedicine System

According to Fong [2], the term telemedicine comes from a simple description of the word 'tele' which means distant or distant from the Greek language, so telemedicine means medical services provided remotely. The term telemedicine refers to the utilization of telecommunication technology for medical diagnosis, treatment, and patient care[3].

Telemedicine and e-health are two areas of biomedical engineering, which are also multidisciplinary, but telemedicine is actually a subset of e-health. Telemedicine can be interpreted as the use of information and communication technology (including electronics, tele-communications, computers, informatics) to transfer (send and / or receive) medical information, to improve clinical services (diagnosis and therapy) and education [4].

Rapid technological developments are a factor supporting the development of telemedicine systems in Indonesia. Indonesia which is an archipelagic country can utilize this telemedicine system as an effort to improve the quality of public health service. The Benefits of Telemedicine include [5]:

a. Improved access to patients
b. Reduce patient costs
c. Reduce remoteness of physician needs
d. Improve the quality of health services.

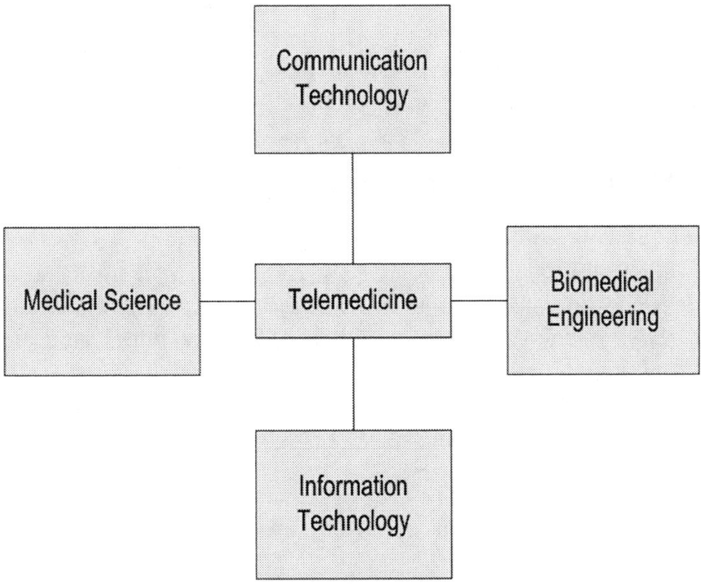

Fig 1. Telemedicine

Along with the pace of development of hardware and software technology as a device supporting telemedicine system has encouraged the development of telemedicine system significantly. One application is the application of telemedicine systems for monitoring and diagnosing patients using wireless technology. The main components of the system as illustrated below:

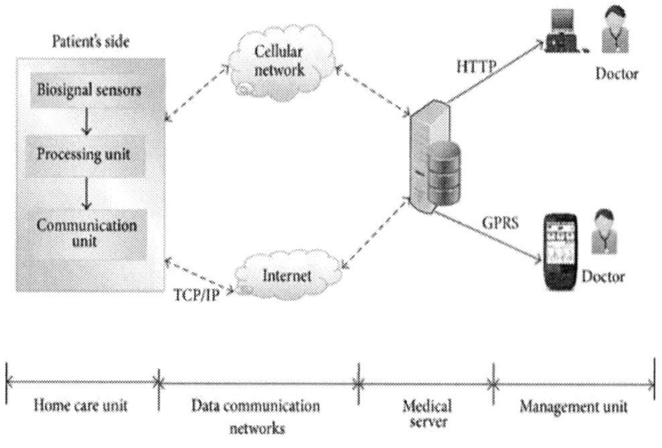

Fig 2. Main Components of Wireless Telemedicine System [6]

Based on the picture above, we can explain the flow of telemedicine system as follows:

a. The patient checks the health condition of the home using a Biosignal Sensor device
b. Further data checks the device is transmitted through a data communication network that can be a cellular network or internet network
c. Through the existing data communication network, data is transmitted to the server for processing and storage
d. Data stored in the server can be accessed by the doctor through the PC in the hospital or through the Smartphone.

In the development of telemedicine this should be noted about the issues that must be considered include: 1) The real effectiveness of the system, 2) Acceptability system, 3) System sustainability, and 4) Further development [7].

3 Android Operating System

The development of Smartphone technology is currently undergoing rapid development. In running the daily activities, humans can not be separated from the use of smartphone devices. This device is operated with an operating system for mobile devices such as iOS, Android, Blackberry, Windows Phone and others. The operating system that is commonly used is the Android operating system.

Android is a Mobile Operating System that grows in the middle of other OS that grows today [8]. The operating system is Linux-based and is for mobile devices. Android provides an open platform for developers to create their own applications for use by a variety of mobile devices. Android is commonly used in smartphones and tablet PCs and it has the same functionality as the Symbian operating system on Nokia, iOS on Apple and BlackBerry OS [9].

Android operating system has attracted the attention of the world community where companies gadgets and mobile phones are competing to create a device that runs under this operating system. The rapid development of the Android operating system seen from the versions that have been launched are as follows :

Table 1. Version of Android Operating System [10]

Name	Version	Launched
Cupcake	1.5	April 27, 2009
Donut	1.6	September 15, 2009
Eclair	2.0 – 2.1	October 26, 2009
Froyo	2.2 – 2.2.3	May 20, 2010
Gingerbread	2.3 – 2.3.7	December 6, 2010
Honeycomb	3.0–3.2.6	Februari 22, 2011
Ice Cream Sandwich	4.0 – 4.0.4	October 18, 2011
Jelly Bean	4.1 – 4.3.1	July 9, 2012
KitKat	4.4 – 4.4.4	October 31, 2013
Lollipop	5.0 – 5.1.1	November 12, 2014
Marshmallow	6.0 – 6.0.1	October 5, 2015
Nougat	7.0	August / September 2016
Oreo	8.0	August 2017

The following diagram shows the major components of the Android platform [11] :

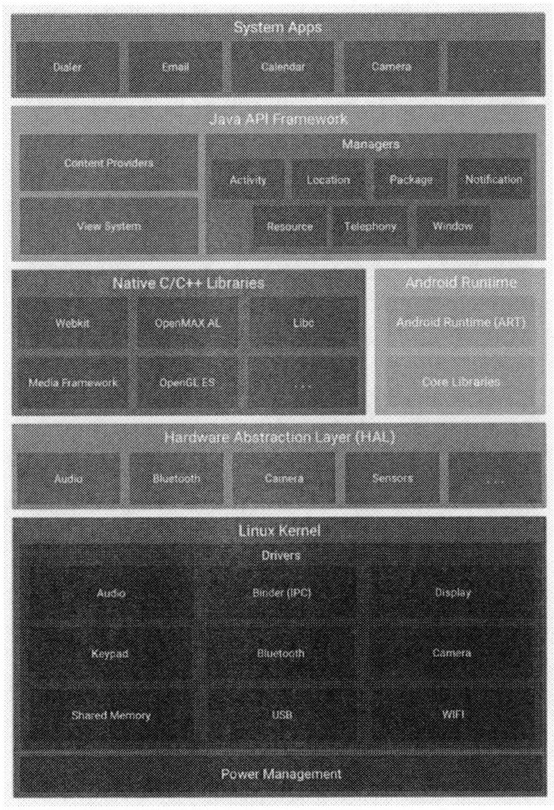

Fig 3. The Android software stack

The major components of the Android platform consists of:

a. Linux Kernel : The foundation of the Android platform is the Linux kernel. Using a Linux kernel allows Android to take advantage of key security features and allows device manufacturers to develop hardware drivers for a well-known kernel.
b. Hardware Abstraction Layer (HAL) : provides standard interfaces that expose device

hardware capabilities to the higher-level Java API framework. The HAL consists of multiple library modules, each of which implements an interface for a specific type of hardware component, such as the camera or bluetooth module. When a framework API makes a call to access device hardware, the Android system loads the library module for that hardware component.

c. Android Runtime (ART) : is written to run multiple virtual machines on low-memory devices by executing DEX files, a bytecode format designed specially for Android that's optimized for minimal memory footprint. Build toolchains, such as Jack, compile Java sources into DEX bytecode, which can run on the Android platform.
d. Native C/C++ Libraries : Many core Android system components and services, such as ART and HAL, are built from native code that require native libraries written in C and C++. The Android platform provides Java framework APIs to expose the functionality of some of these native libraries to apps.
e. Java API Framework : The entire feature-set of the Android OS is available to you through APIs written in the Java language. These APIs form the building blocks you need to create Android apps by simplifying the reuse of core, modular system components and services.

System Apps : Android comes with a set of core apps for email, SMS messaging, calendars, internet browsing, contacts, and more. Apps included with the platform have no special status among the apps the user chooses to install. So a third-party app can become the user's default web browser, SMS messenger, or even the default keyboard (some exceptions apply, such as the system's Settings app).

4 The Design of Mobile Application for Telemedicine System

Telemedicine systems can be developed in accordance with the support of emerging technologies today. The use of Android-based smartphones in almost all levels of society provides a great opportunity for the development of telemedicine systems to be applied to mobile devices.sIn this research, telemedicine system mobile application design to be built is as shown in table below:

Table 2. Use Case Definition of Telemedicine System

No	*Use case*	Description
1.	Register	Patient requests registration on the system
2.	Checking Heart Condition	Pasient do Checking the heart condition
3.	Consult the doctor	Patient conducts cardiovascular consultation with a specialist
4.	Monitoring Patient's Condition	The doctor monitors the patient's condition and provides medical services when needed
5.	Read Patient's Medical Record	Doctors read the patient's medical history to analyze the

		patient's heart health condition
6.	Login	Doctor and Patient enter the user name and password

Based on the above table, the design of mobile application for telemedicine system can be described as follows:

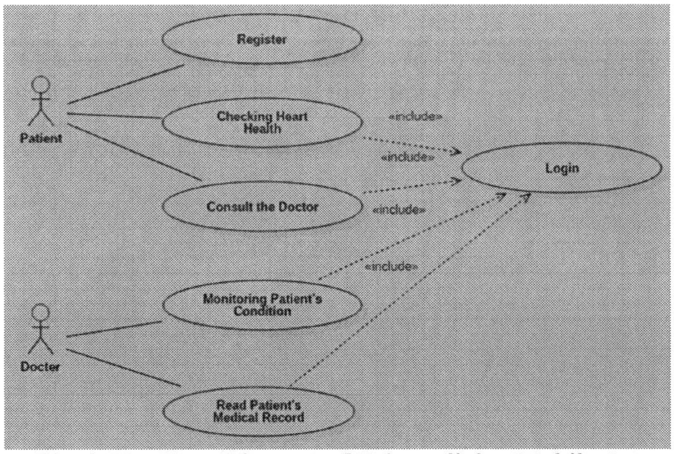

Fig 4. Use Case Diagram of Telemedicine Mobile Apps

5 Conclusion

Telemedicine systems can be further developed with the support of both hardware and software technology. The existence of a telemedicine system can solve health care problems in patients away from health facilities. Patients do not need to come to the hospital regularly to check their physical condition. With this telemedicine system, patients suffering from heart problems can check their heart health condition directly. Thus the physical condition and medical history of the patient can be monitored at any time by a team of cardiologist so that the team can immediately provide medical services.

With the support of mobile operating system technology allows telemedicine system can be applied to Smartphone device. The application of mobile-based telemedicine systems enables health services to be performed by doctors even when they are not in the hospital. With the optimal telemedicine system, medical information processing becomes faster so that efforts to improve the quality of health services to the community can be achieved.

References

1. Indonesian Ministry of Health, *Heart Disease The Highest Cause of Death*. [Online] Available in : http://www.depkes.go.id/article/view/17073100005/penyakit-jantung-penyebab-kematian-tertinggi-kemenkes-ingatkan-cerdik-.html [Accessed date : May[13] 2018]
2. Fong, B., Fong, A.C.M, Li, C.K. 2011.*Telemedicine Technologies : Information Technologies in Medicine and Telehealth (1st edition)*. United Kingdom: John Willey & Sons.
3. J. C. Lin, *"Applying telecommunication technology to health-care delivery,"* IEEE Eng. Med. Biol. Mag., vol. 18, pp. 28–31, 1999

4. Soegijoko, S. 2010. Perkembangan Terkini Telemedika dan E-Health serta Prospek Aplikasinya di Indonesia. National Seminar on Information Technology Applications 2010 (SNATI2010), ISSN: 1907-5022
5. Satyamurthy LS, ISRO,s Experience in Telemedicine with Special Reference to Telemedicine System, 2007
6. M. Abo-Zahhad, SabahM. Ahmed, and O. Elnahas, A Wireless Emergency Telemedicine System for Patients Monitoring and Diagnosis. International Journal of Telemedicine and Applications, 2014
7. Kuat Indartono, Sistem Telemedika Berbasis ICT untuk Manajemen Fasilitas Unit Gawat Darurat. Journal of Electrical Engineering Vol. 5 No. 1, 2013
8. Hermawan S, Stephanus. 2011."Mudah Membuat Aplikasi Android".Yogyakarta : Andi Offset
9. H, Nazruddin Safaat. 2012. Pemrograman Aplikasi Mobile Smartphone dan Tablet PC Berbasis Android Edisi Revisi. Bandung: Informatika
10. https://haiwiki.info/aplikasi/tingkatan-versi-android [Accessed date : May[14] 2018]
11. https://developer.android.com/guide/platform [Accessed date : May[14] 2018].

Design Of Array and Circular Polarization Microstrip Antenna For LTE Communication

Syah Alam[1,*],, *Eddy Wijanto* [2,*] , Budi Harsono[2] , Fidelia Samandatu [2], Markus Upa [2], Indra Surjati [3]

[1]Departemen of Electrical Engineering, Universitas 17 Agustus 1945 Jakarta, DKI Jakarta, Indonesia
[2]Departnent of Electrical Engineering, Universitas Kristen Krida Wacana, DKI Jakarta, Indonesia
[3]Graduate Programe of Electrical Engineering, Universitas Trisakti, DKI Jakarta, Indonesia

Abstract. This paper proposes a high gain microstrip antenna with circular polarization for LTE applications at a working frequency of 2300 MHz. Gain of antenna is optimized by using an array method with 4 elements arranged in plannar using feed line of 50 Ohm, 100 Ohm and 70.7 Ohm. The proposed antenna designed using FR-4 substrate with (εr) = 4.3, substrate thickness (h) of 1.6 mm and loss tangent (tan δ) of 0.0265. Gain of antenna can be optimized by increasing the number of elements in the array of the antenna. Circular polarization can be obtained by cutting the edge of a rectangular patch antenna with an angle of 45^0 . From the simulation results obtained reflection coefficient value of -20.02 dB and VSWR of 1.22 at the working frequency of 2300 MHz. The gain of proposed microstrip antenna is 10.56 dB with an impedance bandwidth value of 714 MHz (1925 - 2639 MHz). The proposed antenna obtained a circular polarization with axial ratio of 1.745 dB at working frequency of 2300 MHz. Array method with four elements increased the gain of antena until 45.07% compared to the single element antenna.

1 Introduction

At present time, the development of mobile telecommunication technology and communication system is growing very rapidly, especially the development in the mobile and wireless telecommunications. The society needs for communication and high transfer data rates are getting higher along with the entry of 4G Long Term Evolution technology. The stability of the connection and the speed of data transfer becomes very important to make the telecom provider optimize the network in order to meet customer needsthere are several telecommunication systems that have been developed, for example is DCS at frequency band (1710 MHz -1885 MHz), PCS (1907 MHz -1912.5MHz), UMTS (1920 MHz – 2170 MHz),WLAN 2.4 GHz and LTE 2.3 GHz [1]. The band frequency of 2300 MHz - 2400 MHz is used for broadband communication system such as Long Term Evolution and Wireless Fidelity in accordance with the Regulation of the Minister of

* Corresponding author: alam_bizkit@yahoo.com

Communication and Information of the Republic of Indonesia Number 28 / PER / M.KOMINFO / 09/2014 [2].

In modern telecommunication devices such as modem usually use an internal antenna that has a small gain so that the reception of signal is not optimal. To solve that problem can be used an external antenna but sometimes the dimension of external antenna is big enough so that it becomes ineffective in its application. In addition, the type of external antenna sold in the market also has a vertical polarization (dipole) while the transmitting antenna used in the BTS (Base Transmission Station) has a horizontal polarization causing a loss on the receiver side due to its cross polarization between the receiving antenna and the transmitter. To overcome this problem microstrip antennas can be developed with circular polarization to reduce the loss caused by cross polarization of the transmitting and receiving antennas. In addition, microstrip antennas has advantages such as small and compact dimensions so the application is more effective beside that microstrio antenna can delovoped to produce circular polarization. It provides both linear and circular polarization. However, microstrip antenna has disadvantages such as low level gain, narrow bandwidth, low efficiency, and low directivity [3]. The circular polarization (CP) is a special case of elliptical polarization and it is realized when antenna configuration excites two equal amplitude, time and space orthogonal resonant modes [4]. Polarization loss may result due to the misalignment in the signal and the receiving antenna. To solve this problem we can used microstrip antenna with circular polarization [5].

To obtain microstrip antenna that has a circular polarization can be used several methods such as fractal [6], truncated corner [7] and proximity coupling [8]. In this research, the truncated corner method is used by cutting the upper and lower edges of the rectangular shape antenna, the selection of this method is because the initial dimension of the designed microstrip antenna is rectangular so that the use of truncated method is much easier and more effective in designing, the truncated corner method also reduces the dimensions of the antenna into smaller ones resulting in a more compact dimension. Beside that to increase the gain of the antenna can be done by using array method [9-12]. In a previous study [13-15] obtained a circular polarization using the truncated corner method.

In this research, the microstrip antenna is designed using truncated corner method to produce axial ratio ≤ 3 dB. After that, gain of the microstrip antenna can be optimized by using an array method using 4 elements. The selection of the trucanted corners method caused by the dimensions used in the initial design is rectangular which is easier and more effective in designing to produce circular polarization. The type of array used is the plannar array because it is more effective to optimizing the gain value of the antenna. The type of substrate used in the design of the antenna is FR-4 Epoxy with (εr) = 4.3, substrate thickness (h) of 1.6 mm and loss tangent (tan) of 0.0265. The design and simulation process of the proposed antenna uses AWR Microwave Office.

2 Design of Antenna

Circular polarization microstrip antenna is designed using double layer substrate of FR-4 Epoxy with relative permittivity (εr) of 4.3, substrate thickness (h) of 1.6 mm and loss tangent (tan δ) of 0.0265. The first stage is to design a rectangular antenna with microstrip line feeder of 50 Ohm that works at working frequency of 2300 MHz for LTE applications. To determine the dimensions of W and L of the initial design antenna we can use the equations below (1), (2), (3) and (4)[16]. The width dimension of the microstrip line feeder of 50 Ohm (Wf) is obtained using equations (6) and (7)[17]. The design of the rectangular microstrip antenna can be seen in Figure 1.

$$W = \frac{c}{2f\sqrt{\frac{\varepsilon r+1}{2}}} \quad (1)$$

$$L = L_{eff} - 2\Delta L \quad (2)$$

$$L_{eff} = \frac{c}{2f\sqrt{\varepsilon_{reff}}} \quad (3)$$

$$\varepsilon_{reff} = \frac{\varepsilon r+1}{2} + \frac{\varepsilon r-1}{2}\left[1 + 12\frac{h}{W}\right]^{-\frac{1}{2}} \quad (4)$$

$$B = \frac{60\pi^2}{Z_0\sqrt{\varepsilon_{eff}}} \quad (5)$$

$$Wf = \frac{2h}{\pi}\left\{B - 1 + \frac{\varepsilon_r-1}{2\varepsilon_r}\left[\ln(B-1) + 0.39 - \frac{0.61}{\varepsilon_r}\right]\right\} \quad (6)$$

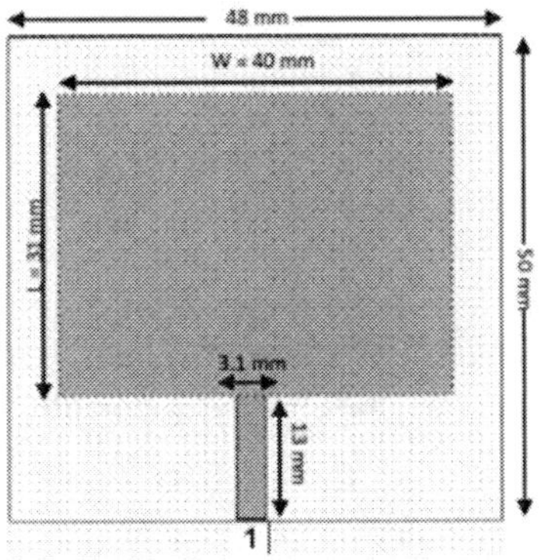

Fig.1 Design of Rectangular Microstrip Antenna

The second stage is to design truncated corner of microstrip antenna with cutting angle of 45^0 at the diagonal position. The main purposes of cutting the patch side of an antenna is to produce a circular polarization with a 3 dB axial ratio. Circular polarization occurs due to interference with the current flowing in the patch affecting the polarization of the antenna. The dimensions of the cut width on the patch antenna can be determined using equation (7) [18]. The design of the truncated corner of microstrip antenna design can be seen in Figure 2.

$$\Delta L = {}^{1}/_{4} \times L \quad (7)$$

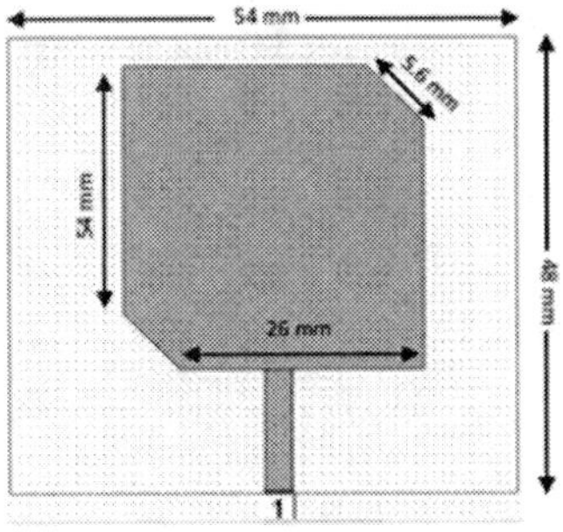

Fig.2 Design of Truncated Microstrip Antenna

The final stage is designing the microstrip antenna using the array method to optimized the gain of antenna by using four patch with feed line of Z_1=50 Ohm, Z_2=70.7 Ohm and Z_3=100 Ohm. Dimension and width of feed line (Wf) of 50 Ohm, 70.7 Ohm and 100 Ohm can be determined using formula (6) and (7) while to determine the distance between patches in the array method is given by (8) [19].

$$d = \frac{\lambda}{2} \tag{8}$$

The design of truncated mirostrip antenna using array with 4 element can be seen in Figure 3.

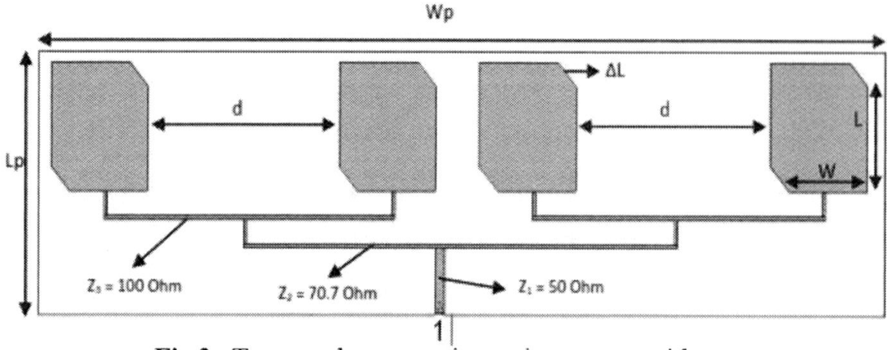

Fig 3. Truncated corner microstrip antenna with array

From Figure 3 above can be seen the dimension of array microstrip antenna with W=26 mm , L = 26 mm, d = 64 mm and ΔL = 5.6 mm. The dimension of the enclosure is Wp = 280 mm and Lp = 65 mm while the width of feeder channel Z_1, Z_2 and Z_3 are 3.1 mm, 1.6 mm and 1 mm.

3 Result and Discussion

After the design process, the next stage performs the simulation process of the design antenna using AWR Microwave Office. The observed parameters are reflection coefficient, VSWR, axial ratio and gain of the antenna. The simulation results of these parameters can be seen in the Figure 4, Figure 5, and Figure 6.

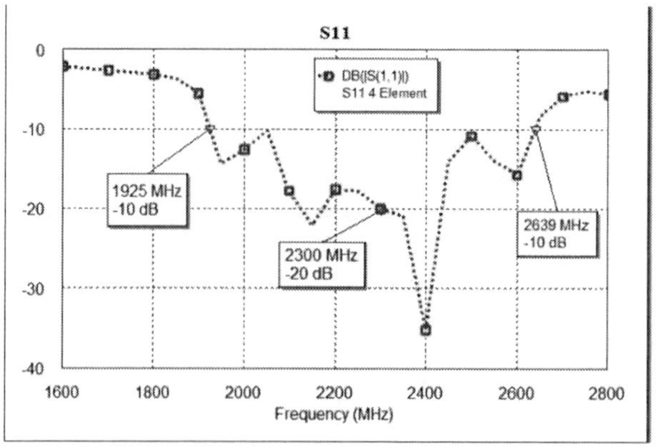

Fig 4. Simulation result of reflection coefficient

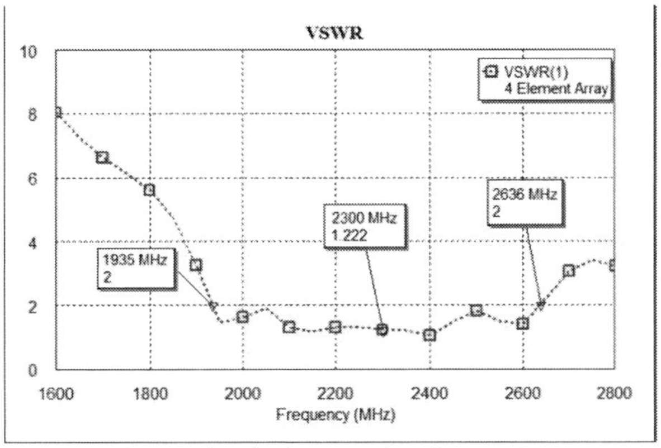

Fig 5. Simulation result of VSWR

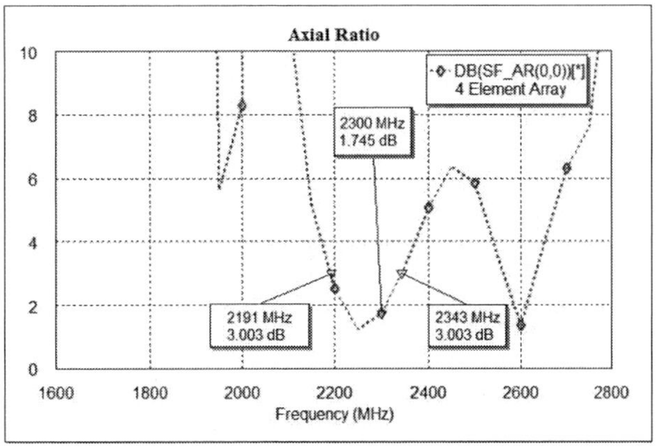

Fig 6. Simulation result of Axial Ratio

From Figure 4 and Figure 5 it can be seen that the proposed antenna has a reflection coefficient of -20 dB and VSWR of 1.22 at a working frequency of 2300 MHz. This is in accordance with the minimum standards set for the reflection coefficient ≤ -10 dB and VSWR ≤ 2 so it can be said that the antenna already has a good reflection coefficient and VSWR. The bandwidth obtained from the proposed antenna is 714 MHz (1925 - 2639 MHz).

From Figure 6 it can be seen that the axial ratio obtained from the proposed antenna is 1.745 dB at a working frequency of 2300 MHz. This proves that the proposed antenna has a circular polarization with an axial ratio of ≤ 3 dB at a predetermined working frequency of 2300 MHz. The bandwidth of the axial ratio obtained from the proposed antenna is 152 MHz (2191 - 2343 MHz) The gain of antena can be optimized by increasing the number of elements in the array of the antenna. The simulation results of gain from optimization process of the proposed antenna are shown in Figure 7.

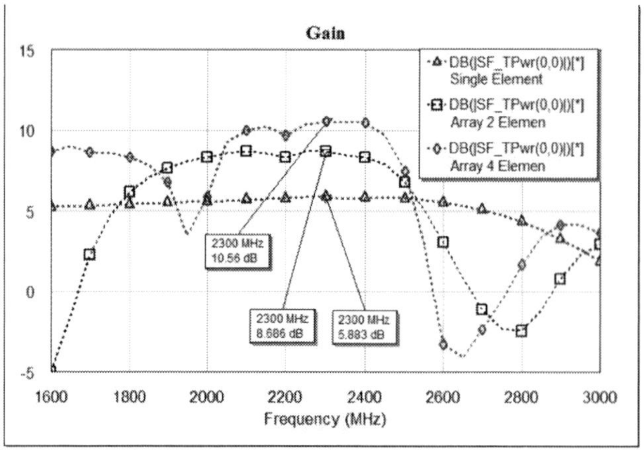

Fig 7. Simulation result of Gain

From Figure 7 it can be seen that the gain increase occurs as the number of elements increases. The gain obtained from the proposed antenna is 10.56 dB at a working frequency of 2300 MHz using 4 elements. The comparison of the overall optimization process of gain on the proposed antenna can be seen in table 1 below.

Table 1. Comparison Gain of Microstrip Antenna

Condition	**Parameter**	
	Gain	***Frequency***
Single Element	5.883 dB	2300 MHz
Array 2 Element	8.685 dB	2300 MHz
Array 4 Element	10.56 dB	2300 MHz

From table 1 above it can be seen that the gain enhancement of the array antenna is directly proportional to the number of elements. The gain obtained from the single element antenna is 5.883 dB, while in the 2 element array it increases to 8.685 dB and in the 4 element array increases to 10.56 dB at working frequency of 2300 MHz. This result proves that the array method on the proposed antenna successfully increases the gain value up to 45.07% compared to the single element antenna. The result of the simulation of the radiation pattern on the proposed antenna can be seen in Figure 6 below.

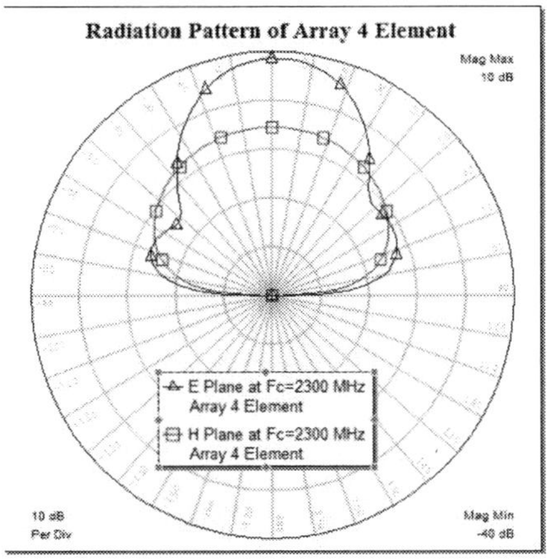

Fig 8. Radiation Pattern of Proposed Antenna

From Figure 8 it can be seen that the radiation pattern generated from the proposed antenna is broadside with the best transmit angle at 0^0. The effect of using arrays is produces a side lobe on the radiation pattern of the proposed antenna.

From the overall results it can be seen that the truncated corner method succeeded in producing circular polarization of antennas with axial ratio ≤ 3dB. In addition, the array method with 4 elements succeeded in increasing gain of antenna from 5.883 dB until 10.56 dB, increasing 47.80% compared to single element antenna.

4 Conclusion

A new design of high gain and circular polarization of microstrip antenna is eventually well proposed. The circular polarization is obtained by cutting the edges of the patch antenna while to increase the gain value is obtained by increasing the number of elements on the antenna. From the simulation results obtained the reflection coefficient of -20 dB, VSWR of 1.22 and axial ratio of 1.72 dB at 2300 MHz working frequency. The gain of the proposed antenna is 10.56 dB or an increase of 40.78% compared to the single elemet antenna.

The authors wish to express their gratitude to the LPPM of Universitas Kristen Krida Wacana for their support through Grant of Internal Researh fiscal of year 2018.

References

1. DITJEN POSTEL , "Structuring the Radio Frequency-Based Service Broadband Wireless Access". Jakarta , 2006.
2. Ministry of Communication and Information of the Republic of Indonesia, PERMENKOMINFO No.28 / PER / M.KOMINFO / 09/2014 "Establishment of Radio Frequency Band For Customers Wireless Broadband Service (Wireless Broadcast) At 2.3 GHz Frequency Band". Jakarta, 2014.
3. Dalvi, S., Jagtap, S., Yadav, V., & Gupta, R. K. (2016, January). High gain wideband 2× 2 microstrip array antenna using RIS and Fabry Perot Cavity resonator. In Microelectronics, Computing and Communications (MicroCom), 2016 International Conference on (pp. 1-6). IEEE.
4. İmeci, T., & Saral, A. (2010). Corners Truncated Microstrip Patch Antenna. 26th Annual Review of Progress in Applied Computational Electromagnetics.

5. Alam, S., Surjati, I., & Ningsih, Y. K. (2017, November). Patch modification and slot arrangement of microstrip antenna for improving the axial ratio. In Broadband Communication, Wireless Sensors and Powering (BCWSP), 2017 International Conference on (pp. 1-5). IEEE.
6. Reddy, V. V., & Sarma, N. V. S. N. (2014). Compact circularly polarized asymmetrical fractal boundary microstrip antenna for wireless applications. IEEE Antennas and Wireless Propagation Letters, 13, 118-121.
7. Rahman, M. A., Nishiyama, E., & Toyoda, I. (2017). A frequency diversity reconfigurable antenna with circular polarization switching capability. In Antennas and Propagation & USNC/URSI National Radio Science Meeting, IEEE International Symposium on (hal. 1367-1368). IEEE.
8. Chen, S. J., Fumeaux, C., Monnai, Y., & Withayachumnankul, W. (2017). Dual Circularly Polarized Series-Fed Microstrip Patch Array With Coplanar Proximity Coupling. IEEE Antennas and Wireless Propagation Letters, 16, 1500-1503.
9. Gupta, A., Choudhary, V., & Tripathy, M. R. (2018). An M-Shaped Microstrip Antenna Array for WLAN, WiMAX and Radar Applications. In Progress in Intelligent Computing Techniques: Theory, Practice, and Applications (pp. 509-515). Springer, Singapore.
10. Amjadi, S. M., & Sarabandi, K. (2017). A Low-Profile, High-Gain, and Full-Band SubArray of Cavity-Backed Slot Antenna. IEEE Transactions on Antennas and Propagation, 65(7), 3456-3464.
11. Faroqi, A., Zaelani, F., Kariadinata, R., & Ramdhani, M. A. (2018, January). On The Design of Array Microstrip Antenna with S-Band Frequency for Radar Communication. In IOP Conference Series: Materials Science and Engineering (Vol. 288, No. 1, p. 012006). IOP Publishing.
12. S. Alam, I. G. N. Y. Wibisana and I. Surjati, "Miniaturization of array microstrip antenna using peripheral slits for wireless fidelity communication," 2017 15th International Conference on Quality in Research (QiR) : International Symposium on Electrical and Computer Engineering, Nusa Dua, 2017, pp. 91-95.
13. Nestoros, M., Christou, M. A., & Polycarpou, A. C. (2017). Design of Wideband, Circularly Polarized Patch Antennas for RFID Applications in the FCC/ETSI UHF Bands. Progress In Electromagnetics Research, 78, 115-127.
14. Lai, H. W., Xue, D., Wong, H., So, K. K., & Zhang, X. Y. (2017). Broadband circularly polarized patch antenna arrays with multiple-layers structure. IEEE Antennas and Wireless Propagation Letters, 16, 525-528.
15. Tang, M. C., Chen, X., Li, M., & Ziolkowski, R. W. (2017). A Bandwidth-Enhanced, Compact, Single-Feed, Low-Profile, Multilayered, Circularly Polarized Patch Antenna. IEEE Antennas and Wireless Propagation Letters, 16, 2258-2261.
16. Wang, M. S., Zhu, X., Guo, Y., & Wu, W. (2018). Compact Circularly Polarized Patch Antenna With Wide Axial Ratio Beamwidth. IEEE Antennas and Wireless Propagation Letters.
17. Alam, S., Surjati, I., Ferawan, A., & Firmansyah, T. (2018). Design and Realization of Compact Microstrip Antenna Using Fractal Sierpenski Carpet For Wireless Fidelity Application. Indonesian Journal of Electrical Engineering and Informatics (IJEEI), 6(1), 70-78.
18. Chen, W. S., Wu, C. K., & Wong, K. L. (2001). Novel compact circularly polarized square microstrip antenna. IEEE Transactions on Antennas and propagation, 49(3), 340-342.
19. Surjati, Indra. "Microstrip Antenna: Concepts and Applications". 2010. Jakarta: Universitas Trisakti

Monitoring Location Prototype Using Lora Module

Mia Rosmiati[1,*], *Mochammad Fachru Rizal*[1], and *Irma Wanti*[1]
[1] Universitas Telkom, School Of Applied Science, 40257 Bandung, Indonesia

Abstract. The number of child abduction cases in Indonesia today is always increasing every year, this has given concern to parents in leaving their children to play outside the home. Therefore, a device that can detect the presence of children when outside the home was needded. At this time, some devices can monitor ones location easily and quickly, such as Smartwatch and GPS Tracker, but the device must be connected to the internet With these conditions, it needs a device that can transmit information in real time without any dependence LoRa module are using low frequency radio as data transmission medium, can be built a Location monitoring prototype without connecting to internet network. From the examination results, the built system successfully shows the real time location coordinate point with a maximum data transmission limit of 2 KM for Line Of Sight (LOS).

I. Introduction

The child is a treasure that is so precious for parents, so the information about its becomes very important for the parents. The abundance of kidnappings makes the parents become more concentrated to always ensure the existence of his child at any time. Based on data obtained from the Indonesian Child Protection Commission (KPAI), that in the range of 2014 to 2017 the number of child abduction cases is increasing every years. As in 2014 the number of child abduction cases according to KPAI is 51 cases, while in 2015 increased to 87 cases, and in 2016 increased sharply to 112 cases. And in early 2017 the number of child abduction cases has reached 23 cases. The number of child abduction cases is motivated by five things. They are an illegal adoption, retaliation or ransom, commercial sex exploitation, organ sales, and forced labor to become street children, beggars and helpers.

With the increasing cases of kidnapping makes parents become always worried at the time when their children are outside the home. So parents need a system that can supervise their children when outside the home. With the development of Internet of Things that integrates the development of computer network technology and sensors, the system can develop smart cities and smart home [1]. Several studies have undertaken the development of smart cities that aim to monitor the existence of children outside the home using several different technologies. Such as the construction of child monitoring system in the school zone area [2]. In this monitoring system, Zigbee and GPS modules are used to monitor the presence of children in the school environments at real time, so by using CCTV and GPS,

* Corresponding author: mia@tass.telkomuniversity.ac.id

the system can monitor the presence of children in the school zone. This system can also alert alarms to data centers when children are not in the school zone. Another study aimed to monitor child activity is automation child monitoring (ACM) system using wireless network [3]. This ACM system used android smartphone to monitor the child's location in real time. By using GPS sensors, accelerator sensors and mobile GIS, this system can provide information to parents through messages displayed on android smartphone in the form of location when the child is lost. This system is controlled by a GSM module that requires an internet connection to transmit its GPS data to related parties. While in the study of An Mobile Safety Monitoring System for Children [4], the child monitoring system uses Bluetooth technology as the child's location transmission media to be displayed on android smartphone. This system is able to provide child's location information on child guard through Bluetooth communication that will give a sense of security to the parents when the children are in the observation area or not. This system integrates GPS sensors, accelerator sensors and mobile GIS (Geographic information system) so that these sensors are able to provide the characteristics of high reliability, short response time and high accuracy and can produce a device that provides a level of security for the child. But this system has a weakness, they are the power of bluetooth and distance limited, because the maximum distance that can be used for children monitoring is 50 m in open areas and 28 m in blocked areas of the walls,

From some of the above research indicate that the problem of distance and power of device become the main weakness in transmission of children data location, so to improve the reliability of child's location monitoring system required a device that can give maximum distance for transmitting child's data location. In addition, devices used for data transmission must require minimal electrical power so that the system can continue to monitor the location of children for long periods of time. According to [5], Low Power Wide Area Network (LPWAN) is a solution to build a network for far enough coverage area and using very small power so that LPWAN concept is widely used to develop the technology of Internet Of Things (IoT). So in this research, the children location monitoring system is built using LPWAN concept that use LoRa module as data location transmission media. And the system which is built do not need the internet connection.

2. Literature

2.1Long Range Radio

LoRa is one of the Low-Power Wide Area Networks (LPWAN) protocol whicdesigned to fulfill IoT requirements. LPWAN act as a wide coverage for IoT devices. LoRa Alliance describer LoRa as the physical layer or the wireless modulation utilized to create the long range communication link[6]. Furthermore, LoRa has significant advantage compared to another communication technology. Its low powered and long range capability exceeds any other standard. A single gateway or base station can cover entire cities or hundreds of square kilometers.[7] For example, the LoRa transmitter chip created by Semtech, SX 1276 could reach 168 dB maximum link budget.[8]

2.2 LoRa Module/GPS HAT

LoRa that is integrated with GPS HAT is an expansion module for building a WAN LoRa that is operated with Raspberry Pi 3. This module is based on SX1276 / SX1278 transceiver made by Semtech. LoRa / GPS HAT is equipped with L80 GPS (base on MTK MT3339), specially designed for applications that use GPS by connecting via serial port on Raspberry Pi. This device can be used for applications such as timing, geolocation, indo or positioning or other applications requiring GPS information. Data transmission and reception on the

LoRa / GPS HAT has a wide spread spectrum as well as the ability to cope with high interference with low power consumption. LoRa / GPS HAT can achieve sensitivity up to 148 dBm at low cost.

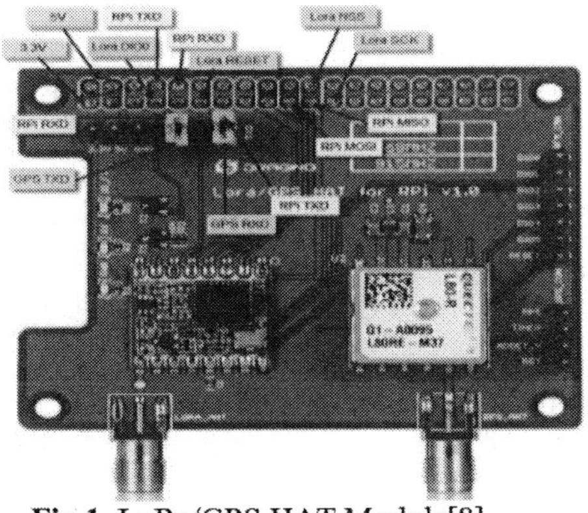

Fig 1. LoRa/GPS HAT Module[8]

At the figure 1, This board can calculate and predict the orbit automatically using ephemeris data stored in the internal flash memory, so that the HAT can improve position quickly even on indoor signal levels with low power consumption. [8] It also can adjust on/off time to achieve a balance between positioning accuracy and power consumption in accordance with environmental and motion conditions. GPS also supports automatic antenna switching function. It can achieve switching between the internal patch antenna and the external active antenna.

2.2 GPS (Global Positioning System)

GPS is an electronic navigation system that can receive information from 4 – 12 satellites so as to indicate position on the surface of the Earth. GPS satellites will transmit satellite positions and GPS receiver distance from satellites. This information is processed and displayed by the GPS receiver. GPS is an open, non-proprietary telecommunications system but a copyright ownership of a rapidly growing and constantly growing company.

GPS is comprised of three segments: satellite constellation, ground-control/monitoring network, and user receiving equipment. The control segment (CS) tracks and maintains the satellites in space. The CS monitors satellite health and signal integrity and maintains the orbital configuration of the satellites. Furthermore, user receiveing equipment provide navigation, timing and several apps.

As seen in Fig. 2, the GPS constellation consists of 24 satellites orbiting the Earth at an altitude of 20,200 kilometers. Orbit is set with the assumption that every point on Earth can see at least four satellites at any time. each satellite orbit the Earth approximately once in 12 hours at a speed of about 11,000 kilometers per hour. GPS satellites have panels Solar energy collectors for generating electrical energy needs.[9]

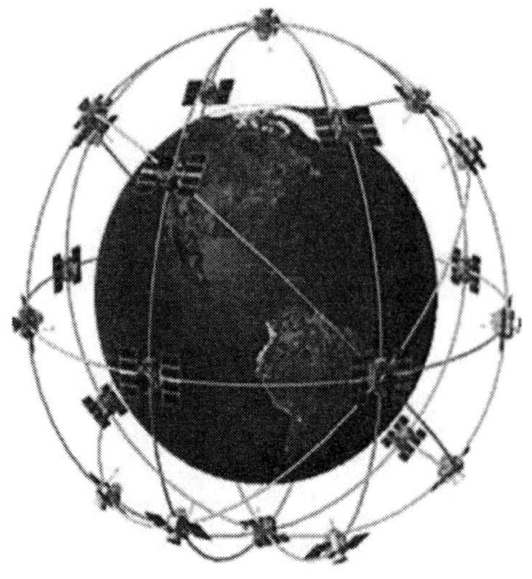

Fig 2. GPS Constellation (9)

3. Methodology

In this research, clild location monitoring system using LoRa with a frequency of 3-300 MHz, assuming LoRa module has a wide range with high efficiency of power usage. Figure 3 below shows an integrated module with Raspberry Pi 3 as processors, this configuration is made with the assumption that the receiver is not connected to the public network.

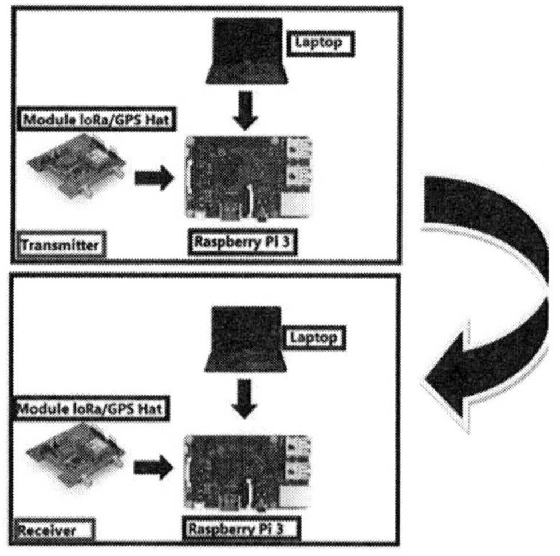

Fig 3. LoRa – Raspberry Pi 3 Configuration

In Figure 3 above, shows the laptop device used as a system controlling on Raspberry Pi 3. Raspberry Pi 3 serves as a microcontroller for LoRa / GPS HAT which is used as a radio packet transceiver.

4. Result & Analysis

To obtain data comparison of system accuracy then the testing process is done for configuration of Line of Sight (LOS) and Non Line Of Sight (NLOS). LOS configuration is a test configuration where the receiver and transmitter positions are in a straight line and are not blocked by a barrier so it is expected that the testing process can take place at a great distance. While NLOS is a test configuration where there is a barrier between transmitter and receiver that can block radio wave propagation, as for the purpose of the NLOS method is to determine the magnitude of the obstacle effect on the number of lost signals.

4.1 Line of Sight (LOS) Testing

The childreen location monitoring system using the LOS scenario is done in Batununggal residence, Bandung, Indonesia. In this test a child carries a device that is equipped with LoRa Transmitter. Then he walks with a maximum distance of 2 Km, the child's location information will be sent to the LoRa receiver, in this scenario a child is in a position facing each other with the receiver and not obstructed by the obstacle. In Figure 4 below shows the location data received by the receiver when a child moves 2 Km from the receiver or observation center point.

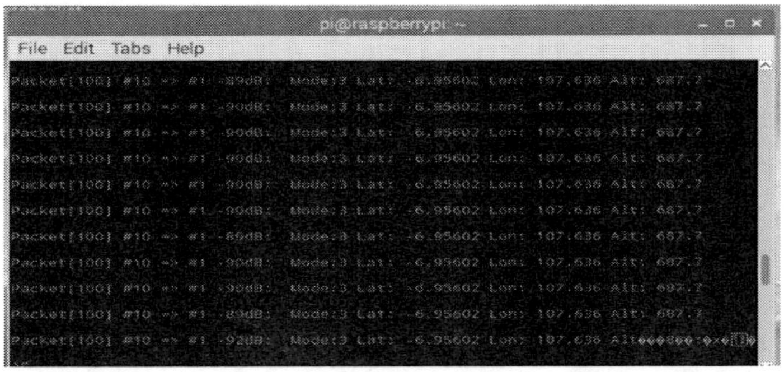

Fig 4. Location monitoring for distance 2 Km of LOS scenario

Figure 4 shows the LoRa transmission configuration capable of transmitting GPS data for transmitter and receiver distance of 2 km. As for the delay of sending data at a distance of 2 km is 5 seconds.

4.2 Non Line Of Sight (NLOS)

The location for child location monitoring system that use NLOS scenario is in the Griya Bandung Asri Residence, Bandung, Indonesia. In this scenario location the child's position with the center of observation (receiver) is blocked by the house or tree.

In the NLOS scenario, a child moves away from the receiver at a maximum distance of 350 m. and the position of transmitter and receiver obstructed by the many obstacles of trees and houses. The monitoing results of the transmit data location using LoRa is shown by figure 5.

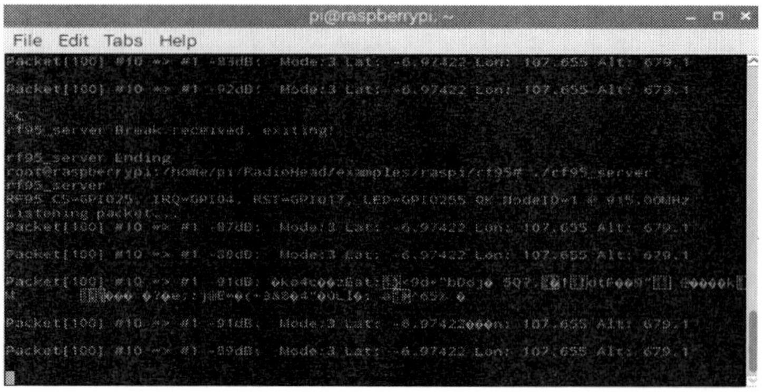

Fig 5. Location monitoring for distance 350 of NLOS Scenario

From Fig. 5 it is seen that the NLOS data communication between the transmitter and receiver is successful. The delay value obtained for this test is 3 seconds.

Below are completed survey from various distance for LOS dan NLOS testing.

Table 1. LOS testing Result for various distance

Distance (KM)	Location	delay (s)	Data GPS		
			Latitude	Longitude	Altitude
1	Batununggal residence	5	-6.95602	107.636	687.7
2	Batununggal residence	5	-6.95602	107.636	687.7
2.5	Batununggal residence	59	-6.97422	107.655	682.4
3	No Location is detected	-	-	-	-

From Table 1 above, the system can send data location for maximum 2.5 km distance with delay is 59 s. The Dragino/Semtech stated LoRa/HAT could reach 168 dB maximum link budget. For Non LOS, see table 2 below.

Table 2. NLOS testing result for various distance

Jarak (m)	Lokasi	delay (s)	Data		
			Latitude	Longitude	Altitude
100	Griya Bandung Asri Residence	3	-6.97422	107.655	679.1
200	Griya Bandung Asri Residence	5	-6.97422	107.655	679.1
300	Griya Bandung Asri Residence	60	679.1	107.655	679.1

From the above NLOS tests, it appears that the obstacles between the transmitter and receiver are very influential in the transmission of radio waves, because they cause the decrease in the intensity of the transmitted radio wave signal. The the more trees that block between the transmitter and receiver make the intensity of the emitted signal decrease and the delay time of radio waves transmission becomes increase. The maximum distance between transmitter and receiver for this scenario is 300 m.. This need further research in spite of maximum link budget stated by Dragino that could reach 168 dB.

5. Summary

Realtime location monitoring and tracking by integrating the loRa module with GPS successfully shows the location in real time. The use of LoRa as a device that functions to transmit and receive radio waves is better when compared to access point, this is because LoRa uses lower power, besides location data can be transmitted with transmitter distance and receiver further that is reach 2 km if using the LOS scheme and 350 m if between the transmitter and receiver there is a barrier (NLOS).

The LoRa further reasearch is to integrate the LoRa module with the processor to be placed on the smartwatch, which aims to detect the location of a person. In network side, to increase coverage and efficiency of LoRa Networking.

Reference

1. M. Elkhodr, S. Shahrestani, H. Cheung, International Journal of Wireless & Mobile Networks, **8,** 67-82 (2016)
2. Y. Song, N. Kim, D.Kim et al, 2010 International Conference on Information and Communication Technology Convergence, ICTC 2010, 579-589 (2010)
3. A.Saranya, C. Venkatesh, International Journal of Computer Sccience and Mobile Computing, **5**, 356-363 (2016)
4. Z.Huang, Z. Gao, H.Lu et al, 2014 10th International Conference on Mobile Ad-hoc and Sensor Networks, 323- 328 (2014)
5. U. Raza, P. Kulkarni, M. Sooriyabandara, IEEE Communications Surveys & Tutorials, **19**, 855-873, (2017)
6. LoRa Alliance, "A Technical Overview of LoRa® and LoRaWAN™," no. November, 2015
7. Semtech, "SX1276/77/78/79 - 137 MHz to 1020 MHz Low Power Long Range Transceiver," 2018.
8. S. Allif A'ssri, F.H.K. Zaman, S. Mubdi, 2017 IEEE 8th Control and System Graduate Research Colloqium (ICSGRC 2017), 127-131 (2017)
9. Lombardi, Michael A, *The Use Of GPS Disciplined Oscillators as Primary Frequency Standars For Calibration and MetrologyLaboratories* , vol. 3, no. 3. 2008

E-Government Interoperability: Architecture Model for Public Information Services of Sub-District Governments

Djalal Er Riyanto[1,*], *Panji* Wisnu Wirawan[1] , and *Kabul* Kurniawan[1]
[1]Departemen of Computer Science, FSM, Diponegoro University, Semarang, Indonesia

Abstract. Sub-District Government as a Public Bodies is such a state administrator in accordance to the law required to apply disclosure, either by publishing information proactively, and or providing information application services. Implementation of e-Government is one solution that can be used to improve performance in running the mandate of the law. E-Government requires collaboration from various Public Bodies, especially in data exchange, information sharing, and processes. E-Government Interoperability is the development of inter e-Government systems for sharing and integrating information using shared standards. The success of e-Government Interoperability is determined by strategies, policies, and architectures that enable data, information technology systems, business processes, and service lines to integrate precisely and efficiently. The architectural model generated from this study illustrates the structure of e-Government Interoperability, the basic organization of system components, the relationship of one component with other components and the environment. Model validation uses result approach/theory analysis for data standardization, solid platforms, easier access to information, and efficient administration and services. The architectural model can serve as a guide for design and evolution, in an effort to create a system for public information services, especially at the sub-district level. It was concluded that the model was able to realize the development of sub-district vertical database integration and single-sign-on.

1. Introduction

The Public Information Disclosure Act gives individuals the right to access information controlled or administered by the Public Agency [1]. It is obliged to apply disclosure, either proactively publishing information, and/or by providing a request for information services. E-Government is broadly defined as the use of information and communication technologies to support government tasks, such as providing or improving public services or managing internal government operations [2].

Benefits of e-Government can be expanded and enhanced with interoperability applied, which is system information technology and communication also business process supported, to be able to exchange data and able to share information and knowledge by

* Corresponding author: erriyanto@undip.ac.id

using common standard [3]. Interoperability involves various parties, inter-agency, cross-sector so that need willingness and mutual understanding to make it happen.

The architectural model called Kenyan e-Government Architecture [4] consists of five layers, ranging from the technology layer to the presentation layer. Another model is a model with four layers, with two layers between them the same as the previous model [5].

In this research, we will examine the non-technical and technical problems to facilitate e-Government Interoperability. Non-technical issues relate to laws and regulations used in order to realize public information services by the government, and from a technical point of view the environment and model of sub-district level information service architecture.

2. Methodology

The three main questions that will be answered in this research are: a) the main information needed to realize public information disclosure in accordance with the law, b) technical and non-technical factors that support the realization of interoperability, and c) architectural model for e-Government Interoperability of public information services the sub-district governments to be used as the basis for implementing the e-Government.

Exploration and study is a preliminary study to obtain materials from literature and observation, related to public agencies, as well as applicants and public information users. As a result, the data and formulation, as well as infra-structure of information and communication technology for public information services of local government (sub-districts) are gathered. The analysis stages produce an environmental analysis covering the technical and non-technical environments of e-Government Interoperability, as well as the factors needed to support the success of e-Government Interoperability. Architectural modelling begins with initial modelling, then validation to ensure that the resulting model meets the model objectives/outputs.

3. Result and Discussion

3.1 Reviews of laws and regulations

Broadly, there are three norms related to public information disclosure, namely: basic norms, general norms, and technical norms [6]. The basic norm contains articles derived from the 1945 Constitution of the Republic of Indonesia [7]. General norms of laws and regulations related to information disclosure [1][8][9][10]. Technical norms contains a provision which provides guidance and information disclosure mechanism to realize public information disclosure [11][12][13]. Everyone has the right to communicate and obtain information to develop his or her personal and social environment; as well as the right to seek, obtain, possess, store, process and convey information by using all kinds of available channels [7].

Rural community has the right to gain access information through village information systems managed by the rural government. Rural Development Information Systems and Rural Area Development shall be developed by the regency/city government, and are designed to ensure the right of information access to the village [10]. Public information services are provided through the Information and Documentation Service Room [10], or the Information Desk [11]. Public Information and Documentation System is provision information system and documentation service, that it is quickly, easily and fairly in accordance with the law on freedom of information [13].

Information Management and Documentation Officer (IMDO) is the official responsible in the field of storage, documentation, provision, and/or information service in the Public

Agency and directly responsible to the supervisor of IMDO [11]. The Information and Documentation Service Manager is the organizer of information and document management services within the Ministry of Home Affairs and Local Government [10]. Rural Information System includes rural data, rural development data, rural area, as well as other information relating to rural development and rural area development.

Rural information through government organizations and departments web sites consist of: a) Potential Rural from The Central Bureau of Statistic, b) Rural and Village Profiles from the Ministry of Home Affairs, c) Rural Community Online from The Ministry of Village, and d) Rural Broadband from the Ministry of Communication and Information.

3.2 Interoperability Environment

Interoperability involves various components, namely: a) data sources and service providers for public, b) applicant user Information, c) supporters, d) laws/regulations, and e) ministry web / government agencies.

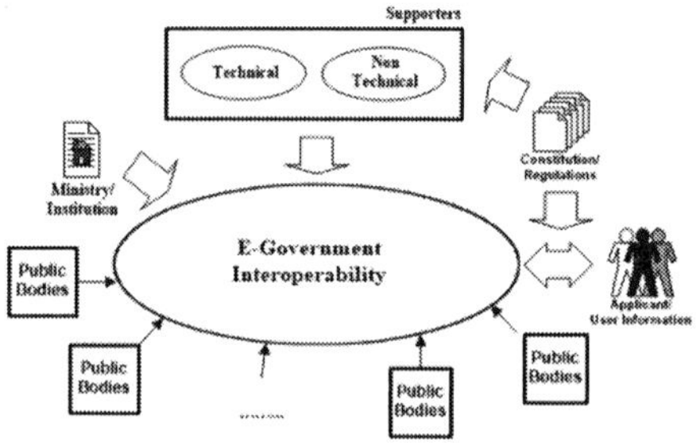

Fig 1. Environmental e-Government Interoperability

Data sources and public service providers come from various public organizations, especially those located within the districts and also stakeholders. Applicant/User Information is those who need data/information or services from E-Government, or those who are targeted to obtain information in relation to government programs. Supporters consist of technical and non-technical components. Technical component consists of information technology and communication infrastructures at each Public Agency. Non-technical supporters include policy and political will as the key success of interoperability. Laws and regulations provide guidance for the development of interoperability, among others, in determining the types of data and information that must be provided and accessible to the public, relations between public agencies, and public service mechanisms. Web Ministries/Agencies are a web related to the management of rural and data village developed, among others by the Ministry of Internal Affairs, Central Bureau of Statistics, and Ministry of Village. The Web is a study material to get the benchmark data of rural and districts in national level.

3.2 Model Architecture for e-Government Interoperability

Architecture is a picture of the structure of a system that describes the basic organization of components, linked to one another with their environment, with principles guide the design and evolution [15]. The architecture is addressed to three things: a) the design is adapted to what is required, not limited to the available technology, b) the effective architecture will accelerate service delivery for the market and lead to cost savings, and c) the architecture should support the overall flexible public services [16].

Architecture development principles of e-Government Interoperability are: a) to use of basic needs in regulations and legislation provided, b) the architectural design has the scalability, thus the implementation and development must be able to adapt to development needs and technological readiness also human resources, and c) to focus on efforts to improve the quality of public services gradually and sustainably.

The e-Government Interoperability architecture is designed to have a number of interconnected components, and the environment of e-Government Interoperability. By adopting the given architecture [4][5], the architectural model for e-Government Interoperability has four layers as a component (see Figure 2).

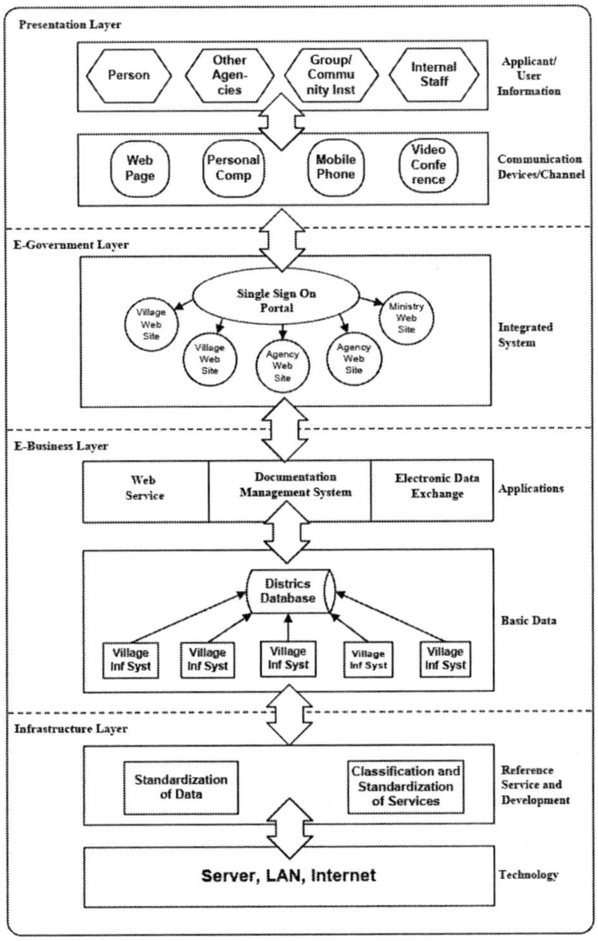

Fig 2. Architecture of e-Government Interoperability

Infrastructure Layer:

Infrastructure Layer consists of two main parts, namely: technology, as well as reference services and development. Technology consists of information and communications technology hardware needed to data capture, perform processing, storage of results, and devices for the distribution of data and information. The Service and Development Reference consist of two sub-sections, namely: a) data standardization, used to eliminate problems caused by multiple independent data processors, and b) Qualification and standardization of services, used to build systems with services integrated and standardized. Both sections are indispensable for interoperability, and subsequent developments.

E-Business Layer

E-Business Layer consists of two parts, namely: basic data, and applications. Basic data is data and processing results that can be accessed through e-Government Interoperability. The data is managed in the sub-district database, and for the long-term period it will form a sub-district data warehouse in the form of historical data obtained from system or operational applications. From various rural information systems an access mechanism will be built to realize "one-door management in obtaining information and analyzing it for decision making"

E-Government Layer

This layer integrates digital data from various organizations into a web portal of government services, in the form of one-stop e-Government portals or single sign on portals, which systems inside different computer networks to provide easy and secure access to multi sub-systems once authenticated, including mandatory and user access permissions. This has resulted in increased access, reduced process service costs, and enabled the availability of services that are easy and fast. To realize e-Government Layer required political policy and the willingness of all stakeholders.

Presentation Layer

The top layer contains facilities for presenting result to applicants and or users of information, whose nature of services are official or non-official. For such service, it needs communication device or channel required, which can be webpage, personal computer, mobile equipment or video conference. By those devices, applicant/user Information can access various information and services of public organization.

Model Validation

The model validation for this research uses the results/theoretical approach to four results/outputs, namely: data standardization, solid platform, easier information access, and efficient administration and services. Data standardization can be realized through the integration of data from various rural information systems into sub-district databases, as well as the adoption of data structures of various information systems developed by various local and central ministries/agencies. A solid platform for realizing inter-system interoperability, document flow efficiency, and inter-company transactions is available through the use of web service applications, document management systems, and electronic data exchange. Easier and cheaper access to public information is available through an open interface with communication or channel equipment that is widely known, owned and operated by the public. Public administration and services are more efficient through "vertical integration", and also by developing Single-Sign-On.

4. Conclusion

Regulations and legislation mandate the need for a broader and more accessible public information service system. The model of the District-level e-Government Interoperability Architecture can be used as a guide in the effort to realize the mandate, through the use and development of an integrated and sustainable information system. Using model validation with the results / theory analysis approach, it was concluded that the model was able to realize the development of sub-district vertical database integration and single-sign-on portal. The implementation of the architectural model of e-Government Interoperability is carried out by optimally utilizing the existing information systems, as well as supporting the willingness of all relevant parties based on the spirit to implement the laws and regulations.

References

1. ___ , Law of The Republic of Indonesia Number 14 Year 2008 concerning Public Information Openness.
2. Novakouski, Marc, and Grace A. Lewis, Interoperability in the e-Government Context, Carnegie Mellon University, 2012.
3. Gottschalk, Petter, and Hans Solli-Saether, e-Government Interoperability and Information Resource Integration: Framework for Aligned Development, IGI Global, 2009.
4. Mose, S.M., Wafula J.M., Kimani S., e-Government Architecture Model for Government-to-Government Deployment of Interoperable Systems (A Case Study of County National Government In Kenya), International Journal of Innovative Research in Technology and Science (IJIRTS), Volume 3, Number 3, ISSN: 2321-1156.
5. Ebrahim, Zakareya, and Zahir Irani: e-Government adoption: architecture and barriers, Business Process Management Journal, Vol. II No. 5, 2005.
6. Prayitno, Dessy Eko, et. al., Module for Public Bodies, Implement The Law Number 14 of 2018 concerning Public Information Openness, Indonesian Center for Enviromental Law, 2013.
7. ___ , The 1945 Constitution of the Republic of Indonesia.
8. ___ , Law of the Republic of Indonesia Number 25 of 2009 concerning Public Service.
9. ___ , Law of The Republic of Indonesia Number 43 of 2009 concerning Archieves.
10. ___ , Law of The Republic of Indonesia Number 6 of 2014 concerning Village.
11. ___ , Information Commission Regulations Number 1 of 2010 concerning Public Information Service Standards.
12. ___ , Information Commission Regulations Number 2 of 2010 concerning Procedures for Resolving Public Information Disputes.
13. ___ , Indonesians Republic's Home Affairs Minister Regulation Number 3 of 2017 concerning Guidelines for Management of Information and Documentation Services of The Ministry Home Affairs and Regional Government.
14. ___, Ministry of Home Affairs Regulation Number 12 of 2007 concerning Guidelines for the Preparation and Utilization of Village and Village Profile Data.
15. Stevanus van Staden, e-Government Interoperability: A Cooperative Architecture Model to Facilitate Information Sharing in Namibia, A Thesis Submitted in Partial Fulfilment of The Requirements of Master of Science in Information Technology of The University of Namibia, 2011.
16. Lallana, Emmanuel C., e-Government Interoperability, United Nation Development Programme, 2008.

Enhancement of Twice Quasi Orthogonal Space Time Block Coded (QOSTBC) Performance System with Zero Forcing EVCM Decoder

Desy Agustin[1], *Nachwan* Mufti Adriansyah[1,1], and *Muhsin*[1,]

[1]Electrical Engineering, the School of Engineering, Telkom University.
Jl. Telekomunikasi No. 01, Terusan Buah Batu, Sukapura, Dayeuhkolot, Bandung, Jawa Barat 40257, Indonesia.

Abstract. In today's modern telecommunications systems, makes the number of studies and development of multiple antennas and multiple-input multiple- output (MIMO) systems to achieve high reliability and low complexity. One attractive approach to improve that performance is using technique transmit diversity which is spacetime block coding and receiver diversity i.e. zero forcing EVCM (ZF EVCM). Although some earlier MIMO standards were develop some space-time codes like (O-STBC)and (Q-OSTBC) to provide high reliability but they are limited able to achieve orthogonality. In this research will be proposed a MIMO system scheme which is an improvement of QOSTBC that used a transmission diversity technique. This improvement from QOSTBC is Twice QOSTBC uses a provision in two codeword matrices to be sent are arranged diagonally so as to have higher levels of orthogonality. In this case Twice QOSTBC highly structured (4x1) can be replaced as an equivalent EVCM channel H. The proposed Twice- QOSTBC's results outperform other QOSTBC techniques with a difference around 3 dB for single-input multi-input (MISO) input configuration at 10^{-6} BER and receiver ZF EVCM has a very similar structure as the code matrix S of the underlying Twice QSTBC which can eliminates the system complexity.

1 Introduction

The needs for achieve high performance in modern telecommunication systems today, such as the 5G standard, makes the number of research and development of multiple antennas and multiple-input multiple-output (MIMO) systems [1]. One attractive approach to fulfill high performance is using transmit diversity technique which is spacetime block coding [2] [3]. Full diversity is achieved while we use different orthogonal constellations for different transmitted symbols. In this research, new paradigm uses the theory of orthogonal designs [4] to design twice quasi orthogonal spacetime block codes.

[1] Corresponding author: nachwanma@telkomuniversity.ac.id

Space Time Block Coding (STBC) [3] [5] is one example of a MIMO technique that utilizes antenna time and dimensions to achieve high reliability with minimum error probabilities [5]. Although STBC excel in minimum desired BER but STBC is limited to [3] two receiver antennas Nt=2 as the antenna configuration is not able to achieve orthogonality. This limitation can be overcome by combining specifically the O-STBC to increase the spatial diversity of the scheme [6]. Unfortunately OSTBC can only applied on BPSK modulation when we want to use complex modulation the code is referred to as QO-STBC [7][8]. Standards of QO-STBC scheme that provides Nt>2 and which can be applied on complex modulation, the full diversity can be achieved. Quasi Orthogonal STBC tends to reduce interference by minimizing the non-orthogonality of the symbol.

A few studies were already proposed and developed on methods QOSTBC some of research are "Two Quasi Orthogonal Space-Time Block Codes with Better Performance and Low Complexity Decoder" [9]. This research trying to exploiting these codes, changed property of the channel from quasi-static fading to block fading and discovered that this alteration improved the proposed code's performance by up to 4 dB in comparison with [6] and [12]. This research can solve the better performance but the orthogonality still needed to be improved by using different coding method.

In this research, we propose constellations for different transmitted symbols to improve the orthogonality property of the codes. With the existence of good coding techniques for MIMO antennas are increasingly massive expected to improve the quality of service to the user. The proposed codes are designed using orthogonal designs which are transmission matrices with orthogonal columns. It is shown how simple decoding which can separately recover transmit symbols using an possible orthogonal design. The emphasis of our study in was to design codes which provide full diversity by generalizing the theory of orthogonal designs with using Twice QOSTBC. Twice QOSTBC is used in 8,16,64 transmitter antenna which will be obtained coded more orthogonal. The ZF EVCM can also be used to apply a low complexity decoder. It has been shown that the ZF EVCM has the same structure as the corresponding QSTBC [13].

The paper is organized as follows, in section II we study on the proposed system along with Twice QOSTBC and Zero Forcing EVCM decoder. Next, in the section III we explore the evaluated performance of Twice QOSTBC. The main conclusion of this research is summarized in section IV.

2 The Proposed System Model

2.1 Twice QOSTBC Encoding

MIMO schemes with full rate and full diversity are met if modulation is in binary. If the complex modulastion used, OSTBC with full rate and diversity cannot be achieved [2]. Another alternative that can be used is QOSTBC which can increase the data rate and maintain the level of diversity of the system [10] [11][12]. QOSTBC tends to decrease interference by reducing the non-orthogonality of symbols. This paper proposed a scheme code word that has better orthogonality to achieve full diversity and full orthogonal. For the proposed Twice QOSTBC scheme can be seen in the following Figure 1.

In [9] a novel QO-STBC scheme for three and four transmit antennas was proposed by reforming the detection matrix of the conventional QO-STBC by using two QOSTBC technique where data matrix of symbol that will be send transformed into a scheme introduce by Tirkkonen, Boariu and Hottinen. This scheme used Alamouti scheme to construct matrix symbol pattern that will be derivative into 4x1, 8x1, 16x1 matrix antenna. We will start sending a code word with $C \in C^{TxM_T}$ where T is time slot and received signal is $Y \in C^{TxM_R}$,

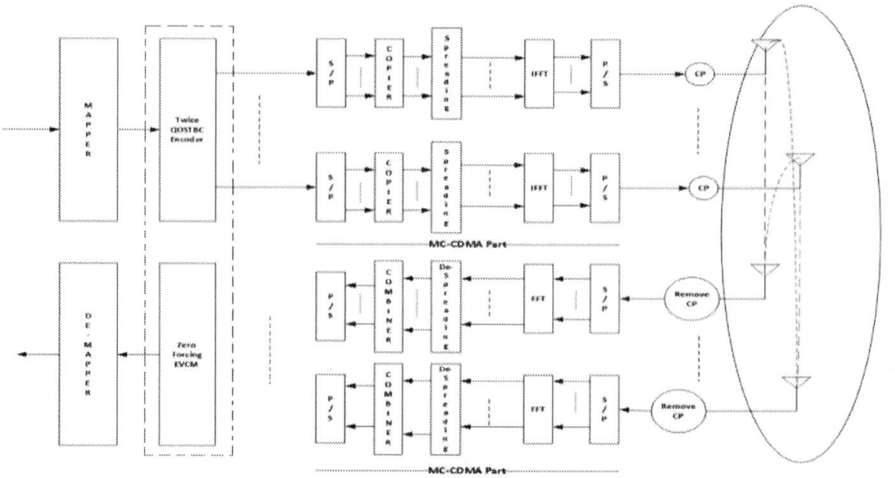

Fig 1. The Proposed Methode Twice QOSTBC transmitter

$H \in C^{T\,xM_R}$ is a fading matrix and $N \in C^{T\,xM_R}$ is Additve White Gaussian Noise (AWGN). For a system that has been designed for two blocks fading the following expression can be written as follows

$$\begin{bmatrix} Y_1 \\ Y_2 \end{bmatrix} = \begin{bmatrix} C_1 & 0 \\ 0 & C_2 \end{bmatrix} \begin{bmatrix} H_1 \\ H_2 \end{bmatrix} + \begin{bmatrix} N_1 \\ N_2 \end{bmatrix} \tag{1}$$

By using Alamouti scheme we can define as matrix generator

$$(x_1, x_2) = \begin{bmatrix} x_1 & x_2 \\ -x_2^* & x_1^* \end{bmatrix} \tag{2}$$

In this paper, a new approach is proposed for the Twice QOSTBC scheme by re-forming the matrix generator using a different technique. The main properties of different technique orthogonal design are simple separate encoding and full diversity. In order to full-diversity codes, we relax the derivation of matrix with better orthogonality. The derivation of matrix Twice-QOSTBC is explained below :

$$S_{4x4} = \begin{bmatrix} U_1 & U_2 & 0 & 0 \\ -U_2^* & U_1^* & 0 & 0 \\ 0 & 0 & U_3 & U_4 \\ 0 & 0 & -U_4^* & U_3^* \end{bmatrix} \tag{3}$$

The aforementioned codewords are the outcome of a case where the careful combination of the two symbols result in the best possible performance to achieve full diversity. In this case, as Eq. (3) is showing we combine S_k and S_{k+4}, where k = 1,2,...4, and replacing U matrix in Eq(4),as follows:

$$C = \begin{bmatrix} S_1 + jS_2 & S_3 + jS_4 & 0 & 0 \\ -(S_3 + jS_4)^* & (S_1 + jS_2)^* & 0 & 0 \\ 0 & 0 & S_1 + jS_2 & S_3 + jS_4 \\ 0 & 0 & -(S_3 + jS_4)^* & (S_1 + jS_2)^* \end{bmatrix} \tag{4}$$

Equation (4) is valid for single rate. Because of the use of zeros in (3), the code has a high peak-to-average power ratio that may increase its cost of implementation. A similar idea can be used to combine transmission matrices (4x1) to build a transmission matrix (8x1) and so on. An example of an 8x1 matrix which provides a rate one code

$$S_{8x8}=\begin{bmatrix} U_1 & U_2 & U_3 & U_4 & 0 & 0 & 0 & 0 \\ -U_2^* & U_1^* & -U_4^* & U_3^* & 0 & 0 & 0 & 0 \\ U_3^* & U_4^* & -U_1^* & -U_2^* & 0 & 0 & 0 & 0 \\ -U_4 & U_3 & U_2 & -U_1 & 0 & 0 & 0 & 0 \\ 0 & 0 & 0 & 0 & U_5 & U_6 & U_7 & U_8 \\ 0 & 0 & 0 & 0 & -U_6^* & U_5^* & -U_8^* & U_7^* \\ 0 & 0 & 0 & 0 & U_7^* & U_8^* & -U_5^* & -U_6^* \\ 0 & 0 & 0 & 0 & -U_8 & U_7 & U_6 & -U_5 \end{bmatrix} \tag{5}$$

For the OSTBC scheme the detection matrix is always diagonal this enables the use of simple linear decoding, but in the two-QOSTBC previous scheme this cannot be done due to the non-orthogonal detection matrix [9]

$$v=\begin{bmatrix} \alpha & 0 & \beta & \beta \\ 0 & \alpha & 0 & 0 \\ \beta & 0 & \alpha & 0 \\ \beta & 0 & 0 & \alpha \end{bmatrix} \tag{6}$$

Where the diagonal elements α represent the channel gains and β represent the interference from the neighbouring signals, for eight transmit antennas. The interference in the detection matrix will cause performance degradation. Now, the encoding matrix codeword Twice QOSTBC result in a free interference detection matrix be given as

$$\begin{array}{ll} < v_1, v_i >= 0, i \neq 1 & < v_2, v_i >= 0, i \neq 2 \\ < v_3, v_i >= 0, i \neq 3 & < v_4, v_i >= 0, i \neq 4 \\ < v_5, v_i >= 0, i \neq 5 & < v_6, v_i >= 0, i \neq 6 \\ < v_7, v_i >= 0, i \neq 7 & < v_8, v_i >= 0, i \neq 8 \end{array} \tag{7}$$

The orthogonality between some columns can help to simplify our ZF EVCM decoder. As mentioned before, this type of coding creates block fading channels [13].

2.2 Zero Forcing Equivalent Channel Matrix (ZF EVCM)

For detection, the interference caused by symbols on each other, related to the non-orthonormality, has to be cancelled [14]. The simple structure of the non-orthonormality matrix results in an easily implementable ZF EVCM decoder [13]. EVCM offers simple detection by ZF in EVCM domain which have better performance compared to conventional ZF, estimation using Zero Forcing EVCM (ZF-EVCM) applied by

$$y(t)=\begin{pmatrix} y_1 \\ y_2 \\ \dots \\ y_N \end{pmatrix} = H_N \begin{pmatrix} x_1 \\ x_2 \\ \dots \\ x_N \end{pmatrix} + \begin{pmatrix} n_1 \\ n_2 \\ \dots \\ n_N \end{pmatrix} \tag{8}$$

The channel coefficients are denoted by $\mathrm{h} = [\mathrm{h}_1, \mathrm{h}_2, \mathrm{h}_3, \mathrm{h}_4]$ and $\mathrm{n} = [\mathrm{n}_1, \mathrm{n}_2, \mathrm{n}_3, \mathrm{n}_4]$ is the noise vector. Zero forcing EVCM has a very similar structure as the code matrix S of the underlying Twice QSTBC which can eliminates the system by

$$\tilde{x} = \left(H_{N_t}^H H_{N_T}\right)^{-1} H_{N_t}^H y \tag{9}$$

with mean value of $x^{\sim}$ are simply taken from each receive antennas as final result of QOSTBC decoding.

3. Simulation Result

This study focused on the performance of the downlink transmission system. The goal is to address the problem of high reliability with full orthogonality. The purpose of this simulation is to demonstrate the superiority of the proposed system compared with QOSTBC system which has higher orthogonolatias value so it can achieve full diversity coding scheme and full orthogonality. The matlab script performs the following : First, generate random binary sequence of +1+j,+1-j,-1+j,-1-j and then group them into pair of four symbols and send four symbols in one time slot (Twice QOSTBC), multiply the Twice QOSTBC with the channel and then add white Gaussian noise. The received symbols will perform zero forcing EVCM and count the bit errors. Repeat for multiple values of SNR and plot the simulation and theoritical results.

Figure 2 shows the result of Twice QOSTBC simulation compared to QOSTBC system in Rayleigh channel for four transmitter antennas. As it is shown from the Figure 2 that the proposed results outperform other QOSTBC techniques with a difference around 3 dB for 10^{-6} BER This is because the Twice QOSTBC coding technique sends two symbols at once which are repeated with the second codeword and the orthogonality of this coding technique is also smaller so the possibility of smaller interference and the reliability of the transmitted data will be more efficient. The simulation results also show the correct data because it does not exceed the predetermined BER result based on diversity order = 4.

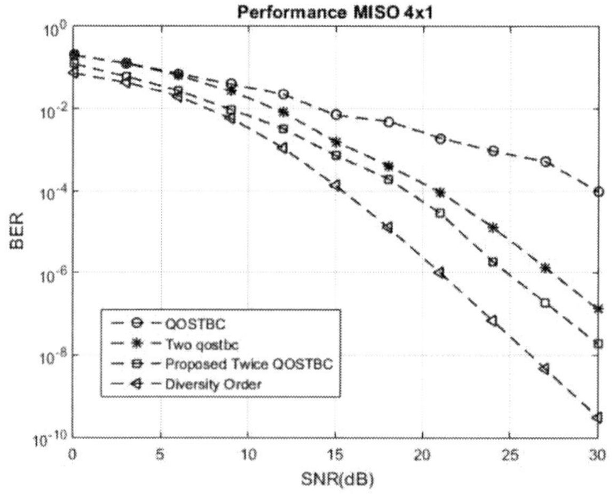

Fig 2. The comparison analysis between MISO 4x1 QOSTBC Scheme, Twice QOSTBC Scheme, and diversity order Scheme

Figure 3 shows the performances of the proposed Twice QOSTBC and the conventional QOSTBC for eight transmit antennas. It is clearly indicated that the proposed scheme achieves better performance than the conventional scheme.We can shown that by carefully choosing the coding scheme, the non-orthogonality can be minimized, and accordingly the performance can be maximized. The simulation results also show the correct data because it does not exceed the predetermined BER result based on diversity order = 8.

Figure 4 shows MIMO 16x16 has a better performance than MIMO 2x2. On the use of 16 antennas, to achieve the BER target of 10^{-5}, the Twice QOSTBC system requires SNR $\pm 3, 5$ dB, while on the use of 2 antennas, SNR larger ± 22 dB, so the improvement performance provided by the number of antennas used is $\pm 18, 5$ dB. It can be concluded,

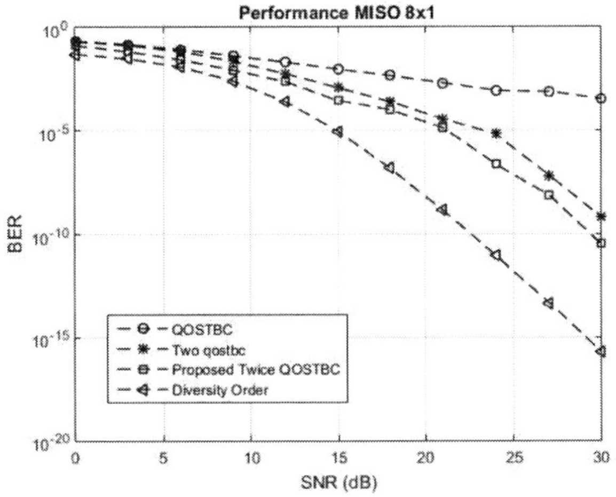

Fig 3. The comparison analysis between MISO 8x1 QOSTBC Scheme, Twice QOSTBC Scheme, and diversity order Scheme

higher number of antennas provides lower BER. In addition, the 16x16 MIMO used the Twice Orthogonal Space Time Block Code (OSTBC) scheme. It was observed that by reforming the detection matrix of the original QO-STBC to eliminate the interference terms, it was able to derive the orthogonal channel matrix that results in simple decoding scheme as in OSTBC. In addition, the more number of antennas used, the faster the process of coding and transmitting data.

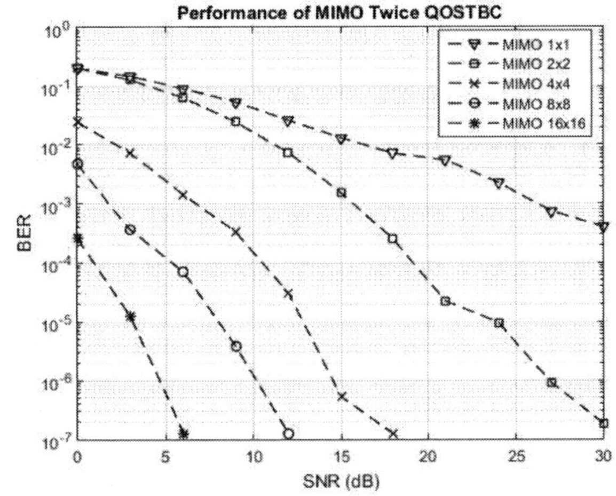

Fig 4. MIMO Twice QOSTBC system

3 Conclusion

From the research results, we can concluded that in general the proposed sytem maintained a better performance compared to other Space Time Coding scheme. Previous research QOSTBC does not have really orthogonal behavior. This non orthogonal behavior leads to increasing error. It has been evaluated in this research using twice qostbc technique coding. Orthogonality rate of full rate Twice-Qostbc by experiment is

$$k = \frac{1}{2} + \frac{2}{n}$$

with n is number of antennas (4,8,16,...). Orthogonality rate is decreasing close to 0,5 by increasing number of antennas. The value of orthogonality of transmitt antenna 4 can provides better SNR at 24 dB to reach the value of BER = $1,617x10^{-6}$.

From the point of view of decoding, the minimal nonorthogonality code presented here may run into trouble in slowly fading channels, when the worst case scenario where all channels are the same may persist for a long time. Alternatively, to prevent this may be applied with algorithm ZF EVCM known to the receiver.

References

1. G. O. Kelvin Anoh and S. Jones, " Full diversity qo-stbc technique for large antenna mimo systems Internet Technologies and Applications",(ITA)Wrexham,UK, (2015)
2. S. Alamouti, "A Simple Transmitter Diversity Technique for Wireless Communications. IEEE Journal on Selected Areas of Communications, Special Issue on Signal Processing for Wireless Communications", vol.16, no.8, pp.1451-1458, Oct. (1998).
3. C. Tarokh.V, Jafarkhani, "Space-time block coding for wireless communications : Performance results" IEEE J.Select Areas COmmun, 17:451460, Mar (1999).
4. V. Geramita and J.Seberry, "Orthogonal designs,quadratic forms and hadamard matrices" ser.Lecture Notes in Pure and Applied Mathematics, 43, (1979).
5. C. Okorafor, Rodriguez, "Performance evaluation of spatial modulation and qostbc for mimo systems" EAI Endorsed Trans.Mobb Commun Appl J.Select Areas COmmun, 44:15, Mar (2015).
6. S. Proakis, J.G. "Digital communications". Mc Graw-Hill:New York,NY,USA,(2008).
7. S. Larsson. "Space time block coding for wireless communications" Cambridge University Press,Cambridge,UK, (2008).
8. H. Jafarkhani. "Space-time coding:theory and practice", Cambridge University Press,Cambridge,UK, (2005)
9. Ali Lotfi-Rezaabad1, Siamak Taleb,"Two Quasi Orthogonal Space-Time Block Codes with Better Performance and Low Complexity Decoder" harif University of Technology (SUT), Tehran, IRAN, (2016)
10. H. Jafarkhani, "A quasi orthogonal space-time block code", IEEE Trans. Comm., vol. 49, pp. 1-4, January, (2001).
11. O. Tirkkonen, A. Boariu, A. Hottinen, "Minimal nonorthogonality rate one space-time block codes for 3+ Tx antennas", IEE International Symposium on Spread Spectrum Tech- niques and Applications (ISSSTA), vol. 2, New Jersey, USA, pp. 429-432, September, (2000).
12. C. B. Papadias and G. J. Foschini,"Capacity-approaching space-time codes for systems employing four transmitter antennas", IEEE Trans. on Information Theory, vol. 49, pp. 726-733, March (2003)
13. Badic, Biljana. "Space-Time Block Coding for Multiple Antenna Systems",November (2005)
14. Patrick Marsch et al., "Quasi-Orthogonal STBC using Streched Constellations for Low Detection Complexity", IEEE Wireless Communications and Networking Conference, (2007)

MATEC Web of Conferences **218**, 03010 (2018) https://doi.org/10.1051/matecconf/201821803010
ICIEE 2018

Location-dependent User Selection Based on Sum Rate Approximation in Large System Regime for Massive MIMO

Vera Noviana Sulistyawan[1], *Rina* Pudji Astuti[1,1] and *Arfianto* Fahmi[1]
[1]Electrical Engineering, the School of Engineering, Telkom University. Bandung, Jawa Barat 40257, Indonesia.

Abstract. Massive MIMO with multiple BS antennas can give simultaneous service for multiple user equipments (UEs) that are widely considered in massive connectivity to meet high data rate requirements. User selection is critical to optimize the overall performance of MIMO systems in various scenarios and has been extensively studied in cellular networks to guarantee service for users. In the previous study, location-dependent user selection (LUS) had extremely low computational complexity which is capable to enhance sum rate performance, but there are many environmental condition assumptions that make this algorithm does not reflect real conditions. In this research, we proposed modified LUS with approximations of sum rate in large system regimes by adding the sum ergodic of the distance from one user to another which enhance sum rate performance. In addition, we vary the user environment that was ignored in previous research by varying the path loss exponent values. In this research, we focus modify on sub-urban areas with each UEs having different environmental conditions. The selection scheme is equipped with spatial correlation fading on the transmitter side MIMO antenna. The simulation shows an increase in sum rate between 0.0012 to 0.3935 in perfect CSI. For the imperfect CSI with antenna correlation coefficient for power at 30 dBm is 0.5 when 32x64 antennas is 14 optimal active UEs with sum rate is 23.4207 bps/Hz. For cases where the user is located in different positions with different environmental circumstances, with 32x64 antennas showing the highest sum rate is 24.8436 bps/Hz with 17 optimal UEs.

1 Introduction

Massive device connectivity is a key requirement for next cellular network technology with many devices but few devices are active at the same time [1]. It is important to know how many optimal user equipments (UEs) that can be served at the same time with few factors to consider. The optimal number of UEs is required to guarantee the rate requirement. To fulfil the requirements of high data rate, massive MIMO with multi antenna BS can simultaneously serve multiple UEs considered to serve a large number of users simultaneously at higher rates[2]. Massive MIMO can achieve very high performance results through signal transmission with a large number of antennas on the transmitter and receiver
sides[3]. It is very important to select the active users to be served simultaneously and determine the optimal number of users to be served to achieve a high rate which is called

[1] Corresponding author: rinapudjiastuti@telkomuniversity.ac.id

user selection. User selection has been extensively studied in cellular networks to guarantee user service.

In a previous research, a massive MIMO user selection was performed using an important instantaneous Channel State Information (CSI) to optimize the overall performance of MIMO massive systems in various scenarios, Semiorthogonal User Selection (SUS) using the channel orthogonality rates among UEs [4][5][6] with instantaneous CSI to improve the sum rates system, but has a high complexity when applied to massive MIMO. While to consider fairness among UEs, round-robin scheduling[4] and RUS (Random User Selection)[7] can be considered. Since they are two simple methods that offer equal opportunities to all candidate UEs. In the study[7], the proposed RUS scheme is unlike conventional Semi-orthogonal User Selection (SUS) that requires instantaneous CSI from small scale fading, but requires only long term CSI and focuses on how many UEs should be selected for simultaneous transmission so have low complexity. Location-dependent User Selection (LUS) [7] had extremely low computational complexity with approximations of sum rate in large system regimes, which are capable of significantly enhances the system sum rate performance. In the previous study, LUS had extremely low computational complexity which capable enhanced sum rate performance, but there are many environmental condition assumptions that make this algorithm does not reflect real conditions.

In this paper, we proposed modified LUS by adding the sum ergodic of the distance from one user to another using approximations of sum rate in large system regimes, which are capable of significantly enhancing the sum rate performance. In addition, we modify the condition of the user's environment which being ignored in previous research. In this paper, we focus modifying users area only for sub-urban areas with each UEs having different environmental conditions. The selection scheme is equipped with spatial correlation fading on the transmitter side MIMO antenna so as to accurately reflect the relevant phenomenon of propagation in real life. Furthermore, this paper is divided into few sections. Design of the algorithm is presented in Section 2. In Section 3, we present results and analysis of the simulation results. Finally, conclusion are drawn in Section 4.

2 Algorithm Design

In this paper, we present the modified LUS where here we add sum ergodic distance which increase performance. User selection is designed with sum rate approximation in a large regime system on the transmitter side. Simulation process focus on downlink transmission in single common circular cell model. From the UEs candidate side, we place the UEs and calculate the distance with the hypergeometric function. We only focus on suburban areas. We analyze the results where UEs are placed in a fixed place with the same environmental conditions and when UEs are placed in locations with different environmental circumstances. Here we vary the value of path loss exponent. This scheme can be seen in the Figure 1.

2.1 Sum Rate Approximation in Large-System Regime

With relying on the random matrix theory based on large system analysis can derive a deterministic approximation of the ergodic sum rate. The deterministic SINR approximations

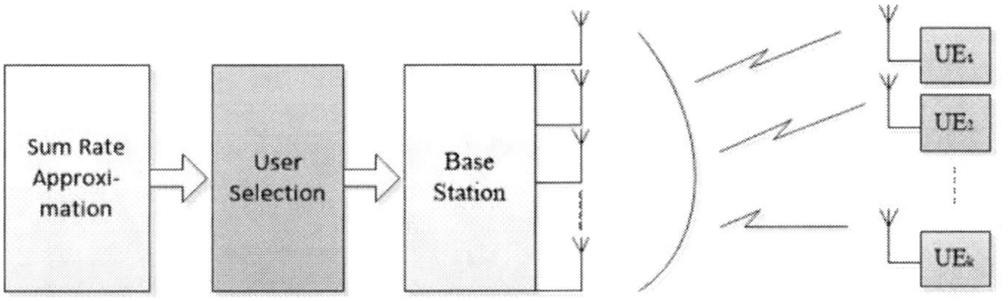

Fig 1. Location-dependent User Selection Based on Sum Rate Approximation in Large System Regime for Massive MIMO

enable us to solve various practical optimization problems[10]. The approximation of the ergodic sum rate is derived. The ($SINR_k$) of each user from the receiver side is defined as follows[7]:

$$SINR_k \xrightarrow{a.s.} \frac{1}{\sum_{k=1}^{K} d_i^{\alpha}(A(K,M)+B(K,M)d_i^{-\alpha})} \quad (1)$$

Each users gives feedback on the best SINR value. Where d_i is the distance between the BS and the active user of K. While the $A(K, M)$ and $B(K, M)$ can be calculated by[7]:

$$A(K,M) = \frac{1}{1-\rho}\frac{\sigma_n^2}{Pc\phi M} \quad (2)$$

$$B(K,M) = \frac{\rho}{1-\rho}\frac{\psi}{M\phi^2-K\psi} \quad (3)$$

In Equation 2 and 3, σ^2_n is the total noise power and c is the path loss in the reference distance of magnitude corresponding to the existing parameter in Table 1. ρ is a channel estimation error where $0 \le \rho \le 1$. P is the power transmit. M is the number of BS antennas and K is active UEs where ($K < M$).

From Equation 2 and Equation 3, ϕ and ψ are the unique solution of the equation that proposed in [7]:

$$\phi = \frac{1}{M}tr\left(R\left(I_M + \frac{K}{M}\frac{1}{\phi}R\right)^{-1}\right) \quad (4)$$

$$\psi = \frac{1}{M}tr\left(R^2\left(I_M + \frac{K}{M}\frac{1}{\phi}R\right)^{-2}\right) \quad (5)$$

I_M is the identity matrix with the (N × M) dimension. The transmit correlation matrix at the BS is modeled by the widely used δ − Kac − Muddock − Szegö matrix R with the (M × M) dimension, in which δ is the antenna correlation coefficient[9]. Note that this model can range from the totally uncorrelated scenario (δ = 0) to the fully correlated scenario (δ = 1)[9].

2.2 Location-dependent User Selection (LUS)

In this research, all UEs candidate (N single antenna) are spread in a single circular cell between inner radius (R_{min}) and outer radius (R_{max}) with Base Station (M antenna) at the center of the cell. Location is denoted in polar coordinates so that the UEs coordinate are (r_k,θ_k) and BS coordinate is (0,0) which can be seen in Figure 2.

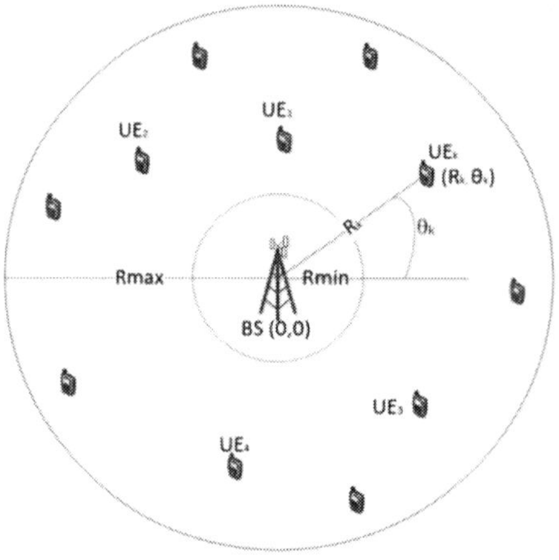

Fig 2. Location-dependent User Selection (LUS) in circular single cell with BS and candidate UEs

At LUS, we select the number of corresponding UEs in the order of large scale fading be dominated by path-loss (i.e., in ascending order of the BS-UE distance) for simultaneous data transmissions. UEs candidates located close to the BS are more likely to be active and selected than UEs located far from the BS. Therefore, with the LUS algorithm can decrease the distance. Meanwhile, due to the shortened distance between base station and terminals, transmission loss will be reduced, thus greatly improving the power efficiency[11]. Therefore $\Re_{LUS}$ which contains sum ergodic values of distances in x and y can formulated as:

$$\Re_{LUS} = \left(1 - \frac{K}{T}\right) \sum_{k=1}^{K} \int_{R_{min}}^{R_{max}} log_2 \left\{1 + \frac{1-\rho}{x+y}\right\} \frac{2r_k \left(R_{max}^2 - r_k^2\right)^{N-k} \left(r_k^2 - R_{min}^2\right)^{k-1}}{B(k, N-k+1)(R_{max}^2 - R_{min}^2)^N} dr_k \quad (6)$$

From Equation 6, ρ is the channel estimate accuracy value whose value is already set at $0 \leq \rho \leq 1$. In the case of massive MIMO where the number of Base Station (BS) antennas is greater than the number of single-antenna users, to estimate the channels, time-division duplex (TDD) operation with the reverse-link pilots are used.

$$x = \frac{\sigma_n^2}{Pc\phi M} \left\{ r_k^\alpha + \sum_{i=1, i \neq k}^{K} R_{min}^\alpha {}_2F_1 \left(k; -\frac{\alpha}{2}; N+1; 1 - \frac{R_{max}^2}{R_{min}^2}\right) \right\} \quad (7)$$

From Equation 7, there are several parameters such as σ_n^2 where is this is noise power, c is the path loss in reference distance, M is the number of BS antennas and N is the number of single-antenna candidate UEs.

$$y = \frac{\rho\psi}{M\phi^2 - K\psi} \left\{ \sum_{i=1, i \neq k}^{K} r_k^{-\alpha} R_{min}^\alpha {}_2F_1 \left(k; -\frac{\alpha}{2}; N+1; 1 - \frac{R_{max}^2}{R_{min}^2}\right) + 1 \right\} \quad (8)$$

The optimal amount of UE presented with the maximal sum rate can be written as:

$$K_{LUS}^* = \text{argmax}_{K \epsilon (1, \ldots, M-1)} \Gamma_2(T, M, K, R_{min}, R_{max}, P, c, \sigma_n^2, \rho, \alpha, \delta) \quad (9)$$

Where $\Re_{LUS} = \Gamma_2(T, M, K, R_{min}, R_{max}, P, c, \sigma_n^2, \rho, \alpha, \delta)$, the value of K_{LUS}^* is obtained when $\Re_{LUS}$ maximum.

3 Result and Analysis

Refers to the cellular model based on [8]. The simulation will be analyzed using the following parameters in the Table 1.

Table 1. Parameters for simulation

Parameters	Value
M-antenna BS	32
N single-antenna candidate UEs	64
Power (P)	30 dBm
Number of symbols in a coherence slot (T)	196
Radius minimum(R_{min})	35 m
Radius maximum (R_{max})	250 m
Path loss in reference distance (c)	$10^{-3.53}$
Noise power (σ_n^2)	-96 dBm
Antenna correlation coefficient (δ)	0.5
Estimated channel accuracy (ρ)	0.2

Using the parameter in Table 1, we obtained the value of correlation matrix on the transmitter side. In this research just considering correlation among transmit antenna BS. Correlation among receive antennas is independent of the correlation between transmit antennas. Therefore the matrix for 32x64 antennas is obtained:

$$R_{32\times32} = \begin{bmatrix} 1 & 0.5000 & 0.2500 & 0.1250 & \cdots & 4.6566\times10^{-10} \\ 0.5000 & 1 & 0.5000 & 0.2500 & \ddots & 9.3132\times10^{-10} \\ 0.2500 & 0.5000 & 1 & 0.5000 & \ddots & 1.8626\times10^{-9} \\ 0.1250 & 0.2500 & 0.5000 & 1 & \ddots & \vdots \\ \vdots & \ddots & \ddots & \ddots & \ddots & 0.5000 \\ 4.6566\times10^{-10} & \cdots & 0.1250 & 0.2500 & 0.5000 & 1 \end{bmatrix}$$

φ and ψ represent the unique solution value in the large-system regime. From simulation, we can get the value of $\varphi = 0.0192$ and $\psi = 3.6868 \times 10^{-4}$. If the value of the antenna correlation coefficient increased, then the value of φ and ψ will also increase but the sum rate will decrease. Therefore, when the correlation is higher then the sum rate decreases.

Simulation results can be seen from Table 2. Where LUS0 is the result of the LUS algorithm without any modification. LUS1 is the result of a modified LUS algorithm by adding sum ergodic from distance which is proven to improve performance. LUS0 and LUS1 are run under the same conditions. While LUS2 is a modification algorithm that runs on imperfect CSI condition where correlation is present in this scheme. LUS3 is the most real condition telling the state of the environment. At this time, we vary the path loss exponent value in

Table 2. Sum rate of simulation results in bps/Hz

K	LUS0	LUS1	LUS2	LUS3	K	LUS0	LUS1	LUS2	LUS3
1	13.9789	14.3724	6.2382	6.2462	17	74.3189	74.3289	22.6756	24.8436
2	24.6197	24.8361	9.7089	9.8131	18	73.3616	73.3705	22.1126	24.0298
3	33.4901	33.6347	12.4374	12.4914	19	71.9682	71.9762	20.3926	23.0858
4	41.0451	41.1493	14.6629	14.6600	20	70.1347	70.1417	20.5157	23.0374
5	47.5355	47.6141	16.5184	16.3958	21	67.8522	67.8585	19.4911	21.9191

6	53.1233	53.1845	18.0819	17.9909	22	65.1067	65.1124	18.3204	21.5407
7	57.9223	57.9710	19.4026	19.5575	23	61.8781	61.8832	17.0024	20.0371
8	62.0162	62.0558	20.5097	20.4332	24	58.1381	58.1427	15.5511	19.6291
9	65.4692	65.5019	21.4241	21.4594	25	53.8485	53.8527	13.9663	18.2946
10	68.3314	68.3588	22.1586	21.7955	26	48.9575	48.9612	12.2563	16.9970
11	70.6426	70.6657	22.7210	23.2224	27	43.3947	43.3979	10.4315	15.3220
12	72.4344	72.4542	23.1169	23.6526	28	37.0624	37.0653	8.5066	12.6050
13	73.7323	73.7494	23.3496	23.7560	29	29.8230	29.8254	6.4877	10.4667
14	74.5563	74.5711	23.4207	24.1961	30	21.4741	21.4760	4.3881	7.5631
15	74.9218	74.9347	23.3310	24.0356	31	11.7044	11.7056	2.2218	4.0176
16	74.8402	74.8515	23.0834	23.8680					

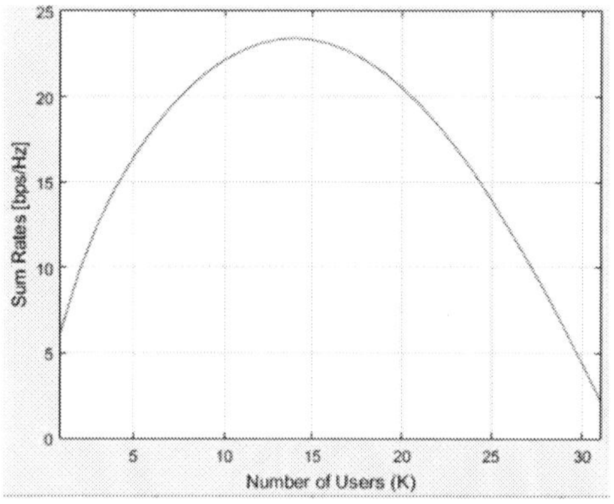

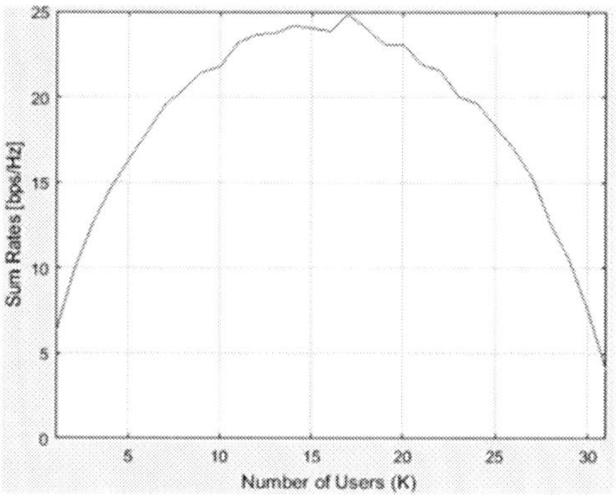

Fig 3. (a) The influence of UEs active (*K*) to the sum rate at the fix condition (b) The influence of UEs active (K) to the sum rate at different environmental for 300 iterations

sub-urban areas where the value is 3 to 3.5. Sum rate values of LUS2 and LUS3 can be seen in Figure 3 a and b, respectively.

The graph of sum rate values can be seen in Figure 3 in point (a) and (b). In Figure 3a, the number of optimal active UEs (K^*) which can be served to reach the optimal sum rate is 14 with sum rate is 23.4207 bit/s/Hz. The sum rate value decreases when correlation is present. In Figure 3a, the candidate UEs have a fix environmental condition with path loss exponent value is 3.76. The number of active UEs candidates that can be served is still similar as in previous studies when imperfect CSI can be seen in Figure 3a. However, the results of this simulation show an increase in the sum rate value than the previous one.

In Figure 3b, for case where the user is located in different positions with different environmental circumstances showing the highest sum rate is 24.8436 bps/Hz with 17 optimal UEs. In this situation, each UEs candidate is dispersed into various areas in the suburban area with a path loss exponent value in the range of 3 to 3.5. This scheme is very different from the scheme in previous research, where in previous studies only provided one condition in all UEs. The number of optimal UEs that can be served depends on environmental circumstances. If the environment is good, then the sum rate will be higher than the bad environmental conditions. The sum rate of 300 iterations is the average value of conditions repeatedly made in 300 times.

4 Conclusion

With the proposed LUS scheme, how many UEs to be served can be determined so that it will guarantee the rate for each UEs. The simulation results show that the optimal number of active UEs candidates is about half of the number of BS antennas. The result of our research have better performance in sum rate compared to previous research. For imperfect CSI conditions for every UEs are put in the same environmental conditions, the number of active UEs that can be served optimally is still the same as in previous research. Our research show an increase in sum rate. As for imperfect conditions and different environmental circumstances resulting in optimal number of active UEs and a sum rate corresponding to the circumstances of the environment. Better environment is needed to achieve optimal served active UEs and higher sum rate. To meet the demand for high data rates with fixed energy, the proposed scheme can be considered for massive device connectivity in 5G mobile communications.

References

1. H. Papadopoulos, C. Wang, O. Bursalioglu, and X. Hou, "Massive MIMO Technologies and Challenges towards 5G," no. 3, pp. 602–621, 2016.
2. N. Krishnan, R. D. Yates, and N. B. Mandayam, "Cellular Systems with Many Antennas : Large System Analysis under Pilot Contamination.", 2013.
3. H. Papadopoulos, C. Wang, O. Bursalioglu, and X. Hou, "Massive MIMO Technologies and Challenges towards 5G," no. 3, pp. 602–621, 2016.
4. T. Yoo, S. Member, and A. Goldsmith, "On the Optimality of Multiantenna Broadcast Scheduling Using Zero-Forcing Beamforming," vol. 24, no. 3, pp. 528–541, 2006.
5. X. Chen, Z. Zhang, and C. Yuen, "Resource Allocation for Cost Minimization in Limited Feedback MU-MIMO Systems with Delay Guarantee," pp. 1–20, 2012.
6. K. T. K. Cheung, S. Member, S. Yang, M. IEEE, and L. Hanzo, "Spectral and Energy Spectral Efficiency Optimization of Joint Transmit and Receive Beamforming Based Multi-Relay MIMO-OFDMA Cellular Networks," pp. 1–20, 2014.
7. H. Liu, H. Gao, S. Yang, T. Lv, and S. Member, "Low-Complexity Downlink User Selection for Massive MIMO Systems," pp. 1–12, 2015.
8. E. Björnson, L. Sanguinetti, and J. Hoydis, "Optimal Design of Energy-Efficient MultiUser MIMO Systems : Is Massive MIMO the Answer ?," vol. 14, no. 6, pp. 3059–3075, 2015.
9. Van Zelst and J. S. Hammerschmidt, "A single coefficient spatial correlation model for multiple-input multiple-output (mimo) radio channels," no. 1, pp. 2–5, 2002.
10. S. Wagner, S. Member, R. Couillet, and D. T. M. Slock, "Large System Analysis of Linear Precoding in Correlated MISO Broadcast Channels under Limited Feedback," pp. 1–28, 2012.
11. Y. Yang, J. Xu, G. Shi, and C. Wang, "5G Wireless Systems Simulation and Evaluation Techniques," 2018.

Switch-Beam Vivaldi Array Antenna Based On 4x4 Butler Matrix for mmWave

Nurlaila Safitri[1] , *Rina Pudji* Astuti[1,1,], and *Bambang Setia* Nugroho[1]
[1]Electrical Engineering, the School of Engineering, Telkom University. Bandung, Jawa Barat 40257, Indonesia.

Abstract. It will be diffcult to use either omnidirectional or fixed beam antenna due to the high propagation losses caused by atmospheric absorption at mmWave for 5G mobile communication. Several studies have been conducted recently using butler matrix which is part of switchable antenna with some advantages such as simple, minimal cost, low loss, etc. Previous studies also have designed vivaldi array antenna at 28 GHz which provides a fix beam directional radiation pattern with narrow beam that requires real phase setting. However, there has been no research using vivaldi antenna with butler matrix, whereas it has some advantages such as wide bandwidth, high gain, high directivity, etc. This paper proposed 4x4 butler matrix integrated with vivaldi antenna by using phase shift of 45 . The design is developed on a single layer of Rogers RT5880 with dielectric constant 2.2 and thickness 0.254 mm. Best results of simulation were picked for overall system at 28 GHz, and the results of antenna as follows: the return loss was below -10 dB, the realized antennas gain was 10.2 dB with unidirectional radiation pattern and bandwidth antenna of 6 GHz that covers from 25 GHz to 31 GHz. The butler matrix average phase di erent between output port are -44.106°, 137.38°, -137.66°, 43.95° with phase error of 0.894°, 2.38°, 2.66°, 1.06°. Antenna array that has been given di erent phase by butler matrix is able to shift radiation pattern on the input port successively with range of beam that can be achieved equal to 185°.

1 Introduction

One of the challenges in making a 5G antenna is to design a high gain antenna to avoid high pathloss due to the atmospheric absorption of electromagnet waves at higher frequencies[1][2]. The much higher gain that is needed to compensate the higher signal attenuation at mmWave introduces the concept of antenna arrays for cellular phones[3]. But the use of antenna arrays to increase the antenna gain causes the antenna radiation pattern to be directional thus requiring the use of beam steering on the antenna [4][5].

There are two ways to steering beam using smart antenna system such adaptive array system and switched-beam [6][7][8]. Switched beam antenna is choosen because it is less cost and simple to be implement and achieve most of the features of adaptive array including capability to steer the antenna beam [6]. Butler matrix is one of the well-known

[1] Corresponding author: rinapudjiastuti@telkomuniversity.ac.id

beam forming which can be used in switched-beam smart antenna system [9]. It is plays an important role in the transmitter and receiver for 5G communication with some advantages such as easy implementation [12], simple, minimal cost[10], the loss involved is very small[8][13]. This study was conducted on the basis of research [5][2] on the design of vivaldi antennas for 5G, which integrated with 4 4 butler matrix [6][11] at mmWave. The working frequency of this 5G antenna is 28GHz [13][5][2].

The proposed design of butler matrix and antenna is applied for 5G mmWave mobile communication at 28 GHz. It must be capable of generating phase di erence between the output ports to provide phase di erence to the input port of antenna array. Thus, the antenna array radiation pattern is capable of shifting in certain directions according to the phase difference assigned to each input. Apart from this introductory remark, this paper has additional three sections. Section II describes the design configuration. Section III discusses the results and analysis, the performance in terms of phase di erent in output port and radiation pattern in every input will be discussed. Finally, a conclusion will be drawn in Section IV.

2 Design Configuration

The proposed butler matrix utilized some elements such as 90 hybrid, phased shifter of 45 and crossover. These element consists of transmission lines and calculated using microstrip feedline. It is created by copper strips separated from the ground plane by a Rogers RT5880 substrate layer with dielectric constant 2.2 and thickness 0.254 mm in 28 GHz.

As indicate in figure 1, the proposed design consists of a butler matrix with four inputs and four outputs, each output of the butler matrix serves as a feeding line for the vivaldi antennas and direct the beam to the desired direction according to the selected input port. The phase di erence which will be generated on each output port with the selection of input ports respectively i.e are 45°, -135°, 135°, -45°.

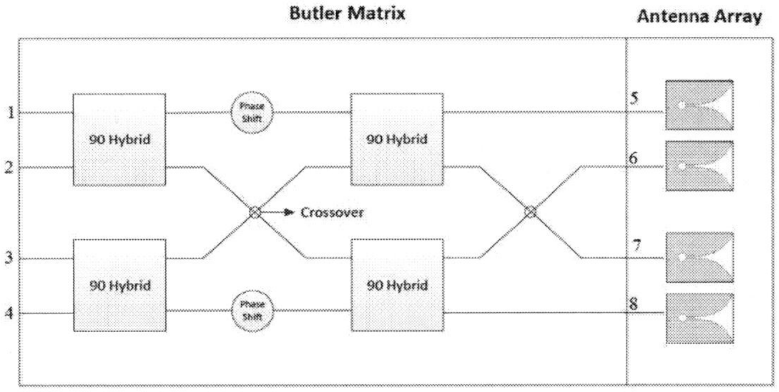

Fig 1. Design of proposed butler matrix and vivaldi array antenna

The first step to design butler matrix is design 90o hybrid which have functions as a power divider with phase margin difference at port 3 and port 4 is 90o caused by the length of each quarter-wavelength transmission line (□/4). The hybrid design consists of two impedance namely Z_o and Zo/□√2 withline length for each impedance of λ/4.

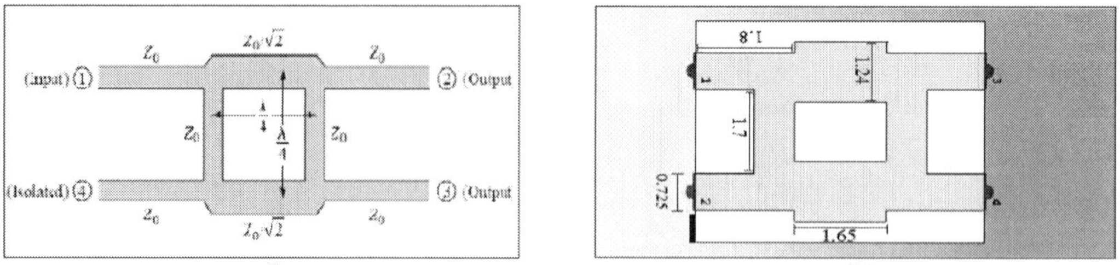

(a) (b)

Fig 2. Configuration of 90 hybrid in (a)geometry (b)simulation

Figure 3 shows that insertion loss S(1,3) and S(1,4) of 90 hybrid not exactly -3 dB because loss of feedline. While the isolation of S(1.2) and return loss S(1,1) magnitude -10 dB at 28 GHz frequency. The phase di erence between the output ports S(1.3) and S(1.4) is approaching 90 which has been in accordance with the desired hybrid specification.

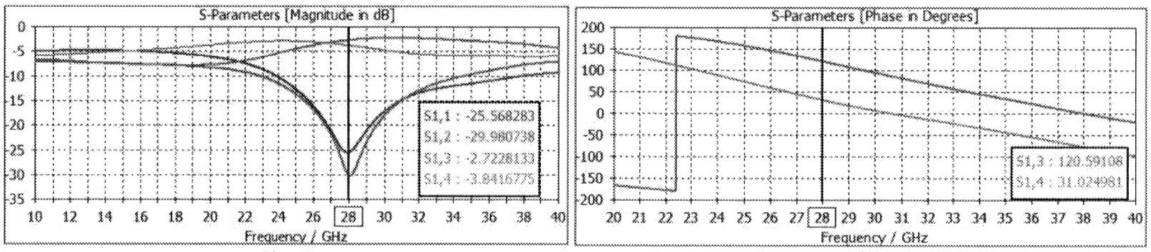

(a) (b)

Fig 3. Simulation result (a)insertion loss, return loss and isolation (b)phase di erent

The second step is design and simulate the crossover which used for obtain high isolation between two intersecting lines in order not to connect electrically by combining the 90 hy-brid in series, but it does not give satisfactory performance[11]. Therefore, the geometry used is shown in Figure 4 where mitred bend is used to prevent discontinuity on the crossover.

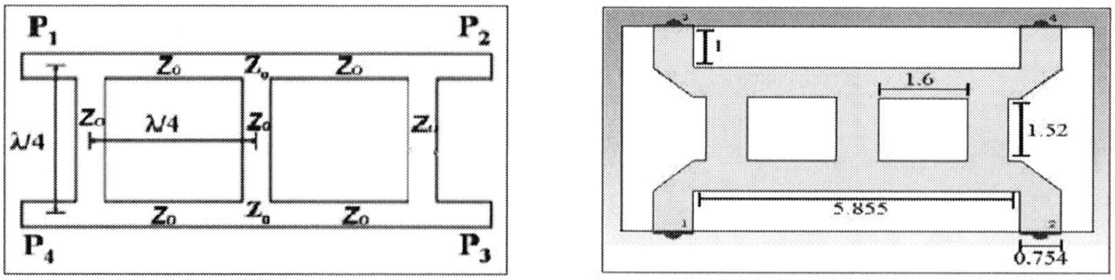

(a) (b)

Fig 4. Configuration of crossover in (a)geometry (b)simulation

The result of crossover simulation after optimized is showed in Figure 5, the insertion loss S(1,4) value is about to 0 dB due to channel loss. While the return loss S(1,1) and isolation

S(2,1) and S(3,1) values are less than -10 dB, which means that almost nothing is missed on those ports.

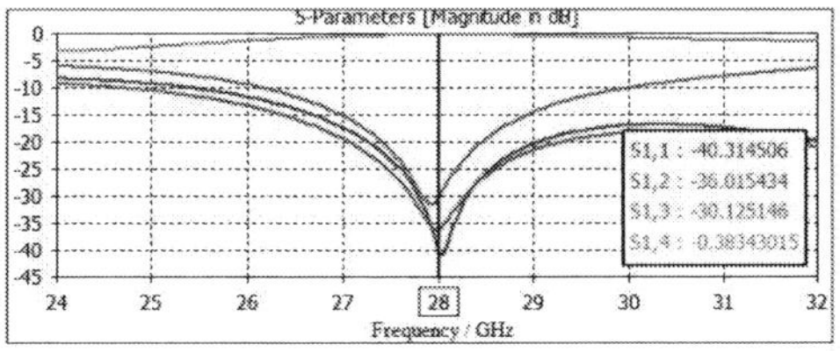

Fig 5. S-Parameter at 28 GHz for crossover

The third step is design and simulate the phase shift which used for steered beam antenna in the desired direction without repositioning the antenna physically [9][10]. The phase out-put at the phase shifter can be seen in figure 6 where at 28 GHz frequency, the phase output is approaching -45°

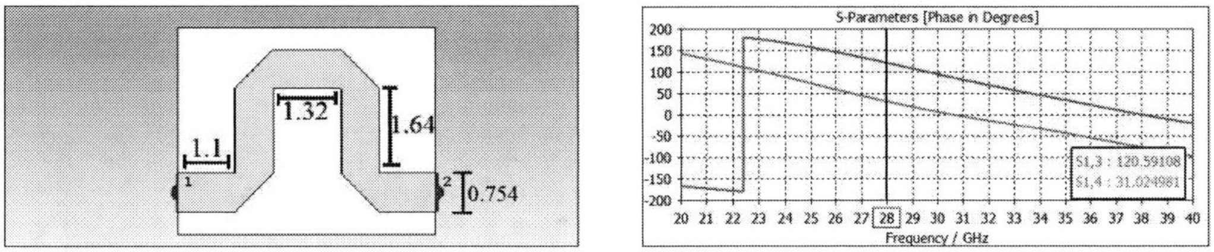

(a) (b)

Fig 6. Geometry of phase shifter after optimization

The final step is design and simulate a vivaldi antenna to an array. Vivaldi antenna config-uration is obtained based on [13] and modified the vivaldi input to make distance of vivaldi input as sama as butler matrix. The optimized vivaldi can be seen in Figure 7 where the vivaldi dimension is 6.6 mm 9 mm.

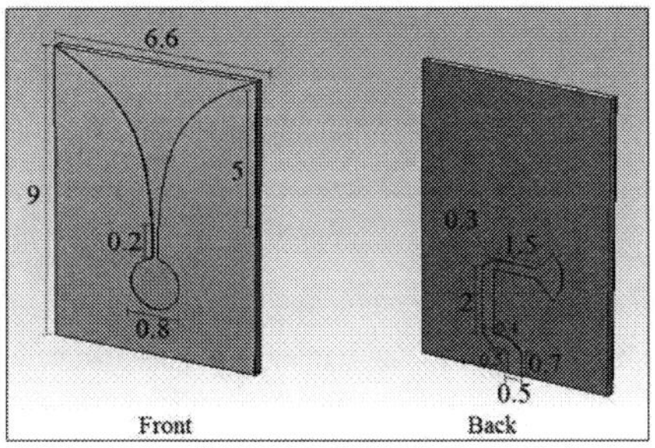

Fig 7. Optimum design of vivaldi antenna at 28 GHz.

The return loss value S(1,1) and radiation pattern of vivaldi antenna is shown in Figure 8. The return loss value at 28 GHz is -24.09165 dB which is less than -10 dB while bandwidth antenna is 6 GHz. It is also has unidirectional radiation pattern with realized gain about 4.13 dB.

 https://doi.org/10.1051/matecconf/201821803011

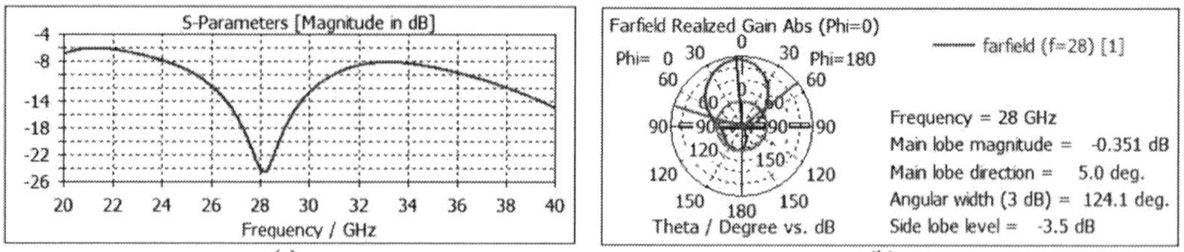

(a) (b)

Fig 8. Result of vivaldi simulation (a)returnloss (b)radiation pattern

3 Result and Analysis

Hereafter, all the individual components of butler matrix are combined on a single substrate together without antenna, it has dimension of 20.52 mm 26.4 mm. After the butler matrix has been designed to provide the output of phase di erence of 45°, -135°, 135°, -45° on the output port, then it was integrated with four vivaldi antennas which has been designed, the final design of butler matrix-antenna array is shown in Figure 9 with dimension of 20.52 mm 29.52 mm.

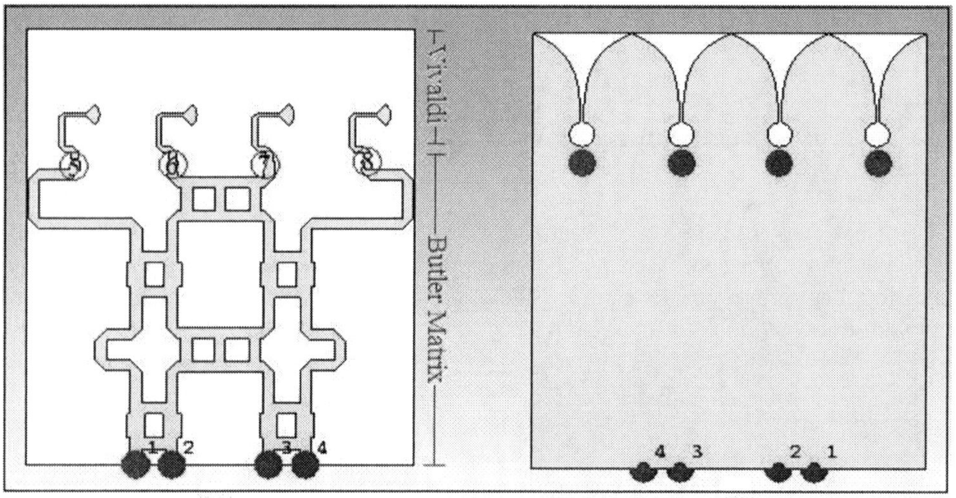

(a) (b)

Fig 9. Simulation of butler matrix array (a)top side (b)bottom side

Table 1. The phase di erence between butler matrix output ports

Input Port	Port(°)			Average(°)	Theoritical Target (°)	Error (°)
	5-6	6-7	7-8			
1	-43.49	-47.02	-41.81	-44.106	-45	0.894
2	150.85	114.63	146.66	+137.38	+135	2.38
3	-146.46	-115.7	-150.83	-137.66	-135	2.66
4	41.82	46.67	43.38	43.95	+45	1.06

It can be seen from Table 1 which obtained through simulation with phase error on the output port of butler matrix respectively are 0.894 , 2.38 , 2.66 , 1.06 . While radiation pattern of butler matrix-vivaldi array antenna in every port are shown in Figure 10.

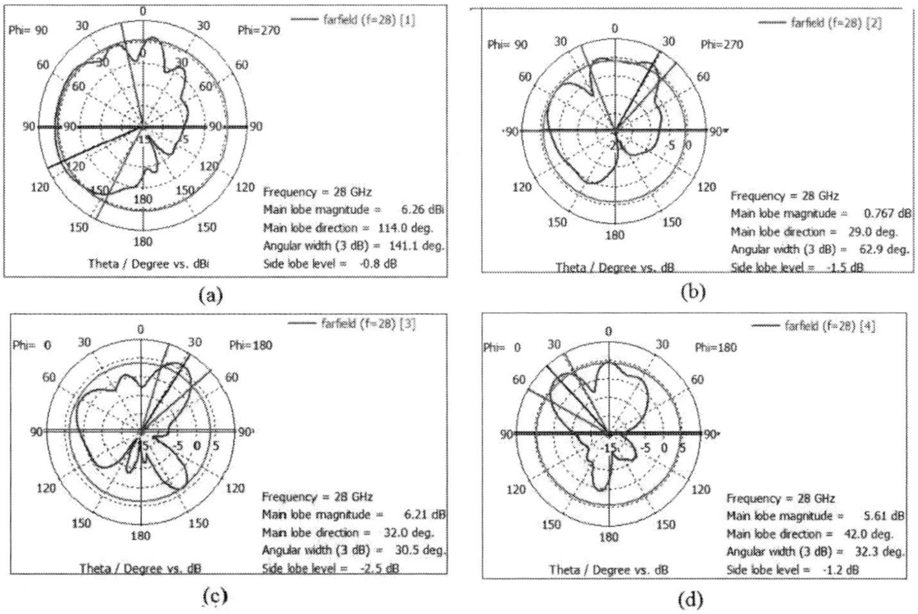

Fig 10. Radiation pattern in every input (a)Port 1 (b)Port 2 (c)Port 3 (d)Port 4

The antenna radiation pattern on each input port when all input ports are combined can be seen in Figure 11. It shows that antenna radiation pattern changes depends on the phase di erence given in each di erent input. The main lobe direction obtained successively on port 1,2,3,4 successively -141.1°, -29°, 32°, 42°. The main lobe direction which is not symmetrical due to the phase di erence in the output port varies in value. To obtain the sequential beam shift from the data, the sequence of input ports selected are 1, 4, 2, 3 with maximum beam range is 185 in azimuth plane.

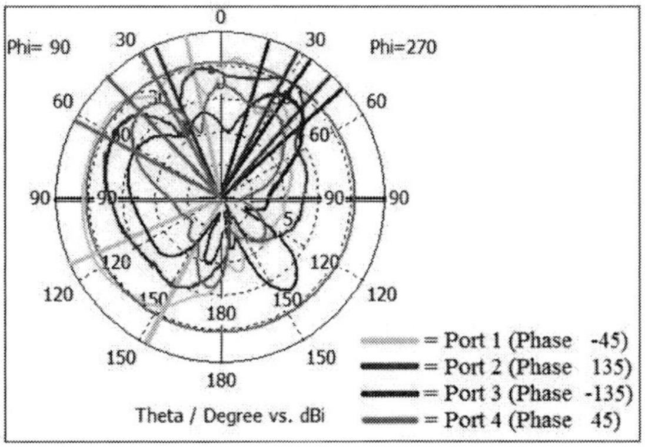

Fig 11. Radiation pattern in every input

4 Conclusion

This paper presents a 4x4 butler matrix using vivaldi array antenna at mmWave for 5G mobile communication. All component of butler matrix such as 90° hybrid, crossover and phase shifter are designed individually. Vivaldi antenna simulation result showed that the antenna can cover from 25 GHz to 31 GHz with the highest array gain performance of 10.2 dBi at 28 GHz. The simulation results of the butler matrix demonstrated a good performance at operating frequency in terms of S-parameters and phase differences in output port. The beam can be switched toward in different directions by switching any of the input ports which will be suitable for 5G mobile communication which would support the future 5G and operate at 28 GHz of frequency. Thus, it has intended to get the butler matrix-vivaldi antennas fabricated and tested to validate the design method. The fabrication will be done in a place where fabrication and measurement equipment are available.

References

1. A. Kumar and P. Kapoor, "Design and Performance Evaluation of a Dual- Band Antenna for the 5G Mobile Communication," pp. 2034–2036, 2016.
2. A. Rahman, A. El, and B. Ahmed, "Design of Vivaldi Antenna Array with Enhancement of Radiation Characteristics for 5G Mobile Applications Design of Vivaldi Antenna Array with Enhancement of Radiation Characteristics for 5G Mobile Applications," no. March, 2017.
3. W. Hong, K. H. Baek, Y. Lee, Y. Kim, and S. T. Ko, "Study and prototyping of practically large-scale mmWave antenna systems for 5G cellular devices," IEEE Commun. Mag., vol. 52, no. 9, pp. 63–69, 2014.
4. N. Ojaroudiparchin, M. Shen, and G. F. Pedersen, "Design of Vivaldi antenna array with end-fire beam steering function for 5G mobile terminals," 2015 23rd Telecommun. Forum, TELFOR 2015, pp. 587–590, 2016.
5. W. Hong, "Solving the Antenna Puzzle," no. October 2017.
6. S. Ifeoma, "A 4x4 Butler Matrix for 28 GHz Switched Multi-Beam Antenna A 4x But-ler Matrix for 28 GHz Switched Multi-Beam Antenna," vol. 7, no. February 2016, pp. 436–442, 2015.
7. H. Nachouane, A. Najid, A. Tribak, and F. Riouch, "Broadband 4X4 Butler matrix using wideband 90; hybrid couplers and crossovers for beamforming networks," 2014 Int. Conf. Multimed. Comput. Syst., no. APRIL, pp. 1444–1448, 2014.
8. W. Bhowmik and S. Srivastava, "Optimum Design of a 4x4 Planar Butler Matrix Array for WLAN Application," vol. 2, no. 1, pp. 68–74, 2010.
9. P. F. I. Shaikh and S. B. Akhade, "Smart Antenna System using 4x4 Butler Matrix switched beam network for 2 . 4 GHz ISM band," vol. 4, no. 3, pp. 278–282, 2015.
10. S. V. Kalam and A. B. Rathi, "Optimum Design of 4X4 Symmetrically Structured Butler Matrix," vol. 5, no. 1, pp. 31–34, 2016.
11. H. Abdul and R. Al, "Design of a 4x4 Butler Matrix for Vehicle Radar Beamforming Antenna systems at 24 GHz By Hani Abdul Rahman Al Habibi."
12. Firmansyah, T. Praptodinoyo, S. et al. "Dual-wideband band pass filter using folded cross-stub stepped impedance resonator". Microwave and Optical Technology Letters.Vol.59(11). pp. 2929-2934. November 2017.
13. Firmansyah,T. et al. "Hepta-band bandpass filter based on folded cross-loaded stepped impedance resonator". Electronics Letters(2017), vol. 53 (16):1119.
14. M. Bouezzeddine, W. L. Schroeder, and T. Kaiser, "On the Design of MIMO Antennas for Cognitive Radio Systems," vol. 2, no. 2, pp. 2–6.
15. G. K. Pandey, H. S. Singh, P. K. Bharti, a. Pandey, and M. K. Meshram, "High Gain Vivaldi Antenna for Radar and Microwave Imaging Applications," Int. J. Signal Process. Syst., vol. 3, no. 1, pp. 35–39, 2014.

Banten Batik Classification with Backpropagation Neural Network

Rian Fahrizal[1]*, *Ridwan Prasetya* Parlindungan Siahaan[2], and *Romi* Wiryadinata[3]

[1,2,3] Department Electrical Engineering, Universitas Sultan Ageng Tirtayasa

Abstract. Batik Cloth is Indonesia heritage of the indigenous cultures. However, the lack of public awareness could made the other nations has claimed that Batik cloth as their heritage. Therefore, We will need serious attention to take the problems. The classification Banten pattern batik cloth had been made to protect the Batik pattern. The classification had been made with Artificial Neural Network with Lavender-Marquant methods. Gray-Level Co-Occurrence Matrix (GLCM) has been used to preprocessing the pattern from Gray scale images. The results showed that the Batik patterns had 97% accuracy and 2% error.

1 Introduction

One of the Indonesian heritage was Batik. It has been recognized as an intangible cultural heritage [1]. However, the lack of public awareness of national culture led to the recent years, batik was claimed and recognized as another nation's culture. Therefore, serious attention was needed to prevent this accident. There were many batik pattern in Indonesia, one of kind was Banten Batik. These batiks had the distinctive color, which is the soft gray color. It showed the character Banten Citizen, that showed the dream, ideas, willing, and the high tempered but the appearance was simple and calm. The Batik color showed the natural environment in Banten. The Batik Banten pattern were Datu Laya, Kapurban, Kawangsan, Mandalikan, Pamaranggen, Pancaniti, Pasepen, Pasulaman, Pejantren, Sabakingking, Srimanganti, and Surosowan. Batik Banten had 10 classification class.

Some methods had been used to clssified Batik Pattern. Batik classification's method were Support Vector Machine [2], Neural Network [3], and Decession Tree Classification [4]. The neural network had great advantages in parallel computation for classified the pattern. Neural network classification could recognized more than 10 class and fast learning convergence for classification. Another advantages were it could adaptive data driven self method that is able to regulate itself against the data that will be processed without requiring an explicit specification function or the form had been distributed, the apportion function of the general public which had accuracy in the classification process, a nonlinear model so that this model became flexible in modeling complex relationships, estimated the subsequent possibilities that provide the basis for classification rules and statistical analysis. Neural networks had been used to classified Batik Banten pattern because of their

* Corresponding author: rian.fahrzal@untirta.ac.id

advantages. The variety of batik pattern that vary in Indonesia made it difficult for the general public to identified batik patterns in each region in the archipelago. The same goes for the Banten batik motif which had approximately 10 batik motifs. This study tried to classify Banten batik motifs that aim to facilitated the public who do not knew the specifications of batik motifs in Indonesia, especially Banten regional batik motifs. Pattern feature extraction that is used as input to the system that is built is GLCM (Gray Level Co-occurrence Matrix) using a gray degree distribution (histogram). The final step is to classify using the artificial neural network method, that is, Backpropagation is used to classify Banten's batik motifs that are more accurate.

Banten Batik is one of the historic batik originating from Banten Province. In this paper, extensive information was needed in processing Banten Batik so that it matched the concept carried without changing the authenticity of Banten Batik. These were batik Banten pattern showed in Figure 1:

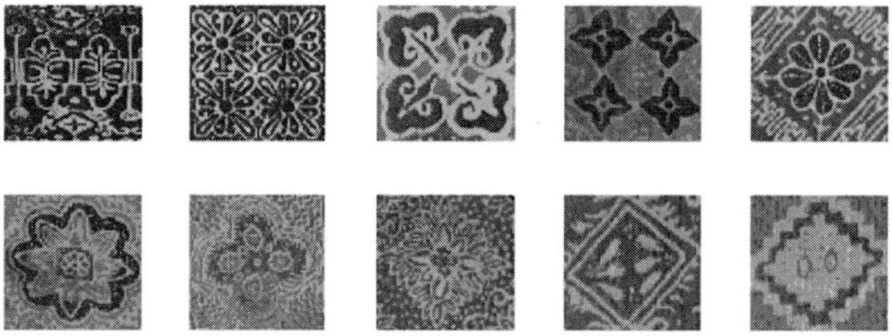

Fig. 1. Batik Banten

2 Research Method

Input image were batik images obtained from batik craftsmen using a Canon D550 camera. The banten batik taken was 300 images consisting of 200 training images and 100 test images. which consists of 10 classes patterns. The methods of artificial neural networks used in this research was backpropagation, in general the algorithm was referred to as back propagation because it can be described as follows, when the network is given an input pattern as a training pattern then the pattern goes to units in the hidden layer to be forwarded to the unit - output layer unit, then the output layer units gave a response called network output. If the output network is not the same as the expected output, the output spreads backward, in the hidden layer is forwarded to the unit at the input layer, therefore the training mechanism is called backpropagation.

Figure 2. explained the algorithm of an artificial neural network design, when the user starts to provide input training data, the training process will be done to get optimal weight on each layer. After optimal weight is obtained, then the optimal weight is stored for the testing process and information on the case of the classification of batik motifs. The results and tests will be compared with the existing targets to determine the system's ability to study the training data provided to the system.

2.1 Preprocessing Image

Batik's images taken from the camera and cropping it with different pixels size. Simplification had been made to resize the images. The resizing image became 640 x 640 pixels with RGB value. GLCM methods needed gray scale image. Resizing images converted to the gray images with *doubles floating point* values. Contrast Limited Adaptive Histogram Equalization (CLAHE) was implemented to enhancement gray images. This method was developed to prevent the over amplification of noise in picture.

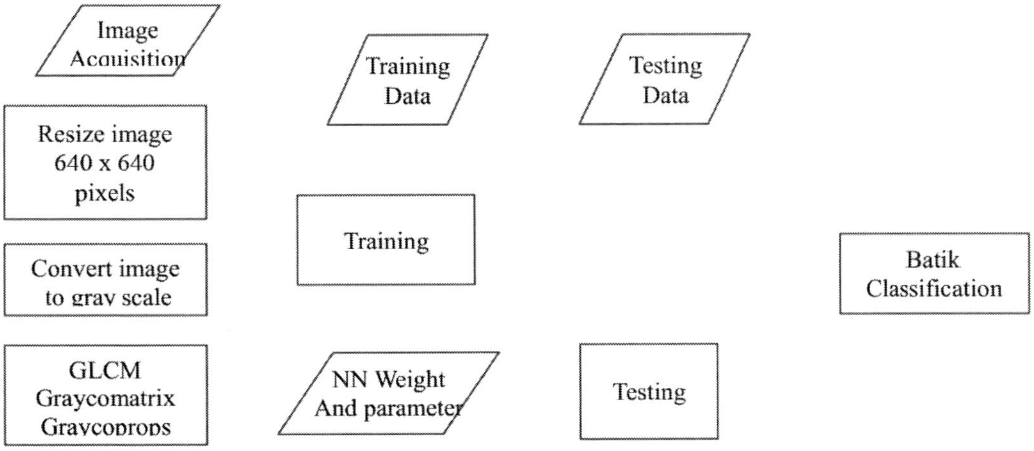

Fig. 2. Batik Banten Classification Algorithms

This method consist the following steps:

1. Collect all the inputs: Image, Number of regions in row and Column paths, Number of bins for the histograms used in image transform function (dynamic range), Clip limit for contrast limiting (normalized from 0 to 1)
2. Preprocess the inputs: Rectify real clip limit from the normalized value if necessary, glorify the image before splitting it into regions
3. Computing each contextual region (tile) thus made gray Level mappings: Thread a single image region, made a Histogram for this region using the specified number of bins, clip the histogram using clip limit, created mapping for this region
4. Computed gray level mapping in order to configure final CLAHE image: Theread cluster of four neighboring mapping Functions, compute image region partly overlapping each Of the mapping tiles, thread a single pixel, apply four Mappings to that pixel, and interpolate between the events to obtain the output pixel; render over the entire image

2.2 Feature Extraction

Feature extraction's method used was Gray Level Co-occurance Matrix (GLCM). GLCM method is a form of removing second order statistical texture features.

A GLCM is a matrix where the number of rows and columns is close to the number of gray levels, G, in the image. The maximum number for the level was 8 and minimum level was 1. The matrix point P (i, j | Δx, Δy) is the relational frequency with which two pixels, isolated by a pixel distance (Δx, Δy), pass within a given neighborhood, one with intensity 'i' and the other with intensity 'j'. The matrix point P (i, j | d, ɵ) included the second order statistical probability values for automatise between gray levels 'i' and 'j' at a specific replacing distance d and at a particular angle (ɵ). Using a large number of intensity levels G showed storing a lot of working data, i.e. a G × G matrix for each collection of (Δx, Δy) or (d, ɵ). Due to their large magnitude, the GLCM's are very reactive to the size of the texture instances on which they are asses. Thus, the number of gray levels is often attenuated. GLCM matrix formulation can be interpreted with the samples illustrated in fig 3. for four varied gray levels. Here one pixel offset is utilized (a reference pixel and its immediate neighbor). If the window was large enough, using a larger offset is mathematical. The top left cell will be full with the number of times the combination 0,0 occurs, i.e. how many time within the image area a pixel with grey level 0 (neighbor pixel) falls to the right of another pixel with gray level 0(reference pixel).

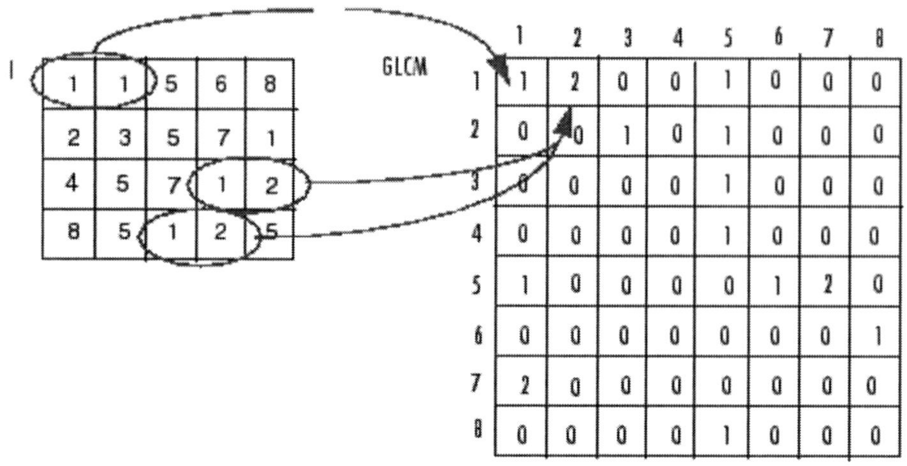

Fig. 3. GLCM Calculation

Four features extraction had been used to get GLCM values which were Contrast, Correlation, Energy, and Homogeneity.

2.2.1 Contrast

The definition of Contrast was the intensity contrast between a pixel and its neighbor over the whole picture. 'Sum of Square Variance' is the other name of Contrast. It defers the calculation of the intensity contrast linking pixel and its neighbor over the whole image. At constant image contrast value is 0.In contrast determine, weight accumulate exponentially(0,1,4,9) as continue from the diagonal.

$$\sum_{i,j} |i-j|^2 p(i,j) \tag{1}$$

Since (i-j) increases contrast proceeded to increase exponentially. When i and j are equal i.e. i-j=0.no contrast is there. When i and j are differ by 1,small contrast is there is 1. When i and j diverged by 2, the contrast is expanding and weight is 4

2.2.2 Correlation

It moved the computation of the correlation of a pixel and its neighbor over the whole picture means it figures out the linear state of gray levels on those of neighboring pixels. On behalf an absolutely positively or negative related image, the correlation value is 1 and -1.On behalf of unvarying image its value is NaN..Range=[-1,1] and the formula is

$$\sum_{i,j} \frac{(i-\mu i)(j-\mu j)p(i,j)}{\sigma_i \sigma_j} \tag{2}$$

2.2.3 Energy

Since energy is utilized for doing work, Thus system. It made use for the texture that computed orders in an image. It gave the sum of square elements in GLCM. It was fully different from entropy. When the window is good orderly, energy value is high .The square root of ASM(Angular Second Moment) texture character is used as Energy. Its range is[0 1].Since constant image its value is 1.The equation of energy is

$$\sum_{i,j} p(i,j)^2 \tag{3}$$

2.2.4 Homogeneity

In brief term it is going by the name of HOM. It moved the value that computed the tightness of sampling of the elements in the GLCM to the GLCM diagonal. For diagonal GLCM its value is 1 and its range is [0,1].Opposite of contrast weight is homogeneity weight values, with weight decreases exponentially loose from the diagonal. The weight employed in contrast is (i-j)^2 and in homogeneity ,it is 1/1+(i-j)^2.The equation is

$$\sum_{i,j} \frac{p(i,j)}{1+|i-j|} \tag{4}$$

2.3 Backpropagation Neural Networks

The method of artificial neural network used in this study is backpropagation, in general, this algorithm is referred to as back propagation because it can be described as follows, when the network is given an input pattern as a training pattern then the pattern goes to units in the hidden layer to be forwarded to the unit and the output layer unit, then the output layer units give a response called network output, if the network output is not the same as the expected output, the output will spread backward, in the hidden layer forwarded to the unit at the input layer, therefore the training mechanism is called backpropagation or back propagation. Lavenberg-Marquardt had been used to classified Batik Banten. This method outperforms simple gradient descent and other conjugate gradient methods in a wide variety of problems. The algorithm provide a solution is called Nonlinear Least Squares Minimization. Figure 4 showed the input layer consisted 160 input, with 20 hidden layers, and 10 output layers, and 10 output layers.

Backpropagation is a supervised learning algorithm and is usually used by multiple layers of perceptrons to change the weights connected to the neurons in the hidden layer. The backpropagation algorithm uses error output to change its weight values in the backward direction. To get this error, the forward propagation stage (forward propagation) must be done first. When forward propagation, neurons are activated by using the sigmoid activation function. The working principle of the backpropagation algorithm is as follows, during the forward feed process, the neurons in the input layer will send information to each neuron in the hidden layer. This hidden layer neuron will do the computation process of the information received and produce output derived from its activation function. Then the output from the hidden layer will be sent again to the next layer, which in this case is the output layer. In this output, the same process takes place as in the hidden layer. However, the output from the activation output of this layer will be used as a response signal from Athe NN. This ANN output response signal will be compared with the desired target and calculated the error magnitude (the difference between the output produced with the target to be achieved). This error was sent back to the previous layers and then used to correct and change the ANN weights based on the learning error given to the ANN.

2.3.1 Training

The training phase is a step in how an artificial neural network is practiced, that is by making a weighing change (the connection between layers that form a network through each unit), while solving new problems will be done if the training process is completed, the phase is the mapping phase or testing process. The training process uses 20 images of each batik, the image is an image.

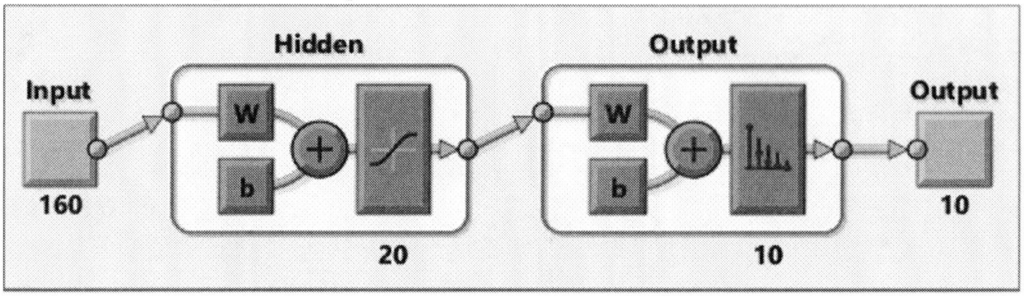

Fig. 4. Neural Networks Backpropagation

The training process by the backpropagation network required several parameters that needed to be set for learning with the trainlm function. The parameters that need to be set were:

1. Maximum epoch is the number of maximum or maximum epochs iterations that may be carried out during the training process.
2. Purpose performance is the target value of the performance function, or used to determine the Mean Square Error (MSE) value limit so that the iteration is stopped. The iteration will be stopped when the MSE value is smaller than the specified limit.
3. Learning Rate is the rate of learning. The greater the learning rate will have implications for the greater the learning step, if the learning rate is set too large, the algorithm will not become stable, and vice versa, if the learning rate is set too small, the algorithm will converge for a very long time.
4. The minimum gradient is the root of the least number of squares (input weight, layer weight, bias weight) that is the smallest allowed.

2.3.2 Testing

The testing process is carried out to determine the success rate of a network in recognizing a new input pattern, where the testing process compares the value of all outputs with the target. The testing processes used 100 images from 10 Batik patterns classification to check the system.Testing is done by entering images that have feature extraction using GLCM. The results of feature extraction are then entered into the neural network and classification is done to validate the results.

3 Result and Discussion

The original image of batik obtained from batik craftsmen in the city of Serang. then the cropping process was done manually to take normal Banten batik as many as 300 images. The obtained image data was divided into 200 images of Banten batik as training data (training) consisting of normal and abnormal Banten batik while 100 images of Banten batik as testing data.

The results of the training are carried out like architectural designs that have been explained previously, the training process will stop when one of the requirements for termination of the network formation process is fulfilled. The conditions that determine the cessation of the formation of this backpropagation network include achieving maximum epoch or maximum iteration, the set MSE value has been reached, the minimum gradient value has been reached, and also the validation check that has been fulfilled. The training process that has been designed.

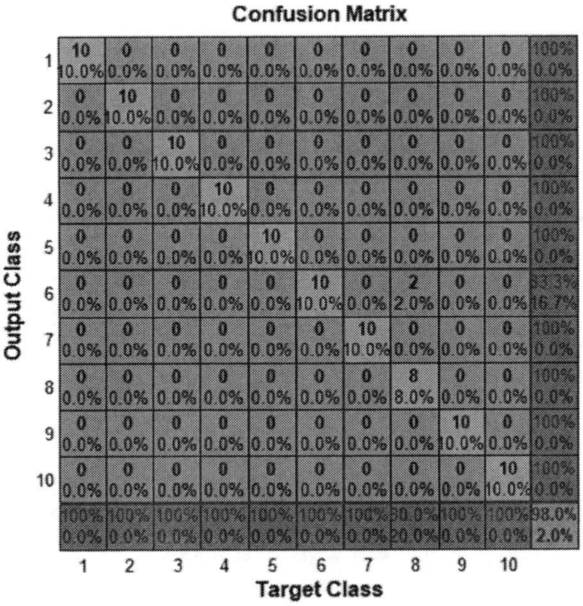

Fig. 5. Confusion Matrix

Figure 5. showed the result of the system. The system presented the 98% in accuracy value. Classification Class were :

1. Datu Laya,
2. Kapurban
3. Kawangsan
4. Mandalikan
5. Pamaranggen
6. Pancaniti
7. Pasepen
8. Pasulaman
9. Pejantren
10. Surosowan

Figure 5. showed the results of testing with the image of Banten batik, it can be seen that not all Banten batik can be recognized by the backpropagation network that has been built. Figure 5. showed that there are still targets that are not or close to the target that has been determined, therefore the network cannot recognize all Banten batik. System validation values for testing are:

$$accuracy = \frac{10+10+10+10+10+10+10+8+10+10}{10+10+10+10+10+10+10+10+10+10} x100\% = 98\%$$

$$error = \frac{2}{10+10+10+10+10+10+10+10+10+10} x100\% = 2\%$$

Batik Banten classification presented the accuracy value 98% and the error rate 2%. The system successfully classified Batik Banten class from 10 different pattern. The system had better value than [5] that had accuracy [3]97.67%,[6] with accuracy value 72%,[7] which had accuracy value 99,5%, but only 2 class,[8]images classification accuracy was 86%, [9] had accuracy value 99,96% but with 5 class had been used ,[4], [10]–[14] had accuracy < 92%,[15] had accuracy 80%, [16] had accuracy value 94,57%.

4 Conclusion

Batik Banten had special texture and shape of specific ornaments that could be used to classification the image of batik. The unique ornament of Batik Banten had size and shape that vary widely in 10 class. Training using Neural network with GLCM to feature extraction had high accuracy such as 98%, and low error as 2%. This method showed promising algorithm to classify Batik Banten, and could be used in face recognation.

References

1. J. Hidayat, "The Art and Sustainable Aspects of Natural Dyeing in KANAWIDA Hand Drawn Batik (Green Batik)," *IPTEK J. Proceeding Ser.*, (2014)
2. R. Azhar, D. Tuwohingide, D. Kamudi, Sarimuddin, and N. Suciati, "Batik Image Classification Using SIFT Feature Extraction, Bag of Features and Support Vector Machine," in *Procedia Computer Science*, (2015)
3. A. H. Rangkuti, "Content based batik image classification using wavelet transform and fuzzy neural network," *J. Comput. Sci.*, (2014).
4. A. H. Rangkuti, Z. E. Rasjid, and D. J. Santoso, "Batik Image Classification Using Treeval and Treefit as Decision Tree Function in Optimizing Content Based Batik Image Retrieval," in *Procedia Computer Science*, (2015)
5. R. Azhar, D. Tuwohingide, D. Kamudi, Sarimuddin, and N. Suciati, "Batik Image Classification Using SIFT Feature Extraction, Bag of Features and Support Vector Machine," *Procedia Comput. Sci.*, (2015)
6. A. Kitipong, W. Rueangsirasak, and R. Chaisricharoen, "Classification system for traditional textile: Case study of the batik," in *13th International Symposium on Communications and Information Technologies: Communication and Information Technology for New Life Style Beyond the Cloud, ISCIT 2013*, (2013)
7. F. Budiman, A. Suhendra, D. Agushinta, and A. Tarigan, "Wavelet decomposition levels analysis for Indonesia traditional batik classification," *J. Theor. Appl. Inf. Technol.* (2016)
8. T. Handhayani, "Batik Lasem images classification using voting feature intervals 5 and statistical features selection approach," *Proceeding - 2016 Int. Semin. Intell. Technol. Its Appl. ISITIA 2016 Recent Trends Intell. Comput. Technol. Sustain. Energy*, pp. 13–16, (2017)

9. I. Nurhaida, R. Manurung, and A. M. Arymurthy, "Performance comparison analysis features extraction methods for Batik recognition," in *2012 International Conference on Advanced Computer Science and Information Systems, ICACSIS 2012 - Proceedings*, (2012)
10. N. Suciati, A. Kridanto, M. F. Naufal, M. Machmud, and A. Y. Wicaksono, "Fast discrete curvelet transform and HSV color features for batik image clansificotlon," in *Proceedings of 2015 International Conference on Information and Communication Technology and Systems, ICTS 2015*, (2016)
11. H. Wijayanto, "Klasifikasi Batik Menggunakan Metode K-Nearest Neighbour Berdasarkan Gray Level Co-Occurrence Matrices (Glcm)," *Klasifikasi Batik Menggunakan Metod. K-Nearest Neighb. Berdasarkan Gray Lev. Co-Occurrence Matrices (Glcm)*, (2015)
12. C. S. K. Aditya, M. Hani'Ah, R. R. Bintana, and N. Suciati, "Batik classification using neural network with gray level co-occurence matrix and statistical color feature extraction," in *Proceedings of 2015 International Conference on Information and Communication Technology and Systems, ICTS 2015*, (2016)
13. F. U. Karimah and A. Harjoko, "Classification of batik kain besurek image using speed up robust features (SURF) and gray level co-occurrence matrix (GLCM)," in *Communications in Computer and Information Science*, (2017)
14. F. Budiman, A. Suhendra, D. Agushinta, and A. Tarigan, "Determination of SVM-RBF kernel space parameter to optimize accuracy value of Indonesian Batik images classification," *J. Comput. Sci.*, (2017)
15. A. A. Kasim, R. Wardoyo, and A. Harjoko, "The selection feature for batik motif classification with information gain value," in *Communications in Computer and Information Science*, (2017)
16. T. Handhayani, "Batik Lasem images classification using voting feature intervals 5 and statistical features selection approach," in *Proceeding - 2016 International Seminar on Intelligent Technology and Its Application, ISITIA 2016: Recent Trends in Intelligent Computational Technologies for Sustainable Energy*, (2017)

Strategy of smart meter infrastructure implementation using LPWAN technology, pilot project PLN Bali Case Study

Gunawan Wibisono[1,*] and *Nofiandri* Badruzzaman[2]

[1,2]Dept of Electrical Engineering, Universitas Indonesia, Kampus Baru UI, Depok, INDONESIA

Abstract. To realize the Bali Eco Smart Grid program, state electricity company (PLN) Bali implemented the advanced metering infrastructure (AMI) through the deployment of a pilot project of 1,000 2 ways smart meters based on LoRa WAN in the Kuta area, Bali. The selection of long range wide area network (LoRa WAN) as smart meter access technology is due to low costs, low power, ease of implementation and technology readiness. LoRa WAN is one of low power wide area network (LPWAN) technology. For deploying smart meter AMI based on LoRa WAN throughout Bali, the right strategy is needed. This research will analyze the strategy of deploying smart meter AMI based on LoRa WAN using strength, weakness, opportunity, and threat (SWOT) analysis. The stages of SWOT analysis are through the collection of internal and external factors obtained through techno-economic studies and surveys to customers, the general public, and PLN employees. Next, matching it by compiling an internal and external evaluation matrix, where an appropriate strategy is a progressive strategy (SO strategy). Components S and W are obtained from the survey while O and T are obtained from observing conditions outside of PLN. It is shown from the SWOT analysis results, obtained S-W = 5.95 and O-T = 2.59, so the strategy is in location (5.95, 2.59) quadrant I. Strategies for deploying smart meters based on WAN LoRa are progressive strategies.

1 Introduction

Decree of the Minister of Energy and Mineral Resources number: 1415K/20/MEM/2017 concerning the Electricity Supply Business Plan (RUPTL) 20117-2026 with one focus on developing new and renewable energy in Indonesia where Bali is used as a center of excellence for energy development. This is in line with the regional government of Bali who wants to make Bali a tourist Eco Green. To start the "Bali Clean and Green" program of the State Electricity Company (PLN) Bali together with the regional government of Bali inaugurated the program "Eco Smart Grid towards the province of Bali Clean and Green" in 2016 and held an advance meter infrastructure (AMI) pilot project on the electric power distribution system to customers based on low power wide area networks

* Corresponding author: gunawan@eng.ui.ac.id

(LPWAN) since 2017 as a substitute for the existing automatic meter reading (AMR). Because, AMR raises several problems including high investment and operational costs, can only be used for prepaid and cannot be for postpaid customers. AMI is a solution for prepaid and postpaid systems, meter data can be accessed by customers, supports the utilization of new renewable energy, on-line billing systems, and many other features. It can be said that the deployment of a smart meter AMI based on LoRa WAN in order to improve customer service and internal efficiency of PLN. Fig 1. shows the evolution of PLN Bali towards Bali Eco Smart Grid, where PLN Bali has gradually developed an electrical system towards the integration of AMI and distribution automation system (DAS). It is seen that AMI will be fully integrated with DAS in 2025.

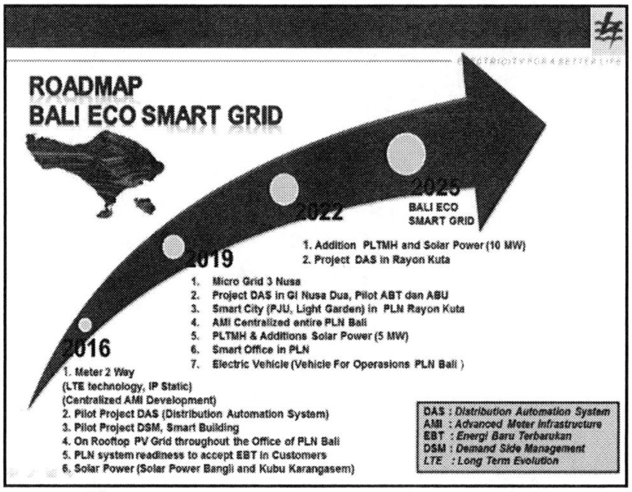

Fig 1. Roadmap Bali Eco Smart Grid [1]

The integration of AMI and DAS requires a large investment and impacts on the management and business of electricity, so a feasibility study that includes technical aspects, social aspects, economic and business aspects is needed. Techno-economic analysis of the application of smart meters based on LoRa WAN in PLN Bali has been carried out by [3]. In this research a feasibility study has been carried out which has resulted in 3 business models that can be used in the implementation of smart meter based on LoRa WAN by PLN Bali including the own build (BO), build operate transfer (BOT), and leasing. If the Bali PLN has sufficient costs, the BO business model is very profitable for PLN, because PLN Bali can utilize more of its network capacity to run smart utility services to PLN customers. This is because every gas and water utility customer is a PLN customer. The LoRa WAN PLN network can be used for other utility services. So that nationally it saves national LoRa WAN network. Research [3] has not reviewed the business model implementation strategy that is suitable for PLN.

Therefore, this research will evaluate the strategy of deploying AMI system based on LoRa WAN by PLN Bali. The first step is to identify internal strengths and weaknesses as well as opportunities and challenges from the external that can affect PLN. In the research used a SWOT analysis approach. Strength (S) and weakness (W) data are obtained from surveys of PLN customers both those who are already and those who have not enjoyed the AMI system based on LoRa WAN and PLN employees. Whereas opportunity (O) and threat (T) were obtained from interviews with PLN leaders and related resource persons. After all internal and external factors are collected and evaluated, then the SO, WO, ST, and WT strategies are compiled.

2 Implementation of smart meter based on lora wan pilot project

In 2017, PLN in the South Kuta region held a pilot project of 1000 smart meter AMI devices based on LoRa WAN in Puri Gading and Nusa Dua Bali. The LoRa WAN network installed has a star configuration, the gateway is installed on an electric pole to cover customers. The results of reading the data obtained in the AMI pilot project at PLN Bali for 1,000 installed devices with meter data readings every 30 minutes with a 100% success rate.

To find out the opinions of users and the general public on the implementation of LoRa-based WAN-based AMI smart meters and the internal readiness of PLN employees, a survey is conducted to customers who use AMI smart meters, the general public, and PLN employees. Of the 104 customers who have not used the AMI smart meter, 60% of customers want to replace the electricity meter with the AMI meter. 24% of customers do not want to change meters and 16% of customers answer normal. Of the 24% who refused to replace their electricity meters the respondents reasoned that they were more accustomed to using an existing electricity meter, while respondents who said they wanted to replace the electricity meter because they wanted to support the government program.

A survey of customers using the AMI smart meter obtained data including 45% of the people felt normal towards AMI, 31% of the people felt dislike of AMI, 21% of the people liked AMI, 2% of the people really liked AMI, and 1 % of the people said they were very dislike of AMI. Of the 45% of respondents who said that they were normal, their reasons were known because at this time there was no noticeable difference to customers using the AMI smart meter. Of the 30% of respondents who said they did not like it, felt that their electricity bill had increased since the change of the AMI smart meter. Of the 21% of respondents who said they liked it because they felt the ease in filling their electricity tokens so that they could regulate their electricity usage needs. A survey of knowledge about smart meter AMI and the enthusiasm of the application of AMI found that 90% of the public did not know about AMI, 8% of the community knew AMI, 1% of the public were very aware, and 1% of the public was normal to AMI. 90% of respondents said there was no socialization of the AMI smart meter from PLN. Of the 107 respondents, 2% strongly agreed, 60% agreed, 26% stated normal, 12% said they did not agree with the plan to implement AMI throughout Bali and Indonesia.

Seeing PLN's readiness in holding the WAN Loans-based smart meter AMI, 100% of the survey data agreed that the implementation of the AMI smart meter would benefit PLN, 100% of PLN employees said that the application of LoRa-based WAN smart meter AMI was more profitable than conventional systems because PLN could receive electricity usage data per customer in real time. Respondents of PLN employees stated that only 52% of PLN had conducted information dissemination to the public about the smart meter AMI system and 48% of respondents stated no. From the survey results to customers and the general public there are still shortcomings in deploying a pilot project for the WAN-based 1000 AMI smart meters in terms of socialization, the impact of using AMI smart meters, limited AMI features, customer resistance to the use of new technologies, concerns over tariff increases for use AMI. In terms of PLN employees, there is still a low level of knowledge about the smart meter AMI, technical capabilities and coordination among employees in working.

3 Swot analysis, results and evaluation

Before deploying a smart meter based on LoRa WAN, it is necessary to evaluate the investment needs and business opportunities that will arise. In the research [3] the needs of capital expenditure (capex), operational expenditure (opex), and investment return rate

(IRR) funds have been submitted for the period of 3 years of deploying the LoRa WAN network to reach the entire region of Bali. Also presented were 3 business models for deploying the Loans Loans Loans to support the implementation of the AMI, namely build own (BO), build operational transfers (BOT), and leasing. Determination of the business model that will be applied by PLN depends very much on the ability of PLN to finance. If PLN has strong financing, the BO model is very suitable. For low financing is rent, but the level of dependence on the party that rents is very large and if there is a problem it takes time to complete it and depends on the willingness of the party who rents out. For BOT, PLN can only have a LoRa WAN network after an agreed period. The BO business model allows PLN to obtain other revenue sources from deploying the WAN LoRa network by utilizing the excess capacity of the LoRa WAN network to serve PLN customers for water and gas utilities. Because water and gas customers are usually PLN customers. So that with one LoRa WAN network can be used for water, gas and electricity utility services.

Once it is known that the business model is suitable for deploying the smart meter based on LoRa WAN and obtaining input from customers, the public, and PLN employees through surveys, PLN needs to develop a strategy through a SWOT analysis to implement the smart meter based on LoRa WAN deployment on a broader and massive scale. Based on the results of the survey and literature study found external and internal factors for the application of smart meter based on LoRa WAN.

External factors include 5% customer growth every year; Bali Eco Smart Grid Roadmap; lack of public knowledge about AMI; Some people refuse, because they feel the price is expensive; error recording KWh Meter; frequency 923 - 925 is intended as SRD; frequency 920 - 925 is a frequency allocation that is widely used for LPWAN in Asian countries. Internal factors include large number of PLN customers; current LPWAN technology readiness; PLN is not a telecommunications company; human resources do not understand about LPWAN management; AMR communication system is not yet reliable; read meter error; inconsistency of billing periods.

These external and internal factors are then evaluated with resource persons and arranged in an external and internal evaluation matrix. The evaluation matrix of external and internal factors is shown in Table 1 and Table 2 respectively.

Table 1. Internal factors evaluation matrix

	STRENGTH	Weight	Ranking	Score
1	PLN is state owned company	0.7.	4	2.8
2	The number of PLN customers is very large	0.6	4	2.4
3	Current LPWAN technology readiness	0.6	3	1.8
4	Coverage of LPWAN is very wide	0.7	3	2.1
				9.1
	WEAKNESS	Weight	Ranking	Score
1	PLN isn't telecommunication company	0.5	2	1
2	pln employees have not understood LPWAN management	0.45	2	0.9
3	AMR communication system is not yet reliable	0.6	1	0.6
4	Billing periode is not consistent	0.65	1	0.65
				3.15

Table 2. External factors evaluation matrix

	OPPORTUNITY	Weight	Ranking	Score
1	LPWAN technology is growing up	0.3	3	0.9
2	LoRa Wan Alliance member	0.65	4	2.6
3	Frequency 920 - 925 MHz is a frequency allocation that is widely used for LPWAN in Asian countries	0.45	3	1.35
4	Customers grow 5% annually	0.35	3	1.05
5	PLN has many subsidiaries	0.35	4	1.4
6	Roadmap Bali Eco Smart Grid	0.55	4	2.2
				9.5
	THREAT	Weight	Ranking	Score
1	Frequency 923 – 925 MHz is allocated for SRD	0.55	3	1.65
2	Low public knowledge about AMI	0.63	2	1.26
3	Error reading KWh Meter	0.6	3	1.8
4	The resistance of some people, because they feel the price is expensive	0.55	4	2.2
				6.91

From the external and internal evaluation matrix, the location of the S-W and O-T quadrants will be determined. From Table 1, the S-W = 9.1 - 3.15 = 5.95 is obtained and from Table 2, the O-T value = 9.5 - 6.91 = 2.59. From the S-W and O-T values are in quadrant I, which is the category of progressive strategies. Strategies that can be developed are the S-O strategy was carried out by testing the implementation of AMI and strategic partnership for the procurement of smart meter modules. Improving HR capabilities through training and outsourcing to third parties is an WO strategy. PLN proposed frequency allocation for LoRa WAN to regulators as ST strategy and PLN conducts ongoing socialization of AMI features and accelerates AMI implementation as an WT strategy.

4 Conclusion

From the survey results found deficiencies in the deployment of a WAN-based smart meter pilot project regarding socialization, the impact of using AMI smart meters, limited AMI features, customer resistance to the use of new technology, concerns over tariff increases for the use of smart meters. In terms of PLN employees, there is still a lack of knowledge on smart meter based on WAN Loans, low technical capabilities and coordination among employees in working. Components S and W are obtained from the survey while O and T are obtained from observing conditions outside of PLN. From the SWOT analysis results obtained S-W = 5.95 and O-T = 2.59, so the strategy is in location (5.95.2.59) quadrant I. Strategies for deploying smart meters based on WAN LoRa are progressive strategies.

Acknowledgements

The authors would like to sincerely acknowledge Universitas Indonesia as this work has been supported by the DRPM Research Grant for PITTA 2018 program No. 2446/UN2.R3.1/HKP 05.00/2018

References

1. Ministry of energy and mineral resources, ministerial decree number: 1415K / 20 / MEM / 2017 concerning ratification of the Electricity Supply Business Plan (RUPTL) PLN 2017-2026 (2017)
2. LoRa Alliance White Paper, "LoRaWAN: What is it?", (2015).
3. Commonwealth of Australia, Handbook of Cost-Benefit Analysis, no. 6. (2006).
4. G P, Saktiadji, "Techno Economic Analysis of Reading Smart Meter Implementation Based on LoRa Technology at PLN Bali " Thesis, Universitas Indonesia (2017)
5. Ministry of Communication and Information Technology, "Ministerial Regulation Number 35/2015 concerning Technical Requirements for Telecommunication Equipment and Short-range Telecommunications" (2015).
6. SigFox White Paper, "One Network A Billion Dreams-M2M and IoT redefined through cost effective and energy optimization connectivity" (2013).
7. Huawei White Paper, "NB-IOT: Enabling New Business Opportunities" (2015)
8. D R, Fred "Strategy Management Concept, 12th Edition" (2010)
9. W G. Sullivan, E Mwick, and J T. Luxhoj, "Engineering Economy 13th Ed. Pearson International Edition" (2006)

Wide ku-band log periodic dipole array microstrip patch antena using defected ground structure for electronic support measure

Immanuel Wicaksono[1*], *Dharu* Arseno[2], *Yuyu* Wahyu[3]

[1]Telecommunication Engineering, Faculty of Electrical Engineering, Telkom University
[2]Indonesian Institute of Science Sangkuriang St., LIPI Office Complex, Building 20, Cisitu, Dago, Coblong, Bandung City, West Java 40135

Abstract, Microstrip antenna as a supporter of wireless communication, which is the development of conventional antenna, has various advantages contained its predecessor, one of which has a patch that can be modified according to the wishes of the user. Microstrip antenna Log Periodic Dipole Array with Defected Ground Structure is one of the patch modification of antenna that allows antenna to work on wideband frequency 12-18 GHz (Ku-Band). Its use is intended for communication on satellites. In this research, will be designed and realized microstrip antenna with modification of Log Periodic Dipole Array patch working on frequency 12-18 GHz (Ku-band). The design will also use modifications to the ground structure with the use of slots using microstrip line feeding techniques. In this research, microstrip antenna design of Log Periodic Dipole Array with Defected Ground Structure works on 12-18 GHz (Ku-Band) frequency using slot using microstrip line feeding technique, which designed by using application of electromagnetic simulator (Integrated Software). . The desired specifications are; gain $\geq$ 6 dBi, VSWR <2, as well as 6 GHz bandwidth, with unidirectional radiation pattern and linear polarization. Substrate materials used in design are Roger 5880 Duroid with a relative permittivity of 2.2 and a dielectric thickness of 1.57mm. Measurement results on the realization of this tool; return loss on each of the 12, 15, 18 GHz frequency markers of -25,457 dB, -12,939 dB, and -11.004 dB; with the value of VSWR 1.112, 1.593, 1.786. Impedance of 44,988 Ω, 34,129 Ω, 27,792 Ω. Gain respectively of 8.907 dB, 8.931 dB, 8.774 dB. Bandwidth 6 GHz. Unidirectional radiation pattern and elliptical polarization.

1 Introduction

The progress of information and communication technology development, especially in the field of telecommunication today is very fast moving. Renewable creations and innovations are essential in maintaining the quality of the technology provided for the customers themselves. Technology devices, especially in the telecommunications sector,

[1] Corresponding author: immancakso@gmail.com

are expected to not only possess qualified technologies but also have the capability, compatibility, and dimensions that support the mobility of the users of the technology itself. The uneasiness that is felt from the high demand of customers is what makes the need of a device provider of a wireless network, the antenna itself.[1]

In this Research, microstrip antenna will be designed which is the development of conventional antenna and has many advantages compared to its predecessor, the dimension of the antenna is small, light and thin, also easy production process because the material is easy to find in the market.[2] However there are some disadvantages such as small bandwidth and small gain. The solution to fix the deficiency is to modify the patch form on the microstrip antenna [3][4].

In the previous research journal, 6 GHz bandwidth was obtained at KU-Band frequency, with a gain of 7.53 dB[5], which used the patch form Log Periodic Dipole Array. Referring to previous research, this Research is done designing and making microstrip antenna Log Periodic Dipole Array with the addition of Defected Ground Structure which is expected to increase the gain of the antenna itself as well as provide improvements in the values of other parameters desired. The design will use the application of electromagnetic simulator (Integrated Software).

2 Antenna design

In this research, the design of microstrip antenna Log Periodic Dipole Array will use copper as groundplane, patch, and feedlinenya. Meanwhile, the substrate will use the Roger 5880 Duroid which has a thin substrate characteristic that is considered suitable for use at high working frequency, wide bandwidth, and supports the use of communications satellites. The characteristics of materials used in this Research are:

1. Copper (copper)
   Used because of its nature as a good conductor. The specifications are as follows:
   a. Relative permittivity (εr): 1
   b. Thickness: 0.035 mm
2. Roger 5880 Duroid
   The selection of this material as a substrate is based on the use of antennas at high working frequency, and thin material characteristics, so that it can adjust to the antenna dimensions according to the specified specifications.
1. Relative permittivity (εr): 2.2
2. Thickness: 1.57 mm

The patch form used is the Log Periodic Dipole Array because of its superiority in overcoming the lack of microstrip antenna that has a narrow bandwidth, by adding several elements in-array, so that large bandwidth can be achieved, depending on the amount of working frequency of the desired antenna. This proposed antenna is designed in the following steps which are explained as:

2.1 Log periodic dipole array MSP antenna without DGS

Firstly, simple rectangular shaped MSP antenna is calculated and designed with the dimension of 39.651 X 27.124 X 1.57 mm^3. While the patch is log periodic dipole array-shaped with the calculation are as table 1. By simulating this log periodic dipole array MSP antenna, we might achieve the frequency of Ku-Band (12-18 GHz). But still, without the optimization and modification of this antenna structure, we were not getting any frequency in it. Geometry of this structure is shown in Figure1. (a-b). Due to this scenario we achieved its return loss is greater than -5 dB. So we needed to move to the next modification scenario

Table 1. Log periodic dipole array MSP antenna without DGS

Element	L(mm)	d(mm)	S(mm)
9	8.427	0.85	5.00225
8	7.5253	0.759	4.467
7	6.7201	0.6778	3.989
6	6.001	0.6053	3.5632
5	5.3589	0.5405	3.181
4	4.7855	0.4826	2.8406
3	4.2734	0.431	2.5367
2	3.8162	0.3849	2.2652
1	3.4878	0.3437	2.0229

2.2 Optimized log periodic dipole array MSP antenna without DGS

The dimension of the rectangular MSP antenna are 124.967 X 79.782 X 1.57 mm^3 and the calculation of the patch which are as follows:

Table 2. Optimized log periodic dipole array MSP antenna without DGS

Element	L(mm)	d(mm)	S(mm)
9	17.42749	5.5	11.5321
8	15.56275	4.9115	10.29816
7	13.89754	4.38596	9.19626
6	12.41054	3.91667	8.21226
5	11.08258	3.49758	7.33354
4	9.89674	3.12334	6.54885
3	8.83779	2.78914	5.84813
2	7.89214	2.49075	5.22238
1	7.04768	2.22423	3.18709

In this scenario, we were achieveing the frequency of Ku-Band, but still the return loss is greater than -10 dB, which is still not a good result yet. So we moved again to the next scenario with the use of defected ground structure. Geometry of this structure is shown in Figure2. (a-b).

2.3 Modified log periodic dipole array MSP antenna with DGS

The dimension of the log periodic dipole array MSP antenna was not changed. We only added the defected ground structure with the calculation are as follows:

Table 3. Modified log periodic dipole array MSP antenna with DGS

DGS parameter	Value(mm)
a	7
b	5
e	4.96793

With the addition of DGS, the return loss of this antenna is less than -10.275 dB, also VSWR is less than 2. Geometry of this structure is shown in Figure3. (a-b).

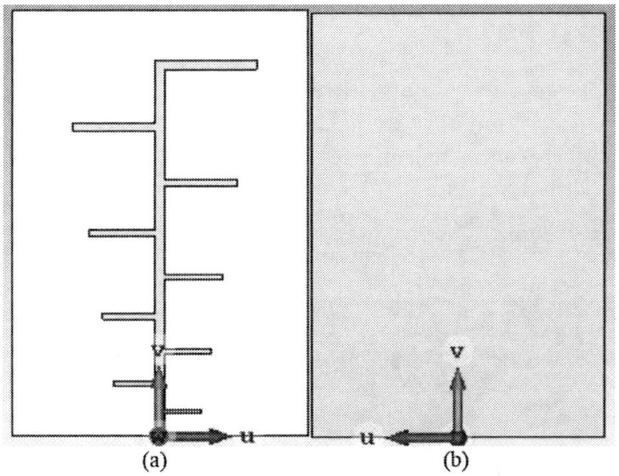

Fig. 1. (a) Log periodic dipole array MSP antenna (b) Back view without DGS

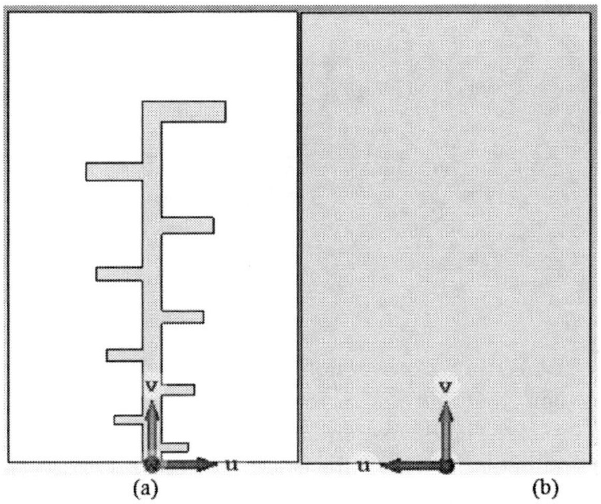

Fig. 2. (a) Optimized Log periodic dipole array MSP antenna (b) Back view without DGS

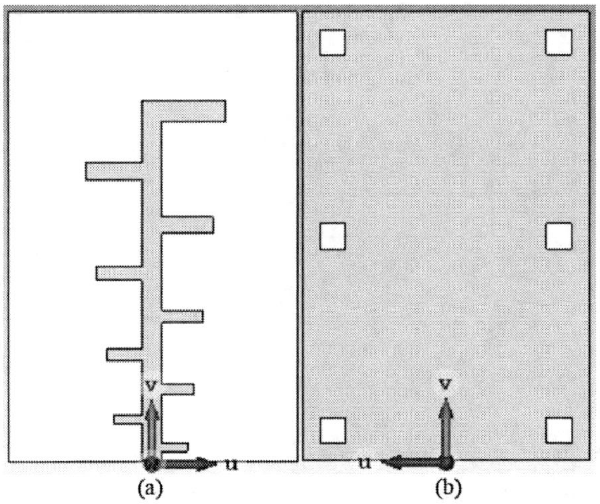

Fig. 3. (a) Modified Log periodic dipole array MSP antenna (b) Back view with DGS

3 Realization results

The proposed antenna is simulated using the Electromagnetic (EM) Simulation Software. By using this simulator, we measured return loss, VSWR, Z-Impedance, and farfield radiation pattern (gain, directivity, and efficiency) of the proposed log periodic dipole array shaped MSP antenna.

Moreover, this antenna is realized at PPET LIPI Bandung. These are the results we got from the realization of the LPDA MSP antenna.

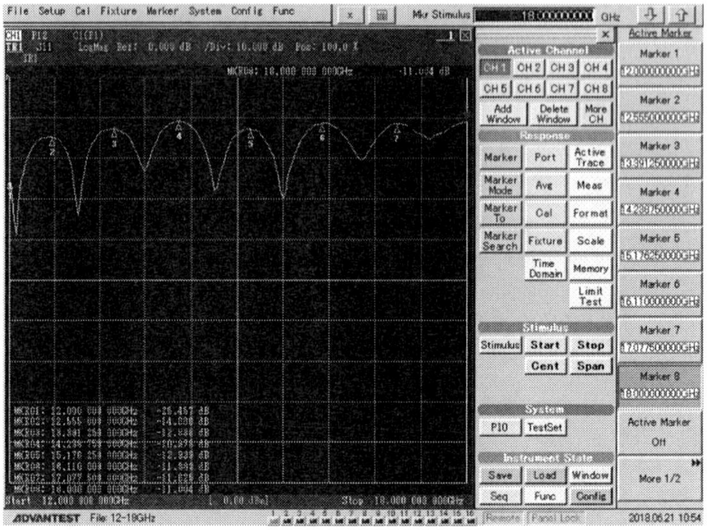

Fig. 4. S11(Return Loss)

The measurement of Return Loss is done by using existing tools in the measurement laboratory at Indonesian Institute of Science (LIPI) using Advantest R3770 Network analyzer 300 kHz - 20 GHz. The measurement results are obtained as the explanation below; namely the 12 GHz frequency marker, the value of Return Loss of -25.457 dB. At the 15 GHz frequency marker the resulting Return Loss is -12.939 dB. Then the Return Loss at the 18 GHz frequency marker shows the result -11.004 dB. The results of this measurement has met the specification with the desired working frequency is also optimal.

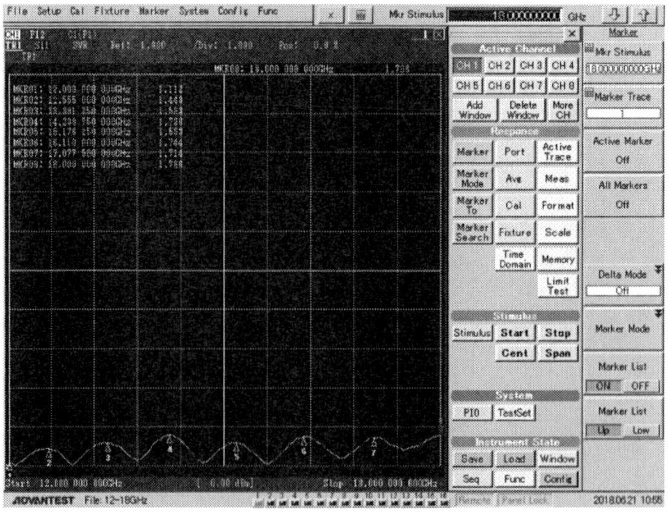

Fig. 5. VSWR

The measurement of VSWR also use the same hardware. The measurement results are obtained as the explanation below; namely VSWR at 12 GHz with a value of 1.112. At the 15 GHz, VSWR generated at the measurement is 1.593. While the VSWR at a frequency of 18 GHz shows the results of 1.786.

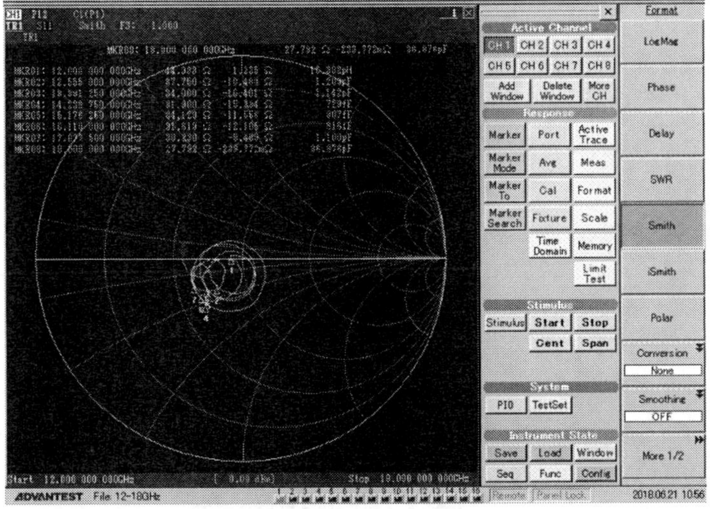

Fig. 6. Z Impedance (Smith Chart)

From Figure6 above, shows that the impedance at 12 GHz frequency marker with value equal to 44.988 Ω. At the 15 GHz frequency marker, the impedance is 34.129 Ω. While the impedance at the 18 GHz frequency marker shows the result of 27.792 Ω. The results are less appropriate with the simulation, which is influenced by the existence of factors affecting the conductor material on the feeding line and the antenna port itself (tin solder material), thus affecting the conduction of feeding.

As the log periodic dipole array using DGS was providing us the Ku-Band frequency, such optimum results of return loss, VSWR, and Z impedance were achieved using this technique.

Pictures below show the comparation of simulation and measurement radiation pattern result:

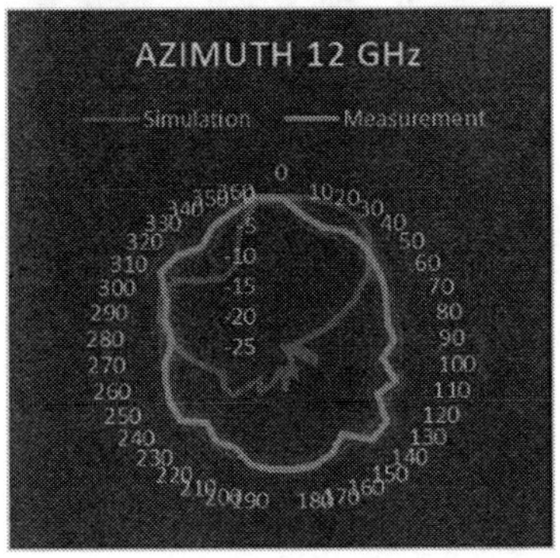

Fig. 7. Azimuth 12 GHz

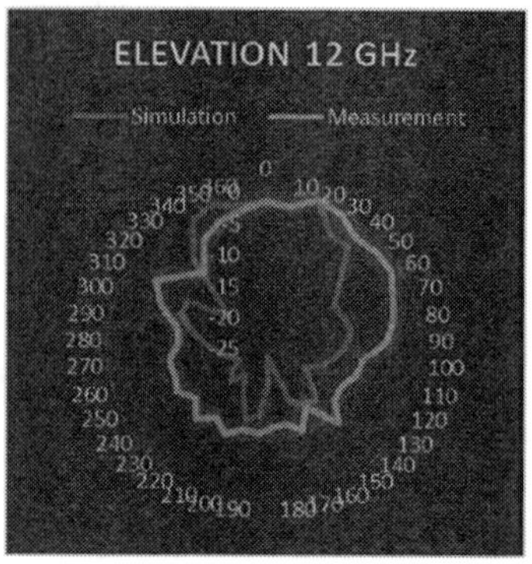

Fig. 8. Elevation 12 GHz

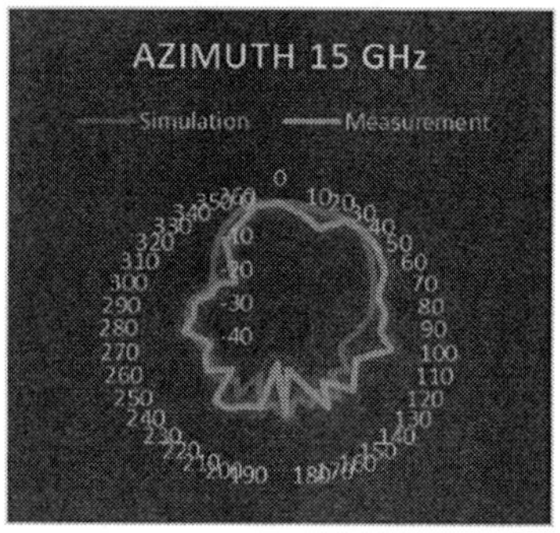

Fig. 9. Azimuth 15 GHz

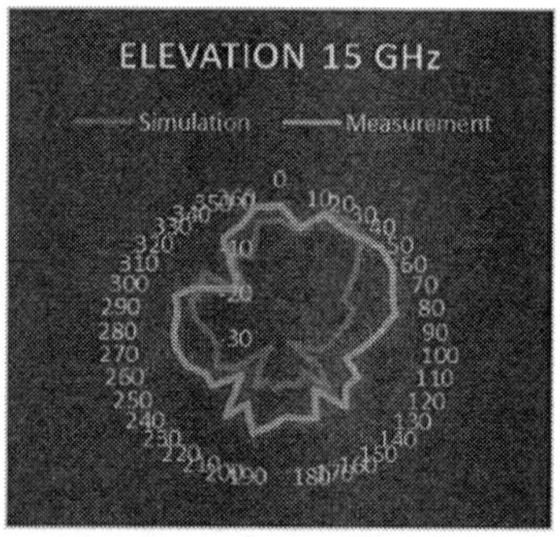

Fig. 10. Elevation 15 GHz

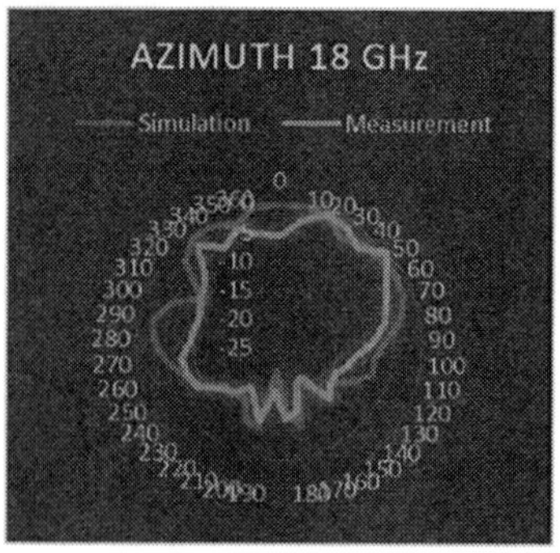

Fig. 11. Azimuth 18 GHz

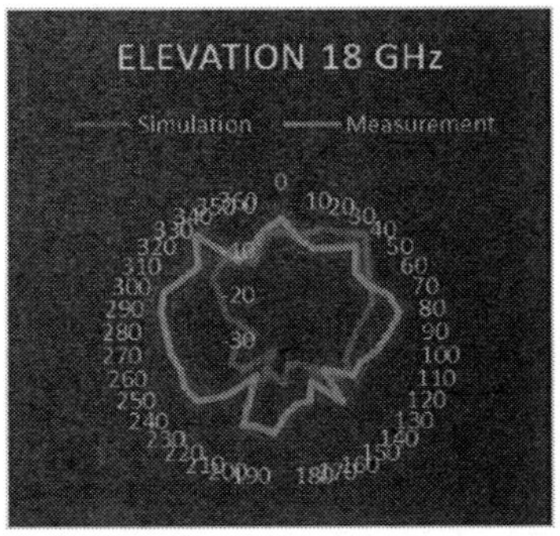

Fig. 12. Elevation 18 GHz

In this measurement, the tools used are spectrum analyzer and function generator / sweep. Where the measurement is done in the anechoic chamber at PPET-LIPI Bandung. By definition, the antenna radiation pattern is the output of the ratio of power levels out of the antenna in from various directions. The 6 picture (Figure7 – Figure12) above shows the radiation pattern comparation of simulated and realized antenna.

4 Conclusions

A compact wide Ku-Band microstrip patch antenna is proposed. Wide Ku-Band (12-18 GHz) is achieved using defected ground structure. Proposed antenna dimensions are 39.651 X 27.124 X 1.57 mm^3. We use Roger 5880 Duroid substrate with dielectric constant 2.2 and thickness 1.57mm is used. For improving the return loss and VSWR of the antenna, Log Periodic Dipole Array MSP antenna with Defected Ground Structure is introduced. All antenna parameters that have been planned in the early stages of the antenna design already meet the desired specifications, ie for Return Loss $\leqslant$ -10.27 dB, VSWR $\leqslant$ 1.883, 6 GHz Bandwidth on Ku-Band frequency, Unidirectional Radiation Pattern, Linear Polarization, 50 Ω Input Impedance, and Gain $\geqslant$ 8.774 dB. The main application of proposed antenna is to make compatible with Ku-Band frequency used-devices which covers maximum satellite applications (Electronic support measure, mobile satellite, fixed satellite, aeronautical radio navigation, radiolocation, space research, etc.)

References

1. SATMARIN. *L-BAND, C-BAND, KU-BAND AND KA-BAND*. (2018, March 22).
2. Ramesh Garg, P. B. *Microstrip Antenna Design Handbook*. Artech House Inc. (2001)
3. Denny, M. KU-Band Log Periodic Microstrip Antenna for Electronic Support Measure. (2017)
4. Firmansyah, T. Herudin., et al. "Bandwidth and gain enhancement of MIMO antenna by using ring and circular parasitic with air-gap microstrip structure". Telkomnika (Telecommunication Computing Electronics and Control). Vol.15(3). Pp. 1155-1163. 2017.
5. Firmansyah, T., Herudin, et a."Multiband microstrip antenna array with slot and array method for GSM, WCDMA, and LTE". International Conference on Broadband Communication, Wireless Sensors and Powering (BCWSP). Pages 1-5. 2017.
6. Fermando, D. Phased Array Microstrip Antenna for 1.27 GHz Synthetic Aperture Radar. (2017)
7. Balanis, C. A. *Antena Theory Analisis and Design 3rd Edition*. United Science: Wiliey Inter Science. (2005)

8. Krauss, J. D. *Antennas*. United Stated: Wiliey Inter Science. (1998)
9. Sujati, I. *Antena Mikrostrip : Konsep dan Aplikasinya*. Jakarta: Universitas Trisakti. (2010)
10. Ardiansyah, R. Ultra Wide Band Log Periodic Misrostrip Patch Antenna for Electronic Support Measure 2-18 GHz. (2017)
11. Bhadouria, A. S., & Kumar, M. Wide Ku-Band Microstrip Patch Antenna Using Defected Patch and Ground. *IEEE International Conference on Advances in Engineering & Technology Research (ICAETR - 2014)*. Unnao, India: Dr. Virendra Swarup Group of Institutions. (2014)
12. Weng, L. H., Guo, Y. C., Shi, X. W., & Chen, X. Q. (2008). An Overview on Defected Ground Structure. In *Progress In Electromagnetics Research B Vol. 7* (pp. 173-189).
13. Lampiran Permen Tasfri (Departemen Komunikasi dan Informatika Republik Indonesia, 2009)
14. *Microstrip Patch Antenna Calculator*. Retrieved from everythingRF. (2015)

https://doi.org/10.1051/matecconf/201821803015

Designing a tri-band microstrip antenna for targetting 5g broadband communications

Ahmad Firdausi [1*], *Galang* Hakim [2] and *Mudrik* Alaydrus [3]

Department of Electrical Engineering, Universitas Mercu Buana, indonesia

Abstract. One of the technologies that has wireless application nowadays was 5G mobile communication.This paper presents the designing of a Tri-band microstrip antenna for targeting 5G broadband communications,This element antenna has 3x3 rectangular patches with feeding line structures are branched. With the use of double feeding proximity coupling structure, we intend to maximize antenna bandwidth, therefore the antenna cover range tri-band frequency from 40 GHz to 70 GHz. The reflection factor comparation between simulation and measurement has a minimum with respective frequency at 45.3 GHz, 57 GHz, and 66 GHz. The total measurement bandwith 11.5 Ghz. With this combination tecnique, the proposed antenna is a promising candidate for 5G communication systems.

1 Introduction

Wireless communication is on one of the frontier technologies that has been widely used in modern application. It was the modern technologies that has evolved in very fast way. One of the technologies that has wireless application nowadays was 5G mobile communication. 5G wireless mobile communication that employs millimeter wave offers high speed wireless data transfer. Because of this wireless data transfer feature, this technologies has been developed into different field such as Internet of Things (IoT), Smart Cities, Etc [1].

The use of milimeter wave frequency has an impact in decrease of the mobile phone antenna dimensions. The conventional antennas was replaced with microstrip antenna structure. With its features such as small and light weight, the microstrip patch antenna has emerged as the best candidate for mobile communication device [2]. Problems arise because the nature of its application, a 5G applications that employs millimeter wave has a very wide frequency spectrum from 30 GHz to 300 GHz [3].

To solve this problem, we proposed a tri-band microstrip antenna design to be used in millimeter wave frequency for targetting 5G application. Several researches have been made for this particular topic but the closes one is by Bondarik-Sjoberg [4].

* Corresponding author: ahmad.firdausi@ymail.com

2 Microstrip antenna design

In this research,we propose a proximity coupling [5] to be used in microstrip antenna design, because from previous research the proximity couple technique produced a significant increase in bandwidth . The Microstrip antenna was to be design at following band, first band was 45.3 GHz, Second Band was 57 GHz, and the third band was 66.6 GHz. This frequency bands was allocated for the mobile communication following the regulation of the allocation usage designation by the National Telecommunications and Information Administration Office of Spectrum Management USA [6]. The tri-band microstrip antenna design is based on double microstrip antenna that has been reasearch as a milimeter wave application [7].

2.1 Design of the Ground Structure

Based on design of the ground structure by Bondarik-Sjoberg, we proposed to applied the same design to coupling the patches on the top antenna side. Figure 1 shows ground structure for antenna design. Geometry for ground structure was $L= 10$ *mm*, $W = 5$ *mm*. Proximity couple combination design patch grid in the radiating element is applied for microstrip structure tri-band antenna characteristics. The antenna substrate is Rogers RO3003 ($\varepsilon_r = 3$, tan δ=0.0013) with a thickness of 0.51 mm.

Fig 1. Ground structure

2.2 Design Feeding Structure

The line feeding structures are branched, therefore for radiating elements there are two feeding point. Figure 2 shows the feeding element structures. To maximize antenna bandwidth the double feeding structure (double-fed proximity) was used [8, 9]. We also applied stub in feeding system to adding the antenna multiband characteritic. The stub at feeding line also function as a capacitive stub that works to reduce inductive that produce in the double feeding system [10]. The stub design length was $l3 = 0.38$ mm and has width $w2$=0.25 mm. See table 1 for another additional dimensions that needed for feeding structure.

Table 1. Dimension of feeding antenna

Symbol	Dimension (mm)
W	5
L	10
w1	1.2
w2	0.25
w3	0.4
l1	1.4
l2	0.25
l3	0.38
l4	6

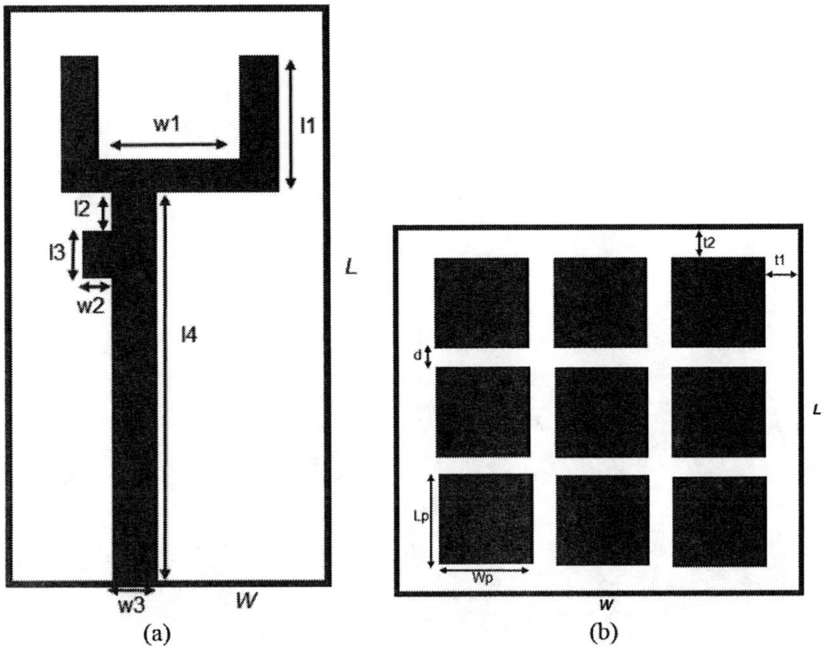

Fig 2. (a) Feeding structure , (b) Radiating elements structure

2.3 Design of the patch layer 2 structure

Table 2. Dimension of radiating elements

Symbol	Dimension (mm)
W	5
L	5
t1	0.45
t2	0.51
d	0.25
Lp	1.16
Wp	1.2

The final antenna structure is constituted by two layers. The first layer (bottom layer), there are feeding structure that connect with 50 Ω microstrip line that feeding the second layer. On the second layer (upper layer), the radiating elements structures are multiple rectangular patch antennas design. This element has 3x3 rectangular patches with no interconnected between patched.

3 Simulation and measured results

In this research, we use ANSY Electronic Desktop HFSS version 16 to simulate the final antenna design. The three dimensional radiation pattern simulated by HFSS is shows at figure 4, it has gain about 5.66 dBi

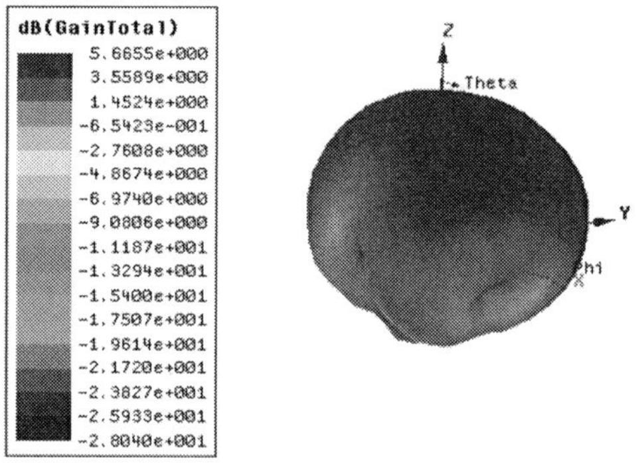

Fig. 4. The microstrip antenna simulation for three dimensional radiation pattern

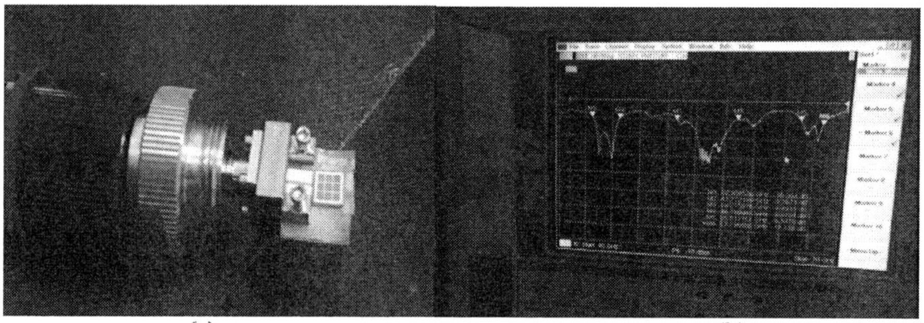

Fig. 5 The final antenna design (the feeding, the rectangular patch structure, and ground at back side substrate) after fabricated. (b) The measured reflection factor of the antenna.

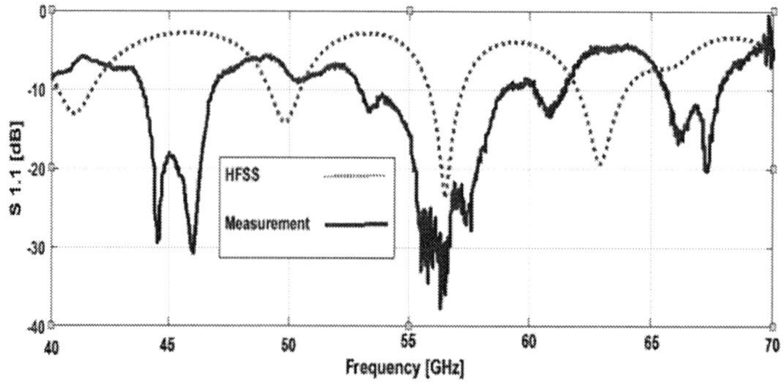

Fig. 6. The reflection factor antenna with HFSS (dashed) and measured (solid).

Figure 6 shows the measurement result for the final antenna design. The minimum reflection factor measurement (solid curve) is at 45.3 GHz, 57 GHz, 66.62 GHz. The total measurement bandwith was 11.5 Ghz. The minimum reflection factor simulation (dashed curve) is at 41 GHz,50 GHz , 57 GHz ,63.2 GHz. There is a shifting in frequency at about 3 GHz from simulation and measurement reflection factor comparison results.

4 Conclusion

This microstrip antenna design was working at 45.3 GHz, 57 GHz, 66.62 GHz frequency range. This is the product of adding the stubs to the feedline design that can produce multi frequency. The tuning stub combine with proximity coupling for patches has produce the required bandwidths at respected frequency, this techniques produce total bandwidth of 11.5 GHz. There is a shifting in frequency at about 3 GHz from simulation and measurement reflection, we believe this is happen because fabrication antenna was not very precision and inaccuracy in placing the upper layer in the prototype during measurement

References

1. Suresh Borkar & Himangi Pande, Application of 5G Next Generation Network to Internet of Things, International Conference on Internet of Things and Applications(IOTA) (2016)
2. Shivangi Verm,Leena Mahajan,Rajesh Kumar,Hardeep Singh Saini,Naveen Kumar 'A Small Microstrip Patch Antenna for Future 5G Applications', 5th International Conference on Reliability, Infocom Technologies and Optimization (Trends and Future Directions) (ICRITO) (2016)
3. P. Roy, R. K. Vishwakarma, A. Jain, and R. Singh, "Multiband Millimeter Wave Antenna Array for 5G Communication," pp. 4–7, (2016)
4. Bondarik and D. Sj, "Gridded Parasitic Patch Stacked Microstrip Antenna with Beam Shift Capability for 60 GHz Band," vol. **62**, no. January, pp. 319–331, (2015)
5. Dian Widi Astuti; Ahmad Firdausi; Mudrik Alaydrus 'Multiband double layered microstrip antenna by proximity coupling for wireless applications',15th International Conference on Quality in Research (QiR) : International Symposium on Electrical and Computer Engineering.,(2017)
6. United States Frequency Allocations The Radio Spectrum,U.S. Department Of Commerce National Telecommunications and Information Administration Office of Spectrum Management, (January 2016)
7. S. K. Goudos, A. Tsiflikiotis, D. Babas, and K. Siakavara, "Evolutionary Design of a Dual Band E-shaped Patch Antenna for 5G Mobile Communications," pp. 17–20, (2017)
8. Bondarik, "Investigation of Reconfigurability for a Stacked Microstrip Patch Antenna Pattern Targeting 5G," pp. 1202–1205.
9. J.-Y. Sze, and K.-L. Wong, "Bandwidth enhancement of a printed wide slot antenna fed by a microstripline with a fork-like tuning stub," Proceeding of ISAP, Japan, pp.1159-1162 (2000)
10. Ahmad Firdausi; Mudrik Alaydrus "Designing multiband multilayered microstrip antenna for mmWave applications", International Conference on Radar, Antenna, Microwave, Electronics, and Telecommunications (ICRAMET) (2016)

Analysis of secondary market policies of radio frequency spectrum in indonesia

Gunawan Wibisono[1*], *Isnur* Widaryanto[2] , and *Cahyo* Nugroho[1]

[1]Department of Electrical Engineering, Universitas Indonesia, Kampus Baru UI, 16424
[2]Ministry of Communication and Informatics, Republik Indonesia, Jakarta, INDONESIA

Abstract. The radio frequency spectrum is an important factor in improving service quality and expanding the network of cellular operators. Operators attempt to obtain additional spectrum allocations through the practice of merging or taking over the radio spectrum, where the process is dependent on ministerial approval. In Indonesia, its implementation produces different treatment in terms of the use of radio frequency spectrum, some are accompanied by divestment another are not. This process is not through the mechanism of the secondary market spectrum. Based on the practice of secondary market spectrum in other countries, it provides business certainty to operators and gives maximum value in the spectrum. In this research, secondary market policy for Indonesia is proposed, where evaluation and analysis will be done using regulation impact analysis (RIA). Three proposed secondary market alternatives include to keep current conditions (alternative 1), secondary market mechanism with direct trading (alternative 2), and broker mechanism with independent supervision (alternative 3). Assessment will be done using cost base analysis (CBA) based on plus minus implications (PMI), and decision will be made based on multi criteria analysis (MCA). According RIA based analysis, the results obtained that alternative 3 provides many advantages compared to 2 other alternatives. The impact of benefits of alternative 3 includes increasing spectrum requirements, frequency spectrum structuring, the need for legal certainty and equality of the process of transferring frequency spectrum.

1 Introduction

Regulation plays an important role in the telecommunications business. There are many regulatory aspects that influence the growth of broadband mobile telecommunications business such as the availability of frequency spectrum, tariffs, interconnection, content, and numbering [1]. Cellular telecommunications regulations have given birth to many cellular operators operating in Indonesia, giving rise to a tight competition climate. The growing number of cellular users as well as the limited number of spectrums owned by operators has led to a decline in service quality. The operator has to have a wide spectrum to provide good services to users and to serve a large system capacity, large number of base

* Corresponding author: gunawan@eng.ui.ac.id

station systems (BTS) to cover a wide service area, and as well as the latest technology to provide services with efficiency levels spectrum and good service quality. Therefore, the operators are trying to get a larger allocation of frequency spectrum.

The use of radio frequency spectrum in Indonesia must obtain permission from the Minister of Communication and Information in the form of radio station permits. The secondary market frequency spectrum policy in Indonesia has not been implemented. Even in Government Regulation (PP) number 53/2000 in Article 25 it has been stated explicitly that holders of radio frequency allocations cannot transfer radio frequency allocations that have been obtained to other parties, unless there is approval from the Minister [2]. Basically, the frequency spectrum belongs to the government. The telecommunications operator is given the right to use the frequency spectrum with the time frame of the permits regulated in PP 53/2000 for 10 years or 5 years and can be extended with permission from the Minister. In addition, to increase the frequency spectrum utilization and avoid the practice of brokering, the Ministry of Communication and Information Technology since 2001 has implemented a modern license to the operator [3]. Modern license is a license given to an operator accompanied by a written network development commitment that must be carried out.

To avoid the frequency spectrum control monopoly through the process of acquisition or merger or consolidation between operators, the government has banned monopolistic practices regulated by Law Number 5 of 1999 concerning Prohibition of Monopolistic Practices and Unfair Business Competition [4], Law Number 36 of 1999 concerning Telecommunications [5], Law 40/2007 concerning Limited Liability Companies [6], and Law 25/2007 concerning Investment [7]. Takeover or transfer of frequencies from one operator to the operator through an auction or a beautiful contest is controlled and determined by the minister of communication and information. In this case, the role and domination of the government is very large. Secondary markets are not permitted yet in Indonesia, even in the takeover or transfer of frequency in practice it is identical to the secondary market. This can be seen from several cases that have occurred and can be classified as secondary market activities in the frequency spectrum in Indonesia, including the merger of PT. Indosat, PT. Satelindo, PT. IM3 and PT. Bimagraha; XL Axiata national roaming cooperation with Axis; cooperation between Mobile-8 and Smart Telecom; network sharing or MVNO cooperation; and the merger of XL with Axis [8] – [9]. The transfer of the frequency spectrum from one operator to another in practice does not get the same approach from the government in an effort to avoid monopolies, maintain a competitive climate, and optimize spectrum values. There are operators who get additional frequencies accompanied by divestments but there are operators who get additional frequencies without divestment.

On the other hand, the application of a pure secondary market that has been carried out by the United Kingdom, the United States, Australia, New Zealand and Guatemala has been proven to create a healthy competition climate, increase the economic value of the spectrum, provide certainty of telecommunications business law, reduce government dominance, and give equal treatment to all parties [10 – 13]. To ensure legal certainty and equal treatment in the implementation of the secondary market, the government makes strict rules to protect interference with the next operator and maintains maximum emission level to avoid outband interference for all operators. The guarantee of business law certainty in the implementation of secondary markets makes the frequency spectrum trade receive the attention of many operators.

Therefore, it is very interesting to analyze the application of the secondary frequency spectrum market in Indonesia. There are 3 alternatives of the secondary market spectrum policy that will be evaluated, namely alternatives 1, 2, and 3. To get good policy results, then in this study an approach of regulatory impact analysis (RIA) will be conducted. To assess the 3 alternative policies, cost-based analysis (CBA) is based on plus minus implication (PMI). Whereas to decide the best policy alternative, multi-criteria analysis (MCA) evaluation is used. From the results of the CBA based on PMI approach, an alternative 1 has score = -16 is obtained, alternative 2 = 4 and alternative value 3 = 22. With an MCA, the alternative 1 has score = 206, alternative 2 = 250, and alternative 3 = 321. It is seen that with BCA based on PMI and MCA obtained the highest score achieved by alternative 3. The impact of alternative benefits 3 includes increasing spectrum requirements, frequency spectrum structuring that continues to experience constraints and obstacles, and the need for legal certainty and equality of the process of transferring a clear frequency spectrum

2 Secondary market spectrums based on RIA

In order to ensure that secondary market spectrum regulation is indispensable to the public interest, encouraging an evidence-based approach and ensuring that all practical options for addressing the problem have been considered and providing the highest benefits that exceed the costs incurred, the RIA approach is used. As a process, the implementation of the RIA method is carried out through various stages, namely:

a. Identification and analysis of problems related to policies

According to Government Regulation 53/2000, it is stated that holders of frequency allocation cannot be transferred to other parties, unless there is approval from the Minister. This is supported by different treatments in granting frequency spectrum usage rights when there is a merger or acquisition of an operator's ownership.

b. Setting goals

The government's secondary market spectrum policy aims to streamline the use of frequency spectrum by operators, ensure equality and a clear process in transferring the license to use the frequency spectrum, disclosure of information on the use of frequency spectrum, supervision and strict sanctions.

c. Development of various policy options / alternatives

There are 3 proposed alternative formulations of the secondary market spectrum policy, namely: alternative 1 which keep with the current conditions, does nothing, alternative 2 implements a secondary market with a direct trade mechanism, and alternative 3 is to implement a secondary market with a broker mechanism supervised by an independent supervisory body.

d. Assessment of alternative policy choices

Three alternatives for the preparation of the secondary market spectrum policy, the feasibility of the aspects of benefits and costs is proposed called as cost based analysis (CBA). Aspects of benefits include both direct and indirect regulators, operators, and users. While the cost aspect includes the cost of the selection process, buying and selling, controlling, operating and using the frequency spectrum which will be the obligation of the licensee. One effective method to be applied in CBA is plus-minus implication (PMI).

e. Selection of the best policy

To get a policy selection of the secondary market spectrum application, a multiple criteria analysis (MCA) approach is used. The MCA shows all the criterias that are decisive in formulating secondary market policy objectives, given weighting for each criterion and given an assessment.

3 Results and discussions

To be able to decide on the best policy alternative, an evaluation is done using CBA bsed PMI. Whereas the selection of the best policy alternatives is carried out with an MCA that involves all the factors that show the achievement of the policy objectives compiled, Table 1 shows CBA based PMI the proposed alternatives 1, 2 and 3. It can be seen from Table 1, an alternative 1 gets score = -16, alternative 2 = 4, and alternative 3 = 22. This means that alternative 3 provides a higher (plus) benefit than alternatives 1 and 2. It is indicated that the all criterias at alternative 3 get score 1 as beneficial.

Table 1. CBA based pmi of the proposed alternative 1, 2, and 3

Stakeholder	**No**	**Plus**	**+1/-1 Alternative**			**Minus**	**+1/-1 Alternative**		
			1	**2**	**3**		**1**	**2**	**3**
Regulator	1	Frequency spectrum arrangement	1	-1	1	Regulatory socialization costs	-1	1	1
	2	Frequency spectrum control	1	-1	1	Monitoring and evaluation costs	-1	1	1
	3	Regulatory arrangement	-1	-1	1				
	4	Economic value of frequency band	-1	1	1				
Operator	1	The need of additional frequency spectrum	-1	-1	1	Refarming costs	-1	1	1
	2	Use of frequency spectrum	-1	1	1	OPEX	1	-1	1
	3	Competision	-1	-1	1	Licence Fee	-1	1	1
	4	Equality of	-1	1	1				
	5	Legal certainty of the right to transfer the use of frequency spectrum	-1	1	1				
	6	Secondary market frequency spectrum	-1	1	1				
	7	Information disclosure of the use of frequency spectrum	-1	-1	1				
User	1	Network access	-1	1	1				
	2	Choice of products and services	-1	1	1				
	3	Producs and services price	-1	1	1				
	4	Quality of Service	-1	-1	1				
	5	Technology change	-1	1	1				
	6	Customer protection	-1	-1	1				
Subtotal			**-13**	**1**	**17**		**-3**	**3**	**5**
Total							**-16**	**4**	**22**

The CBA based on PMI evaluation is insufficient to decide that alternative 3 provides many benefits. Therefore, to strengthen in deciding the best alternative proposed, the three alternatives were evaluated and analyzed by the MCA. Table 2 shows the comparison of the results of MCA for alternatives 1, 2, and 3. From Table 2 shows the alternative value 1 = 206, alternative 2 = 250, and alternative 3 = 321. This means that the factors in alternative 3 strongly support the purpose of applying the market Secondary frequency spectrum in Indonesia and have a beneficial effect. The impact of alternative benefits 3 includes increasing spectrum requirements, frequency spectrum structuring that continues to experience constraints and obstacles, and the need for legal certainty and equality of the process of transferring a clear frequency spectrum. By implementing alternative policy 3, it is expected that constraints on spectrum requirements can be overcome by configuring permits based on geographic, bandwidth and time, where it can provide flexibility in radio frequency spectrum management, but still based on certain parameters through an independent oversight body evaluation process. It aims to be able to provide legal certainty and equality in the transfer of the radio frequency spectrum.

Table 2. MCA of the Proposed Alternative 1, 2, and 3

No	Indikator	Weight	Alternative 1		Alternative 2		Alternative 3	
			Score	Total	Score	Total	Score	Total
A	**Regulator**							
1	Frequency spectrum arrangement	5	4	20	2	10	3	15
2	Frequency spectrum control	5	4	20	2	10	3	15
3	Regulatory arrangement	4	2	8	3	12	5	20
4	Economic value of frequency band	5	3	15	3	15	4	20
5	Regulatory socialization costs	2	3	6	3	6	3	6
6	Monitoring and evaluation costs	2	2	4	3	6	4	8
	Subtotal			**73**		**59**		84
B	**Operator**							
1	The need of additional frequency spectrum	4	3	12	2	8	4	16
2	Use of frequency spectrum	5	2	10	3	15	4	20
3	Competision	5	2	10	3	15	4	20
4	Equality of transfer frequency spectrum	5	1	5	3	15	4	20
5	Legal certainty of the right to transfer the use of frequency spectrum	5	3	15	3	15	3	15
6	Refarming costs	2	2	4	3	6	4	8
7	OPEX	2	4	8	2	4	3	6
8	Licence Fee	2	2	4	3	6	3	6
9	Secondary market spectrum	5	1	5	4	20	4	20
10	Secondary market spectrum benefits	4	3	12	3	12	4	16
	Subtotal			**85**		**116**		147
C	**User**							
1	Network access	3	2	6	3	9	4	12
2	Choice of products and services	3	2	6	4	12	3	9
3	Products and services price	3	2	6	3	9	3	9
4	Quality of service	5	2	10	3	15	4	20
5	Technology change	5	1	5	5	25	3	15
6	Customer protection	5	3	15	1	5	5	25
	Subtotal			**48**		**75**		90
Total				**206**		**250**		321

4 Conclusions

In this study secondary market spectrum policy has been analyzed with the RIA approach for the three proposed alternatives. Three alternatives are determined, calculated and compared between the benefits and costs of each alternative using CBA based on PMI.

It is shown that alternative 1 had scores = -16, scores for alternatives 2 = 4 and scores for alternatives 3 = 22. By using MCA, a score of alternative 1 = 206, score for alternative 2 = 250 and score for alternative 3 = 321. Based on the results of the analysis using RIA to 3 alternative choices, alternative 3 is obtained which has benefits that are greater than the costs (losses) incurred. This can be seen from the results that impacts related to the increasing spectrum requirements, frequency spectrum structuring that continues to experience constraints and obstacles, and the need for legal certainty and equality of the process of transferring a clear frequency spectrum. By implementing alternative policy 3, it is expected that the constraints of spectrum requirements can be overcome by configuring permits based on geographic, bandwidth and specific time, where it can provide flexibility in the arrangement of radio frequency spectrum, but still based on certain parameters through the supervisory body evaluation process independent.

Aknowledgment

The authors would like to sincerely acknowledge Universitas Indonesia as this work has been supported by the DRPM Research Grant for PITTA 2018 program No. 2445/UN2.R3.1/HKP 05.00/2018

References

1. B. Elian, Analysis of Radio Frequency Spectrum Needs for Indonesian Wireless Broadband Radio Communication in 2010-2025, Thesis, Universitas Indonesia (2015)
2. Republic of Indonesia, Government Regulation (PP) No. 53/2000 concerning Use of Radio Frequency Spectrum and Satellite Orbit (2000)
3. Ministry of Transportation, Decree of the Minister of Transportation No. 30/2004 concerning Amendments to the Decree of the Minister of Transportation No. 21/2001 concerning the implementation of Telecommunications (2004)
4. Republic of Indonesia, Law Number 5 of 1999 concerning Prohibition of Monopolistic Practices and Unfair Business Competition (1999)
5. Republic of Indonesia, Law of the Republic of Indonesia Number 36/1999 concerning Telecommunications (1999)
6. Republic of Indonesia, Law 40/2007 concerning Limited Liability Companies (2007)
7. Republic of Indonesia, Law 25/2007 concerning Investment (2007)
8. Ministry of Communication and Information Technology, Decree of the Minister of Communication and Information of the Republic of Indonesia Number 1192/2013 concerning Stipulation of Block Allocation of Radio Frequency Bands of Comprehensive Arrangement of 2.1 GHz Radio Frequency Bands (2013)
9. Ministry of Communication and Information, Regulation of the Minister of Communication and Information of the Republic of Indonesia Number 4/2015 concerning Operational Provisions and Procedures for Licensing to Use Radio Frequency Spectrum (2015)
10. Electronic Communications Committee (ECC), Refarming and Secondary Trading in a Changing Radiocommunication World, in the European Conference of Postal and Telecommunications Administrations (2002).
11. M. Barrie, S. Delaere and P. Ballon, Classification of business scenarios for spectrum sensing, in Proc. PIMRC, 2010 IEEE 21st International Symposium on (pp. 2626-2631) (2010)

12. S. Bridge, The secondary spectrum market: a primary licensing and leasing, Florida: Spectrum Bridge, Inc., (2009)
13. J. W. Mayo and S. Wallsten, Enabling efficient wireless communications: the role of secondary spektrum markets, Washington DC: Technology Policy Institute, pp. 1-30 (2009)

Applications of temperature and humidity monitoring system at aerophonic plants based on IoT

Irma Saraswati[1*], *Vera* Dwi Puspitasari[1], *Anggoro* S Pramudyo[1], Alimuddin[1], *Teguh* Firmansyah[1], *Rida* Oktorida Khastini[2] and *Untung* Mardono[3,4]

[1]Electrical Engineering, University of Sultan Ageng Tirtayasa Indonesia
[2]Biology Education, University of Sultan Ageng Tirtayasa Indonesia
[3]Mechanical Engineering, University of Sultan Ageng Tirtayasa Indonesia
[3]Industrial Engineering, University of Engineering

Abstract. Indonesian agricultural system has improved significantly along with technological developments. However, as a result of technological developments, agricultural land becomes narrow. To overcome the increasingly narrow farming land, start developing systems suitable for planting with aeroponic system. Unpredictable environmental climate changes can disrupt productivity growth and create low quality crops. To solve this problem, monitoring system based on Android needed. This system is used to monitor in parameters change,the parameters include temperature and humidity. Android-based monitoring system uses java programming and is open source, besides the user can monitor the development of spinach plants, by looking at data that has been stored into the database so that users can know the development of the plant. This application can be accessed via smartphone with android operating system as a client server.

1 Introduction

Indonesia's agricultural system has begun to develop cultivation systems with various systems. One example of a system that began to be developed is the aeroponic system, using the air media by hanging plants on the air and then the nutrients sprayed on the roots of plants [1]. Plants suitable for aeroponic systems are vegetables [2]. The object used in this study is spinach, because spinach requires a relatively short cultivation time of 25-35 days after planting [3]. Spinach plants require temperature, humidity, carbon dioxide levels and sufficient lighting to grow well. If one of these parameters is not met or excessive then the crop is not maximal and cause the spinach plants to dry will not even grow [4].

The effects of climate change in agriculture are characterized by floods, droughts and rainy seasons. Climate change leads to reduced productivity growth and low quality crops [5]. It requires a continuous monitoring system from seed planting to harvest time. Manual monitoring of plant growth has limitations caused by human physical factors including

* Corresponding author: irma.saraswati@untirta.ac.id

fatigue, non-continuity, uniformity and inaccuracy. The solution of the problem is by applying technology in the form of remote monitoring. Remote monitoring is very effective because the results of direct monitoring data are stored in the database [6].

The monitoring system research on spinach plant is focused on monitoring system only with parameters of temperature, humidity, light intensity and carbon dioxide levels. The differentiator of this study with previous research [8-10] is to use android application for remote monitoring on spinach plants with aeroponic system. Android is one of the operating systems for linux-based mobile devices that include operating systems, middleware, and applications. Android is made by using Java programming and is open source so it can grow quickly and also provide an open platform [7]. Application of aeromosis can monitor the development of spinach plants, by looking at data that has been stored into the database so that users can know the development of the spinach plant. This application is accessed via smartphone with Android operating system and is client server.

2 Research methods

2.1 Research methods

The design of research monitoring system (aeromosis) is divided into 4 stages namely, system design, architecture design (linking between the database with android), designing the application and programming view on android studio, and testing applications against multiple smartphones. Flow diagram design research can be seen in Figure 1.

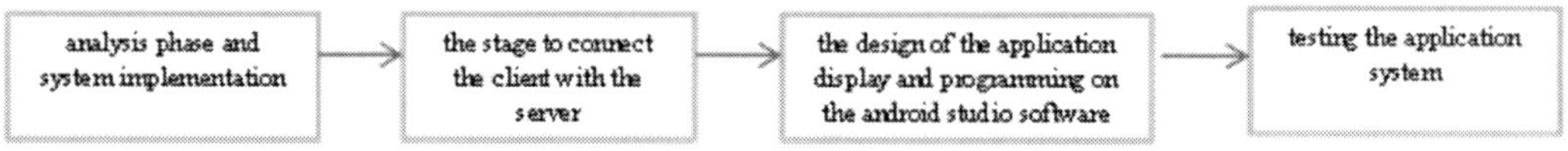

Fig. 1. Diagram block of designing research

Figure 1 describes the steps to be used in this study. The first phase begins with the analysis phase and system implementation. The second stage is the stage to connect the client with the server. In the process of connecting the database to the application monitoring system using JSON (JavaScript Object Notation) format. The third stage discusses the design of the application display and programming on the android studio software. The fourth stage is the last stage of testing the application system, whether the system runs according to the design or not as well as on a variety of smartphones.

2.2 Design Architecture

The architecture design of the monitoring system application research aims to connect the client with the server via the internet. Client will request server to get information about the last condition parameter. The depiction of the architectural design flow can be seen in Figure 2.

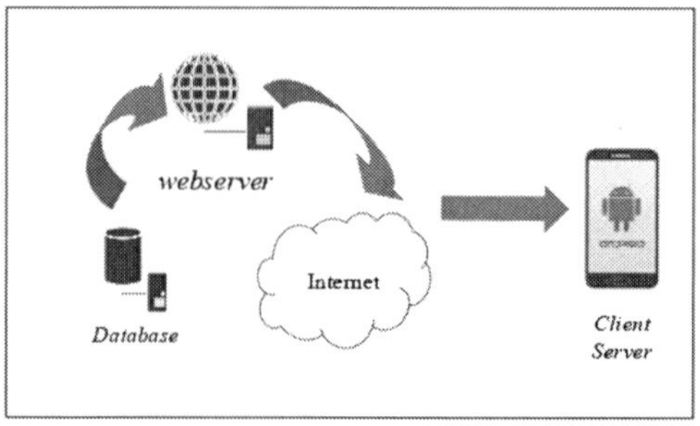

Fig. 2. The architecture design of the monitoring system application

Figure 2 is an architectural design of aeroponic plant monitoring system applications. Which consists of database, web server, internet and client server ie smartphone. The database is created using MySQL, which connects to the webserver. In order to connect between database, web server, and smartphone required internet network.

2.3 Application and Programming Display Design on Android Studio

This design begins by creating a block diagram of the display sketch of the application content (description of the application content). The block diagram can be seen in Figure 3.

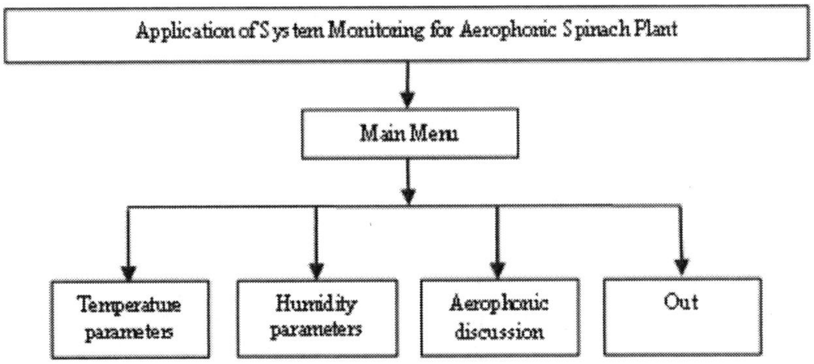

Fig. 3. Diagram Block of Aplication System Monitoring

Figure 3 illustrates the description of the contents of aeroponic monitoring system application (aeromosis) consisting of a main menu containing temperature and humidity content, aeroponic discussion (containing questions and answers about aeroponics), and exit application). Figure 3 can run with any programming on the android studio software. The programming stage of android studio is very important in making the application of aeromic monitoring system (aerophonic monitoring system). In android studio there are two programming languages that is xml to make the application and java display to run the application display in order to run in accordance with the design.

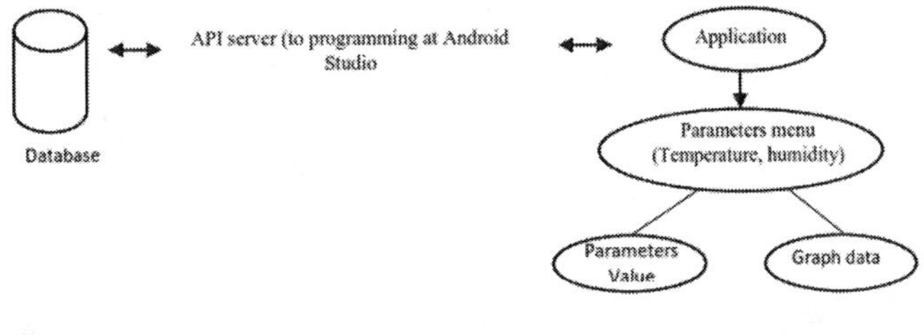

Fig. 4. How to Get Data of Aplication System Monitoring

Figure 4 describes the steps in data retrieval for the application of this monitoring system (aeromosis). Data stored on the database, connected with php files that have been uploaded on the web server. Connecting data from the database into this aeromosis application by calling the database url in the java file in android studio. Once the url is entered into java programming, the application can access the data with the internet connection.

3 Results and discussion

All menus can run because of coding. Coding is a process of translation of design or application design from the main menu to the existing menu in it into the java programming language to be a unity in order to become a medium that can be used in accordance with the design. Testing is very important because at this stage aims to determine whether the menu in the application can run correctly and to determine the shortcomings or errors that must be corrected. Testing the application is done using a smartphone.

3.1 Running application

The most important stage for aeromosis applications to be used is the stage of running the application. At this stage the application is run using a smartphone (USB Debugging), this stage is also a stage to test the application can run in accordance with the design or not. The application runs on Lenovo brand A6600d40 smartphone with android 6.0 operating system, API 23 and 5-inch screen size. Here are the steps in running the application using a smartphone:

1. Preparing smartphones as well as Micro USB to USB cable. Then open Developer Mode (Smartphone Mode) on your smartphone. If Developer Mode does not appear on the smartphone, enter the settings first, then enter the about phone menu, select the build number (press 5 times) until the message appears if the smartphone has become a developer.
2. Enable USB debugging on smartphone by connecting smartphone with PC using USB cable. Enable developer mode through the settings menu, by searching for USB debugging.
3. Run the application by Run on the android studio software, and search auto smartphone will become emulator and appear on the screen android studio software. Once the smartphone that we use appears then click OK, and wait until the compile process is complete. The application will appear on the smartphone screen.

3.2 Monitoring system application view

The aeromosis application displays the values of those parameters, the graph data that serves to see the values of the parameters that have been stored into the database in the form of line graphs, some questions and answers about aeroponics. Interface testing is done on the main menu to make sure the menus in the main menu interface can run according to the design. Display applications in accordance with the design using a smartphone with 5-inch screen size, operating system android 5.0.0 as the brand Lenovo A6600d40 hp.

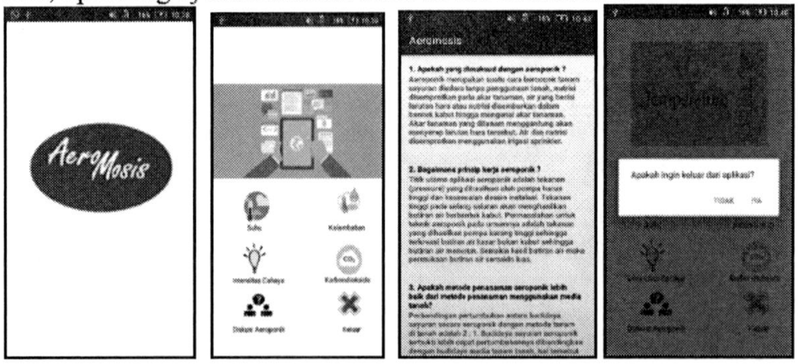

Fig. 5. Display of Aeromosis

Figure 5 is an aeromosis display, a splashscreen menu, a main menu, an aeroponic discussion menu and an exit menu. In the splashscreen display contains a logo that describes the application with aeroponic writing (in aeroponic monitoring system).

Splashscreen duration of about 5 seconds, then after automatically going to the main menu view.

The main menu consists of imageslider and some menu options. Imageslider on this main menu contains some animations that can shift automatically with delay every two to three to move from one image to another. After the three image will automatically move to the first image. In addition to the main menu imageslider also displays some menu options are: temperature menu, and humidity menu aeroponic discussion menu and exit menu.

3.2.1 Menu of temperature

Aeromosis is an aeroponic plant monitoring application that focuses on spinach. In aeromosis applications, this temperature menu displays values in real time graphs and lines. Display the menu menu temperature can be seen in Figure 6.

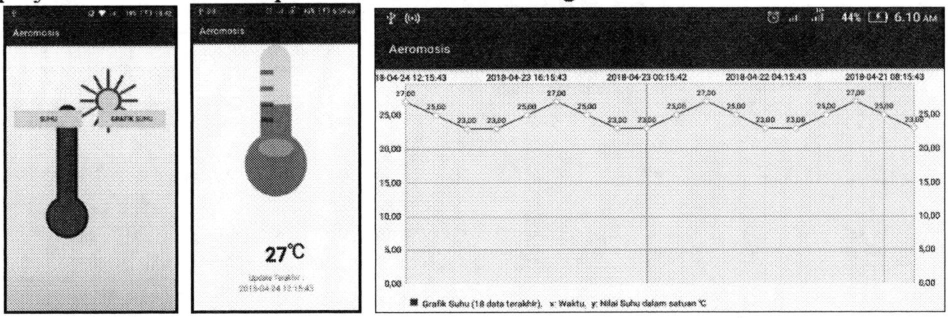

Fig. 6. Display of Temperature Menu in Smartphone

3.2.2 Menu of humidity

The humidity parameter menu display does not vary much with the temperature menu menu view. The humidity parameter menu display can be seen in Figure 7.

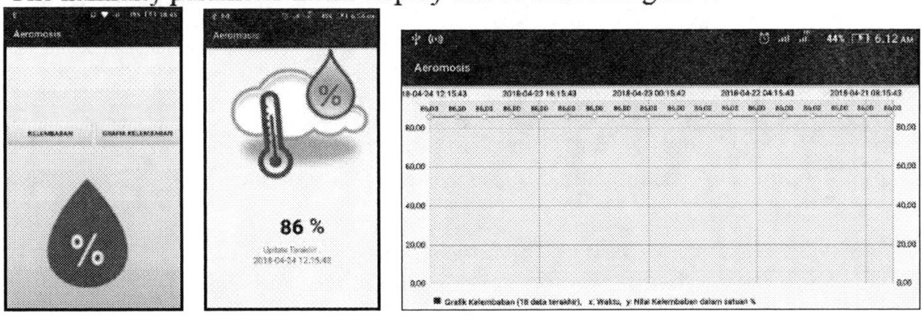

Fig. 7. Display of Humidity Menu in Smartphone

3.3 Testing of Device Screens

Tests on various screen devices done to check the system can work well or not on other smartphones. In this study using 10 kinds of smartphones with different types and specifications like can be seen in Table 1.

Test results from various screen device sizes in Table 1, obtained successful measurements 9 and 1 failed on the smartphone. Failure on this test is caused by screen display that is not in accordance with the design. This failure occurs because the application is run on a smartphone with a screen size of the device under 4 inches and the operating system on the smartphone 4.1.2. The size of the screen and operating system on the smartphone is very influential in the application of this monitoring system

Tabel 1. List Of Various Devices Used

No.	Devices Name	Screen Size	Operation System	API	Result
1.	Lenovo A6600d40	5 inci (720x1280 piksel)	Android 6.0	23	Succes
2.	Samsung GT-S6310	3.27 inci (320x480 piksel)	Android 4.1.2	16	Failed
3.	Xiaomi Redmi 4A	5 inci (720x1280 piksel)	Android 6.0.1	23	Succes
4.	Xiaomi Mi A1	5.5 inci (1920x1080 piksel)	Android 8.0	26	Succes
5.	Asus Z00RD	5 inci (720x1280 piksel)	Android 5.0.2	23	Succes
6.	Lenovo A7010a48	5 inci (720x1280 piksel)	Android 6.0	23	Succes
7.	Xiaomi Mi Max	6.44 inci (1920x1080 piksel)	Android 6.0.1	23	Succes
8.	Xiaomi Redmi Note 4	5.5 inci (1920x1080 piksel)	Android 6.0	23	Succes
9.	ASUS-X0011D	5.5 inci (1920x1080 piksel)	Android 7.1.1	25	Succes
10.	Freetel FTJ161A	4 inci (480x800 piksel)	Android 5.1	22	Succes

4 Conclusion

Based on the results of the application of temperature monitoring systems, humidity, light content and carbon dioxide in aeroponic system can be explained as follows:

- Application of aeroponic spinach system monitoring system has been completed in accordance with the design. All functions are used on phones with minimal size on 4-inch smartphones and 5.0.0 Android operating system and the system already meets the needs of the plant.
- Application of remote monitoring system can display parameter value; temperature and humidity content by displaying value and time. In addition, this application can also display graph data in the form of line graph display as much as 18 data obtained from the database. This application is very useful to develop aeroponic spinach plants.

References

1. Wahono, S., Sugiyanto, Elfita, "Eksperimen Pengaturan Suhu dan Kelembaban pada Rumah Tanaman (Greenhouse) dengan Sistem Humidifikasi", J. Teknik Mesin S-1, Vol. **2**, 1, (2014)
2. Ferdiansyah, "Rancang Bangun Sistem Kontrol pada Budidaya Tanaman dengan Teknik Aeroponik", (Institut Teknologi Sepuluh November, pp. 11-17, 2012)
3. Ramdani M., Andrian R., Novian Anggit, "Perancangan Sistem Monitoring Tanaman Menggunakan Zigbee dan Platform M2M", (Universitas Telkom Bandung, pp. 1-7, 2014)
4. Y. Sutiyoso, "Aeroponik Sayuran, Budidaya dengan Sistem Pengabutan Swadaya", (Jakarta, 2003)
5. Ruminta, "Analisis Penuruna Produksi Tanaman Padi Akibat Perubahan Iklim Di Kabupaten Bandung Jawa Barat", Universitas Padjajaran, J. Kultivasi Vol. **15**, 1, (2016)
6. T. Budioko, "Sistem Monitoring Jarak Jauh Berbasis Internet of Things Menggunakan Protokol MQTT", SemNas Riset Teknologi Informasi, Vol. **1**, (2016)

7. Syarief, Syafrizal, "Sistem Monitoring Suhu dan Kelembaban Tanaman Cabai pada Greenhouse Berbasis LabVIEW", Vol. **15**, No. 2 (2016)
8. Yahwe P. Caesar, Isnawaty, L.M. Fid Aksara, "Rancang Bangun Prototype System Monitoring Kelembaban Tanah Melalui SMS Berdasarkan Hasil Penyiraman Tanaman", semanTIK, Vol. **2**, No. 1, pp. 97-110, (2016)
9. Fahmi, Amanda, Wijaya, Ikhlal Aldhi Ghani, Nurulli Abdul Sugiharto, Aldhyth, "Sistem Monitoring dan Controlling Air Nutrisi Aquaponik Menggunakan Arduino Uno Berbasis Web Server", Vol. **1**, No. 1, pp. 39-46, (2016)
10. Andri, Rachmad, "Sistem Monitoring dan Pengendalian Suhu dan Kelembaban Ruang pada Rumah Walet Berbasis Android, Web dan SMS", Institut Teknologi Sepuluh November, pp. 1-9, (2013)

MATEC Web of Conferences **218**, 03018 (2018) https://doi.org/10.1051/matecconf/201821803018
ICIEE 2018

A Study on internet addiction among teenagers in selangor state of malaysia

Raja Kumar Murugesan[1,*], *Siti Khadijah* Binti Mohd Hamzaht[2] , and *Supriyanto*[3]

[1]Taylor's University, Malaysia
[2]International University of Malaysia Wales, Malaysia
[3]University of Sultan Ageng Tirtayasa, Indonesia

Abstract. Internet addiction is becoming a growing phenomenon across the world due to the tendency of Internet overuse for varied reasons. The objective of this study is to identify the extent of Internet use and level of addiction among teenagers with the age ranging from 13 to 17 years old. This also include the risk factor and impact to their social life. The sample data were taken from several schools covering 500 students from the state of Selangor in Malaysia. The survey was done using an instrument from Internet Addiction Test (IAT) developed by Dr. Kimberly Young. This test had been modified to meet the requirements of the youngster's mind-set in Malaysia. Demographics factors such as age, gender and race were examined apart from family and personal factors. The results of the study and Chi-square test summarizes that the level of internet addiction among teenagers in Selangor is $p = 0.161$ and tends to be minimal. We need to be aware of the level of Internet usage by teenagers so that measures could be made in place to deter or mitigate Internet addiction.

1 Introduction

The Internet since its inception for public use in the late 1980s has grown exponentially connecting people globally turning the world in to a global village. The Internet has become an essential part of our life. It cuts across geographical boundaries facilitating new opportunities, and learn new things faster. Today, the Internet has led to many disruptive innovations, disruptive business, and has grown in to one of the biggest economy in the world. Today, the Internet has become an integral part of everyday life especially among the teenagers.

Table 1 shows the world Internet usage statistics where, Asia tops the list in terms of highest number of Internet users with an Internet penetration rate of 48.1%. Malaysia has an Internet penetration of 78.3% with 68.7% penetration in terms of Facebook subscribers as of 31 December 2017 [2]. This statistics shows that Malaysia is one of the countries in Asia with a very high Internet penetration rate and substantially high number of users active on social media.

* Corresponding author: rajakumar.murugesan@taylors.edu.my

With the growing use of the Internet, excessive use of the Internet that could interfere daily routine activities may become a concern. The higher Internet penetration rate could positively influence the Internet overuse or addiction occurrence rate. This establishes a raising concern with especially youngsters. Studies from the University of Hong Kong done in December 2014 shows that the Internet addiction around the world is about 6% while, in Asia it is about 7.1% [3].

Table 1. World Internet usage statistics [1]

WORLD INTERNET USAGE AND POPULATION STATISTICS DEC 31, 2017 - Update						
World Regions	Population (2018 Est.)	Population % of World	Internet Users 31 Dec 2017	Penetration Rate (% Pop.)	Growth 2000-2018	Internet Users %
Africa	1,287,914,329	16.9 %	453,329,534	35.2 %	9,941 %	10.9
Asia	4,207,588,157	55.1 %	2,023,630,194	48.1 %	1,670 %	48.7
Europe	827,650,849	10.8 %	704,833,752	85.2 %	570 %	17.0
Latin America	652,047,996	8.5 %	437,001,277	67.0 %	2,318 %	10.5
Middle East	254,438,981	3.3 %	164,037,259	64.5 %	4,893 %	3.9
North America	363,844,662	4.8 %	345,660,847	95.0 %	219 %	8.3
Oceania / Australia	41,273,454	0.6 %	28,439,277	68.9 %	273 %	0.7
WORLD TOTAL	7,634,758,428	100.0 %	4,156,932,140	54.4 %	1,052 %	100

Internet Addiction refers to the overuse of the Internet that affects a person's daily routine life. Internet addiction is the inability to control Internet use that leads to depression, anxiety and abnormal behaviour if deprived of Internet use. Internet Addiction Disorder (IAD) is described by Dr Ivan K Goldberg as the symptoms of giving up or reduced social or occupational activities because of the use of the Internet [4]. Internet with cheaper and fast bandwidth has made easy access especially now freely available in most restaurants. Repercussions and undesirable outcomes may happen due to negative self-perception among users with Internet addiction. This could lead to poor mental health, stress, low self-esteem, anxiety, depression, and loneliness that would affect the lifestyle of a person. The above reasons had motivated this study on Internet addiction among teenagers in Malaysia.

This main objectives of this study are; to investigate on the number of teenagers who have Internet dependency based on demographic characteristics, Internet usage, family and other related factors such as personal characteristics and psychology of an individual; Identify the type of internet addiction and the factors that influence this disorder among teenagers; and to develop a mitigation framework on the issues of Internet addiction.

This study was conducted with teenagers in the age of 13 to 17 years old studying in the schools located in Hulu Langat and Kajang in the state of Selangor in Malaysia. Reason being that Selangor is the state with highest Internet penetration rate in Malaysia. A sample size of 430 students was used. The next section of this paper reviews critically the existing literature on Internet addiction and assessment methods. Section 3 discusses the research method, Section 4 discusses the results and findings. Section 5 concludes the paper.

2 Internet Addiction

2.1 Internet Addiction Disorder

Excessive use of the Internet possibly for online relationships, socialising, chat rooms, gaming, gambling, shopping or browsing would result in Internet addiction. Internet addiction as a disorder was matured by Dr Young which used to be classified as pathological gambling [5]. Internet Addiction Disorder (IAD) is also synonymously called as Problematic Internet Use (PIU) [6], Internet dependence [7], and compulsive Internet use [8]. According to Beard [9], the following five criteria needs to be satisfied to be diagnosed as Internet Addiction: Internet preoccupancy, Need to use the Internet most of the time to get satisfaction, Efforts to manage, control, reduce or stop using the Internet were not successful, Gets restless, depressed, moody or irritated when trying to reduce or stop using the Internet. In addition to the above, he says that any one of the following three criteria must be met: Has risked losing an important relationship, Lies to the family or others excessive use or dependency of the Internet, and use the Internet to escape from problem or difficult situation.

2.2 Internet Addiction Prevalence

The Internet has become an integral part of our life and as essential as water and electricity. Today there are many people who quip that they cannot live without the Internet. A study on public attitude over using the Internet conducted by Ipsos across 23 countries involving about eighteen thousand people shows that two-third of the respondents cannot imagine life without the Internet. The top ten countries with Internet addiction prevalence was India 82%, Great Britain 78%, China 77%, Germany 73%, United States 73%, Russia 66%, Spain 65%, France 64%, Italy 62% and Japan 62% [10]. The same study also reveals the frequency of use of the Internet. The top ten countries that has highest frequency of use of the Internet are Canada 97%, Great Britain 95%, Russia 96%, Sweden 96%, Australia 96%, Indonesia 95%, Poland 95%, S Korea 94%, and U.S., Belgium, Germany and S Africa 94% each. Most of these studies focused on younger population. From the above study it can be observed that the Internet addiction prevalence is more in Asian countries. It is observed that young age, early exposure to the Internet, frequent use, and environmental factors influences Internet addiction [11-13].

2.3 Internet Addiction Assessment

There are many instruments that have been developed for the Internet addiction assessment. The study done by Laconi et al on the measurement of Internet addiction identifies forty five such tools [14]. All these instruments measure Internet addiction through scales, interviews or diagnostics done through questionnaires. Among these, the Internet Addiction Test (IAT) [15], the Chen's Internet addiction scale (CIAS) [16], the Compulsive Internet Use Scale (CIUS) [17], and the Generalized Problematic Internet Use Scale (GPIUS) [18] were found to be most commonly used. Studies shows that IAT exhibits excellent test, retest reliability and excellent internal consistency of $r \geq 0.75$, and $a = 0.92$ respectively [19-21].

3 Research Methodology

This study in terms of a theoretical framework tries to understand the Internet addiction among teenagers by studying the student's personal characteristics and their cyber behaviour. The key dimension used to conceptualize the concerns include losing control of time in online activity, escaping from social activities, and compulsive response to environment. As for students personal characteristics the key dimensions used are gender, self-esteem, adjusting to family, adapting to learning, and interpersonal relationship. For Cyber behaviours, the key dimensions are entertainment, social communication online, virtual emotions and browsing for information search.

The research conceptual framework for this study is developed based on the modification of IAT [15, 20]. This conceptual framework to study the factors that influence Internet addiction tendency is shown in Fig. 1.

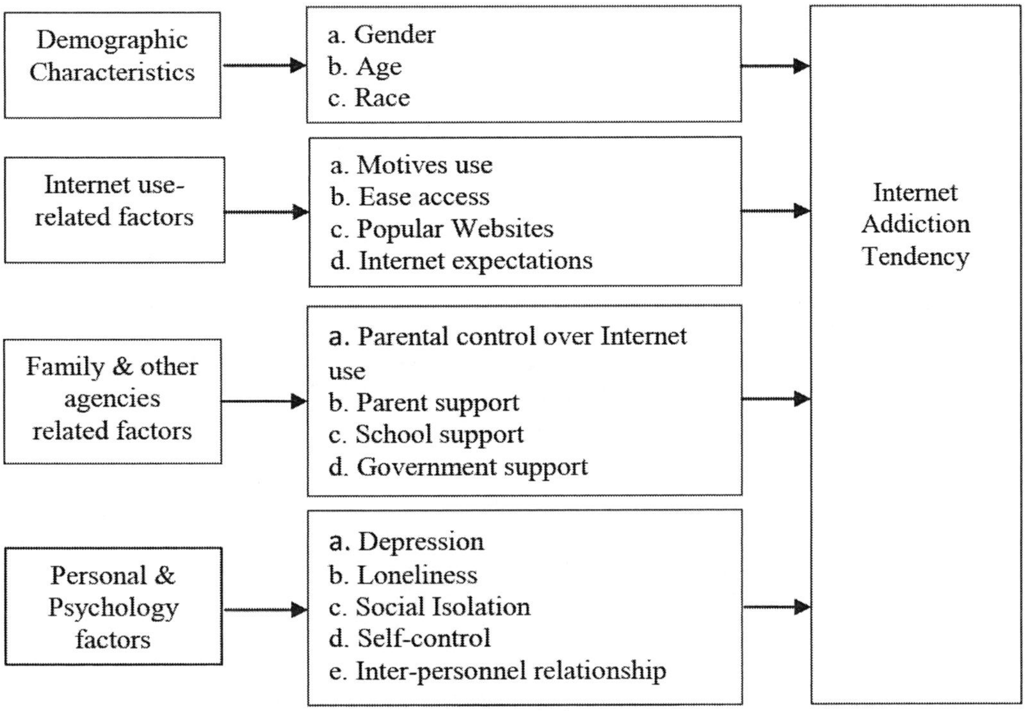

Fig. 1. Conceptual frameworks for internet addiction tendency.

Questionnaires were used to collect data from the teenagers. The questionnaire comprises of three sections, where the 1st section had twenty questions to assess the level of Internet addiction from the frequency of Internet usage. 2nd section had five questions to assess the type of Internet addiction, and 3rd section had six questions. Out of 500 questionnaires distributed, 430 were received completed. The study was conducted in Kajang and Hulu Langat in the Selangor state of Malaysia. Four out of forty five schools were used for this study.

4 Results and Analysis

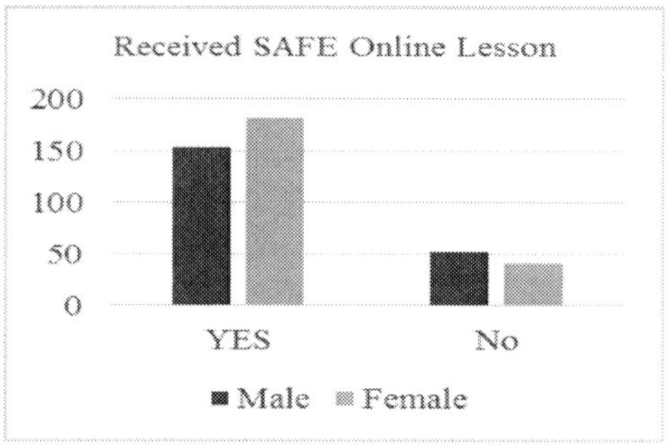

Fig. 2. Number of teenagers who received safe online lesson.

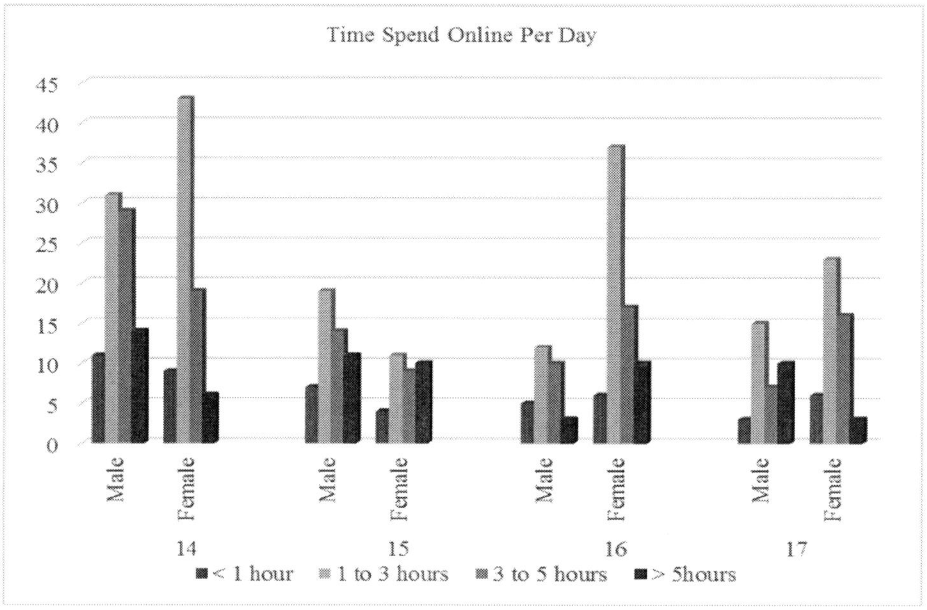

Fig. 3. Time spent online per day.

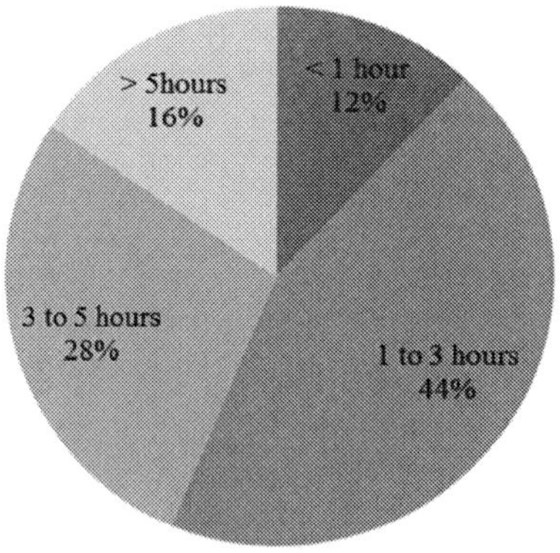

Fig. 4. Percentage of Time spent online per day.

Among the respondents, 31.6% were in the age group of 14, 30.7% were in the age group of 16, 21.86% were in the age group of 17, and 16.28% were in the age group of 15. Some of the results obtained through the questionnaire is shown in Fig. 2 to 5.

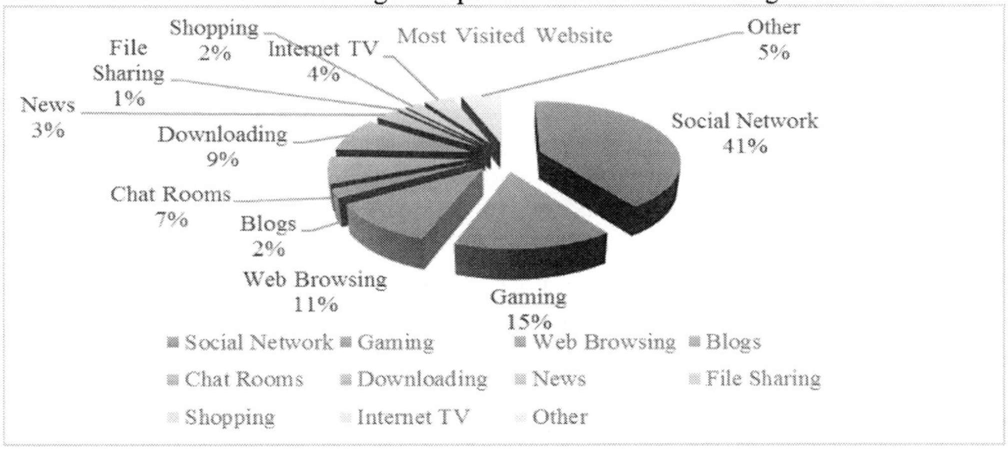

Fig. 5. Most visited website on a day.

From the data obtained for the time spend on Internet per day by teenagers in the age of 14 to 17 years old are that chi-square = 27.3, degrees of freedom = 21, and probability = 0.161. The probability of Internet addiction to happen is only 0.161 which, is minimal. The limitation of this research is that the data collected is only from four different schools in the state of Selangor and the conclusion made may not represent as a whole for the entire state of Selangor or Malaysia.

5 Conclusion

Increased and frequent use of the Internet could lead to Internet addiction that would be of major concern especially for adolescents or teenagers. This study was done to assess the level and type of Internet usage among Malaysian teenagers with the sample of data taken from four schools in the state of Selangor. IAT made and statistical analysis shows that the level of Internet addiction is minimal and is not of a concern. Though Internet Addiction is not observed to be a concern from the sample data taken it will be good to be aware of the level of Internet use periodically so that proper measures could be made in place to deter or mitigate Internet Addiction.

References

1. World Internet Usage and Population Statistics (Dec 31 2017 update).
2. Asia Marketing Research, Internet Usage, Population Statistics and Facebook Subscribers Dec. 31(2017)
3. Cecilia Cheng and Angel Yee-lam Li. Cyberpsychology, Behav., and Soc. Networking. **17**, 12 (2014)
4. D. Wallis, Just Click No. *The New Yorker* January 13, 28 (1997)
5. K.S. Young, Internet addiction: The emergence of a new clinical disorder. *104th annual meeting of the American Psychological Association*; (1996)
6. R. A. Davis, A cognitive behavioural model of pathological internet use (PIU) Comput. in Hum. Behav. **17**(2), 187–95 (2001)

7. N. A. Dowling and Quirk KL, Screening for Internet dependence: do the proposed diagnostic criteria differentiate normal from dependent Internet use? Cyberpsychol Behav. **12**(1), 21-7 (2009)
8. S. E. Caplan, Problematic Internet use and psychosocial well-being: development of a theory-based cognitive-behavioural measurement instrument. Comput. in Hum. Behav. **18** (5), 553–75 (2002)
9. K. W. Internet addiction: a review of current assessment techniques and potential assessment questions. Cyberpsychol Behav. **8**(1):7-14 2005
10. Life without the Internet. Technology – Attitudes and Use, Charts. Ipsos Global Trends. (07 Aug 2018)
11. D. J. Kuss, O. L. Lopez-Fernandez, Internet addiction and problematic Internet use: A systematic review of clinical research. World. J. Psychiatr **22** (6-1), 143-176 (2016)
12. E. Stip, A. Thibault, Beauchamp-Chatel Ab, et al. Internet addiction, hikikomori syndrome, and the prodromal phase of psychosis. Front. Psychiatry **7**: 6 (2016)
13. S. Krishnamurthy, S. K. Chetlapalli, Internet addiction: prevalence and risk factors: a crosssectional study among college students in Bengaluru, the Silicon Valley of India. Ind. J. Public. Health **59** (2), 115-121 (2015)
14. S. Laconi, R. F. Rodgers & Chabrol H: The measurement of Internet addiction: a critical review of existing scales and their psychometric properties. Comput. Hum. Behav. **41**, 190–202 (2014)
15. K. S. Young, Internet Addiction: The Emergence of a New Clinical Disorder. CyberPsychology. Behav. **1**, 237–244 (1998)
16. S. Chen, L. Weng L, Y. Su, H. Wu, and P. Yang, Development of a Chinese Internet addiction scale and its psychometric study. Chinese J Psychol. **45**:279, (2003)
17. G. Meerkerk, RJJM. Van Den Eijnden, A. A. Vermulst, and H. F. L. Garretsen, The Compulsive Internet Use Scale (CIUS): some psychometric properties. Cyberpsychol Behav. **12**, 1–6 (2009)
18. Caplan S: Problematic internet use and psychosocial well-being: Development of a theory-based cognitive behavioural measurement instrument. Comput. Hum. Behav. **18**, 553–575 (2002)
19. A. Barke, N. Nyenhuis, and B. Kroner-Herwig, The German version of the internet addiction test: a validation study. Cyberpsychol Behav. Soc. Netw. **15**, 534–542 (2012)
20. K. Lee, H. Gyeong, Yu B, Y. M. Song, H. K. Lee and D. Kim, Reliability and validity of the Korean version of the internet addiction test among college students. J Korean Med Sci. **28,** 763–768 (2013)
21. H. Osada, Internet addiction in Japanese college students: Is Japanese version of Internet Addiction Test (JIAT) useful as a screening tool. Bull Senshu Univ Sch Hum Sci. **3**, 71–80 (2013)

Genetic algorithm based protocols to select cluster heads and find multi-hop path in wireless sensor networks: review

Mohammed Al-Shalabi[1,*], *Mohammed* Anbar[1,*] , and *Tat-Chee* Wan[1,2]

[1]National Advanced IPv6 Centre, Universiti Sains Malaysia, 11800 Penang, Malaysia
[2] School of Computer Sciences, Universiti Sains Malaysia, 11800 Penang, Malaysia

Abstract. A wireless sensor network (WSN) is a modern technology in radio communication. A WSN comprises a number of sensor nodes that are randomly spread in a specific area for sensing and monitoring physical attributes that are difficult to monitor by humans, such as temperature, fire, and pressure. Many problems, including data transmission, power consumption and selecting cluster heads, may occur due to the nature of WSNs. Various protocols have been conducted to resolve these issues. Most of the proposed protocols are based on the Genetic Algorithm as an optimization technique to select the Cluster Heads (CHs) or to find a multi-hop path for sending the data from the CHs to the Base Station (BS). This paper presents a comprehensive study of the protocols for WSNs that are proposed to come up with these issues. This study emphasises on CHs selection protocols and multi-hop path finding protocols and their strengths and weaknesses. A new taxonomy is presented to discuss these protocols on the basis of different classes. A complete comparison of the main features and behaviors of the protocols is conducted. This study will give basic guidelines for the researchers those have a motivation to develop a new CHs selection protocol or a multi-hop path finding protocol.

1 Introduction

Data Transmission is the process of sending the data from the Cluster Heads (CHs) or the sensor nodes to the Base Station (BS) [1]. By contrast, the CH selection is the process of selecting the proper nodes to be CHs. There are many protocols have been proposed to deal with these issues in WSNs using different algorithms [1,2]. The evolutionary based algorithms play important role in solving such problems due to their efficiency in finding the optimal solutions. The most popular and familiar evolutionary algorithm is the Genetic Algorithm (GA) which is adapted in this research to solve the direct transmission problem [3,4].

This paper will review and compare between the protocols that have been conducted in the last few years. This review is realized as a clear understanding for the researchers of

* Corresponding author: moshalabi@nav6.my, anbar@nav6.usm.my

different CHs selection protocols' and multi-hop path protocols' vision, their strengths and weaknesses, and the used parameters in them.

The rest of the paper will be organized as following; Section II will clarify the Genetic Algorithm. Section III reviews the related protocols based on the new taxonomy. Section IV presents a comparison between the protocols. Finally, the paper is concluded in Section V.

2 Genetic algorithm

In this section, a background of Genetic Algorithm will be presented because it is considered as an optimization method used to find optimal solutions.

GA is an optimization algorithm that simulates a nature evolution, and it uses the same methods in the natural selection to find optimal solutions [5]. GA is used in optimization and search problems to find the best solution. The main operations in GA are selection, crossover, and mutation [6]. At first, initial population is randomly generated, and then a new population will be selected from the initial population depending on the fitness function and will contribute in the generation of the population [7]. The fitness function is very important in GA and depends on the problem, it is used to evaluate the quality of individuals and select the best of them to contribute in the next generation [8]. The main operators in GA are crossover and mutation, crossover operator is used to generate new children from the parents by combination them, and these children will be better than the two parents [9]. In mutation, random swap is applied on the parents in order to form new children. The algorithm stops when a specific condition is reached [10].

One of the GA operators is crossover which is applied on two selected individuals to produce new offspring in order to vary the programming of the chromosomes [9]. The individuals are selected depending on the fitness value. There are different types of crossover such as one-point crossover and two-point crossover which can be used to exchange portions; from these individuals; that are separated by a crossover point. The crossover point is randomly selected and then the crossover operation will be applied on the two selected individuals.

The mutation operation is applied on each gene of an individual with the probability of mutation rate. Mutation affects only a single chromosome, and the chromosome which is selected for mutation will have a randomly selected gene that is called mutation point. There are many ways for the mutation operation, flipping the value of a gene is one of the ways, the value of the gene that has a position equals to the mutation point will be changed from 0 to 1 or vice versa to give a new possible path.

In the selection process, some individuals are selected from the population in the current generation based on the values of the fitness function. The individuals with smaller fitness values are selected as the individuals of the population in the next generation. The algorithm stops if the number of generations equals a specific number which represents the maximum generations.

The main drawback in the WSNs is ignoring the residual energy of the nodes when selecting them as CHs. Moreover, WSNs consider the direct transmission of data between a CH and the BS, which consumes a lot of energy and negatively affects the lifetime of a network. For these reasons, there are many routing protocols that have been proposed to enhance the WSNs respecting to maximize the network's lifetime based on a proposed taxonomy, as will describe in the following section.

3 Proposed taxonomy of GA-based protocols in wsns

In this paper, the focus is on the CHs selection and the multi-hop path finding as designated criteria to present the classification of the GA-based protocols in the WSNs. In Fig 1, the GA-based routing protocols are categorized into two categories: CHs selection-based and multi-hop path finding-based protocols.

Each main category is further classified into different classifications based on the main factor in the proposed approaches. The next subsections describe each main category in detail. Then, existing familiar protocols that belong to each main category are presented in detail.

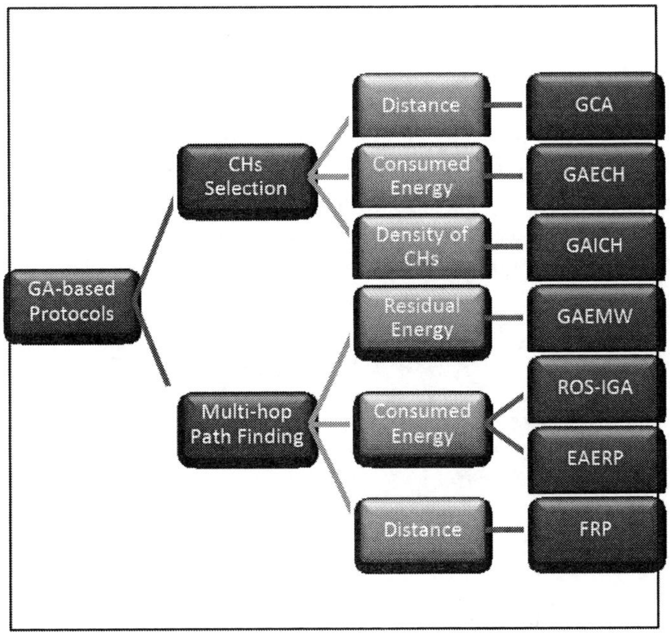

Fig. 1. Proposed taxonomy of LEACH-based routing protocols in WSNs

3.1 Multi-hop path finding protocols

This subsection presents the protocols that have been proposed to find the multi-hop path for transmitting the data from the CH to the BS based on the GA as will be described in the following paragraphs.

3.1.1 FRP (Fitness-base routing protocol):

Authors in [11] have been proposed a new protocol to improve the efficiency of data transmission by using GA. In order to find an efficient rout to transmit data from a sensor node to the BS, the fitness function uses parameters such as the distance from the sensor node to the BS, based on the fitness value, the optimal nodes are selected and then the data packet is routed in an optimized route to the BS [11]. The fitness value of a node is calculated as follows:

$$Fitness = dist(i,j) + dist(j,BS) \tag{1}$$

Where the distance between two sensor nodes or between any sensor node to the BS is calculated by Equation 1. The packet is transmitted from a sensor node to the BS through an efficient path, the intermediate nodes between the source node and the BS are selected depending on their fitness values, and these nodes route the packet to other nodes until reach to the BS [11].

In FRP protocol, the energy consumption in the transmission phase is reduced, and also the packet loss percentage is minimized by finding alternative nodes in case of node failure.

3.1.2 GAEMW (Genetic algorithm for energy-entropy based multipath routing in wsns):

In GAEMW protocol [12], authors proposed a GA for multipath routing in WSNs; the idea is to find the minimal node residual energy of each route in selecting multipath routing.

To initialize population, a routing path is encoded by a set of integers represent IDs, energy and other node's information. The first gene represents a source node and the last gene represents the BS, then several paths (chromosomes) will be created from the source to the destination. Fitness function is used to evaluate the quality of each chromosome (path) by finding the least cost path between the source and destination.

The fitness function of a chromosome Ci is defined as follows [12]:

$$f(C_i) = \frac{1}{G}\sum_{k=1}^{path}\left((E_{tr} + E_{rec}) \times P_{ki} \times \left|P_{ki}^{s,d}\right| - E_{avg}\right) \quad (2)$$

After that, the two operators in GA (crossover and mutation) are applied on the population and the next generation is created. This protocol provides efficient protocol for evaluating the route stability in WSNs.

3.1.3 ROS-IGA (Routing optimization strategy based on improved GA):

The literature that has been proposed in [13] describes an improved protocol using GA to find a suitable route between a sensor node and the BS, the new protocol called ROS-IGA. In this protocol, authors tried to solve the problem of generating invalid individuals; this problem negatively affects the efficiency of GA. In the ROS-IGA protocol, the locations of nodes and other information such as energy consumption between adjacent nodes, accumulated distance, and the remaining energy of nodes are considered in the fitness function to find suitable and practical path. The fitness function is maximization function, thus, the higher fitness value means more paths [13].

The crossover and mutation operations are improved in this protocol based on the neighboring information and the structure of the network. This improvement of crossover and mutation operations provides better routing and practical routes in WSNs.

3.1.4 EAERP (Energy-aware evolutionary routing protocol for dynamic clustering of wireless sensor network):

The idea of this protocol is to increase the stability period of the network and to increase the lifetime until the last node dies. Authors in [14] have used the concept of the centralized evolutionary routing protocol to optimize the CH selection process by the BS. The proposed fitness function includes the intercluster and the intracluster energies. The limitation in this protocol is the consideration of the single-hop transmission only without taking into account the multi-hop transmission.

3.2 Cluster heads selection protocols

This subsection presents the protocols that have been proposed to select the nodes to be CHs based on the GA. The following paragraphs describe each protocol in details with the formulated fitness functions and the considered parameters.

3.2.1 GAICH (Genetic algorithm inspired clustering hierarchy):

In this protocol, authors [15] have used the GA at the BS in order to select the suitable cluster heads and find the optimal number of them by applying a proposed fitness function which is based on the following parameters: density of cluster heads, distances from cluster heads to the BS, the centrality of cluster heads, and the residual energy of cluster heads. In this protocol, the authors proposed three levels of cluster heads percentage depend on the average energy of the network, and the population size depends on the number of candidate cluster heads. The limitation in this protocol is that the stability period is not achieved. In addition, this protocol doesn't expose the data transmission problem.

3.2.2 GAECH (Genetic algorithm based energy efficient clustering hierarchy in wireless sensor networks):

In this protocol, the authors [16] have enhanced the LEACH protocol by using the GA (run at the BS) with proposed fitness function in order to form balance clusters. The chromosomes in this protocol are presented using the binary presentation, and the single point crossover is used. The parameters which are used in the fitness function are the total energy consumption, the standard deviation among clusters in basis of energy consumption, the energy consumption in the cluster heads, and the cluster head distribution. The results showed that the performance of this protocol is better than other protocols. On the contrast, this protocol suffers from some drawbacks, such as it did not consider the distance when selecting the cluster heads, and it did not consider the routing process.

3.2.3 GCA (A new robust genetic algorithm for dynamic cluster formation in wireless sensor networks):

In order to increase the lifetime of the network, the authors in [17] have proposed a robust genetic clustering algorithm to balance the consumed energy within the clusters by controlling the number of them. The proposed fitness function in this protocol is presented in the following equation:

$$Fitness = W * CH + (1 - W) * D \quad (3)$$

where CH is the number of cluster heads, D is the total distance from cluster members to the CH in each cluster and W is a value based on the application. The authors of [17], based on this fitness function, tried to minimize the consumed energy by minimizing the number of CHs in the network and minimizing the total distance between the cluster members and the CH. The stability period of the network is not considered in this protocol.

4 Comparative analysis

Table 1 shows a comparison among the above mentioned protocols which proposed to select the CHs or to find the multi-hop path for transmitting the data from the CH to the BS using different techniques based on the GA. In Table 1, some information such as the year of when the protocol is proposed, the modeling parameters, and the drawbacks and limitations of each protocol are highlighted.

Table 1. Comparison among GA-based Protocols

Protocol	Modeling Parameter	Drawbacks and Limitations
GCA (2007)	Number of CHs and distance to the BS.	-Does not consider the routing problem.
EAERP (2011)	Inter and Intra clusters energies to select the CHs.	-Consider just the single-hop transmission.
FRP (2014)	Distance from the sensor node to the BS	Fitness function just considers the distance to the BS.
GAEMW (2015)	Residual Energy of nodes	Fitness function concentrates only on the residual energy.
GAECH (2015)	The Total and standard deviation of energy, and CHs distribution.	-Does not consider the distance in selecting the CHs. -Does not consider the routing problem.
ROS-IGA (2016)	Locations of nodes, Energy consumption between adjacent nodes, Accumulated distance, Remaining energy of nodes	-Concentration on the invalid individuals, it is based on the locations of nodes which got during the deployment. -Does not consider the CH selection.
GAICH (2016)	Density of CHS, distance and residual energy.	-Stability period is not achieved. -Does not consider the routing problem.

According to Table 1, it is clearly seen that the parameters which used in the fitness function in order to optimize the data transmission problem in many protocols were not enough to find the best solution; which is the optimal multi-hop path between the CH and the BS; as the protocols that have been proposed in [11] and [12], which the authors just considered the distance from the sensor node to the BS in addition to the residual energy of nodes respectively. Otherwise, the authors [17] used GA to select the best CH without considering the data transmission path.

Other proposed protocols to select the CHs also did not consider suitable and enough parameters to select the proper nodes to be CHs. In general, gaps still exist in the proposed protocols in terms of the CH selection and the data transmission issues.

4 Conclusions

Several protocols have been discussed and proposed to solve the CHs selection and the data transmission problems in the hierarchical-based WSNs. In this paper, the common protocols are discussed and reviewed with considering the advantages and disadvantages for each protocol. The GA evolutionary algorithm was used to handle the data transmission between a CH and the BS by finding a multi-hop path between them based on some parameters such as the distance and the energy, or to select the CHs based on some parameters such as the number of CHs and the centrality of the CHs. Each of these protocols is analyzed in order to clarify the guidelines when the researchers want to propose mechanisms; these guidelines consider the gaps in the previous protocols.

References

1. M. Al-Shalabi, M. Anbar, T.C. Wan, A. Khasawneh, Variants of the Low-Energy Adaptive Clustering Hierarchy Protocol: Survey, Issues and Challenges. Electronics **7**, 136, (2018).
2. A. Abbasi, M. Younis, A survey on clustering algorithms for wireless sensor networks. Computer Communications **30**, pp. 2826-2841, (2007).
3. A. Norouzi, H. Zaim. Genetic Algorithm application in optimization of wireless sensor networks. The Scientific World Journal, (2014).
4. M. Kumar, M. Husian, N. Upreti, D. Gupta. Genetic Algorithm: Review and Application. International Journal of Information Technology and Knowledge Management **2**, pp. 451–454, (2010).
5. Q. Yang, X. Yang, Y. Wang, D. An, A strategy based on cooperative transmission for minimizing delivery delay in WSN. International Journal of Distributed Sensor Networks **2013**, (2013).
6. R. Zhou, M. Chen, G. Feng, H. Liu, S. He. Genetic clustering route algorithm in WSN. *Natural Computation* (**8**, pp. 4023–4026, 2010).
7. W. Liu, Y. Wu. Routing protocol based on genetic algorithm for energy harvesting-wireless sensor networks. IET Wireless Sensor Systems **3**, pp. 112–118, (2013).
8. S.N.Sivanandam, S. N. D. Genetic Algorithm Implementation Using Matlab. Introduction to Genetic Algorithms, pp. 453, (2008).
9. M. Warrier, A. Kumar. An Energy Efficient Approach for Routing in Wireless Sensor Networks. Procedia Technology **25**, pp. 520-527, (2016).
10. Y. Song, C. Gui, X. Lu, H. Chen, B. Sun. A Genetic Algorithm for Energy-Efficient Based Multipath Routing in Wireless Sensor Networks. Wireless Personal Communications **85**, pp. 2055, (2015).
11. A. Balamurugan. An Energy Efficient Fitness Based Routing Protocol In Wireless Sensor Networks. Ictact Journal On Communication Technology **5**, (2014).
12. Y. Song, C. Gui, X. Lu, H. Chen, B. Sun. A Genetic Algorithm for Energy-Efficient Based Multipath Routing in Wireless Sensor Networks. Wireless Personal Communications **85**, pp. 2055, (2015).
13. G. Yao, Z. Dong, W. Wen, Q. Ren. A Routing Optimization Strategy for Wireless Sensor Networks Based on Improved Genetic Algorithm. Journal of Applied Science and Engineering **19**, pp. 221-228, (2016).
14. E. A. Khalil and B. A. Attea. Energy-aware evolutionary routing protocol for dynamic clustering of wireless sensor networks. Swarm and Evolutionary Computation **1**, pp. 195-203, (2011).
15. Y. Lu, V. Wong. An energy-efficient multipath routing protocol for wireless sensor networks. International Journal of Communication Systems **23**, pp. 633-652, (2006).
16. B. Baranidharan, B. Santhi. GAECH: Genetic algorithm based energy efficient clustering hierarchy in wireless sensor networks. J. Sensors **2015**, (2015).
17. S. R. Mudundi, H. H. Ali. A new robust genetic algorithm for dynamic cluster formation in wireless sensor networks. *Wireless and Optical Communications*, (pp. 360–367, Montreal, Canada, June 2007).

Critical Factor of E-Learning Component using HELAM and AHP

Generosa Lukhayu Pritalia[1] *, *Djoko Budiyanto* Setyohadi[1], *Luciana* Triani Dewi[2] and *Sri* Kusrohmaniah[3]

[1] Department of Informatics Engineering, Faculty of Industrial Technology, Atma Jaya Yogyakarta University, Yogyakarta – Indonesia
[2] Department of Industrial Engineering, Faculty of Industrial Technology, Atma Jaya Yogyakarta University, Yogyakarta – Indonesia
[3] Department of Psychology, Faculty of Psychology, Gadjah Mada University, Yogyakarta – Indonesia

Abstract. Today's e-learning is an essential component of an education process, especially in higher education. The performance of education process can be improved significantly when higher education applies e-learning. Furthermore, much higher educations use e-learning, however many of them fail to use e-learning since they do not investigate the critical component of e-learning which suitable for their institution. This paper proposes how to identify the critical factor of e-learning in higher education using *Hexagonal E-learning Assessment Model* (HELAM) and *Analytic Hierarchy Process* (AHP). Using two type of student regarding the use of e-learning, i.e., the student who everyday use e-learning and not. Our study shows that the most critical factor chosen by intense students using e-learning is Course Management factor, while the most critical factor for not intense students is Knowledgeable. From the results of this study can be concluded that the students already have the confidence and ability that are high in using e-learning sites, but support and encouragement are also needed to motivate them in the use of e-learning. Such motivation can be their private lecture site management services as well as support and training so that students can be well-informed about e-learning sites.

1 Introduction

Education is the unification of some elements of teaching, namely: a source of information, ideas, educational material as well as students. Some parts of the element integrate the information technology, this precipitating the birth of ideas about e-learning. E-learning is a teaching and learning system using internet technology to enhance knowledge and performance [1]. Learning using electronic media or e-learning had been already started in the 1970 's [2]. In Indonesia, the use of e-learning started in 1999 and continued to grow to this day. The selection of e-learning interface in the learning media is

* Corresponding author: djoko.bdy@gmail.com

useful to improve the effectiveness and efficiency of the teaching and learning process, the previous study show that interface is able to influenced user emotion [3]. Social aspects or users such as students and teacher attitude as well as technical aspects such as the quality of the system, the quality of the information and content as well as the quality of service are the important aspects in e-learning. Among these aspects, an identification should be done to investigate which aspect that is worth the careful attention for the e-learning site.

This study adopts the Hexagonal E-learning Assessment Model (HELAM) developed by Ozkan and Koseler [4]. HELAM model is one evaluation model to assess user satisfaction towards a system of e-learning. HELAM is chosen because this model tries to offer judgments on all aspects of the faculty, students, institutions, and systems. The method of Analytic Hierarchy Process (AHP) is the method selected in the decision making on this research. Analytic Hierarchy Process (AHP) is a method used in solving problems that are multi-criteria (MCDM) [5]. The AHP method is a method of decision making which is developed to give priority when several criteria are considered in decision making. AHP method is widely used in earlier studies because of its ability to generate optimal decisions [6]. This study will classify the factors elements based on interests. Furthermore, the result element groups will get a major concern in e-learning.

2 Literature Review

Hexagonal E-learning Assessment Model (HELAM) which was conceived by Ozkan, Koseler [4] is a model to evaluate or measure the success of e-learning. Many scoring models e-learning are successful but they did not consider the importance of each aspect of e-learning. An assessment of e-learning in general only focuses on some aspects of the course. Hexagonal E-learning Assessment Model (HELAM) combines all aspects such as the faculty, students, institutions, and systems. Hexagonal in E-learning Assessment Model (HELAM) describes that the success of the e-learning system due to two concepts: social issues and Technical Issues. E-Learning process is open system so it will be influenced by social issue. In the other side, the issue is the technical factors affecting e-learning that emerge within the system itself and the supporting factors are complementary factors of using of e-learning.

According to Ozkan, Koseler [7], HELAM has been developed to evaluate e-learning through six dimensions with two important issues namely (1) social issues: social issues: a perspective (the perspective of) student (7 factors), the attitude of the teaching staff (9 factors) and (2) Technical Issues: the quality of the System (11 factors), quality of information (4 factors), quality of service (11 factors) as well as a supporting Factor (4 factors)

Analytical Hierarchy Process (AHP) is one of multi-criteria decision-making models that can help the human's mind frame in which the factor of logic, experience, knowledge, emotions, and sense are optimized into a systematic process [8-10]. AHP (Analytical Hierarchy Process) method is a method of decision making which is developed for granting priority of some alternatives when there are several criteria that should be considered, as well as allowing the decision makers to compose a complex problem into a hierarchy or an integrated set of levels. The AHP method is used to solve complex problems. It is performed by setting the group into a hierarchy. And then, a numeric value is used as a substitute for human perception in doing a comparison to deal with the priority

3 Research Methodology

Research methodology in the flowchart shown in the figure below.

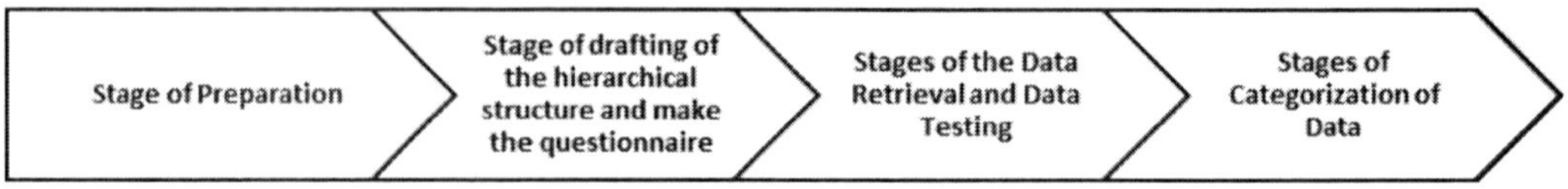

Fig. 1. Research flowchart

3.1 Stage of Research

3.1.1 Stage of Preparation

Preparation phase includes the study of literature, the design research, the research's time schedule and location, the population and sample. Population samples are grouped into intensive and not intensive students using e-learning website (http://kuliah.uajy.ac.id).

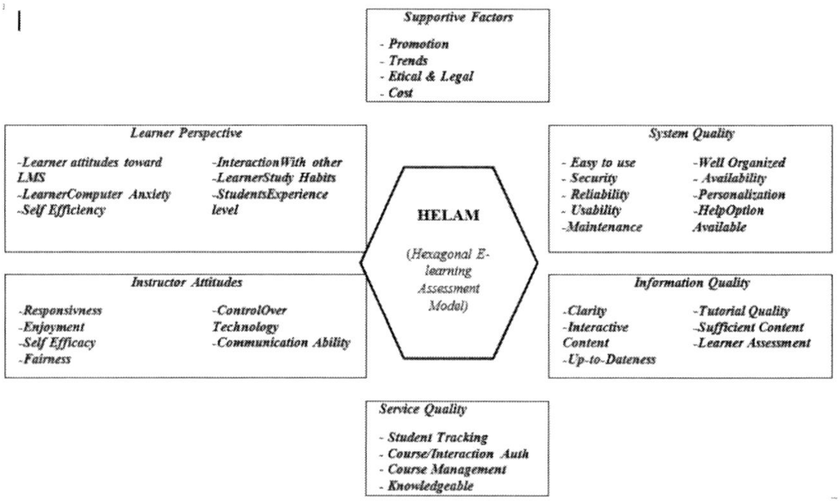

Fig. 2. Hexagonal E-learning Assessment Model

Dimensions and criteria owned HELAM contain some elements that have dependencies or relationships of one criteria to the other. It is contradictory from AHP method because one of the conditions of the method of AHP requires the independent criteria and sub criteria. To overcome these problems, the variable is selected so the used variable is only independent variable. It is shown in Fig 2.

3.1.2 Stage of drafting of the hierarchical structure and make the questionnaire

The next step is to create of a hierarchical structure. The creation of hierarchical structure is the basic stage which will be used to solve the problem. After the structure is formed and then a Questionnaire is compiled based on the use of reference questionnaires from AHP (Analytical Hierarchy Process) which was introduced by the Saaty [9]. Preparation of questionnaires is carried out based on the obtained hierarchical structure.

3.1.3 Stages of the Data Retrieval and Data Testing

The stage of data retrieval is done by filling out the questionnaires, where the factor priotiry of quiestionare is adopted from Saaty [11]. The sampling method used sampling purposive.

Determination of the number of samples is 100 relying the common descriptive research [12].

3.1.4 Stages of Categorization of Data

Based on the stage of analysis and presentation of data, the next step is to distribute the elements into a class which aims to divide the range into top priority data categories and additional priority categories [13].

4 Result and Discussion

4.1 Data of Respondents

The data was collected from 146 respondents who student which use e-learning. There were 20 questionnaires that did not return. Based on the demographic data obtained from respondents, the data can be seen in the following table.

Table 1. Respondent demographic data

Demographic	Total	Demographic	Total
Gender		**Use of the internet for task**	
Male	45	More than 6 hours	35
Female	98	4-6 hours	35
Faculty		2-4 hours	60
Industrial Technology	29	Less than 2 hours	16
Economy	21	**Use of the computer for task**	
Biology	25	More than 6 hours	42
Social and Political Science	22	4-6 hours	34
Legal Science	25	2-4 hours	55
Engineering	24	Less than 2 hours	15
Daily use of internet		**Number of courses that use e-learning**	
More than 6 hours	62	0	18
4-6 hours	48	1	24
2-4 hours	31	2	13
Less than 2 hours	5	3	20
Daily use of computer		4	10
More than 6 hours	31	>4	61
4-6 hours	34		
2-4 hours	53		
Less than 2 hours	28		

Based on the results above,this study indicated that the internet has become part of the respondents' everyday life. It is reflected that 75% of respondents have been accustomed to using the internet for everyday purposes above 4 hours per day. However the use of the internet for the purposes of completing the coursework only gets allocation fewer than 4 hours per day. Thedata reveals that students' attention to use e-learning are likely to be distracted by social media which are more familiar for them and more often to use. The total of respondents can be said to have been wonted to using the sites College because 61 respondents have used the site in College for more than four courses this semester.

4.2 Data Analysis

Analysis of questionnaire data from respondents using the AHP method made by Saaty [9]. It can be seen in table 2 that the quality of the system and the quality of information basically have a very strong interaction with the quality of the relevant information in the e-learning and it is supported by the quality of a good system. The quality of the services that are responsive and appropriate needs is also required so that e-learning can be used optimally by the user. The e-learning system must be effective and efficient [14-16]. The effectiveness and the efficiency of the process are measured from the of user, and it can be used for developing e-learning which is useful to enhance user satisfaction.

Table 2. Dimension ranking between intensive and not intensive students

Dimension	Intensive		Not Intensive	
	Weights	Ranking	Weights	Ranking
Learner Characteristics	0.1471	4	0.1435	5
Instructor Characteristics	0.1405	5	0.1452	4
Institution and Service Quality	0.1896	3	0.1980	3
Infrastructure and System Quality	0.2330	1	0.2158	1
Course and Information Quality	0.1907	2	0.1985	2
Extrinsic Motivation	0.0988	6	0.0988	6

Table 3 shows that the most important factor of the intensive student is course management. For students who are not intensive, a factor that is considered the most important one is knowledge, the factors that are considered most important for not intensive or student are a student's anxiety towards the system.

Table 3. Factor global ranking between intensive and not intensive students

Factor	Not intensive				Intensive			
	Local weight	Rank Local	Global weight	Rank Global	Local weight	Rank Local	Global weight	Rank Global
Clarity	0.1656	3	0.0358	7	0.2035	2	0.0474	4
Sufficient Content	0.2030	1	0.0439	4	0.2070	1	0.0482	3
Student Tracking	0.2335	3	0.0465	3	0.2389	3	0.0455	5
Course Management	0.2435	2	0.0484	2	0.2870	1	0.0547	1
Knowledgeable	0.3443	1	0.0685	1	0.2798	2	0.0533	2

4.3 Categorizing Data

Data categorization is done through the stages of determining the number of intervals class followed by determining the range of the data and determining the length of the class intervals.

Table 4. Frequency distribution of global weights for intensive and not intensive students

class	Not intensive			Intensive		
	Range	Frequency	Rank	Range	Frequency	Rank
1	0.0594– 0.0685	1	1	0.0480– 0.0547	2	1-2
2	0.0502– 0.0594	0	-	0.0413– 0.0480	4	3-6
3	0.0410– 0.0502	3	2-4	0.0346– 0.0413	1	7
4	0.0319– 0.0410	7	5-11	0.0279– 0.0346	9	8-16
5	0.0227– 0.0319	13	12-24	0.0212– 0.0279	6	17-22
6	0.0135– 0.0227	11	25-35	0.0145– 0.0212	13	23-35

The category of priority is gained from data at the top of the class. The results in table 4 major priority Category obtained from data on class 1. The table show that the top priority for intensive students are: Knowledgeable while the top priority for non intensive students are Course Management and Knowledgeable.

4.1 Discussion

The result of table 4 implies that Knowledgeable Service is in the form of instructions and special training given to students to increase the ability of students in utilizing computer-based learning systems and e-learning. This is inline to previous studies of Bhuasiri [17] that computer users and e-learning in developing country institutions have not so acquainted with technologies such as institutions in developed countries. It should be noted that e-learning is not the main need. Therefore, usually e-learning is not well performed [18]. Raising awareness of the importance of technology users, computer usage skills and knowledge is an important factor in the success of e-learning at institutions in the developing countries.

Course Management services should be provided in program settings or materials to serve the needs of students in the education process [19]. Learning materials management services is a service that should be given by the developer or the owner of the system to the users of the system. Material management services are provided for teachers and users to organize and to manage the material needed for a learning activity.

Based on the results of the elements deliberations that became the top priority for students who are intensive and not intensive, there are some suggestions that can be given to the owner and developer of the system in order to optimize the site e-learning, such as :

- Socialization features in the e-learning are increasingly optimized so that the user can find out every feature of the e-learning site. Socialization can be performed by informaton disemanitation from the owner of the system or teachers and fellow students. It will be useful to encourage the students who are not intensive to use e-learning can be done by providing them the required knowledge and training. It will play the importance of using e-learning site in learning.
- Optimizing service management courses and materials courses taken by students are available in the system. Increasing the use of e-learning can be encouraged by faculty and students who routinely or even obligated use e-learning in the learning process.

In addition, the most important factor for students foundin the study is a surprising thing Computer anxiety of the system becomes the most unnoticed factor for students who are intensive and not intensive using e-learning. They pay more attention to the technical aspects or factors that are derived from the system rather than the social aspects or factors that is originated from users. The study found that factors that can affect the success of e-learning in higher education are related to the improvement of technology and attitude towards e-learning, the increased knowledge and skills of basic technology, content learning improvement, computer training and motivation.

5 Conclusion

The results of this study showed that college students have the confidence and ability in using e-learning sites, but they need support and encouragement in motivating. The motivation comes from the owner or developer of the system, teachers and even from fellow students. The study found six dimensions to implement e-learning system in college. These dimensions include the characteristics of student Attitude, the teachers attitude, systems quality, information and content quality as well as the service quality. The most important factor is the course management factor or services provided by the system to manage the courses which are being taken by students and factor in knowledge or services of the owner and developer of the system so that users can become well-informed about the e-learning system. Further research should examine and focus on stakeholders' different contexts such as system developer and administrator. The results of this research may change from time to time by the interests of stakeholders. This insight can reveal additional information that is useful for the e-learning application of in college.

References

1. Ruiz, J.,Mintzer, M., & Leipzig, RThe Impact of E-Learning in Medical. . (2006).
2. Waller, V., & Wilson, J. A Definition of E-Learning. (2001).
3. Djoko Budiyanto Setyohadi, Sri Kusrohmaniah, Efrans Christian, Luciana Triani Dewi. M-learning Interface Design Based on Emotional Aspect Analysis. In Paper presented at International Conference on Intelligent Human Computer Interaction, 276-287. (2017).
4. Ozkan, S., & Koseler, R. Multi-Dimensional Evaluation of E-Learning Systems in the Higher Education Context: An Empirical Investigation of a Computer Literacy Course. (2009).
5. Lee, MC. The Analytic Hierarchy and the Network Process in Multicriteria Decision Making: Performance Evaluation and Selecting Key Performance Indicators Based on ANP Model. Convergence and Hybrid Information Technologies. (2010).
6. Singh, Rana Pratap., Nachtnebel, Hans Peter., Analytical hierarchy process (AHP) application for reinforcement of hydropower strategy in Nepal. Renewable and Sustainable Energy Reviews. (2016).
7. Ozkan, S., Koseler, R., & Baykal, N. Evaluating learning management systems. (2009).
8. Chen, M. K., & Wang, S.-C. The critical factors of success for information service industry in developing international market: using analytic hierarchy process (AHP) approach. Expert Systems with Applications, 37(1), 694–704. (2010).
9. Saaty, T.L. *Fundamentals of Decision Making and Priority Theory with the Analytic Hierarchy Process.*Pittsburgh USA: RWS Publications. (1994).
10. Saaty, T. L. How to Make a Decision The Analytic Hierarchy Process. European Journal of Operational Research, 48, 9-26. (1970).
11. Bozbura, F. T., Beskese, A., & Kahraman, C. Prioritization of human capital measurement indicators using fuzzy AHP. Expert Systems with Applications, 32(4), 1100–1112. (2007).
12. Fraenkel, J., Wallen, N. How to Design and evaluate research in education (2nd ed). New York: McGraw-Hill Inc. (1993).
13. Sturges, Herbert. (The Choice of a Class Interval. *Journal of American Statistica Association.* 1926).
14. Holsapple, C. W., & Lee-Post, A. Defining, assessing, and promoting e-learning success: An information systems perspective. Decision Sciences Journal of Innovative Education, 4(1), 67–85. (2006).
15. Selim, H. M. Critical success factors for e-learning acceptance: Confirmatory factor models. International Journal of Technology Marketing. (2007).
16. Sun, P. C., et al. What drives a successful e-learning? An empirical investigation of the critical factors influencing learner satisfaction. Computers Education, 50(4), 1183–1202. (2008).
17. Bhuasiri, Wannasiri; Xaymoungkhoun, Oudone; Zo, Hangjung. Critical Success factors for e-learning in developing countries: A Comparative Analysis between ICT experts and faculty. Journal Computer & Education. (2012).
18. Djoko Budiyanto Setyohadi, Michael Aristian, Benyamin Langgu Sinaga and Nor Aziati Abdul Hamid. Social Critical Factors Affecting Intentions and Behaviours to

Use E-Learning: An Empirical Investigation Using Technology Acceptance Model. Asian Journal of Scientific Research, (2017). 10: 271-280.
19. Khan, B. Managing e-learning strategies: Design, delivery, implementation and evaluation. Hershey, London, Melbourne, Singapore: Information Science Publishing. (2005).

Time Series Traffic Speed Prediction Using k-Nearest Neighbour Based on Similar Traffic Data

Bagus Priambodo[1*], and *Yuwan* Jumaryadi[1]
[1]Universitas Mercu Buana, Faculty of Computer Science, Jakarta, Indonesia

Abstract. During the past few years, time series models and neural network models are widely used to predict traffic flow and traffic congestion based on historical data. Historical data traffic from sensors is often applied to time series prediction or various neural network predictions. Recent research shows that traffic flow pattern will be different on weekdays and weekends. We conducted a time series prediction of traffic flow on Monday, using data on weekdays and whole days data. Prediction of short time traffic flows on Monday based on weekdays data using k-NN methods shows a better result, compared to prediction based on all day's data. We compared the results of the experiment using k-NN and Neural Network methods. From this study, we observed that generally, using similar traffic data for time series prediction show a better result than using the whole data.

1 Introduction

Due to population increase and numbers of private cars in this modern era, traffic congestion has become significantly worse, not only leading to economic losses but also causes environment damages, human stress and environmental pollution [1]. Drivers need traffic information which can affect their driving to a certain option [2] and thus lead to chain changes in traffic flow state in the upstream and downstream of relative road section and other road segments in the network. Some studies show that a road always has a similar traffic state on same working day or weekends at the same time interval [5]. Adjacent roads have a similar history in terms of road traffic condition during work days or weekend [6]. Investigate similar traffic condition pattern can lead us to increase accuracy and reduce high computation in predicting traffic flow by using only similar traffic data rather use the whole data.

Related Work and Background

Time series [3], and neural network models[4][5], are often applied to prediction traffic flow and predict traffic congestion based on vehicle speeds, weather, incident, and special days of historical data. Some studies use a Bayes classifier to predict traffic congestion

* Corresponding author: bagus.priambodo@mercubuana.ac.id, yuwan.jumaryadi@mercubuana.ac.id

[6][7]. Other used non-parametric regression *k*-NN to predict short-term traffic flow[8][9][10][11].

Traffic flow is uncertain, nonlinear and complex, it is difficult to predict the traffic flow effectively and accurately by the predict method based on traditional mathematics and physics models [12], although the neural network can train based on historical data to modelling complex relationships of traffic flow, its unavoidable because its need lengthy amount of time required to train the network [8]. K-nearest neighbour non-parametric regression method [10] is a reliable method for short-term traffic flow forecasting, which can reflect the nonlinear characteristics.

Some studies have conducted predict short-term traffic flow using k-NN, using three-layer k-NN [10], using k-NN with composite method [8], using k-NN with weight[12], using k-NN with tensor for calculating similarity traffic [11], and k-NN with connected link upstream or downstream[9].

Previously, we have conducted a prediction of traffic flow in normal condition based on all day's data using multiple regression method[13]. Other research show there is a difference in congestion pattern between weekdays and weekend [14]and between regular and irregular days [15]. Based on this information, we found that traffic flow pattern on roads will be different on weekdays and weekends. We assumed time series prediction using the k-NN method on weekdays using only weekday's data will have a better result if compared to using all days' data.

2 Dataset and Method

2.1 Dataset

For our experiments, we used data set from IoT traffic sensor in Aarhus, Denmark [16][17]. The total number of IoT sensors is 449 and their location is shown in Figure 1.

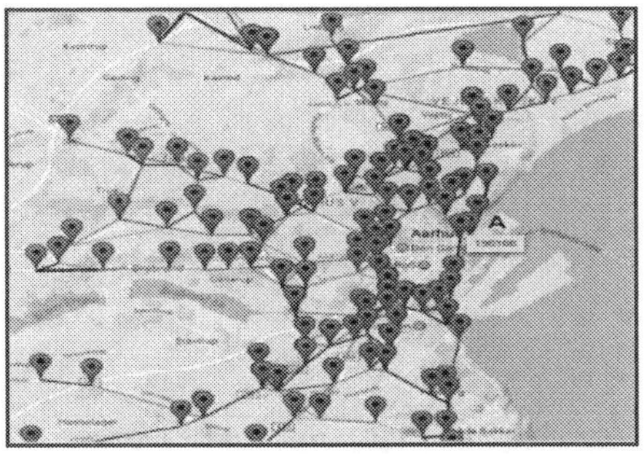

Fig. 1. Map of 449 IoT traffic sensors in the city of Aarhus, Denmark

2.2 Methodology

In order to achieve our main research objective that is to do time series traffic speed prediction using k-NN based on similar traffic data. We conduct our experiment as follows. In the beginning, we observe traffic average speed pattern on road 158415 and road 158595, on weekdays and in whole days. Second, we conduct time series prediction on road 158415 and road 158595 using week day's data and using whole days data. Third, we calculate an error using MAD, MAPE and RMSE.

2.3 Observe traffic average speed pattern

Our aim of the research is predicting average speed using a k-NN method based on similar traffic data. We observed traffic flow pattern on two roads and found there are different patterns of traffic flow between weekdays and the whole days. Weekdays' data is traffic data from Monday to Thursday. Whole week data is traffic data from Mondays to Sundays. We observed that traffic pattern on weekdays looks sharper than traffic pattern on whole days, as seen in figure 2 and figure 3 below.

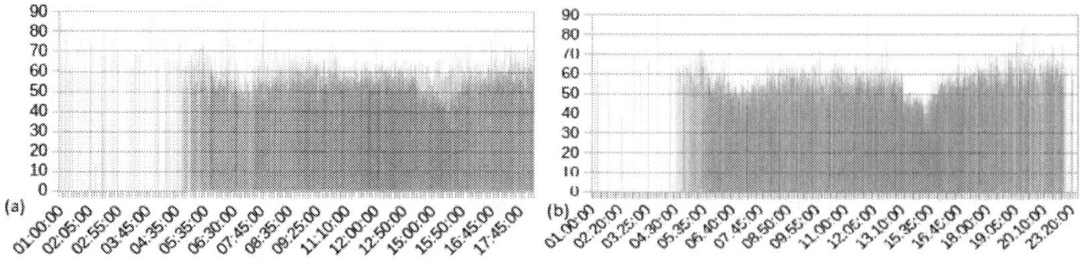

Fig. 2. Traffic pattern on road 158415 (a) all day's data, (b) week day's data

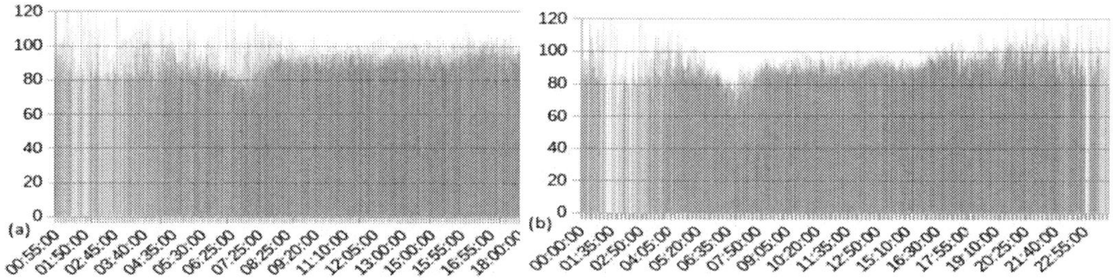

Fig. 3. Traffic pattern on road 158595 (a) all day's data (b) week day's data.

2.4 Time series prediction

For time series prediction using the k-NN method, we set lags = 10 and k = 20. For time series prediction using elm neural network, we set lags = 4, layers = 3 (input, hidden and output), with 4 input neurons, 100 hidden neurons and one input neuron.

3 Result and Discussion

3.1 Dataset

Results of the experiment for short time prediction (25 minutes) on road 158415 are shown in Table 1 and the errors are displayed in Table 2. Results of the experiment for short time prediction (25 minutes) on road 158595 are shown in Table 3 and the errors are displayed in Table 4.

Table 1. Results of short time prediction using K-NN and ELM neural network on road 158415

TIME	Actual	K-NN Weekdays	NN Weekdays	K-NN All days	NN All days
15:05:00	45	50,15	47,91	49,50	48,24
15:10:00	44	54,2	48,84	46,85	49,49
15:15:00	53	45,35	49,60	41,05	50,25
15:20:00	48	50,7	49,59	44,85	50,27
15:25:00	52	51,8	49,31	47,75	50,11

Table 2. The error of short time prediction on road 158415

Error	K-NN Weekdays	NN Weekdays	K-NN All days	NN All days
MAD	5,180	3,087	5,340	3,129
RMSE	6,268	3,264	6,311	3,374
MAPE	11,01	6,48	10,75	6,65

Table 3. Results of short time series prediction using K-NN and ELM neural network on road 158595

TIME	Actual	K-NN Weekdays	NN Weekdays	K-NN All days	NN All days
06:05:00	73	69,2	74,69	70,95	74,46
06:10:00	69	72,9	75,79	64,55	75,45
06:15:00	74	67,55	77,29	62,7	77,16
06:20:00	78	73,45	78,03	68,8	78,08
06:25:00	71	69	78,58	70	78,74

Table 4. The error of short time series prediction on road 158595

Error	K-NN Weekdays	NN Weekdays	K-NN All days	NN All days
MAD	4,140	3,880	5,600	3,780
RMSE	4,381	4,848	6,890	4,770
MAPE	5,64	5,47	7,55	5,33

3.2 Discussion

From observation of traffic speed, we gathered information that traffic flow pattern on weekdays looks sharper than traffic flow in whole days. This show traffic flow on weekdays has a similar pattern. From experiment result, time series prediction using the k-NN method on Monday using weekdays data show a better result than prediction using whole days data. As seen in Figure 4 and Figure 5. This shows us that time series prediction using a k-NN method based on similar traffic flow data produced a better result than based on whole data.

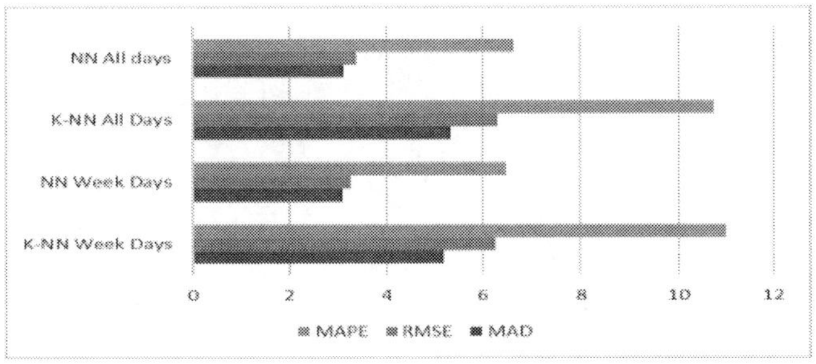

Fig. 4. Bar chart of error of short time series prediction on road 158415

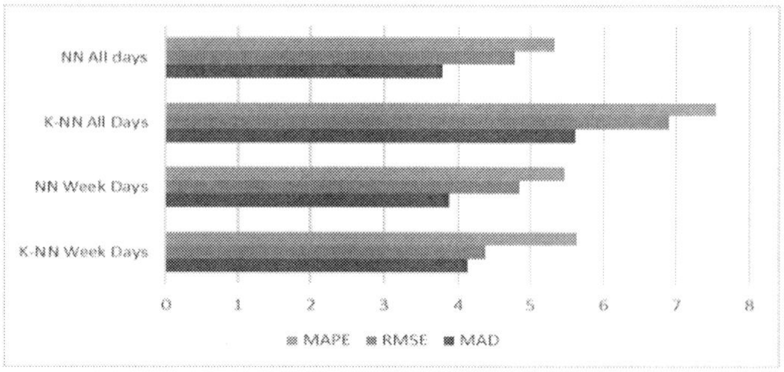

Fig. 5. Bar chart of error of short time series prediction on road 158595

4 Conclusions

Generally, our results show that time series prediction on Monday using a k-NN method based on weekdays' data or based on similar traffic data produced better results compared to prediction using whole day's data, also from our experiment, prediction using time series using neural network produced better results when compared with the k-NN method. Further investigation is needed to understand the results clearer. We propose to cluster traffic condition in each day on weekdays and weekends to obtain better results.

Acknowledgement

The authors gratefully acknowledge use of service and facilities at Information System Department, Universitas Mercu Buana, funded by Universitas Mercu Buana.

References

1. N. Petrovska and A. Stevanovic, "Traffic Congestion Analysis Visualisation Tool," *IEEE Conf. Intell. Transp. Syst. Proceedings, ITSC*, vol. 2015–Octob, pp. 1489–1494, (2015).
2. G. Zhu, K. Song, P. Zhang, and L. Wang, "A traffic flow state transition model for urban road network based on Hidden Markov Model," *Neurocomputing*, vol. **214**, pp. 567–574, (2016).
3. H. Dong, L. Jia, X. Sun, C. Li, and Y. Qin, "Road traffic flow prediction with a time-oriented ARIMA model," in *NCM 2009 - 5th International Joint Conference on INC, IMS, and IDC*, no. 1, pp. 1649–1652 (2009).
4. E. E. E. Lee, J. J. J. Kim, and W. W. Yoon, "Traffic speed prediction under weekday, time, and neighboring links' speed: Back propagation neural network approach," in ... *Theories and Applications. With Aspects of* ..., no. Mic, pp. 626–635 (2007).
5. K. Kumar, M. Parida, and V. K. Katiyar, "Short Term Traffic Flow Prediction for a Non Urban Highway Using Artificial Neural Network," in *Procedia - Social and Behavioral Sciences*, vol. **104**, pp. 755–764 (2007).
6. J. Kim and G. Wang, "Diagnosis and Prediction of Traffic Congestion on Urban Road Networks Using Bayesian Networks," in *Australasian Transport Research Forum 2016*, no. 2595, pp. 1–21 (2016).
7. S. S. Aung and T. T. Naing, "Naïve Bayes Classifier Based Traffic Detection System

on Cloud Infrastructure," *2015 6th Int. Conf. Intell. Syst. Model. Simul.*, pp. 193–198, (2015).

8. M. D. Kindzerske and D. Ni, "Composite Nearest Neighbor Nonparametric Regression to Improve Traffic Prediction," *Transp. Res. Rec.*, no. 1993, p. pp 30-35, (2007).
9. B. Yu, X. Song, F. Guan, Z. Yang, and B. Yao, "k-Nearest Neighbor Model for Multiple-Time-Step Prediction of Short-Term Traffic Condition," *J. Transp. Eng.*, vol. **142**, no. 6, p. 04016018 (2016).
10. X. Pang, C. Wang, and G. Huang, "A Short-Term Traffic Flow Forecasting Method Based on a Three-Layer K-Nearest Neighbor Non-Parametric Regression Algorithm," *J. Transp. Technol.*, vol. **6**, no. 4, pp. 200–206 (2016).
11. Y. Wu, H. Tan, P. Jin, B. Shen, and B. Ran, "Short-Term Traffic Flow Prediction Based on Multilinear Analysis and k- Nearest Neighbor Regression," *15th Int. Conf. Transp. Prof.*, pp. 557–569 (2015).
12. L. Zhang, Q. Liu, W. Yang, N. Wei, and D. Dong, "An Improved K-nearest Neighbor Model for Short-term Traffic Flow Prediction," *Procedia - Soc. Behav. Sci.*, vol. **96**, no. Cictp, pp. 653–662 (2013).
13. B. Priambodo and A. Ahmad, "Predicting Traffic Flow Based on Average Speed of Neighbouring Road Using Multiple Regression BT - Advances in Visual Informatics: 5th International Visual Informatics Conference, IVIC 2017, Bangi, Malaysia, November 28–30, 2017, Proceedings," pp. 309–318 (2017).
14. K. Lee, B. Hong, D. Jeong, and J. Lee, "Congestion pattern model for predicting short-term traffic decongestion times," in *2014 17th IEEE International Conference on Intelligent Transportation Systems, ITSC 2014*, pp. 2828–2833 (2014).
15. F. Rempe, G. Huber, and K. Bogenberger, "Spatio-Temporal Congestion Patterns in Urban Traffic Networks," *Transp. Res. Procedia*, vol. **15**, pp. 513–524 (2016).
16. G. Wang and J. Kim, "Prediction of Traffic Congestion and Incidents on Urban Road Networks Using Naive Bayes Classifier," no. February (2017, 2016).
17. S. Bischof, A. Karapantelakis, A. Sheth, A. Mileo, and P. Barnaghi, "Semantic Modelling of Smart City Data," in *W3C Workshop on the Web of Things Enablers and services for an open Web of Devices*, pp. 1–5 (2014).

The Performance Analysis of Wireless Distribution System Using Point to Multipoint Network Topology

Rizal Munadi[1,2*], *Fardian*[1,2] , *Zeri* Falmiza[2], *Ernita Dewi* Meutia[1,2] and *Roslidar*[1,2]

[1]Wireless and Networking Research Group (Winner), Computer Network Laboratory, Electrical and Computer Engineering Department, Syiah Kuala University, Banda Aceh 23111, Indonesia
[2]Electrical and Computer Engineering Department, Syiah Kuala University, Banda Aceh 23111, Indonesia

Abstract. Current information technology development continues to evolve along with the needs in obtaining data or information either audio, images or a combination of both which is growing rapidly. Therefore information technology must be able to provide services that could fulfill user expectations in terms of both quality and quantity. One of the prominent changes in information exchange field is the use of wireless technologies, including its network configurations. Besides having to function optimally, the performance of the wireless technology is also needed to be maintained. This research is about the technology performance of Wireless Distribution System (WDS) with point to multi-point network configuration using 802.11g access point. The research was conducted in two phases. The first phase was building a WDS network model using four access points that are connected to each other. The second one was examining WDS network performance on server side. Based on the results obtained seen throughput on each segment WDS tested in gradual decline and the sharpest occurred in the last segment. On the other side the value of jitter increased from WDS-1 to WDS-2 and WDS-3 in a normal number, as well as the value of packet loss which increased in a very good category. The results obtained indicate good quality and generally very appropriate WDS technology to expand wireless access.

1 Introduction

The development of information device technology continues to increase along with the needs of users in communicating and using data communication applications. The growth of users who switch to communication with IP based services needs to be accompanied by good service quality. To maintain data transmission requirements, equipment integrated in the network must be designed according to the desired expectations. The need to give satisfaction in accessing the internet is what motivates the motivation of this research.

* Corresponding author: rizal.munadi@unsyiah.ac.id

Nowadays, computer network infrastructure technology shows remarkable development on wireless networks compared to wire networks. Various variants of wireless network technology with the 802.11 standards are now starting to dominate with the 802.11n variant. Although the 802.11n variant offers a 2.4 GHz and 5 GHz frequency band, it turns out that hardware manufacturers have even installed network cards on the latest computer products by adopting the most recent variant, 802.11ac. However, all installed hardware wireless network is intended to provide optimum results but with network configuration and unpredictable traffic conditions, can produce an impact value from the output that is beyond expectations.

Wireless Distribution System (WDS) is a coordination technology for wireless network (WLAN) networks to optimize the use of network resources and service area coverage. The use of WDS in network implementation can expand the coverage area and provide flexibility and mobility for users. Limitations of service coverage of a network by using only single Access Points have been solved. Although the problem of coverage is resolved, the use of WDS as a system also has drawbacks, namely a decrease in capacity. In terms of capacity, the increasing number of a node according to users, the traffic collision, the channel utilization rate, the total throughput is among the research focus in system performance.

Network performance issues have provided challenges for researchers to conduct studies and provide solutions. The research conducted by Jeong [1], which proposed a new medium access scheme, showed that the results of throughput performance were better than the existing schemes. Sung, et al. have studied about signal quality and packet delivery and purposed an algorithm name Node Degree Algorithm for achieving throughput performance [2]. Another research which done by Tahar using multiple interfaces per participating Access Point (AP) in order to improve the end to end performance[3].

This paper is focused on network performance and describes an implementation of Wireless Distribution System point to multi point network in AP 802.11g. The purpose of the implementation is to figure out the performace of the network at the client site by evaluating the parameter of throughput, jitter, and packet loss of the WDS network that is connected to APs. The rest of this paper is organized as follows: section two explains the literature review; section three describes the research method; section four discusses the data analysis; and finally, the conclusion is summarized in section five.

2 Literature Review

Wireless Local Area Network (WLAN) is a wireless technology that uses air as a propagation medium to transmit data from the sender to the receiving device. In the operation of equipment, WLANs with 802.11 standards use frequencies allocated for Industrial, Scientific, and Medical (ISM) purposes that get exemption from usage permits or are identified with unlicensed bands. Equipment that works and meets the designation of the ISM band is free to be used by anyone. In addition to the impact on this simplification of convenience for users and increasing popularity, on the other hand it provides opportunities and challenges for researchers to get better results. In its development, various types of 802.11 standards try to provide high data transmission speeds in order to provide satisfying services for users.

To connect between 802.11 WLAN devices so that they can deliver information between interconnected entities, network topology needs to be planned. Theoretically, there are two topologies in a wireless network which are Ad-Hoc topology and infrastructure. Following is the short explanation of the topology:

a. Ad-hoc mode

Ad-hoc mode is a simple mode of the WLAN network. Ad-hoc topology is similar to peer to peer network, where the network is built only by applying the component of wireless device. Ad hoc wireless networks, consisting of a collection of wireless nodes, all of which may be mobile, dynamically create a wireless network among themselves without using any such infrastructure or administrative support [4] The drawback of this mode is the computer is unable to communicate with the computer in the wired network.

b. Infrastructure mode

A more complex topology of the wireless network is the network topology with Basic Service Set (BSS) infrastructure mode. This configuration places the AP as a device that functions as data traffic in forwarding information in the network. As the agent, AP in addition to serving connected nodes also plays a role as a bridge with conventional networks that are physically connected via UTP cable.

2.1 Wireless distribution system

Wireless Distribution System (WDS), which is also named as Wireless Repeater, is a system to develop a wireless network without using cable as the media of data transfer but applying the wireless path of each AP equipment. It is very contrast to conventional communication network, where the media of distribution system is cable (wired LAN). All of the base stations in WDS have to be configured using the same radio channel, the same encryption methode (without encryption, WEP, or WAP), and the same encryption keys. And the connection is a configuration using different Service Set Identifiers (SSID) on each user identity [5].

WDS allows the interconnection of each AP equipment in a wireless environment. By applying WDS, the connection in a wireless network can be expanded using some APs or base station without the requirement of backbone cable network. WDS can be applied in two features/modes connectivity between APs, which are:

a. Wireless Bridging; where the communication between APs in the WDS network is limited and does not allow other wireless client or station (STA) to access the network.
b. Wireless repeater; where AP can communicate with each other and to the wireless Station (STA).

Zikri conducted a research using 2 APs for Wireless Distribution System (WDS) [6]. In this research, *throughput, jitter, packet loss and Signal to Noise Ratio (SNR)* are used as evaluation parameters to measure the quality of the test network. The results obtained on the receiving-end from the test network throughput obtained showed that the average results decline of 51.3% from the maximum bandwidth. In the measurement of other parameters, indicating that the value of the jitter that occurs on average 1.95 ms and the number of packet loss of 0.1%. Another research evaluation showed that accessing a network with WDS system by a variation of client number might reduce the number of channel allocation that may be used by the client [7]. The calculation uses the same client that exists in the network with the WDS system. In the initial stages of testing with 1 client, it was known that the channel allocation in throughput degraded 40.3%. This percentage gradually decreased to 21. 2% on the test with clients, and increased again to 25.2% on the test with 12 clients. The last, the research conducted by using 3 APs with WDS technology [8]. As for the use of such research, WDS technology produces an average percentage decline in the value of the throughput generated by using one client on the AP-2 (WDS-1) is 31.27%, while the average percentage of the throughput on the AP-3 (WDS- 2) is 54.92%. Tests were performed using point-to-point network topology with the distance between the AP range of 20 meters conducted in open space. The results show a gradual decrease in throughput for each AP used.

2.2 Network performance

2.2.1 Throughput

Throughput is the real capability of a network in sending data. Throughput is always related to bandwidth even in some context it is also called bandwidth. The principal different is on their nature, while bandwidth is static of fixed, throughput, on the other hand, is dynamic. Globally speaking, through put is a measure of data traffic in a communication, with the unit of megabit per second (mbps). The higher the value, the better the network performance. Actual throughput will vary. Network conditions and environmental factors, including the volume of network traffic, building materials and construction, and network overhead, lower the actual data throughput rate.

2.2.2 Jitter

Jitter is a delay variation due to time difference or interval of the arriving of receiving the packet. In data communication, packets delivered are varied and traffics that occur in the network are very dynamic in accordance with the volume of the data transaction. The amount of jitter that occurs is dependent on the number of collisions between packets that pass through in the network. The number of jilter in a reasonable tolerance is between *0* to *100 ms* as shown in Table 1. As jitter indicates the package collision in the internet protocol, the heavier the traffic load, the bigger the possibility that collision will occur. This will have an impact on the value of the jitter will become larger.

Table 1. Jitter criteria [9]

Category	Peak Jitter
Excellent	0 ms
Good	0-75 ms
Poor	76-125 ms
Worst	126-225

2.2.3 Packet loss

The failure in the transmission of a data packet on its way to the destination is called packet loss. There are some factors that may cause the packet loss, including collision, congestion, error transmission of physical media, and the failure at the receiving side which may due to overflow on the buffer. Attempts to minimize the number of packet loss are a challenge that must be faced and very difficult to avoid. Therefore, if the smaller value of packet loss in a network the better the performance of the network. Performance categories of the packet loss are shown in Table 2.

Table 2. Packet Loss Criteria

Category	Packet Loss
Excellent	0 %
Good	1-3 %
Poor	4-15 %
Worst	16-25 %

3 Research Methodology

In this research, four APs have been configured using point to multipoint topology approach to test their WDS performances. All APs placed in open space and examine their capabilities to deliver a message to the client under three scenarios as listed in Table 3. We propose throughput, jitter and packet loss as the parameters to measure their performances by using the jperf software. Fig. 1 shows the tested topology model in this research.

Table 3. Testing scenarios

Skenario	AP-1 (*main*)	AP-2 (WDS-1)	AP-3 (WDS-3)	AP-4 (WDS-4)
1.	*Server*	*Client*	-	-
2.	*Server*	-	*Client*	-
3.	*Server*	-	-	*client*

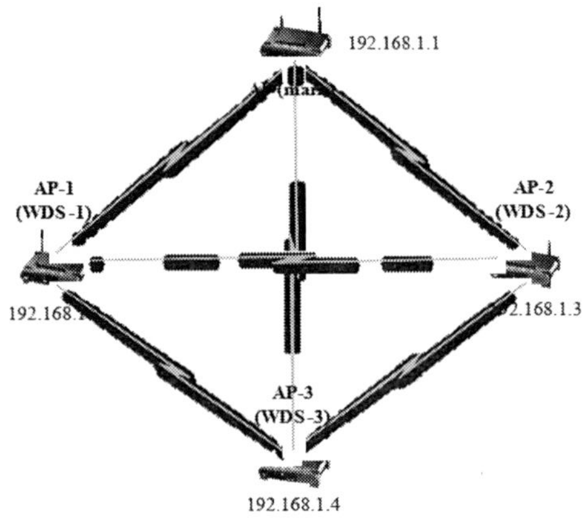

Fig. 1. The model of topology

4 Discussion

This section discusses the result from the average value of throughput, jitter, and packet loss during the test. The average value from every result test will be compared with the result from throughput, jitter, and packet loss on WDS network that using point to multipoint topology.

4.1 Throughput

This test uses one client for every AP-2 (WDS-1), AP-3 (WDS-2), AP-4 (WDS-3) and 1 server for AP-1. Port used for every client is 5001 (WD-S-1), 6001 (WDS-2), and 8001 (WDS-3).

As shown in Fig. 2, throughput measurements are carried out involving 3 APs that work together as a WDS network. From the resulting test, it is found that the average throughput value of WDS-1 is 6360.06 Kbps, WDS-2 is 6281.54 Kbps, and WDS-3 is 5535.04 Kbps. In aggregate, there is a decrease in throughput when data traffic passes between APs gradually

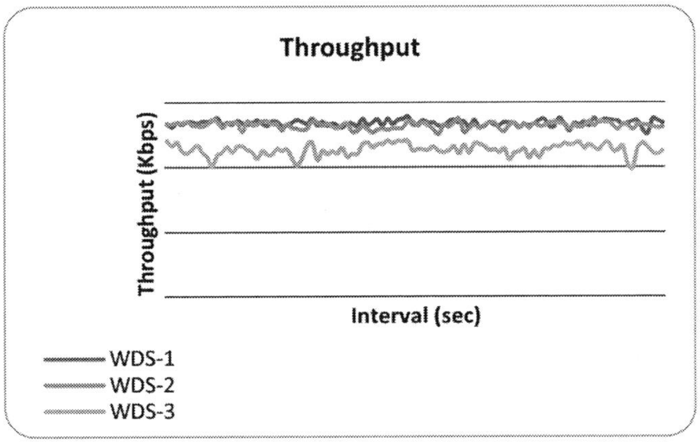

Fig. 2. Throughput average value

4.2 Jitter

The results of Jitter's measurement are shown in Fig. 3. The result shows that WDS-1 is 2,215 ms for its average value and the highest value is 4,574 ms. On WDS-2, its average jitter value is 3,417 ms and the highest value is 3,966 ms. On WDS-3, its average value is 3,897 ms with the highest value is 5,660 ms. Unlike the results of throughput for each WDS stage, jitter measurement shows a gradual increase between each WDS.

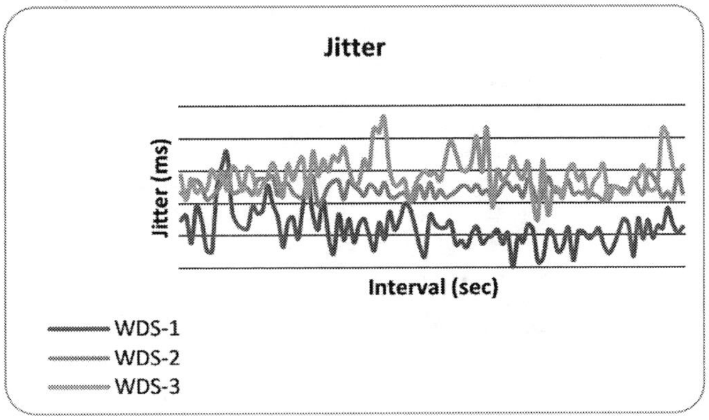

Fig. 3. Jitter average value

4.3 Packet Loss

The amount of packet loss on every WDS client is figured out in Fig. 4. The percentage of packet loss resulted from WDS-1 is 0,15%, WDS-2 is 0,18%, and WDS-3 0,35%. From this data, it can be seen that the amount of the packet losses is greater for every distance addition.

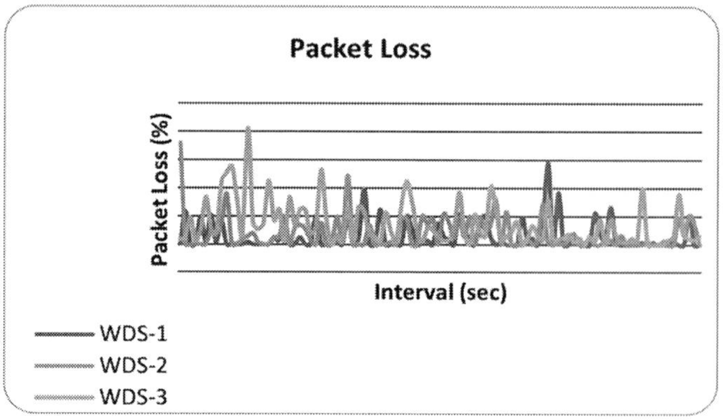

Fig. 4. The graphic of packet loss average value

4.4 Data Analysis

4.4.1 Throughput Comparison

Fig. 5 shows that the average value of WDS-1 is 6360,05 Kbps, WDS-2 is 6281,54 Kbps, and WDS-3 is 5535,04 Kbps. The percentage of decreasing value from WDS-1 to WDS-2 is 1,23%, from WDS-2 to WDS-3 is 11,88%, and from WDS-1 to WDS-3 is 12,97%.

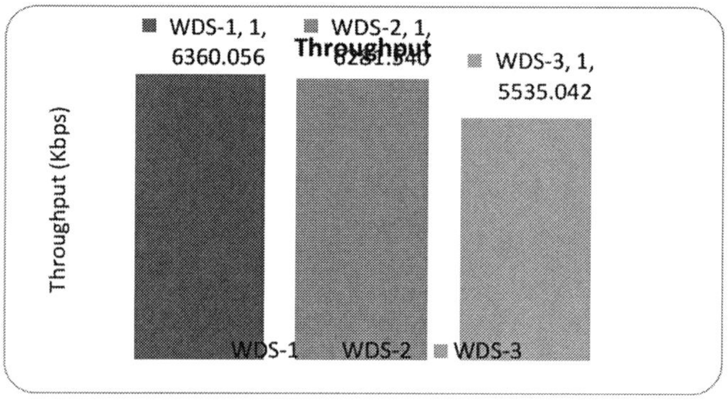

Fig. 5. The comparison result of throughput average value

4.4.2 Jitter Comparison

Fig. 6 shows that the average value of jitter on WDS-1 is 2.215 ms, WDS-2 is 3.417 ms, and WDS-3 is 3.897 ms. The percentage of jitter average value on the WDS-1 network is 1,20%, WDS-2 is 0,47%, and WDS-3 is 1,68%. The highest jitter value is found on WDS-3, it is because there is packet queueing during data distribution process. All the jitter results are still categorized as good according to ITU-T G.8251, 2001 standard where the maximum value of allowed jitter is 75 ms.

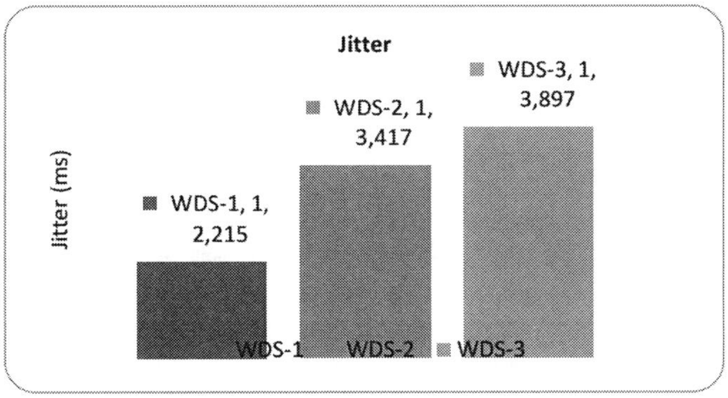

Fig. 6. The comparison result of jitter average value

4.4.3 Packet loss average value

Fig. 7 shows the percentage of packet loss average values on WDS-1 is 0,15%, WDS-2 is 0,18%, and WDS-3 is 0,35%. The losses percentage occured during the test between WDS-1 and WDS-2 is 0,03%, WDS-2 and WDS-3 is 0,17%, and WDS-1 and WDS-3 is 0,20%. But according to ITU TIPHON, 1998 standard, above losses are still categorizing good. The losses amount of data packet is influenced by throughput and jitter value on the network. The greater packet loss resulted from the low throughput value and the high jitter value.

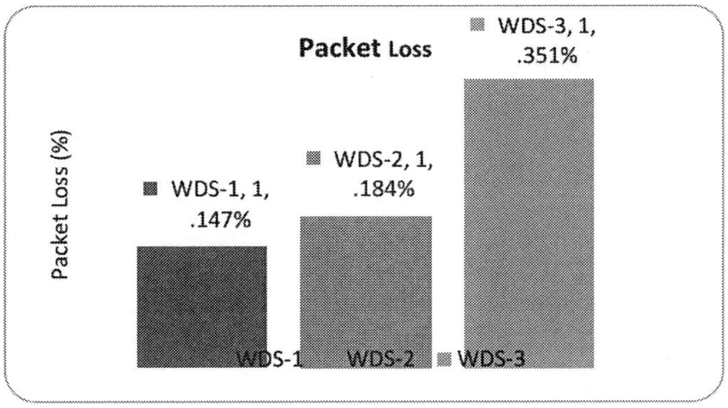

Fig. 7. The comparison result of packet loss average value

5 Conclusion

This research has shown the evaluation of throughput, jitter, and packet loss based on *point to multi point* topology and the WDS network. The results obtained indicate good quality and generally very appropriate WDS technology to expand wireless access.

References

1. J. Jeong, W. Lim, and Y. Suh, "Throughput Enhancement for WDS-Based WLANs," *22nd International Conference on Computer Communication and Networks (ICCCN)*, Nassau, pp. 1-6 (2013).

2. Sung, T.-W. *et al.* An Algorithm for Improving Network Throughput of Wireless Distribution System. in *2016 Third International Conference on Computing Measurement Control and Sensor Network (CMCSN)* 178–181.
3. R. Tahar, A. Belghith, and R. Braham, "Performance evaluation of IEEE 802.11 multi-interface based wireless distribution system (MI-WDS)," *2009 IEEE/ACS International Conference on Computer Systems and Applications*, Rabat, pp. 764-770 (2009).
4. J. Haas et al. , "Guest Editorial," IEEE JSAC, Special Issue on Wireless Networks, vol. **17**, no. 8, (1999).
5. Sulistyo, W., Bayu, T. I., & Sathibi, A. (2012). Implementasi Dan Analisis Penggunaan Dd-Wrt Firmware Untuk Membangun Jaringan Wireless Distribution System Pada Jaringan Hotspot (Studi Kasus : FTI UKSW), *2012*(Snati), 15–16.
6. K. Zikri, "Analisis Performansi dan Kualitas Penerimaan Sinyal Pada Multi Access point Menggunakan Wireless Distribution System (WDS) Pada Jaringan WLAN IEEE 802.11g. Skripsi. Universitas Syiah Kuala, (2012).
7. Putra, D. L. A., and Subhan, A. "Analisa Kinerja Implementasi Wireless Distribution System Pada Perangkat Access Point 802.11g Menggunakan OpenWRT". EEPIS Final Project (2011)..
8. R. Munadi, R. Wahyudi, Fardian, Ardiansyah, Analisis Kinerja Jaringan Wireless Distribution System Menggunakan Access Point 802.11g, Konferensi Nasional ke-7 Fortei (2013).
9. Sugeng, W., Istiyanto, J. E., Mustofa, K., & Ashari, A. The Impact of QoS Changes towards Network Performance. *International Jurnal of Computer Networks and Communications Security*, *3*(2), 48–53. Retrieved from http://www.ijcncs.org/published/volume3/issue2/p5_3-2.pdf (2015).

The Application of Structural Equation Modelling for Predicting the Link between Motivation and Experience Quality in Creative Tourism

Dwi Suhartanto[1,*], *Regina Agustina* [2], *Nono* Wibisono[1], *Gundur* Leo[1]
[1]Department of Business Administration, Politeknik Negeri Bandung, Bandung, Indonesia
[2]Department of Business Administration, Widyatama University, Bandung, Indonesia

Abstract. The Indonesian creative tourism industry is growing and competitive. In this competitive industry, the tourism attraction capability in delivering a quality of experience and motivating customers to visit is imperative. However, research to explore tourist motivation impact on experience quality in creative tourism is limited. This paper is designed to predict the influence of motivation on experience quality in the creative tourism by applying Structural Equation Modelling (SEM). Data were collected through survey from 260 tourists visiting creative tourism attractions in Bandung. The result shows that push motivation has a significant impact on experience quality dimension of escape, learning, and recognition; while the impact of pull motivation is not significant. This finding improves the theoretical and managerial understanding on the link between motivation and experience quality as well as the application of SEM in creative tourism industry.

1 Introduction

Tourism industry, especially creative tourism, is one of the growing and a competitive industry [1]. In this dynamic and competitive environment, creative tourism destinations face a stiff competition as numerous destinations present similar attractions [2]. Further, in this unique tourism industry, tourists expect more than a good service experience [3], for both the new and the returning tourists. As returning customers yield higher than new customers due to less marketing expenditure, acquiring new tourists should be a continuing task. For this task, creative tourism managers have to comprehend the factors that create such a favourable behaviour [4].

The literature in services context has reported that there are several factors affecting tourist experience with the attraction, including, among others, customer expectation, past experience, customer image on the attraction, and tourist characteristics [5]. Further, recent studies have identified that motivation has an impact on tourist perception on the quality of the service in various service industries [6]. However, limited studies have assessed motivation as an experience quality determinant in creative tourism. Additionally, a limited

[*] Corresponding author: dwi.suhartanto@polban.ac.id

study in creative tourism attraction applies structural equation modelling (SEM) as a comprehensive analysis method.

This paper extends the existing experience quality literature by focusing on creative attractions, which can be challenging as such an experience may initially be considered as a once-in-a-lifetime experience. More specifically, this study is intended to (1) examine the effect of tourists' motivation on their experience with the quality of the attraction, and (2) apply structural equation modelling to predict the association between motivation and experience quality.

2 Literature review

The service experience concept has attracted attention among scholars [7], however, studies related to the service experience suggest that what the meaning of service experience is still lack of agreement [8]. As part of the service experience, the consumption experience is articulated by a sense of fun and fantasy [7]. This notion suggests that the experience of the service is not only during the customers encounter with the service, but also when the customers communicate and interact with the service providers. In tourism, experience quality refers to tourists' psychological and social response towards the consequence of the service performance during their visit and consume the service attraction [9]. This notion implies that experience quality is different from service quality which reflects the evaluation of the service consumption [10].

The literature identifies various dimensions of experience quality. Pine and Gilmore [7] introduce a measurement scale of 4E (educational, entertainment, escapist, and aesthetic) to assess experience in the accommodation industry. While in the tourism context, Xu and Chan [11] present dimension of peace of mind, relaxation, involvement, recognition, escape, and hedonics. These researchers use different dimensions but have a relatively similar concept that renders tourist participation as the main factor. Past studies report that tourist intention to revisit and recommend an attraction is an impact of customer experience with the performance. Further, literature also informs that tourist experience quality has a positive consequence on tourist behaviour toward attraction such as their level of satisfaction and loyalty as well as their memories [12, 13]. While past studies were conducted in the general attraction arena, as far as the authors' knowledge, no research has been undertaken in the creative attraction context.

The reason why people travelling is one of the earliest focus of study in tourism research [14, 15]. Past studies identify that motivation is not only enabling to rationalize a certain tourist behaviour, but it is also a pivotal factor in influencing tourist post-purchase behaviour [4, 16]. There are many definitions given to the term of motivation in tourism context which one of generally accepted refers to a combination of traveller wants and need that influence a person propensity to visit and enjoy the destination attractions [17]. This propensity is the only factor rationalizing the reason for a certain traveller behaviour [4]. Many theories have explained motivation, but the push-pull theory is the most commonly applied theory to explain tourist travel behaviour [18]. As named, the push-pull theory explains motivation using a two-dimensional perspective; a push factor is a person's internal wants such as the need to have fun, relax, or break from routine life; while a pull factor is the attributes of destination such as attractions, entertainment performances, restaurants, art galleries and so on which attracts a person to visit. In short, push motivation represents what tourist is seeking when visiting a destination, while pull motivation represents the destination offering to satisfy tourists' needs in travelling. It is expected that in a tourist attraction, both pull and push factors affect tourist experience with the creative tourism attraction, leading to hypothesis (H1): Pull motivation influences the experience quality dimension and (H2): Push motivation influences the experience quality dimension.

3 Method

The data were collected by using a self-administrated questionnaire from 260 tourists at four creative tourism destinations in Bandung. This number of data is adequate to test the simple proposed model [19]. Specifically, the destinations are Saung Angklung Udjo, Batik Komar, Daarus Sunnah Horse Riding and Archery, and Eco Bambu Cipaku. The data consists of females (56.2 %) and males (43. 8%) respondents with the majority (62.7%) were under the age of 40. The experience quality consists of three variables (escape, learning, and recognition) with 12 indicators (see Table 1). The motivation consist of two constructs, pull and push motivation with each consists of three indicators adopted from preceding researches [3]. All variables were gauged by 5-point Likert measurement scale, from "strongly disagree" to "strongly agree". To test the hypotheses, this research uses Structural Equation Modelling (SEM). The validity and reliability of the scale measurement constructs were examined by applying confirmatory factor analysis. This method was also used to verify the proposed model. SEM enables a researcher to assess latent construct and testing the relationships between the variables simultaneously [19]. Additionally, SEM is a noted technique to estimate coefficient paths of structural models.

4 Result

The measurement model of the link between motivation and experience quality was evaluated by employing confirmatory factor analysis (CFA) with the assumption that all of the variables tested co-vary each other. The CFA test result indicates that the goodness-of-fit index of the model is poor (CMIN/DF: 3.484, GFI: 0.879, CFI: 0.876, and RMSEA: 0.083).

Table 1. Reliability and Validity Check.

Construct/Item	CR	AVE
Escape (EQ1)	0.764	0.52
- Forget daily activity (E1)		
- Escape from routine (E2)		
- Comfortable (E3)		
- Relaxing (E4)		
Learning involvement (EQ2)	0.783	0.55
- Involve in the activity (L1)		
- Get a unique experience (L2)		
- Can choose the activity (L3)		
- Increase knowledge (L4)		
- Increase expertise (L5)		
- Understand something new (L6)		
Recognition (EQ3)	0.799	0.67
- Treated as an important person (R1)		
- Treated wholeheartedly (R2)		
Push Motivation (PUSH)	0.786	0.65
- Find a new experience (M1)		
- Enjoy the holiday (M2)		
- Refresh from daily activity (M3)		
Pull Motivation (PULL)	0.822	0.7
- Accompany friend/family (M4)		
- Passing by the attraction (M5)		
- Attracted by the promotion (M6)		

In order to make the model fit, the next step modifies the variables by re-parameterizing based on the standardized residual, insignificant path, and modification index value with theoretical concept consideration [19]. Using this process, indicators from motivation variables (M1 and M4), Escape variable (E3), and Learning variable (L1, L2, L3) we excluded from the model. Excluding these items resulting that the constructs are valid and reliable, indicated by the value of composite reliability higher than 0.7 and the value of average variance extracted also higher than 0.5 [19]. Additionally, the requirement for discriminant validity is also met as the value of average variance extracted is higher than the constructs squared correlation [19].

Following the validity and reliability check, the test of the proposed model was performed to analyse the hypothesized causal paths between variables. The result of the modified model shows that the requirements of goodness-of-fit indexes are satisfied (CMIN/DF: 3.15, GFI: 0.930, CFI: 0.931, and RMSEA: 0.077). In addition, the result of model testing indicates that the estimation model by using bootstrapping is satisfactory as there are no estimation problems occurred.

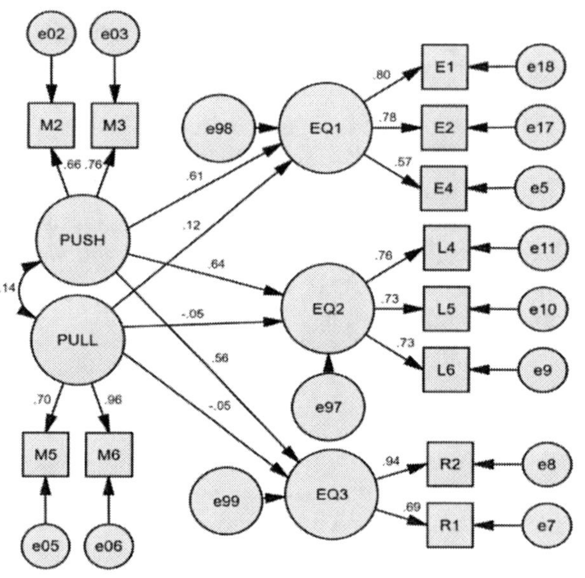

Fig 1. Result of testing the model.

The testing model results notes that this study reveals the association between push and pull motivation and three dimensions of experience quality. The effect of push motivation on all of the experience quality dimensions, escape, learning, and recognition (0.610, 0.648, and 0.568 respectively) is significant at $p<0.01$. In contrast, the effect of pull motivation on experience quality dimension is not significant. Thus, the results of testing hypothesis show that hypothesis H1 is supported, while hypothesis H2 is not supported.

The study results extend the result of past studies [20, 21] which report the importance of motivation on quality, satisfaction and loyalty. The result reveals the effect of push motivation factor in influencing tourist experience with the performance of the attraction. The positive association between push motivations on tourist experience quality implies that the more tourists are motivated by the push factor, the more they perceived a high quality of the attraction performance; while less pull motivation will not impact on tourist experience with the attraction performance. In addition, this finding uncovers that the effect of push motivation toward the learning dimension of experience quality is the highest compared to that of the other dimensions. Thus, this study stresses the push factors as a critical determinant of experience quality, mainly on the learning dimension.

The testing of proposed model on the link between motivations on experience quality in creative tourism shows the strength of using the SEM to examine a complex model including constructs with many indicators. Using this method, a researcher can simultaneously inspect the validity, reliability, fitness of the model and relationships between the variables. Compared to other methods such as multiple regression which is commonly used in analysing the variables relationship in creative tourism study, this method clearly can help the researcher understand the relationship between the variables tested comprehensively.

5 Conclusion, Implication, and Future Research

Although creative tourism is a promising sector, studies to scrutinize tourist motivation in relation to tourist experience with the creative tourism attraction is limited. This research examines the creative attraction motivation and its impact on three experience quality, namely escape, learning, and recognition. The data analysis reveals a positive effect of push motivation on three dimensions of experience quality. Theoretically, this finding offers a solid support for the argument of the relationship between motivation and customer experience with the service, where motivation is an important determinant of customer service experience. However, in terms of pull motivation, this research suggests that this factor has less importance as a determinant of experience quality. Further, the positive linkage between push factor and experience quality dimension highlights the effect of push factor on attraction experience. Considering other studies reporting a positive effect of motivation on satisfaction and loyalty [6, 16], these findings implies that the experience quality consequences of motivation extends our understanding that, besides experience quality, satisfaction, and perceived value, motivation, especially push factor is also one of the tourist loyalty building blocks in creative tourism attraction. Lastly, this study offers an evidence of a better analysis method of using SEM compared to regression analysis to examine a research model in creative tourism.

The results of this study bear implications for managing and marketing creative attractions. It reveals that push motivation is an important determinant of experience quality. To attract and develop visitor experience, the attraction managers need to develop customer pushing motivation strategy. Developing appropriate pricing and marketing strategies may not be necessary. To attract customers, it is important that the creative attraction managers to stressing on the benefit that the attraction to offer, that is to provide a new learning and enjoying the experience. Further, collaborative promotions with other tourist attractions and destination organizations to attract visitors to the destination by stressing the benefit of the experience with attraction offered are important, as tourists generally also enjoy visiting other city' attractions.

While this paper reveals some important findings, it does have limitations. First, the data was gathered from four cultural-based attractions located in Bandung, Indonesia which essentially hampers the generalization of the findings. This issue of generalization can be solved with replication of this research in a different destination with a different culture. Replication such a study can also focus on different tourists, for instance between domestic and foreign tourists. Second, in addition to experience quality, motivation has a direct consequence on visitor satisfaction and loyalty, it is also critical to include these variables on the model of the link between motivation and experience quality, such as tourism involvement which is an important factor for the employee to serve tourists better [22].

References

1. D. Suhartanto, "The role of store coopetition and attractiveness on the performance of tourism destination and its retail stores," *International Journal of Tourism Policy,* **7**, pp. 151-165 (2017)
2. H. Zhang, X. Fu, L. A. Cai, and L. Lu, "Destination image and tourist loyalty: A meta-analysis," *Tourism Management,* 40, pp. 213-223 (2014)
3. D. Suhartanto, A. Brien, N. Sumarjan, and N. Wibisono, "Examining attraction loyalty formation in creative tourism," *International Journal of Quality and Service Sciences,* **10**, pp. 163-175 (2018)
4. L.-L. Chang, K. F. Backman, and Y.-C. Huang, "Creative tourism: a preliminary examination of creative tourists' motivation, experience, perceived value and revisit intention," *International Journal of Culture, Tourism and Hospitality Research,* **8**, pp. 401 - 419 (2014)
5. R. Dodds and L. Jolliffe, "Experiential Tourism: Creating and Marketing Tourism Attraction Experiences," in ***The Handbook of Managing and Marketing Tourism***

Experiences, M. Sotiriadis and D. Gursoy, Eds., ed Bingley, UK: EmeraldInsight, pp. 113-129 (2016)
6. A. Subrahmanyam, "Relationship between service quality, satisfaction, motivation and loyalty: a multi-dimensional perspective," *Quality Assurance in Education,* **25**, pp. 171-188 (2017)
7. B. J. Pine and J. H. Gilmore, "Welcome to the experience economy," *Harvard Business Review,* pp. 97-105 (1998)
8. P. Deshwal, "Customer experience quality and demographic variables (age, gender, education level, and family income) in retail stores," *International Journal of Retail & Distribution Management,* **44**, pp. 940-955 (2016)
9. C.-F. Chen and F.-S. Chen, "Experience quality, perceived value, satisfaction and behavioral intentions for heritage tourists," *Tourism Management,* **31**, pp. 29-35, 2 (2010)
10. D. Suhartanto, "Tourist satisfaction with souvenir shopping: evidence from Indonesian domestic tourists," *Current Issues in Tourism,* **21**, pp. 663-679 (2018)
11. J. Xu and A. Chan, "Service experience and package tours," *Asia Pacific Journal of Tourism Research,* **15**, pp. 177-194 (2010)
12. H. S. Kim and B. Choi, "The effects of three customer-to-customer interaction quality types on customer experience quality and citizenship behavior in mass service settings," *Journal of Services Marketing,* **30**, pp. 384-397 (2016)
13. D. Suhartanto, D. Dean, A. Sosianika, and T. Suhaeni, "Food souvenirs and their influence on tourist satisfaction and behavioural intentions," *European Journal of Tourism Research,* **18**, pp. 133-145 (2018)
14. M. Sundstrom, C. Lundberg, and S. Giannakis, "Tourist shopping motivation: Go with the flow or follow the plan," *International Journal of Quality and Services Sciences,* **3**, pp. 211-224 (2011)
15. C. Antón, C. Camarero, and M. Laguna-García, "Towards a new approach of destination loyalty drivers: satisfaction, visit intensity and tourist motivations," *Current Issues in Tourism,* **20**, pp. 238-260 (2017)
16. C. White, "The impact of motivation on customer satisfaction formation: A self-determination perspective," *European Journal of Marketing,* **49**, pp. 1923-1940 (2015)
17. F. Meng, Y. Tepanon, and M. Uysal, "Measuring tourist satisfaction by attribute and motivation: The case of a nature-based resort," *Journal of Vacation Marketing,* **14**, pp. 41-56 (2008)
18. N. K. Prebesen, E. W. Woo, J. S. Chen, and M. Uysal, "Motivation and Involvement as Antecedents of the Perceived Value of the Destination Experience," *Journal of Travel Research,* **52**, pp. 253-264 (2013)
19. J. E. Hair, G. T. Hult, C. M. Ringle, and M. Sarstedt, ***A Primer on Partial Least Squares Structural Equation Modeling (PLS-SEM)***, 2 ed. Thousand Oaks: Sage (2017)
20. M. Devesa, M. Laguna, and A. Palacios, "The role of motivation in visitor satisfaction: Empirical evidence in rural tourism," *Tourism Management,* **31**, pp. 547-552 (2010)
21. Y. Yoon and M. Uysal, "An examination of the effects of motivation and satisfaction on destination loyalty: A structural model," *Tourism Management,* **26**, pp. 45-56 (2005)
22. D. Suhartanto, D. Dean, R. Nansuri, and N. N. Triyuni, "The link between tourism involvement and service performance: Evidence from frontline retail employees," *Journal of Business Research,* **83**, pp. 130-137 (2018)

The Dimension of Food Souvenir: An Exploratory-Confirmatory Factor Analysis

Adila Sosianika[1] , *Tintin* Suhaeni[1], Nono wibisono[1], and *Dwi* Suhartanto[1*]
[1]Department of Business Administration, Politeknik Negeri Bandung, Indonesia

Abstract. Food souvenirs as the tangible reminder that encounters the travelling experience are considered as an essential element in the tourism industry. However, the studies exploring the dimension of food souvenirs, are limited, and most of them applying single factor analysis, either exploratory or confirmatory. In response to the identified gap, this research scrutinizes the aspects of food souvenirs by utilizing a mixed method of exploratory-confirmatory factor analysis. The data was collected from 252 domestic tourists in Bandung, Indonesia. The exploratory factor analysis test identifies four dimensions of food souvenirs: tangible, brand and packaging, value, food quality. Although confirming the food souvenir dimension, the confirmatory factor analysis notes that the elements of the dimensions are different from that identified in exploratory factor analysis. Besides offering a theoretical understanding on the food souvenirs dimension, this study extends our understanding of a methodological issue that the exploratory-confirmatory factor analysis offers a better explanation on the food souvenir dimension compared to that of single factor analysis.

1 Introduction

A souvenir is a tangible product that encounters the perceptual images of travelling experience and its effects on a favourable tourist future behaviour. Many tourists feel that they need a souvenir in their travelling to keep their unforgettable experience and also to complete their trips [1, 2]. A souvenir might even have a social function [3]. By sharing and giving of souvenirs to their families, friends, and colleagues, the travellers might have values of self-esteem, prestige, and pride. Moreover, a souvenir also can give unforgettable retention of the cultural value of the destination visited. Thus, a souvenir consider as an essential element for the tourism industry since it has a good impact not only on the industry but also for the economy of the destination [4, 5]. As a result of its essential role in the tourism industry, souvenirs have been studied from different perspectives, including cultural, behavioural, and retailing [4, 6, 7].

The recent studies have considered the food which a tourist bought in a tourist destination as an important souvenir [2, 7]]. Food souvenirs, mainly local delicacies and specialty foods, are recognized as a tangible product to capture the unique nature of the destination. Therefore, travellers buy a particular food to enjoy physically and mentally and relate its identity to the destination visited [3]]. Food souvenirs also consider as an important aspect on the development of economic, other industries, such as agriculture,

* Corresponding author: dwi.suhartanto@polban.ac.id

dairy, and fishery [8]] as well as cultural and environmental sustainability of the tourist destination [9, 10]]. However, the role of food souvenir and its essential meaning in the tourism industry, especially the study exploring the dimension of food souvenirs, are still limited in the literature. Moreover, previous studies assessing the dimension of food souvenirs mostly use exploratory factor analysis or confirmatory analysis only [3, 5, 11]. The usage of a single factor analysis very often provides different findings [12]]. Thus, using a single factor analysis approach to assess the dimension of food souvenir could potentially result in un-comprehensive dimension. This study argues that exploring the dimension of food souvenirs using a mixed method, exploratory-confirmatory factor analysis, will provide a better result.

Based on the research gap identified, this research aims to scrutinize the dimension of food souvenir by applying an exploratory-confirmatory factor analysis approach. In the competitive tourism industry, understanding the dimension of food souvenirs is critical for both retail managers as well as tourism destination managers to develop a proper marketing strategy to satisfy tourists' need of food souvenir.

2 Literature review

A souvenir is a tangible object and intangible memorable experience for tourists during and after their travelling [1, 6]]. Past studies categorize souvenirs into three types: local products, products distinct with graphics, and outfits and jeweller [[13]]. The researchers have also identified local food and beverage products in some countries that are related to cultural tourism, such as candy, alcohols, spices, and tea souvenirs [10]. Food souvenir is considered as a tool to express the local culture as well as assign tourists with a unique culture and way of life of the residents [14]].

Recently, researchers classify the local food as a souvenir and suggest its critical role in improving the competitive advantage of a tourist destination [15]]. Although there are many studies conducted on local food as a souvenir, previous research focusing on the dimension of food souvenirs is limited [3, 8]. A survey in Chinese tourism identifies four dimensions of food souvenir, i.e. quality, symbol, innovation, and practicality [8]. Altintzoglou et al. [9]' study on Norway food souvenirs and tourist innovativeness propose the key elements affecting travellers in selecting and purchasing food as souvenirs, including taste, quality, authenticity and local origin. Suhartanto [6, 16]] report five dimensions of food souvenirs in the Indonesian domestic tourist context, namely value, store, collectability, display, and functionality. A different result is indicated by Paraskevaidis and Andriotis [17]] who identify exchange value, use value, spiritual value, and sign value as the dimension of food souvenir.

To sum up, the previous literature has indicated a significant growth in understanding souvenirs as an element of a tourism destination's contributor. Recently, researchers have also reported the critical role of food souvenirs in influencing tourist buying decisions as well as highlighting the local tourist destination values and community creativities. Much of past researchers tend to emphasize on the meaning of food as a souvenir, but limited attention has been given on the dimensionality of food as a souvenir.

3 Method

The measurement of food souvenirs scales in this research was drawn from the previous literature on souvenir and specialty food [3, 8, 9] (see table1). All variable elements were anchored using a 5-point Likert's scale (from 1 strongly disagree to 5 strongly agree). Self-Administrated questionnaires were distributed to 252 tourists. This number of data is considered as adequate to test the dimension of food souvenir using both exploratory and confirmatory factor analysis [12, 18]. The respondents consist of females (39.68%), and males (60.32%) with the majority (48.02%) were at the age of 17-25 years. The data was gathered from tourists who visited Bandung City. To obtain a robust data, the respondents were approached after buying a food souvenir. As it is impossible to use random sampling, the purposive sampling method was applied. To test the proposed model of food souvenir, this sampling method is considered as appropriate [19]. To examine the dimension of food souvenirs, this study uses the mixing method of exploratory-confirmatory factor analysis approach. The confirmatory factor analysis was executed by applying Structural Equation Modelling (SEM) as suggested by scholars [12, 18].

4 Result

To scrutinize the dimensions of food souvenirs, an exploratory factor analysis (EFA) was conducted with the 12 items to measure tourist perception of a food souvenir. This study utilizes principal component analysis as a method of extraction and a varimax rotation with Kaiser normalization as recommended by past researchers [18] for a parsimonious description of the dimensions.

Table 1. Rotated Component Matrix

Items	Component			
	Tangible	Packaging an Brand	Value	Food Quality
1. Taste	.098	-.123	.389	.515
2. Natural	.084	.145	-.017	.789
3.Healthy food	.210	.313	.034	.702
4. Interesting food display	.085	.821	-.038	.264
5. Good packaging design	.244	.752	.102	.199
6. Practical	.695	.123	.035	.209
7. Eco-friendly	.772	-.027	.132	.240
8.Value product	.103	-.015	.820	.146
9.Reasonable price	.029	.174	.866	.006
10.Brand memorable	.277	.494	.425	-.153
11.Interesting packaging	.612	.479	.144	-.159
12.Trusted producer	.613	.261	.029	.034

The result (Table 1) exposes the 12 items can be grouped into four factors covering 62.316% of the variance. All of the four dimensions identified have Cronbach's Alphas above 0.70, specifying the reliability of the identified dimensions[18]. The result of Kaiser–Meyer–Olkin test (0.754) and Bartlett's test of sphericity ($p< 0.01$) support for the sampling adequacy and convergent validity of the instruments. Further, the results suggest that the variance of the original values is relatively supported by the common factors as indicated by their commonalities; ranging from 0.441 to 0.781. The construct discriminant validity is satisfied with factor loadings exceeding the cut-off value of 0.5[18].

As exposed in Table 1, the exploratory factor analysis extracts four underlying factors of the food souvenir. The first factor consists of four items (with eigenvalue 3.63) and explains 17.044% of the variance, namely practical, eco-friendly, interesting packaging and trusted producer. Thus, the first factor is labelled as Tangible. The second factor consists of three items (with eigenvalue 1.512) which explain 16.352% of the variance of the data. This factor is loaded with the item of interesting food display, good packaging design, and brand memorize able. So, this factor is labelled as Packaging and Brand.

The third factor consists of two items (eigenvalue 1.316) and accounts for 15.070% of the variance. The third factor is loaded with items of value product and reasonable price; thus it is labelled as Value. The fourth factor contains three elements (with eigenvalue value 1.20) explains 13.849% of the variance. The three items connected with this factor consist of taste, natural and healthy food; so this factor is labelled Food Quality[20].

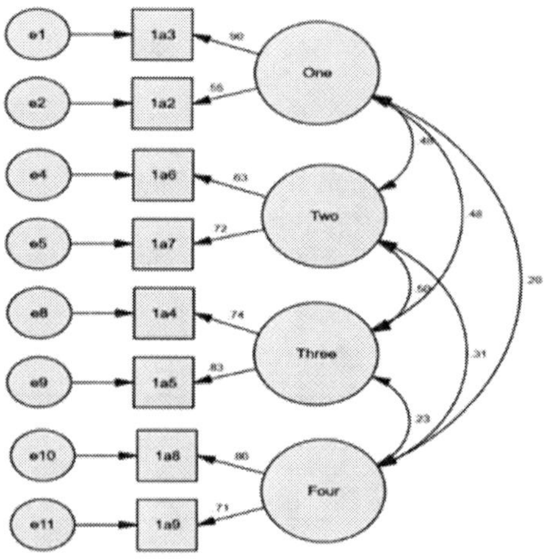

Fig. 1. The modified model

The measurement of food souvenirs model was assessed using confirmatory factor analysis (CFA), where the dimensions of food quality were supposed to co-vary with one another. The outcome of CFA test indicates that the goodness-of-fit index of the model (CMIN/DF: 5.390, GFI: 0.920, CFI: 0.846, and RMSEA: 0.096) identified is inadequate level of model fit. To increase the level of model fitness, the adjustment of the food souvenir attributes was directed by re-parameterizing the model on the basis of the insignificant path, standardized residuals, modification index value, and hypothetical considerations [21-23]. Using this process (Table 2 and Figure 1), the best food souvenir dimension consists four dimensions with each consist of two items. The exclusion of item 1a1 (Dimension One), 1a11 and 1a12 (Dimension Two), and item 1a10 (Dimension Three) did not considerably modify the hypothesized content of the construct. The reason was the remaining items had a robust statistical similarity and theoretically consistent with each dimension.

The result of modifying the CFA test exhibited that the goodness-of-fit indexes of the model (CMIN/DF: 4.109, GFI: 0.970, CFI: 0.951, and RMSEA: 0.081) specifies an excellent of model fit. Further, Table 2 indicates that the values of the composite reliability of all four dimensions are higher than 0.7 and the values of average variance extracted are also higher than 0.5 [5].

Table 2: Reliability and Validity Check

Variable	CR	AVE	Tangible	Brand & Packaging	Value	Food Quality
Tangible	0.626	0.501	**0.676**			
Brand & Packaging	0.759	0.613	0.502	**0.783**		
Value	0.730	0.575	0.314	0.232	**0.758**	
Food Quality	0.707	0.560	0.482	0.484	0.205	**0.749**

Additionally, the requirement for discriminant validity is also met as average variance values extracted are higher than the constructs formed correlation[5]. Thus, the requirement of construct reliability and validity is satisfied. The finding of this research indicates that tangible aspect is a critical element of food souvenir. Thus, this study suggests the souvenir retail managers need to consider the tangible attributes of the food, including practicality, environmentally friendly, attractive packaging, and a good producer image [6, 10]. Besides, this study also suggests the usage of a unique brand as well as interesting packaging design to attract tourist to buy food souvenir[20]. Moreover, consistent with previous studies[6, 7], the result of this study indicates that value and price are important elements of food souvenirs. Finally, this study highlights the food quality as an essential attribute of food souvenir. It suggests that the food souvenir should meet with the food health standard[8], have a good taste, and contain natural ingredients.

From the methodological perspective, this study shows that although the dimensions identified in exploratory, as well as confirmatory factor analysis, are similar, the items are different. Thus, this research gives a new perspective that examining the dimensionality of food souvenir, as well as another souvenir, should use exploratory-confirmatory factor analysis rather than only a single factor analysis. This finding is important as none of past studies has discussed such issue in food souvenir context.

5 Conclusion, Implication, and Future Research

The previous scholars have examined the dimension of food souvenir, but they are only applying exploratory factor analysis or confirmatory factor analysis. Thus, their results are considered as less robust. This study focuses on the usage of exploratory-confirmatory factor analysis to identify the dimension of food as a souvenir. The results show that this method identifies four dimensions of food souvenirs, but the elements of the dimensions are different. Besides offering useful implications for understanding tourist behaviours towards the food souvenir, this study extends our understanding that in assessing the dimension of food souvenir researchers should use both factor analysis instead of only a single factor analysis. Although this study offers significant findings, this study bears drawback concerning the sample which is collected in Bandung. As a result, this study cannot be generalized in other location. Thus, the future research could replicate this study in a different location, culture, and souvenir type.

References

1. W. H. Kong and T. Z. Chang, "The Role of Souvenir Shopping in a Diversified Macau Destination Portfolio," *Journal of Hospitality Marketing and Management,* **21**, pp. 357-373 (2012)
2. K. K. Swanson and D. J. Timothy, "Souvenirs: Icons of meaning, commercialization, and commoditization," *Tourism Management,* **33**, pp. 489-499 (2012)
3. L. Lin and P. C. Mao, "Food for memories and culture - A content analysis study of food specialties and souvenirs," *Journal of Hospitality and Tourism Management,* **22**, pp. 19-29 (2015)
4. C. Tosun, P. Temizkan, D. Timothy, and A. Fyall, "Tourist shopping experience and satisfaction," *International Journal of Tourism Research,* **9**, pp. 87-102 (2007)
5. D. Suhartanto, D. Dean, A. Sosianika, and T. Suhaeni, "Food souvenirs and their influence on tourist satisfaction and behavioral intentions," *European Journal of Tourism Research,* **18**, pp. 133-145 (2018)
6. D. Suhartanto, "Tourist satisfaction with souvenir shopping: evidence from Indonesian domestic tourists," *Current Issue in Tourism,* **21**, pp. 663-679 (2018).

7. D. Suhartanto, D. Dean, R. Nansuri, and N. N. Triyuni, "The link between tourism involvement and service performance: Evidence from frontline retail employees," *Journal of Business Research,* **83**, pp. 130-137 (2018)
8. L. Lin, "Food souvenirs as gifts: tourist perspectives and their motivational basis in Chinese culture," *Journal of Tourism and Cultural Change,* pp. 1-16 (2016)
9. T. Altintzoglou, M. Heide, and T. Borch, "Food souvenirs: Buying behavior of tourists in Norway," *British Food Journal,* **118**, pp. 119-131 (2016)
10. K. Buczkowska, "Local food and beverage products as important tourist souvenirs," *Turystyka Kulturowa,* **1**, pp. 47-58 (2014)
11. D. Suhartanto, B. T. Chen, Z. Mohi, and A. Sosianika, "Exploring loyalty to specialty foods among tourists and residents," *British Food Journal,* **120**, pp. 1120-1131 (2018)
12. [12] J. E. Hair, G. T. Hult, C. M. Ringle, and M. Sarstedt, ***A Primer on Partial Least Squares Structural Equation Modeling (PLS-SEM)***, 2nd ed. Thousand Oaks: Sage (2017)
13. H. Wilkins, "Souvenirs: What and why we buy," *Journal of Travel Research,* **50**, pp. 239-247, (2011)
14. H.-W. Joung, E.-K. Choi, and E. Wang, "Effects of Perceived Quality and Perceived Value of Campus Foodservice on Customer Satisfaction: Moderating Role of Gender," *Journal of Quality Assurance in Hospitality & Tourism,* **17**, pp. 101-113 (2016)
15. T. T. Trinh, C. Ryan, and J. Cave, "Souvenir sellers and perceptions of authenticity - The retailers of Hội An, Vietnam," *Tourism Management,* **45**, pp. 275-283 (2014)
16. D. Suhartanto, "The role of store coopetition and attractiveness on the performance of tourism destination and its retail stores," *International Journal of Tourism Policy,* **7**, pp. 151-165 (2017)
17. P. Paraskevaidis and K. Andriotis, "Values of souvenirs as commodities," *Tourism Management,* **48**, pp. 1-10 (2015)
18. [18] J. F. Hair, W. C. Black, B. J. Babin, and R. E. Anderson, ***Multivariate data analysis: A global perspective,*** 7th ed. Upper Saddle River: Pearson Education (2010)
19. N. L. Reynolds, A. C. Simintiras, and A. Diamantopoulos, "Theoretical justification of sampling choices in international marketing research: Key issues and guidelines for researchers," *Journal of International Business Studies,* **34**, p. 80-95 (2003)
20. S. Mynttinen, L. Logren, M. Sarkka-Tirkkonen, and T. Rautiainen, "Perceptions of food and its locality among Russian tourists in the South Savo region of Finland," *Tourism Management,* **48**, pp. 455-466 (2015)
21. J. C. Anderson and D. W. Gerbing, "Structural equation modeling in practice: A review and recommended two-step approach," *Psychological Bulletin,* **103**, p. 411-423) (1988)
22. P. M. Bentler and C.-P. Chou, "Practical issues in structural modeling," *Sociological Methods & Research,* **16** (1987)
23. J. C. Anderson and D. W. Gerbing, "Structural equation modeling in practice: A review and recommended two-step approach," *Psychological Bulletin,* **103**, pp. 411-423 (1988)

Prototype Design Of Chicken Meat Cutting Tool

Laurence[1,*], *Natalia* Hartono[1] , and *Clara Alverina* Setiawan[1]

[1]Industrial Engineering, Faculty of Science and Technology, University of Pelita Harapan, Indonesia

Abstract. The main thing the chicken skewers owners do is cut the meat into blocks. However, to produce a cubical meat shape is not an easy task and takes a long time. Therefore, it is necessary to design a tool to help the chicken skewers business owner to cut the chicken breast faster. The purpose of this research was to design an ergonomically chicken meat cutting tool to produce cubical chicken meat in large quantities so that the work can be completed in a shorter time. The planning and product development stage of this research was based on Ulrich and Eppinger steps from phases 0 through 4. After that, a prototype evaluation was performed. The conclusion of this research was that the tool successfully worked as the design intended. The maximum cutting yield is 225 pieces, 15 times faster, suitable for half-frozen chicken meat, and the evaluation results show a good performance rating. Overall, most of the respondents (87.5%) were pleased with this tools, shown by 57.5% choose a smiling face and 30% show a big smiling face as their feeling of satisfaction after trying the gamma prototype.

1 Introduction

Chicken skewer business categorized as a micro business where the maximum profit is Rp.300 million per year according to the constitutional law number 20, the year 2008 and also categorized as a home industry because at average there are 1 to 4 workers. This study starts with the observation in Tangerang area where there is a lot of chicken skewers business. When the process of making skewers was observed, it is clear that the main process is to cut the meat and put onto skewers. However, to produce a cubical meat shape is not easy and takes a long time. The preliminary information gathering was done with talking informally with several skewers business owner in Tangerang area and it is found that they need a tool to cut the chicken breast fillet faster.

There is an automatic chicken cutting machine that can cut chicken meat into blocks but the price of such machine is expensive and it means a large amount of investment needed to start the chicken skewer business. Small companies have scarce resources that disrupt them to innovate [1]. Tools and equipment help to perform a task that not easily perform in more efficient way and ergonomics should be involved in the design to ensure safe and effective operation [2]. Therefore, it is necessary to design a manual chicken meat cutter to help micro business and home industry so they can reduce the investment of expensive machine to cut chicken meat faster than using knife only. The purpose of this research was to design

* Corresponding author: laurence.fti@uph.edu

ergonomically chicken meat cutting tool to help the work of skewer business owner to produce cubical chicken meat in large quantities so that the work can be completed with a shorter time.

2 Scope and Constraint

This research collected data from observation and interview to skewer business owner at Tangerang area from August to November 2017. The knife size was designed to cut chicken breast meat to 2x2 cm size and can be only used to cut boneless chicken meat or chicken breast fillet. The steps from Ulrich and Eppinger until gamma prototyping due to financial constraint.

3 Methodology

The research starts with an identification problem with skewer business owner who needs a lot of preparation and the most time consuming was cutting the chicken meat to shape cubical. Product ideas are based on customer problem and become the heart of the concept generation process [3]. Then, the cutting tools for this needs were searched and continue with an interview with several skewer business owners to gather preliminary data and preliminary voice of customer (VOC). After that, the purpose of this research was established.

Theory of product design and development from Ulrich and Eppinger was used for this research because it provides a step-by-step approach thus the methods are structured and suitable as a starting point for continuous improvement of the product. Ulrich and Eppinger's methods consist of phase 0 to phase 5 [4]. Phase 5 is production ramp-up, and this research until prototyping, so this research were from phases 0 to 4, which depict in Fig 1. Finally, the conclusion and suggestion for further research were made.

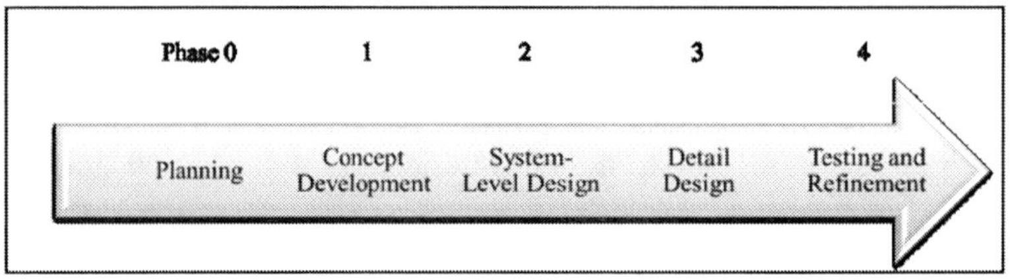

Fig 1. Phases in this research

4 Product Design and Development

The phase 0 to phase 4 were explained with detailed in this section.

4.1 Phase 0

At zero phase, there is the five-step process: identify opportunities, evaluate and prioritize projects, allocate resources and plan timing, complete pre-project planning, and reflect on the results and the process. Identify opportunities has been done through observation and interview to skewer business owner. The second step is to evaluate and prioritize the project. As this project has focused on designing cutting tools, the competitive strategy used was customer focus, where researchers assess the customer needs and preference of end users. Resource allocation and plan timing in this research were using budget under 2 million rupiah and the prototype plan to finish in November 2017.

The next step was complete pre-project planning where the guidance for product development was written in a mission statement that informs the product description, benefit proposition, key business goal, target market, assumption and constraints, and stakeholders. Product description dan benefit proposition was a manual cutting tool that can cut cubical chicken breast meat in large quantities. Not only skewer business owner but also catering can use the same tool to cut vegetables and fruit. The key business goal that the product will be launch at 2018 with primary target market is skewer business owner and secondary target is catering business owner. Assumption and constraints in this product development are this product is designed to cut boneless half-frozen chicken meat and other meats like duck, fish, beef and can cut soft vegetable and fruit. The stakeholders are all the groups of people who are affected by the product's success or failure [4], that are skewer business owner, catering business owners and the design team. The final step in this phase is reflected in the results and the process and it is decided that this idea can be continued to phase 1.

4.2 Phase 1

Phase 1 is a concept development that consists of identifying customer needs, product specification, concept generation, concept selection, and concept testing. The first steps in phase 1 are identifying customer need with gathering raw data from customers (skewer business owner in Tangerang area) using questionnaire and interpret the raw data in terms of customer needs, followed by organize the needs into a hierarchy, establish the relative importance of the needs and finally reflect on the results and the process. The product specification use data from the previous step and interpret the raw data into the voice of customer (VOC) and establish the relative importance of the needs using Quality Function Deployment (QFD) that consists of House of Quality (HOQ) 1 and HOQ 2. The next step is comparing the design with competitor products and reflect on the results and the process.

From the HOQ 1 table, the highest importance level was tool durability test and product usage test 29.1% and 24% respectively. The result of HQO 2 shows that the most important part of this tool is the knife shown by the level of importance by 53.2%. Two benchmark competitor product is skewered machining chicken meat lamb and automatic chicken skewer machine. The weaknesses of first competitor products are it needs several cuts and the results are too dense between the meat, where the skewer business owner usually want the meat not too dense. The second competitor needs electricity and expensive.

The next steps are concept generation. A product concept is an approximate description of the technology, working principles, and form of the product [4]. This research develops several concepts that follow the customer needs and target specification from the previous steps. The concept is usually expressed as a sketch or a rough three-dimensional model [4]. The concept generation steps are clarified the problem, search externally, search internally, explore systematically, and reflect on the solutions and the process. The first step is to make a description of the product, that the cutting tool will cut chicken meat faster, in large quantity and has a 2x2 cm shape, made from food-grade stainless steel to ensure safety in kitchenware, equipped with a handle to grip the tools, will have square shaped container. The tools designed to cut the half frozen chicken meat because based on the search on the interviews with skewer business owner and catering business owners, a half frozen meat will give better cutting results because of its solid meat structure. In addition, data from the USDA (United States Department of Agriculture Food Safety and Inspection Service) said that bacteria in chicken meat will grow rapidly at temperatures between 4.4 ° C and 60 ° C [5]. The external search steps were a search for the patent of chicken meat cutting tools and the results are there is no similar design. The search was at the Indonesian government website of Directorate General of Intellectual Property [6]. The internal search step using brainstorming to explore the idea and make sketches. The idea is to combine the cutting board and a kitchen knife. The process of external and internal search results in many concept fragments and need to organize and synthesized using concept combination table. This process resulting 4 alternative concepts and can be seen in Fig 2.

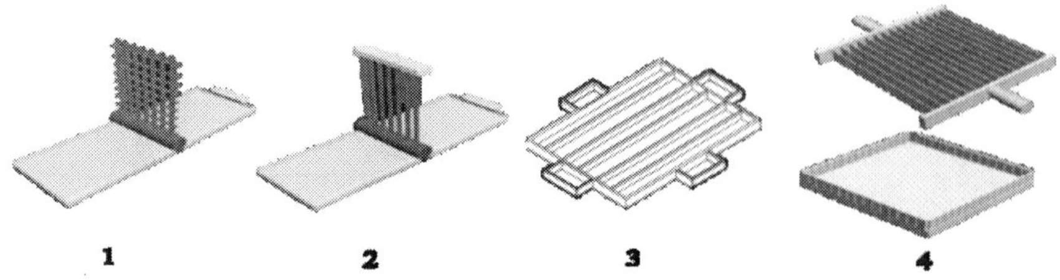

Fig 2. Four alternative concepts

After that, reflect on the solutions and the process shows that this concept can continue to the next step. The next step is concept selection and concept testing, and in this research using external decision to choose among concepts. The four concepts were showed to the skewer business owner and catering business owners and 50% of them choose number 4, so this concept was chosen.

4.3 Phase 2

Phase 2 System Level Design consists of product architecture. Product architecture is the assignment of the functional elements of a product to the physical building blocks of the product [4]. The cutting tools are simple without many parts so that the product architecture is the same as the product description.

4.4 Phase 3

Detail design in this third phase consists of a specification of the shape and material of all components of the product, material and planning of product process making, and prototyping. Based on interviews and observations, the skewer business owner age ranged from 21-45 years. This data used to determine the size of the handle to get an ergonomic product. The dimensional data was obtained from the web antropometriindonesia.org [7]. To determine the size of the handle, the data use the bottom of the percentile, that is the 5th because the handle is a part that must be able to be grip so the size used is the minimal dimension size so the people with small or large hand sizes will be able to hold the handle. It is known that the size of the hand for the fifth percentile is 16.5 cm and for the width of the hand is 9.09 cm. The handle should be designed to have large contact surfaces so the force distributes over a larger area [8]. Therefore, the size of the handle is 10 cm in length with a diameter of 2.5 cm. The handling also has a similar resemblance to the size of a general motor handlebar. Fig 3 shows the summarize of product detail design. This cutting tool would suitable for both right and lefthanders person and for both male and female because it is designed with two handles so the force will be smaller compared to just one handle. The design follows the principle of hand tool design that is design for use by either hand, for an individual user, and design for a power grip [2].

Parts	Function	Material	Dimension	Weight	Make /buy	Picture	Quantity
Cutting board	as a place to lay chicken meat	cutting board wood	34 x 34 x 5 cm	1.5 kg	make		1 pcs
Side of the cutting board	become a support for the cutting board	cutting board wood	36 x 36 x 7 cm	0.25 kg	make		4 pcs
Knife	to cut	stainless steel	38 cm with 3mm thickness	0.25 kg	make		14 pcs
Blade connection	as connection between knifes	stainless steel	33 cm	0.25 kg	make		2 pcs
Holder	for easy handling at the cutting process	stainless steel	10 cm long with diameter 2.5 cm	0.25 kg	make		2 pcs

Fig 3. Refining Specification and Parts

After the specification, the next step is to make a prototype. Prototyping is the process of developing such an approximation of the product [4]. There are two prototypes in this phase. Prototype alpha was the first prototype made from the plastic board. In making this prototype, the handle was different than the planning due to the difficulty to make rolling handlebar. This prototype making was used to know the big picture of the product designed. After analyzing the alpha prototype, the beta prototype is made from 3 mm medium-density fibreboard (MDF) using laser cutting to the actual dimension of the design. Using prototype beta, an internal evaluation is performed to identify the required engineering changes for the final product.

4.5 Phase 4

Evaluation of the previous prototypes shows that the cutting tools can work as the design intended, and the step progress to phase four, testing and refinement. The gamma prototype was build to be used as trial and evaluation to the end users. In this prototype, the material used for the knife and handle was iron plates while for the container was made from wood. The shape of this prototype is the same with a beta prototype but this prototype had different sizes and shapes of the handle. The handle was shaped like a tube with the length of 10 cm and diameter of 2.5 cm as the design in the previous phase for easier grip. The process of making the gamma prototypes will be the same with the actual product, but with different material for the knife. Prototype knife made from iron while the actual product from stainless steel. The process starts with pattern making, measurement, cutting the materials, polishing, made the knife blade using milling and grinding machine, assembling and final inspection. Figure 5 shows the picture of the gamma prototype.

Alpha prototype Beta prototype Gamma prototype

Fig 5. Prototypes

The testing involved 40 users (29 skewer business owner and 11 catering business owner) to test the product and followed by evaluation using a questionnaire. The results are the cutting board container can hold to 10 – 12 boneless chicken breast in one cut and produces 225 cubical 2x2 cm shaped. This test was compared with cutting with a knife and the results show that to cut with knife 10 boneless chicken breast need 685 seconds and with this gamma prototype only need 45 seconds, which is 15 times faster using this cutting tools. The questionnaire adapted from Ulrich and Eppinger to evaluate the quality of the design with rate 1 to 5 from low to high valuation. At the end of the questionnaire was a question about the overall satisfaction feeling of the respondent to this gamma prototype. The results show that on average, respondents give a performance rating of 4 that shows that overall the tools have good performance rating. The respondents felt about the tools showed that 12.5% feels neutral, 57.5% happy, and 30% feel very happy with the products.

5 Conclusion

The process of making the cutting tools was done using Ulrich and Eppinger phases until prototyping. The evaluation of gamma prototype shows that this tools can cut as the design intended, which are 15 times faster, cut in cubical shaped with maximum cutting yield is 225 pieces, has performance rating of 4 which means good performance and most of the respondents were pleased with the products shown by 57.5% choose happy face and 30% choose big smiling face as their feeling of satisfaction after trying the gamma prototype.

6 Recommendations

This research shows that not only a sophisticated product can use the phase from Ulrich and Eppinger, but also simpler products. The phases direct the product designer to focus and in practical implication will be useful to people without product design background who wants to make products based on the voice of the end customers.

The research can add economic analysis to evaluate if this gamma prototype could be developed further to be sold.

References

1. L. Löfqvist, Product Innovation in Small Companies: Managing Resource Scarcity Through Financial Bootstrapping, Int. J. of Innovation Management **21** 2 1750020 (2016)
2. M.R. Lehto, J.R. Buck, *Introduction to Human Factors and Ergonomics for Engineers* (Taylor & Francis Group, New York, 2008)
3. M. Crawford, A. Di Benedetto, *New Product Management* (McGraw-Hill Irwin, New York, 2011)
4. K.T.Ulrich, S.D.Eppinger. *Product Design and Development.* Sixth Edition (McGraw-Hill Education, New York, 2016)
5. United States Department of Agriculture Food Safety and Inspection Sevice. *Chicken from Farm to Table.* https://www.fsis.usda.gov/wps/wcm/connect/ad74bb8d-1dab-49c1-b05e-390a74ba7471/Chicken_from_Farm_to_Table.pdf?MOD=AJPERES (retrieved September 2017)
6. *Direktorat Jendral Kekayaan Intelektual* (Directorate General of Intellectual Property) www.pdki-indonesia.dgip.go.id (retrieved August 2017)
7. *Antropometri Indonesia* (Indonesian Anthropometry) http://antropometriindonesia.org/index.php/detail/artikel/4/10/data_antropometri (retrieved September 2017).
8. M.S. Sanders, E.J. McCormick, *Human Factors in Engineering and Design* (McGraw Hill, Singapore,1992)

MATEC Web of Conferences **218**, 04004 (2018) https://doi.org/10.1051/matecconf/201821804004
ICIEE 2018

Prediction of Grinding Work Roll Demand in a Job Shop Company By using Artificial Neural Network and ARIMA Method

Yusraini Muharni[1*], *Ade* Irman[1], *Muhammad* Ilhamsyah[2]
[1]Department of Industrial Engineering, Engineering Faculty, Universitas Sultan Ageng Tirtayasa, Banten, Indonesia 42435
[2]Production System Laboratory, Department of Industrial Engineering, Engineering Faculty, Universitas Sultan Ageng Tirtayasa, Banten, Indonesia 42435

Abstract: This study concern about forecasting grinding work roll demand in a job shop company located in industrial area in Cilegon. This factory main production is fabrication, which accepts various orders from other companies especially from the company around. Grinding Work Roll is one of those products that frequently request by customer. Although the order is frequent but the volume is fluctuation month by month. This situation drives the company to face the problem in preparing the resources required in fabrication process specially in scheduling the operators. To cope with this problem, we proposed to apply two robust forecasting methods, Artificial Neural Network and ARIMA to help in prediction the grinding work roll demand so as the company could make a good plan for the production process. The best architecture for ANN is obtained through applying Taguchi Method which applies Levenberg-Marquardt algorithm as Training Function. The best number for hidden layer is 10, while Momentum is 0.9. The Prediction result shows that ANN predicts better than ARIMA Method according to the lower Mean Square Error (MSE). MSE Value for ANN is 0.002 while for ARIMA MSE is 0.0043. From this study, we experienced that by applying Taguchi method could improve the performance of Artificial Neural Network.

1. Introduction

The role of forecasting in today's business world is counting essential as company focus on improving customer satisfaction while reducing the cost of fulfill demand and running production process [1]. Forecasting is crucial input to wide range of business decision making process. This research was undergone in a job shop company with fabrication as main activity. Located in industrial area, this company receives various orders mainly from manufacturer nearby. One regular order comes every month from a loyal customer to fabricate Grinding Work Roll. The problem exist in this order is the variation in number of orders that changeability each month so at the time order could be increased while decrease in the next month. This situation influences the completion time of the order, since it over the due date in the contract causing losses for companies and clients. For example is the demand for April 2008 there is a demand of 12 units but the output produced only 10 units then the next month is May 2008 demand decreased

* Corresponding author: yusmuharni@gmail.com

dramatically to one unit but the output produced exceeds the demand of two units. This situation drove the company into losses because if production exceeds demand. It will still be given to the client because there is no place to accommodate it, whereas if the production does not reach the request it will be fined. Therefore the company intends to predict the number of requests in the Grinding Work Roll for the upcoming year in order to prepare the resources better and avoid further losses in the future.

Time series forecasting method has been known for years such as simple linear regression, exponential smoothing etc. However the simple linear regression required causal variables that explain the observed time series [1]. On the other hand, with only time series data with monthly basis, without any explanatory variables we directed into ARIMA models since it has advantage of determining valuable information from time series. There are three types of model could be applied: (1) Autoregressive (AR) models, (2) Moving Average (MA) Models, (3) Mixed Autoregressive-Moving Average (ARIMA) model. Whereas in fact, fluctuating data is difficult to meet the assumptions that exist in time series analysis as well as regression analysis, so that becomes an obstacle, therefore, an alternative method required which is free from assumptions with good estimation ability, i.e. Non parametric method, one of which can use Artificial Neural Network because this method is able to identify pattern from an input data (input) by using learning method to further trained to study pattern of past data and trying to find function that connect pattern of past data with output that desirable at this time.

Therefore, this study focus on applying Artificial Neural Network and ARIMA method for predicting the number of Grinding Work Roll demand since the data used is fluctuated and has a large scale. However, Artificial Neural Network Method is a more suitable method to use. The objective of this study is the company able to anticipate an increase in the number of orders, and can prepare the resources precisely

2. Literature Review

Forecasting has come to play an important role in managing business. To support sales and operations planning process, forecasting can be integrated effectively by involved data from previous sales [2]. The sufficient data in a time series basis can support the time series forecasting process especially for sales forecasting [3].

Forecasting is the activity of predicting or predicting what will happen in the future with a relatively prediction requires accurate data in the past, so that can be seen prospects of future situations and conditions. In general, the use of forecasting is as follows:

1. As a effective and efficient tool in planning.
2. To determine future resource requirements.
3. To make the right decision.

The Usefulness of forecasting is seen in a decision making. A good decision is a decision based on consideration of what will happen at the time of the decision in the various activities of the company. Nevertheless note that the forecast is always an element of error, so that needs to be considered is an attempt to minimize the error of the forecast [1,5]. One aspect of making good forecast is deeply understand what comprises the previous data to be used in preparing forecast [5].

Artificial neural networks (ANN) originated from modeling neural networks in the human brain with similar work characteristics with biological neural networks. It can also design the basic structure of a network that is a neuron into a definite design such as pattern recognition faster. This network resembles the human brain in two ways, namely, knowledge acquired by the network through the learning process, and the connection strength between neurons is known as the synaptic weight used to store knowledge [6]. ANN algorithms are commonly used in pattern classification, pattern completion, approach function, prediction, optimization, and control system applications [7]. Artificial neural networks is an information processing system that has certain performance characteristics similar to that of human biological neural networks that have been developed as generalizations of mathematical models of human cognition or biological nerves. Generalized mathematics formsANN from biological neural networks, assuming [4]:

1. Information processing occurs in many simple elements (neurons)
2. The signal is sent between the neurons through the link
3. The liason between neurons has a bonded weight which in the network multiplies the transmission signal.
4. Each neuron implements an activation function (usually nonlinear) on its input to determine the output signal.

Forecasting time series data using ARIMA model is encouraging to develop as long data series available (at least 50 data points). The advantages of this model are the ability to handle variability in the data as long as the series has stationary pattern. Besides ARIMA model can be applied for short, medium, and long term forecast horizon. [1]

3. Research Method

A. Data

The forecasting process begins with the collection of historical data on a monthly basis. Data are collected from the year 2008 to 2016 from a job shop company which fabrication as main business.

B. Artificial Neural Network

In this study, we apply the Artificial Neural Network Model with multilayer architecture consist of three layers: Input layer, Hidden layer, and Output layer as seen in Figure 1.

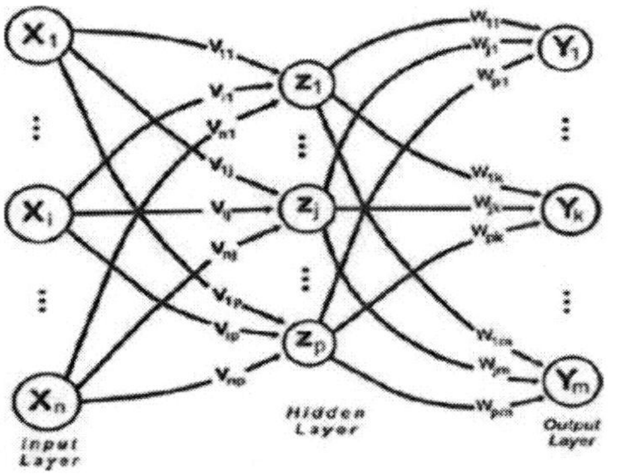

Fig 1. Multilayer Architecture of ANN [4]

We conduct a Taguchi Method in order to select the best ANN Architecture design. Three factors are considered influenced to the accuracy such as learning function, number of hidden layer and the momentum.

C. Taguchi Method

First introduced by Genichi Taguchi, this method is recognised as an effective way to conduct experiments since it could improve the efficiency by reducing experimental costs. Originally the method is applied with main focus to improve product quality through designed. Taguchi method introduces orthogonal arrays as a technique to find the best combination of manufacturing factors while reducing the experiments [12,13]. Taguchi method applies the orthogonal array is noted as $L_a(b^c)$, where a is the number of factor combination in the experiment, b is the number of level for factors, and c represents the number of factors in the experiment. The selection the number of factors and the number of levels for each factor in the experiment is based on the proper orthogonal array. In this study, we applied an orthogonal array as $L_9(3^3)$ meaning that we have 9 of factor combinations in the experiment, three level for factors, and three factors in the experiment which are learning function, number of hidden layer and the momentum.

4. Result And Discussion

Table 1 is a recapitulation of the best architectural design results. Based on the combination Taguchi method used is the combination of number 3 that is with the training function Levenberg-Marquardt, hidden layer 10, and momentum of 0.9. This combination yields an average MSE of 0.00105.

Table 1. The Recapitulation of ANN's Best Architecture Design with Taguchi Method

No	Fungsi Pelatihan	Hidden Layer	Momentum	MSE			MSE
				Replikasi 1	Replikasi 2	Replikasi 3	Rata-rata
1	Levenberg-Marquardt	1	0.1	0.00113	0.00322	0.00151	0.00195
2	Levenberg-Marquardt	5	0.5	0.00124	0.00231	0.00157	0.00171
3	Levenberg-Marquardt	10	0.9	0.00096	0.00099	0.00119	0.00105
4	Bayesian Regularization	1	0.5	0.00130	0.00271	0.00151	0.00184
5	Bayesian Regularization	5	0.9	0.00115	0.00581	0.00323	0.00340
6	Bayesian Regularization	10	0.1	0.00118	0.00121	0.00141	0.00127
7	Gradient Descent	1	0.9	0.00129	0.00235	0.00379	0.00247
8	Gradient Descent	5	0.1	0.00152	0.00967	0.00912	0.00677
9	Gradient Descent	10	0.5	0.00143	0.00371	0.00195	0.00236

The best performance results of the Levenberg-Marquardt training function can be seen in Figure 2.

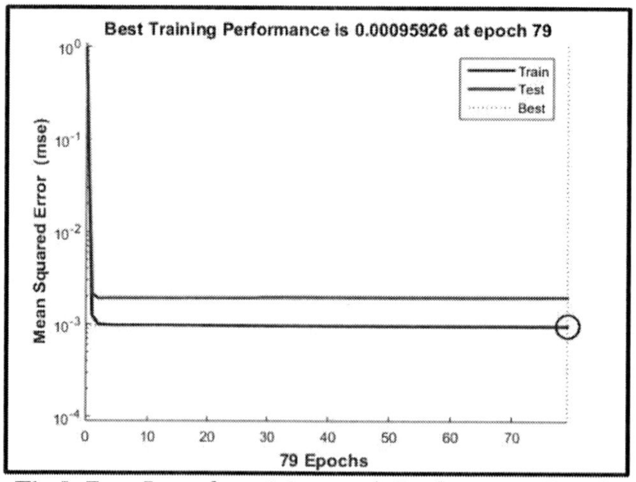

Fig 2. Best Levenberg-Marquardt Performance Result

Figure 2 shows the performance results of the third alternative design with Levenberg-Marquardt training function, 10 hidden hidden layers, and Momentum of 0.9. The design produced the best MSE of 0.00095926 when it reached the 79th iteration. Testing process is done by using the best architectural design that has been obtained by using Taguchi method, network structure used consists of training function using Levenberg-Marquardt, hidden layer amounted to 10, and momentum of 0.9.

Comparison between target and network output aims to see the proximity between data. If getting in between the target and the network output then the difference error is small or not much different between the target with the output network. The regression test graph shows the result of 0.96948 can be seen in Figure 3.

Comparison of ANN forecasting result is done to ascertain whether ANN method is better than other method, for comparison used ARIMA method, based on previous research used ARIMA method (1,1,0) (2,1,0) 12. Comparison of ANN method with ARIMA can be seen in Table 2.

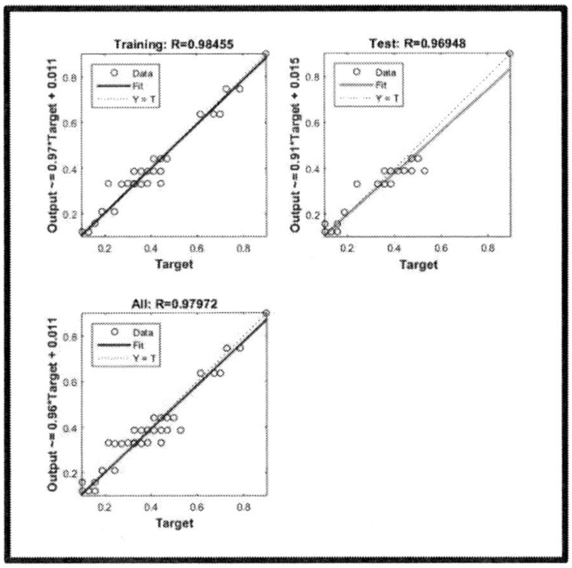

Fig 3. Regression

Table 2. ANN and ARIMA Method Comparison

No	Metode	MSE
1	Artificial Neural Network	0.0020
2	ARIMA	0.043

Based on the above table can be seen ANN method has a lower MSE compared with ARIMA method where ANN method has MSE 0.002 and ARIMA method has MSE of 0.043. Thus the best result for this research is using ANN method. Here is the ARIMA and ANN comparison plot to the actual data:

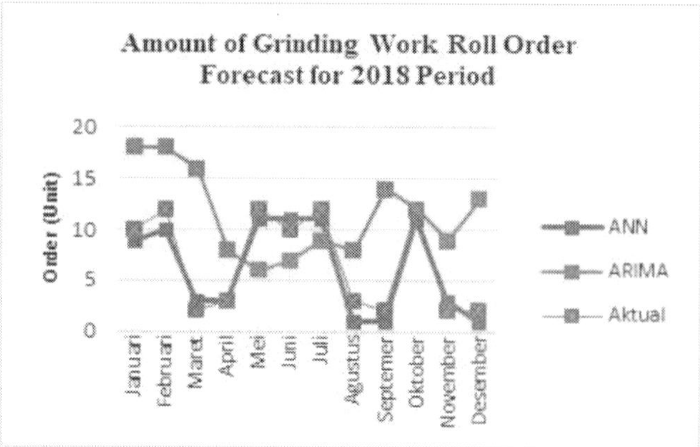

Fig 4. Comparison Between ANN and ARIMA Method Result over Actual Data

Comparison of forecasting the number of order Grinding Work Roll for period 2018 can be seen in Figure 4 that the forecasting with ANN method shows no significant difference to actual data, whereas for forecasting using ARIMA method seen in plot of graph there is a significant difference compared with ANN method. Therefore can be inferred based on MSE, that the ANN method in this work outperform the ARIMA method.

5. Conclusion

Based on the results of research, it can be drawn conclusions, as follows:

1. The best Artificial Neural Network architecture design for demand for Grinding Work Roll is to use Levenberg-Marquardt training function, 10 hidden layers, and 0.9 momentum.
2. The results of the demand forecast for Grinding Work Roll using Artificial Neural Network for the period of 2018 are 9 units, 10 units, 3 units, 3 units, 11 units, 11 units, 11 units, 1 unit, 1 unit, 11 units, 3 units, and 1 unit.
3. Forecasting result using ANN method produce MSE equal to 0.0019668 while for ARIMA method produce MSE equal to 0.0043. Based on the comparison of MSE values of ANN method is outperform the ARIMA method.
4. The suggestion for further research is to conduct further research on forecasting production demand on other products as well as the addition of influencing factors to determine the best network architecture to produce more accurate forecasting, resulting in more precise forecasting results for subsequent production.

References

1. J. H. Wilson, B. Keating, J.G. Solution Inc, Business Forecasting With ForecastXTM, Mc Graw Hill, Singapore (2009)
2. T. Alhadeff, J. of B.. Forecasting 23.1. (2004)
3. S. Harris, J. of B.. Forecasting (1995-96)
4. Fausett, L. Fundamental of neural Network : Architecture. Algorithms and Applications. Prentice Hall Inc., New Jersey. (1994)
5. M.A. Moon, T. Mentzer, J. of B. Forecasting.18.2.(1999)
6. J. Lawrence, Introduction to neural networks. Design, theory, and applications. Nevada City, CA: California Scientific Software; 1994.

7. L. Medsker, J. Liebowitz, Design and development of expert systems and neural networks. New York: Macmillan College Publishing Company; 1994.
8. L.Wang, Z.-G. Wang, H. Qu, S.Liu. Appl. Soft Computing. (2018) https://doi.org/10.1016/j.asoc.2018.02.004
9. D. Matsypura, R. Thompson, A.L. Vasnev. Omega. (2017). Doi:10.1016/j.omega.2017.06.010
10. D. Nibbering, R. Paap, M.V.D. Wel. Int. J. of Forecasting 3 (2018)
11. S. Benabderrahmane, N. Melloouli., M. Lamolle. Knowledge-Based Systems 000. 1-19 (2018)
12. G. Tkacz. Int. J. of. Forecasting. 17.1.(2001)
13. M.-L. Huang, Y.-H. Hung, Z.-S. Yang, Measurement. 54.(2016)
14. G. Taguchi, E.A. Elsayed, T.C. Hsiang, Quality Engineering in Production System, Mc Graw Hill Book company (1989)
15. K. Ouyang, H.-W. Wu, S.-C. Huang, S.-J. Wu. Energy (2017)

Effect of Zn Interlayer Particles on Mechanical Properties and Microstructure of Friction Stir Spot Welding Aluminum Alloy

Lingga Arti Saputra[1,1,] Nurul Muhayat[2], and Triyono[1]
[1]*Graduate School of Mechanical Engineering, Sebelas Maret University, Surakarta, Indonesia*

Abstract. Global warming can be reduced by controlling emissions in motor vehicles. Lightweight aluminum materials can lower the engine work so as to reduce fuel consumption.The effect of dwell time on mechanical properties and microstructure friction stir spot welded AA1100 with particle interlayer Zn were investigated. The result shows that the particle interlayer Zn affect to the mechanical properties. The tensile shear load friction stir spot welded AA1100 with particle interlayer Zn is higher than hart of without particle interlayer Zn. In line with tensile shear load that the hardness of friction stir spot welded Al with particle interlayer Zn got the higher hardness than of without particle interlayer Zn. The addition of particle interlayer Zn reduce the hook defect and spread in the aluminum matrix as solid solution.

1. Introduction

Global warming can be controlled by reducing the amount of emissions in vehicles[1-3]. The use of lighter and stronger materials is one of method to improve fuel efficiency[4-5]. Welding aluminum material using a resistance spot welding (RSW) on the vehicle frame has many shortcomings. For example, the costs that used to be quite expensive, joint strength is low[6]. Kawasaki developed the method of welding friction is a single point of Friction Stir Join (SFJ) and Friction Stir Spot Welding (FSSW) [7]. FSSW is one of the development of Friction Stir Welding (FSW) process using a system where the friction[7]. FSSW which has the characteristics of solid-state properties that have many advantages over traditional fusion welding techniques[8]. for example, the welding process easier, a low porosity level, is a mechanical joint, can be used for different materials, weight reduction joint, welding low energy and produce less pollution[9].

FSSW can save 90% energy and can save the useof 40% work tools of RSW and RW[10]. FSSW problems began when their decreased ability hooking joint because there are defects[11-14]. Hooking defects can be solved by adding a second interlayer on the surface[15]. Use of Zn interlayer can eliminate hook defects, ICM formation, and grain size changes thereby strengthening the joint[16-18]. Interlayer Zn can be diffused to form structures eutectoid significantly improve the joint[19]. Use of the tool is very influential in the formation of a weld and hook defects[20-22]. Hook defects not completely lost the use pinless tool, but it can increase the usage of the cylinder pin joint and fix defects hook formation[23-24]. Therefore, this study will examine the use of aluminum

[1] Corresponding author: linggalampu@gmail.com

welding FSSW on the addition of Zn in the form of particles with the *interlayer* parameters influence dwell time, welding depth with the use of the pin on the tool size variations shoulder.

2. Experimental

Aluminum material with a thickness of 1.6 mm. materials with chemical composition 0.12%Si, 0.4963%Fe, 0.099%Cu, 0.1265%Mg, 0.2001%Zn, 0.0159%Ti, 0.001% Cr, 0.0005% Ni, 0,0191% Pb, 0,0055% Sn, 98,78% Al. Interlayer using pure Zn particles with a thickness of 0.2mm. **Fig.1** shows the use of pure Zn interlayer particles of 0.2mm thickness. **Fig.2a** Shows the process of placing the Zn particles by using the stopper tool. **Fig.2b** Zn interlayer particles have been composed. **Fig.2c** shows the FSSW welding process. The use of welding tools has the diameter of solder 10mm and 2mm pin diameter. Use of dwell time parameter variation for 20, 30, 40, 50s. FSSW welding specimen with length 125mm x wide 40mm. After the specimen is welded, then cut to see metallographic using Keller's reagent (4ml HF, 6ml HCl, 10ml HNO_3, 180ml H_2So4). Violence value is done by using Micro Vickers using ASTM E384 standard. Tensile/shear load (Universal testing machine). Macrostructure (Olympus/ SZ 1145) scanning electron microscope (SEM, Phenom ProX) energy dispersion spectrometer (EDS). X-ray diffraction (XRD, Shimadzhu 7000).

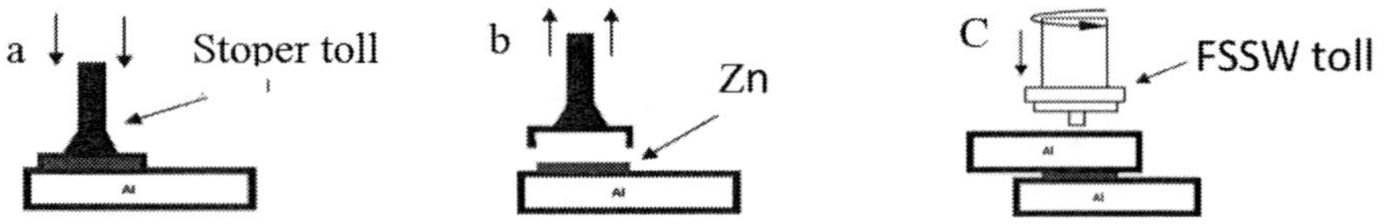

Fig.1. Position placement of Zn interlayer: (a) stopper tool Zn interlayer (b) position Zn particle interlayer (c) welding process

3. Results and discussion

3.1 Mechanical properties and macrostructure of FSSW joints without the addition of Zn interlayer

Fig.2 Show FSSWjoint without using Zn particle interlayer. On the surface of the joint there is no flow of Zn particles. Using pins on the tool can fix the joint [24-25]. But still looks hook defect formed around the joint area. **Fig.4** is the results of the shear tensile testing for the FSSW joint and without the use of a Zn particle interlayer. The highest result of tensile test of FSSW without using Zn interlayer particles has a value of 2.7kN at dwell time 20s with the lowest value of 2.4kN at dwelling time 50s. **Fig.3** Hardness on the surface of the marked joint. Starting point of testing starts from center key hole. Distribution of penetration is the same in every variation. Profile of hardness level distribution with W shape [26]. Hardness value on each part shows a varied result. The highest hardness value was 46.8 HV in the SZ region.

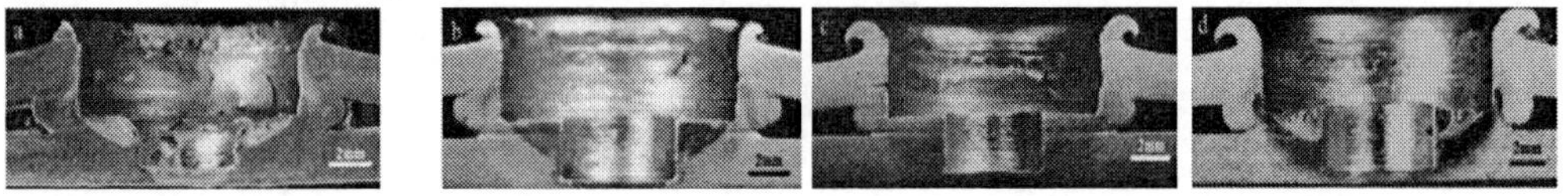

Fig 2. Mackrostruktur of FSSW without the Zn interlayer 10mm(a.10, b.20, c.40s, d 50s)

3.2. Mechanical properties and microstructure of FSSW joints with the addition of Zn interlayer

3.2.1. Mechanical properties

3.2.1.1. Tensile/shear test

Fig.4 showed tensile / shear load in joint without Zn interlayer particlesand with Zn Interlayer particlesFSSW.The addition of FSSW Zn Interlayer particles has a higher tensile / shear load than with Zn Interlayer FSSW particles. An increase in dwell time having results different each variations with Zn Interlayer particles FSSW tensile / shear load highest 3.8kN in dwell time 20s and tensile / shear load lowest 3.1kN in dwell time 50s. Without Zn interlayer particlesFSSW tensile / shear load highest 2.7kN in dwell time 20s and tensile / shear load lowest 2.4kN in dwell time 50s. FSSW welding with Zn Interlayer particlesas a whole provides better joint results than without Zn interlayer particles[25,3, 27].

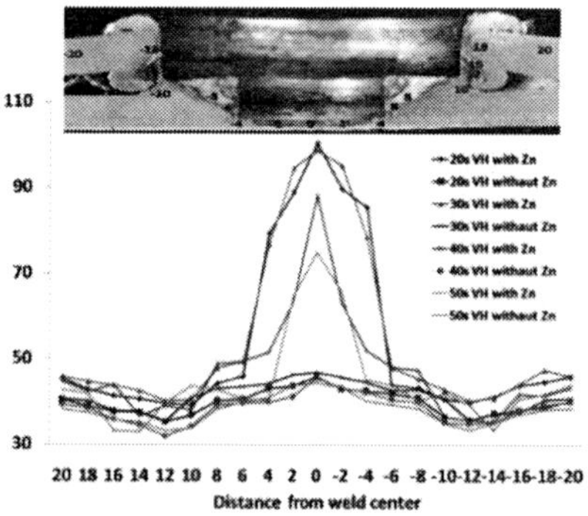

Fig 3. Microhardness of the FSSW with interlayer and without interlayer

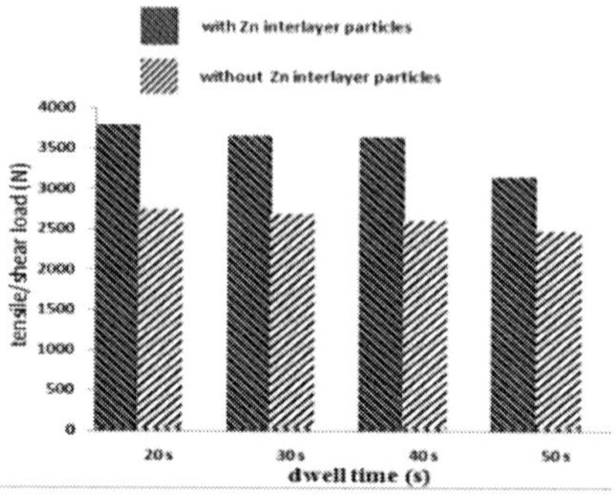

Fig 4. Tensile-shear loads of FSSW

3.2.1.2. Hardness test

Fig.3 overall FSSW with Zn particle interlayer has the highest hardness value of 100.3HV. While the FSSW without Zn interlayer particles only has the highest score of 46.8 HV.The SZ area is the hardest point. Hardness values begin to decrease in the HAZ area and reach the minimum value between HAZ and TMAZ. The value of hardness rises again in the SZ region[14].

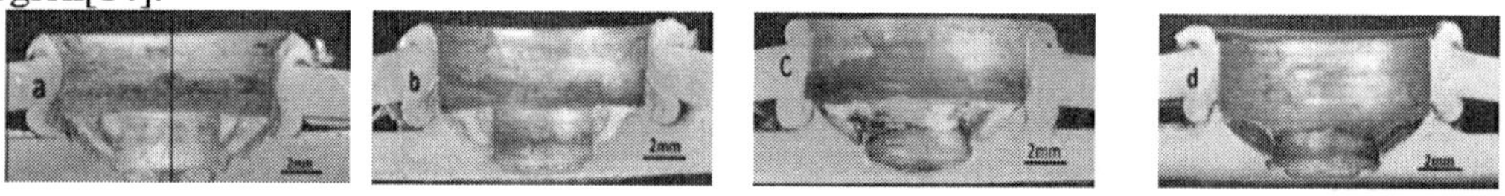

Fig 5. Mackrostruktur of FSSW with the Zn interlayer 10mm (a.10s, b.20s, c.40s, d.50s)

3.3 Microstructure of FSSW Al alloy joints with the addition of without Zn interlayer particles

Microstructure testing is performed to determine the symptoms that arise during the welding process. **Fig.5** shows the difference in dwell time. the mechanical process of welding produces a hot input. Material flow is formed between substrate Al with Zn thus affecting the result of joint[26]. **Fig.5** shows the variation of FSSW joint by using Zn particle interlayer. **Fig4.a**. shows the state of the joint between Al material and the Zn particle interlayer. Zn particles with Al form a reaction zone on the surface of the joint. **Fig 6.a** flow of Zn material in the nugget area appears dominant. Material flow formed due to extrusion process.This process occurs because the pin penetration force pushes the material back toward the toll and rotate the toll that stirs the material [28-29]. **Fig.6.b** shows mapping areas that have different views.

Zn particles are seen scattered around the joint region [16, 20]. The flow of Zn particles has a tendency to move away from the center of the mortar due to the centrifugal effect. **Fig.7.(d).** analysis with EDS. **Fig.7.(d).** shows a material composition comprising 1% Zn of

78at% Al and the remainder of the other material. **Fig.7** The use of Zn interlayer particles gives different effects on each joint[15, 17, 19, 25]. The dispersion of Zn particles reacting with Al forms a reaction area. The direction of dispersion of Zn particles has different characteristics in each welding variation. The dwell time effect makes a difference on each joint. **Fig.5** shows the wide-ranging Zn spread on each variation.

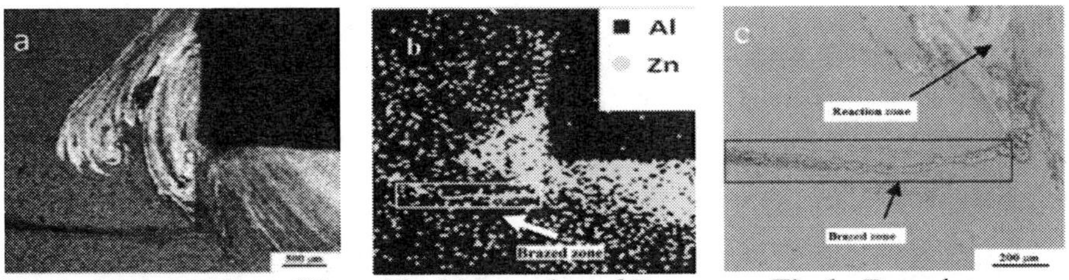

Fig.6.a Material Flow of Zn **Fig.6.b** mapping of Zn **Fig.6.c** Brazed zone

The effect of mechanical rotation on the tool and the heat input produces a reaction on the substrate. **Fig.6.a** Zn dispersion rate in welding results tends to decrease. The effect of dwell time results in a material that is softened longer so that the Zn particles tend toward the middle of the mortar. **Fig.6.c** shows the graph of bonded zone formation. Increased bonded zone due to mechanical pressure process of toll and heat input from friction effect. This process gives a longer time on Zn to soften and fill the gap in the joint. **Fig.7.(f).** XRD results in the transition region show the formation of Al Zn and Al.

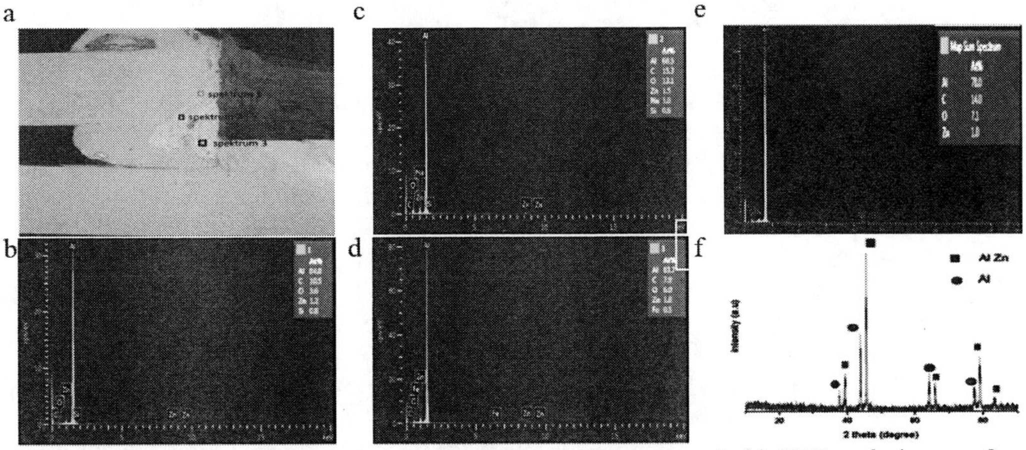

Fig.7.(a) EDS section. (b).spectrum 1. (c). spectrum 2. (d). spectrum 3. (e). EDS analysis on surface mapping. (f). XRD pattern of the interface between Al substrate and Zn interlayer

EDS analysis addresses the distribution of Zn interlayers[17, 19, 25]. **Fig.7.a** addresses the EDS section. The Zn spread differs in each section. **Fig.7.b.** shows the of 1.2% Zn percentage of Al 84.0%. **Fig.7.c** shows a of 1.5% Zn percentage of AL 68.5%. **Fig.7.d** shows a 1.8% Zn percentage of Al 83.7%. Zn material has increased toward the center of the stir zone.

3.3.1 Distribution of Zn interlayer particles

During the welding process occurs mechanical friction that causes heat. The work of the tool produces a mix between Al and Zn. Zn substrate reacts with Al to form a material stream. **Fig.6.b** shows the extent to which Zn is spreading in the HAZ area. This process occurs the mechanism of pressure and rotation of the tool. An increase in the dwell time variable leads to a decrease in the length of Zn. This is due to the tendency of Zn material to flow towards SZ. **Fig.6.c** shows the direction of the Zn substrate undergoing brazing process and closing the hook defect. The increased dwell time and pressure of the tool give the Zn substrate time flowing between the material gaps.

4. Conclusion

In this study, the effect of Zn particle interlayer with dwell time variation on Mechanical Properties and Microstructure of Stir Spot Friction Welding Aluminum Alloy. The following conclusions were drawn.
(1) Highest tensile strength at FSSW joint without using Zn particle interlayer at dwell time 20s with tensile strength of 2.7kN. lowest tensile strength 2.5kN at 50s dwell time. The level of violence in the SZ region with a value of 46.8HV the tensile strength of the joint without the Zn interlayer has a value of 2.7kN.
(2) The highest tensile load at the FSSW joint with the Zn particle interlayer at the right time 20s with a value of 3.8kN. Tensile load increases with time. The highest hardness value in SZ with value 100.3HV.
(3) Zn particles react with Al to form Al-Zn. Hook defects can be reduced so the joint strength increases. The addition of Zn interlayer reduces defects in the joint area. Area mapping Zn particle spreads around the joint area. Zn diffuses with Al formed the material and form of Al Zn.

References

1. Marcia McNutt, "Climate Change Impacts," vol. 341.
2. Y. Liu, Q. Sun, J. Liu, S. Wang, and J. Feng, "Effect of axial external magnetic fi eld on cold metal transfer welds of aluminum alloy and stainless steel," *Mater. Lett.*, vol. 152, no. 2, pp. 29–31, 2015.
3. P. Xue, D. R. Ni, D. Wang, B. L. Xiao, and Z. Y. Ma, "Effect of friction stir welding parameters on the microstructure and mechanical properties of the dissimilar Al – Cu joints," vol. 528, pp. 4683–4689, 2011.
4. J. M. Piccini and H. G. Svoboda, "Effect of pin length on Friction Stir Spot Welding (FSSW) of dissimilar Aluminum-Steel joints," *Procedia Mater. Sci.*, vol. 9, pp. 504–513, 2015.
5. H. Bisadi, A. Tavakoli, M. T. Sangsaraki, and K. T. Sangsaraki, "The influences of rotational and welding speeds on microstructures and mechanical properties of friction stir welded Al5083 and commercially pure copper sheets lap joints," *J. Mater.*, vol. 43, pp. 80–88, 2013.
6. J. Peng, S. Fukumoto, L. Brown, and N. Zhou, "Image analysis of electrode degradation in resistance spot welding of aluminium," vol. 9, no. 4, pp. 331–336, 2004.
7. M. P. Mubiayi and E. T. Akinlabi, "Friction Stir Spot Welding between Copper and Aluminium : Microstructural Evolution," vol. II, 2015.
8. X. Liu, S. Lan, and J. Ni, "Analysis of process parameters effects on friction stir welding of dissimilar aluminum alloy to advanced high strength steel," *Mater. Des.*, vol. 59, pp. 50–62, 2014.
9. K. P. Babu, S., Sankar, V. S., Ram, G. J., Venkitakrishnan, P. V., Reddy, G. M., & Rao, "Microstructures and mechanical properties of friction stir spot welded aluminum alloy AA2014," *J. Mater. Eng. Perform.*, vol. 71–84, 2013.
10. G. Buffa, P. Fanelli, L. Fratini, and F. Vivio, "Influence of joint geometry on micro and macro mechanical properties of friction stir spot welded joints," *Procedia Eng.*, vol. 81, no. October, pp. 2086–2091, 2014.
11. W. Li, J. Li, Z. Zhang, D. Gao, W. Wang, and C. Dong, "Improving mechanical properties of pinless friction stir spot welded joints by eliminating hook defect," *Mater. Des.*, vol. 62, pp. 247–254, 2014.
12. S. Z. Badarinarayan, H., Q. Yang, "Effect of tool geometry on static strength of friction stir spot-welded aluminum alloy," *Int. J. Mach. Tools Manuf. 49.2*, vol. 142–148, 2009.

13. A. Buffa, G., Campanile, G., Fratini, L., & Prisco, "Friction stir welding of lap joints: Influence of process parameters on the metallurgical and mechanical properties," *Mater. Sci. Eng.*, vol. 19–26, 2009.
14. J. Y. Cao, M. Wang, L. Kong, and L. J. Guo, "Hook formation and mechanical properties of friction spot welding in alloy 6061-T6," *J. Mater. Process. Technol.*, vol. 230, pp. 254–262, 2016.
15. M. Miyagawa, K., Tsubaki, M., Yasui, T., Fukumoto, "Spot Welding between Aluminum Alloy and Zn Coated Steel by Friction Stirring," *J. Japan Inst. Light Met. 23*, pp. 131–136, 2008.
16. J. M. Piccini and H. G. Svoboda, "Effect of the tool penetration depth in Friction Stir Spot Welding (FSSW) of dissimilar aluminum alloys," *Procedia Mater. Sci.*, vol. 8, pp. 868–877, 2015.
17. R. Z. Xu, D. R. Ni, Q. Yang, C. Z. Liu, and Z. Y. Ma, "Journal of Materials Science & Technology Pinless Friction Stir Spot Welding of Mg – 3Al – 1Zn Alloy with Zn Interlayer," vol. 32, pp. 76–88, 2016.
18. S. Lazarevic, S. F. Miller, J. Li, and B. E. Carlson, "Experimental analysis of friction stir forming for dissimilar material joining application," *J. Manuf. Process.*, vol. 15, no. 4, pp. 616–624, 2013.
19. R. Z. Xu, D. R. Ni, Q. Yang, C. Z. Liu, and Z. Y. Ma, "Influencing mechanism of Zn interlayer addition on hook defects of friction stir spot welded Mg – Al – Zn alloy joints," vol. 69, pp. 163–169, 2015.
20. S. Hirasawa, H. Badarinarayan, K. Okamoto, T. Tomimura, and T. Kawanami, "Analysis of effect of tool geometry on plastic flow during friction stir spot welding using particle method," *J. Mater. Process. Technol.*, vol. 210, no. 11, pp. 1455–1463, 2010.
21. P. Su, A. Gerlich, T. H. North, and G. J. Bendzsak, "Intermixing in Dissimilar Friction Stir Spot Welds," *Metall. Mater. Trans. A*, vol. 38, no. 3, pp. 584–595, 2007.
22. S. S. Yin, Y. H., Sun, N., North, T. H., & Hu, "Influence of tool design on mechanical properties of AZ31 friction stir spot welds," *Sci. Technol. Weld. Joining.*, 2013.
23. M. K. Bilici and A. I. Yükler, "Influence of tool geometry and process parameters on macrostructure and static strength in friction stir spot welded polyethylene sheets," *Mater. Des.*, vol. 33, no. 1, pp. 145–152, 2012.
24. V. Malik, N. K. Sanjeev, H. S. Hebbar, and S. V. Kailas, "Investigations on the effect of various tool pin profiles in friction stir welding using finite element simulations," *Procedia Eng.*, vol. 97, pp. 1060–1068, 2014.
25. R. Z. Xu, D. R. Ni, Q. Yang, C. Z. Liu, and Z. Y. Ma, "Influence of Zn interlayer addition on microstructure and mechanical properties of friction stir welded AZ31 Mg alloy," *J. Mater. Sci.*, pp. 4160–4173, 2015.
26. P. Xue, D. R. Ni, D. Wang, B. L. Xiao, and Z. Y. Ma, "Effect of friction stir welding parameters on the microstructure and mechanical properties of the dissimilar Al-Cu joints," *Mater. Sci. Eng. A*, vol. 528, no. 13–14, pp. 4683–4689, 2011.
27. A. Boucherit and R. Taillard, "AC PT," *Mater. Des.*, vol. 124, pp. 87–99, 2017.
28. O. Lorrain, V. Favier, H. Zahrouni, and D. Lawrjaniec, "Understanding the material flow path of friction stir welding process using unthreaded tools," *J. Mater. Process. Technol.*, vol. 210, no. 4, pp. 603–609, 2010.
29. Y. C. Lin, J. J. Liu, and J. N. Chen, "Material flow tracking for various tool geometries during the friction stir spot welding process," *J. Mater. Eng. Perform.*, vol. 22, no. 12, pp. 3674–3683, 2013.

MATEC Web of Conferences **218**, 04006 (2018) https://doi.org/10.1051/matecconf/201821804006
ICIEE 2018

Quality Improvement for A Shoe Laundry Using Integration Model of Service Blueprint, Failure Mode and Effect Analysis, and Fault Tree Analysis

Natalia Hartono[1,*], *Andry M Panjaitan*[1], and *Abram Noel*[1]

[1]Industrial Engineering, Faculty of Science and Technology, University of Pelita Harapan, Indonesia

Abstract. Nowadays, shoes are not just a casual footwear. Certain shoes can tell the social class of a person. The increase of shoe prices and social status of wearing expensive shoes became a trigger for the development of shoe laundry services. There were a service quality problems in a shoe laundry in Tangerang, Indonesia. Several methods to improve service quality was studied and it is decided to propose a new model, which is integration model of Service Blueprint, Failure Mode and Effect Analysis (FMEA) and Fault Tree Analysis (FTA). The research starts with identifying the problem with observation and interview, then build Service Blueprint. Based on the fail point that has been identified in Service Blueprint, the FMEA used to find which process is the most dominant cause of failure and the most urgent for improvement. The next step is using FTA to find the root cause of the failure of the dominant cause. After analyzing the FTA, the improvement was proposed and implemented. Service failure before and after implementation was compared to see the improvements. There are 6 suggestion and implemented. After the implementation, the error in each process was measured and it is found a decrease in error in each process.

1 Introduction

Nowadays, shoes are no longer just being footwear because certain shoes are able to show a person's social class and provide value for their own satisfaction because of the high price. For example, Nike Air Vapormax Flyknit iD [1] with the price of 3.8 million rupiah and Adidas Ultraboost X shoes is 3.4 million rupiah [2]. Not to mention the price of limited edition shoes that can be twice the price. These prices are close to regional minimum wage in a month in Jakarta, Indonesia. With the increasingly expensive price of shoes, shoe laundry began to grow like mushrooms in the rainy season in Indonesia. Shoe laundry provides shoe washing with a different method with cleaner and preserves the shoes compare than washing your own shoes.

* Corresponding author: natalia.hartono@uph.edu

Shoe laundry which is become the object of this research is a Small-Medium Micro Company that only has 4 workers including the owner. This shoe laundry has the quality of service problems. One of the obvious problems is mistaken in the process of washing shoes. Based on initial observations and interview, these errors occur due to several factors, such as human factors and tool factors. Errors that occur are varied, such as input errors on queuing registration system, soap used does not match the shoe material and wrong washing technique due to an error in deciding the shoe material. These errors cause delay and raise complaints from the customer, which is a service quality problem of this shoe laundry.

Literature study shows that to improve the quality of service, there are many tools. Service blueprinting is a well-known model of planning the service process as a system and this model was originally introduced by Lynn Shostack [3]. The components in Service Blueprint are able to analyze each service process in detail. A service blueprint is a picture or map that portrays the customer experience and the service system, so that the different people involved in providing the service can understand it objectively, regardless of their roles or their individual points of view [4]. In the service blueprint, sub-processes and activities in a service process are depicted and the relationship between them is established and related to the customer's contact points so it is possible to analyze how a service process can best be designed to support good customer perceived service quality [3]. The blueprint can be analyzed to determine likely failure points or bottlenecks in the process, as well as customer pain points [4]. Service Blueprint needs to be combined with other methods to give more directed quality improvement analysis because there is no analysis of the cause of service failure in Service Blueprint.

There are two methods that widely and commonly used in manufacturing as a quality tool, namely Failure Mode and Effect Analysis (FMEA) and Fault Tree Analysis (FTA). Both of these methods can also be used in the service field. FMEA is a quality tool that allows the user to examine the process from the viewpoint of its potential for failure and plan for its control through service, product or process design requirements so the failure is designed out of the process before it reaches the customer [5]. FTA can be simply described as an analytical technique, whereby an undesired state of the system is specified, depicts the logical interrelationships of basic events that lead to the undesired event [6].

The application of FMEA in service and FTA in service has been done, for example, is research of Geum [7] and Carvalho [8]. Previous studies on service improvements using Service Blueprint, FMEA, and FTA were used separately or combination of two of the three tools. For example, Geum et al using FMEA to improve service productivity, Carvalho et al using FTA in service, Chuang's research that combines Service Blueprint with FMEA, and Bitner et al research using Service Blueprint [7,8,9,10].

This research is the first that integrates Service Blueprint, FMEA, and FTA as an effort to improve the quality of service. Therefore, this research aims to improve quality service for a shoe laundry using an integrated model of Service Blueprint, Failure Mode, and Effect Analysis, and Fault Tree Analysis. The research was conducted from August – December 2017.

2 Methodology

The research starts with identifying the problem with observation and interview in a shoe laundry and after literature study, it is found that there are several tools to improve the quality of service. It is decided to integrate Service Blueprint, FMEA and FTA because of the strength of the tools. Service Blueprint will give a picture of the process and identify fail point and risk of excessive wait of the service. Based on the fail point that has been identified in Service Blueprint, the FMEA used to find which process is the most dominant cause of failure and the most urgent for improvement. The next step is using FTA to find the root cause of the failure of the dominant cause. The proposed integrated model of Service Blueprint, FMEA and FTA depict in Fig 1. After analyzing the FTA, the improvement was proposed and implemented. Service failure before and after implementation was compared to see the improvements.

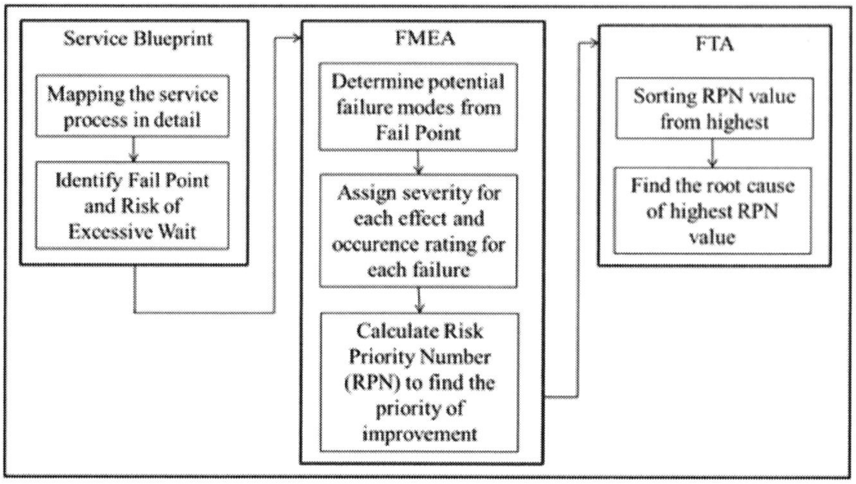

Fig 1. Proposed Integrated model of Service Blueprint, FMEA, and FTA

3 Data Collection and Analysis

A brief description of the shoe laundry that used in this research is that its start in 2016 with only 4 employees, and its location in Lippo Village, Karawaci, Tangerang, Indonesia. The shoe laundry offers a full cleaning service, unyellowing (whitening), recoloring, and repaint.

The first step is taken data to depict all of the processes and identify the fail points and risk of excessive wait where each process is described in the five Service Blueprint components: customer action, physical evidence, the front of stage interaction, back of stage interaction and support processes. Service Blueprint very useful to see how customer involves in the service and identify the fail point and risk of an excessive wait. The results show there are 11 fail points. There is no time data on service processes, so there is no standard time for every process. There are five processes that need to have time standard, so this research collect the data and calculate the standard time. The steps to calculating standard time took the time of the process, validate the data with a normality test, uniform test, and adequacy test, then determine the normal time and standard time. The standard time becomes standard for the shoe laundry to determine their service target time. The fail point used as input in FMEA to find the most urgent process that needs to improve first. Severity and detection value determined based on theory and discussion with the shoe laundry owner while occurrence value from observation. The FMEA table depicted in Table 1

In order to choose which failure mode that will get analyze further, the RPN value is sorted descending from largest to smallest value. The first quartile represents 25% of the highest RPN value in FMEA. The first quartile shows there are 4 processes with highest RPN, but because the 5th in the list has the same RPN value with the 4th, then there are 5 processes that analyze further using Fault Tree Analysis (FTA).

Table 1. Failure Mode Effect Analysis

Process	Failure Mode	Failure Effect	Severity	Occurence	Detection	Risk Priority Number
Input data (customer data, shoes data, shoes service)	Wrong input data in the system	Wrong printed receipt	7	8	3	168
Give receipt that also as slip to took the shoes	Give the wrong receipt	Causing confusion to customer and employee	9	1	3	27
Brushing the shoes (upper with soap "A" and sole with soap "B")	Damaged shoes	Customer dissappointed/angry	7	1	2	14
	Shoes still dirty	Rework	4	10	2	80
Removing spot from rubber surface (usually the sole)	There is still spot	Rework	5	1	3	15
Brushing inner sole and shoelace with soap "C"	Insole or shoelace broken	Customer dissappointed/angry	7	1	2	14
Quality control	Insole or shoelace still dirty	Rework	4	9	3	108
	Dirty shoes pass the inspection	There is still spot	2	9	4	72
Call the customer (mostly Whatsapp)	Call the wrong customer	Causing confusion to customer and employee	1	1	4	4
Give the finished shoes	Give the wrong finished shoes	Causing confusion to customer and employee	5	1	2	10
Write the address	Input the wrong address	The courier got lost so the delivery time is longer than it should be	5	8	2	80
Packing the finished shoes	Misplacement the shoes when packing	Shoes damaged or dirty	5	1	2	10
Send the shoes	The courier got lost	Delay in delivery	5	10	5	250
	Time delivery not appropriate with agreement	Customer is not at home/place	5	1	4	20

FTA used to find the root cause of failure to generate a logical solution to solve the problems. Each FTA made based on the results of discussions with the owner of the shoe laundry and the employee.

1. FTA for The courier got lost

   There is 4 basic events (root cause) of the courier got lost. The first is the lack of knowledge of employees of the intended address. The second root is a human error caused by having health problems or personal problems that interfere with his thinking while working. Through the results of interviews with all employees, personal problems that have occurred is a family problem that ends up becoming a financial problem. Next root is because the previous process inputs the wrong address. The fourth root is a guided error in the GPS application. Every courier who delivered always using GPS applications so that this GPS application becomes crucial. There are 3 causes of error in the GPS application, the first one is because the GPS does not lead to the appropriate path, this can be due to any repair or closure or change of the road because the GPS is not updated. Guidance errors in the GPS application can also be caused by a human error due to incorrect enter the address or wrong destination address.
2. FTA for Wrong input data in the system

   The next type of error is an error entering data into the system. This can lead to a lot of errors in subsequent processes, one of which is the shoe washing process that will be discussed on the next FTA. There are several causes of the process error, that is because it is caused by the employee mishears or the customer is giving the wrong info or not complete info, otherwise, it could be employees not focus when entering data to the system due to physical health problems or personal problems. Another cause is the lack of employee knowledge of the types of shoes. This can be due to lack of training from the owner or also because of the lack of guidance for the employee.

3. FTA for Insole or shoelaces still dirty
   The root cause is the same with shoes still dirty, the difference is that the shoelaces and insole have the same washing treatment, so there is no cause of wrong input data.
4. FTA for shoes still dirty
   This is caused by two major factors namely internal and external. The external factor is from the customer itself, where the shoes already damaged or too dirty and have a permanent stain. This can cause all efforts to clean these shoes is not optimal. Internal factor is an error in the process of washing the shoe itself, this is caused by several factors. The first factor is the lack of supervision, this has the potential to make employees do with perfunctory. The next cause is the focus of employees disturbed due to health problems or personal problems. Next is because the employee is poorly trained in brushing shoes. The cause of the next brushing error is the error of the brushing method, in the process of Full Cleaning Service, the shoes are divided into three, namely general shoes, leather shoes, and suede shoes. The three shoes have their own brushes and slightly different washing ways, this error can be caused by two things, the first is incorrect data entry to the system, because when viewing the Service Blueprint, data input to the system are name and type of shoes, if this is wrong, it cause the wrong choice of washing method. The second error is a mistake to read the SOP order that becomes a guide for washing employees to wash shoes or the employee misread. The next cause is a brush tool that does not work optimally, this could be due to a tool that has been damaged or the tool has never been check periodically.
5. FTA for Input the wrong address
   Errors in the process write the address may cause errors in shoe delivery processes that have been discussed in previous FTAs. Record errors can occur because employees are less focused in recording, even though the info provided is correct and the employee heard correctly. This is because employees have health problems or personal problems that interfere with his mind that has been discussed in the previous FTA. The second cause is the conventional method of recording. Viewing Service Blueprint in attachment 1, the process of recording the address is done via phone, this causes 2 errors that lead to wrong in recording the address, the employee is mishearing the info and the customer is giving wrong information. The manual input can cause confusion, lost, and untidy.

There is 6 suggestion based on the root cause that has been identified with FTA and after discussed with the owner, and they agree to implement the suggestion. After 2 weeks of implementation, the error was measured in each process. The implementations are :

1. Briefing and feedback between owners and employees every day at 9.30 AM
2. Money lending policy for the urgent needs of the employee. This policy is welcomed by employees, but in the period of implementation, no employee has used this facilitation so the effect is not yet known.
3. Method of recording the customer address is changed to send location method using Google maps via WhatsApp application. This method is deemed to be effective and making the delivery process is running smoothly. The problem arises if the customer does not understand how to use send location feature and the employee need to take time to explains. In the implementation, from 50 times observation, this only happens 2 times.

4. Cleaning and maintenance of the tools regularly. The initial plan is to clean the brush after cleaning each pair of shoes, but after implementation, the owner decides to clean the brush every 4 pairs of shoes because it deemed to be more effective.
5. Guidance for the employee about the shoe material. The sample of shoe material placed in a cardboard with the name of the material underneath. the sample. In the implementation, the employee never asked the owner to ensure the shoe material and there is no mistake in deciding the methods for cleaning shoes.
6. Motivation cards containing informative and persuasive sentences about keeping shoes clean for customers. The cards are 9x5 cm in size and put into packaged shoes. The results of this implementation not yet known.

After the implementation, the error in each process was measured and it is found a decreased error in each process. There were 64% error free process after implementation.

4 Conclusion

The proposed model can describe the fail point in the service process in a shoe laundry using Service Blueprint can identify the most urgent process that needs to improve first and identify the root cause using FMEA and find the root cause using FTA, so the solution can be proposed. The implementation shows that there is a decreased of error in the service process, but due to the short time in implementation, it is suggested to measure after some period of time to do a quality check. Further research should use the model in other service setting to give this model generalizability since this is the first research.

References

1. Adidas Shop. https://shop.adidas.co.id/pria/sepatu/training.html (retrieved September 2017)
2. Nike Store. https://store.nike.com/id/en_gb/pw/mens-shoes/7puZoi3?ipp=120 (retrieved September 2017)
3. C. Grönroos, Service Management and Marketing: Managing the Service Profit Logic (John Wiley & Sons, United Kingdom, 2015)
4. V.A.Zeithaml, M.J.Bitner, D.D.Gremler, Services Marketing: Integrating Customer Focus Across the Firm (McGraw-Hill, New York, 2013)
5. C. McCain, Using an FMEA In a Service Setting. Quality Progress **39**, 9 (2006)
6. W.E.Vesely, F.F.Goldberg, N.H.Roberts, D.F.Haasl, Fault Tree Handbook, US Nuclear Regulatory Commission (Washington, DC, 1981)
7. Y.Geum, H.Seol, S.Lee, Y.Park, Application of fault tree analysis to the service process: service tree analysis approach, Journal of Service Management **20**, 4:433-54 (2009)
8. A.A.P.de Carvalho, C.A.Scapin, Service fault tree analysis: its use for improving the efficiency of service processes: Annual Reliability and Maintainability Symposium, Proceeding p. 186-9 (2002)
9. P.T.Chuang, Combining Service Blueprint and FMEA for Service Design, The Service Industries Journal **27** 2 p.91-104 (2007)
10. M.J.Bitner, A.L.Ostrom, F.N.Morgan, Service Blueprinting: A Practical Technique for Service Innovation, California Management Review (2007)

Sustainability in Accommodation Service Industry: The Role of Quality and Value

Tintin Suhaeni[1,a], Adila Sosianika[1]
[1]*Department of Business Administration, Politeknik Negeri Bandung, Bandung, Indonesia*

Abstract. Hotel industry as a provider of tourist accommodation is an essential pillar of the tourism industry. This study is designed to identify how non-star budget accommodation industry sustain its business, through providing service that enables to satisfy and make loyal their customers. Referring to the existing literature, three predictors included in the model of developing guest loyalty, i.e. service quality, service value, and customer satisfaction. Data were collected from 298 domestic tourists in five tourism destinations in West Java; there are Cianjur, Cirebon, Bandung, Sukabumi, and Ciamis. The findings indicate that tourists tend not loyal to budget accommodation because of the service quality that they received does not meet their expectations. Similarly, tourists' perception on the perceived value and level of satisfaction are not as customer expected. In terms of how domestic tourist develop loyalty, the result confirms that service quality and perceived value are important determinants of tourist satisfaction and loyalty. Thus, it is suggested that creating a high service quality and value are imperative for the non-star hotel industry to develop its sustainable business.

1 Introduction

Literature [1, 2] concludes that accommodation service availability that suitable for tourist needs can give a positive impact on tourists intention to revisit a tourist destination. To meet the tourist expectation, hotels should provide a high quality of service that at least matched with their expectation in order they revisit to the same hotel for the next trip and attract more tourists to come. One of the main factors to develop such a quality service is by employing competent and loyal employees [3, 4].

There are many types of the hotel to cater the customer needs of accommodation services with broadly can be classified into hotels with star-rating and non-star-rating or budget hotel. Based on the number of employees, capital, and its sales turnover, non-star-rating hotels can be categorized into small-medium enterprises. Like other small-medium enterprises, limited human resource and capital make these non-star hotels difficult in providing excellent service to the customers [5]. Meanwhile, star-rating hotels which do not face those problems tend to be able to provide a high-quality service and delivering program to create guest loyalty. On the other hand, statistical data [6] shows that tourists who stay in star-rating hotels are less than the number of tourist in non-star-rating hotels, 22,6 million compared to 37,6 million. Further, the data indicates that 75% tourists who

[a] Corresponding author : tintin.suhaeni@polban.ac.id

stay in non-star-rating hotels are domestic tourists. Therefore, domestic tourists are the potential market for non-star hotels.

Due to the high potential of Indonesian tourism and a large number of domestic tourists, the development of non-star hotels should be an important issue for the hotel industry. This research is intended to explore guest satisfaction and guest loyalty as a sustainable business strategy for non-star hotel.

2 Literature review

2.1 Concept of loyalty

Arguably, having satisfied and loyal customer is one of the most fundamental issues for flourishing and sustaining the business, including in the accommodation industry [7]. Customer loyalty concept has been developed and tested by many researchers in the various industry context. Oliver [8] defines loyalty as a psychological commitment to keep buying the same product or services in the future, even though there is a persuasion from another competitor to switch buying their products. This definition emphasizes that the proper attitude towards a product or service not only driving customers to rebuy in the future but also make customers resistance to the competitors marketing efforts [9].

There are three aspects of customer intention to be loyal: cognitive, affective, and conative. Cognitive refers to customer trust and knowledge regarding a phenomenon [10]. Scholars [11] define cognitive loyalty as customer attitude towards a product or service based on their trust and knowledge about a product being preferred than competitor product. Affective loyalty is defined as a liking based on satisfaction. Scholars [8] maintain that loyalty is built based on affection concept, meaning that the whole of customer evaluation towards a brand consist of involvement, liking, and caring. Those factors occur because of satisfaction in using a product or service. Conative loyalty or behavioral intention is a loyalty state that triggers a commitment to buy something that appears first. Commitment to purchase a product will be influenced by customer preference to the product. As a consequence, having customers who commit is very important for every business and become a key for its sustainability, because they will be resistant to competitor promotion. Scholars [2, 12] report that commitment will differentiate customer who keep buying the product with loyalty and who is not.

2.2 Loyalty determinant factors

Literature has widely recognized that the main driver of loyalty is not only service quality but also value and satisfaction with the product or service. The quality of service refers to customer opinion about the superiority of service as a whole [13]. Past studies have verified the association between quality and satisfaction, value, and loyalty [2, 12, 14]. Basically, scholars have agreed that quality of service is a critical factor that influences on customer perceived service value, level of satisfaction, and customer favorable behavior. Perceived service value is a customer evaluation towards something that they receive based on what they already give [15]. Many studies have been conducted to analyses service value because of its favorable effect on both of customers and employees. Most of the researchers confirm a positive effect of service value towards the satisfaction level and loyalty behavior. Customer satisfaction is a significant factor that will influence to customer purchasing behavior, profit rate, and company stock value. The recent studies, using multivariate models, have analyzed and reported the positive consequences of satisfaction on customer loyalty behavior [16, 17].

2.3 Loyalty in accommodation industry

The latest research about client loyalty in hotel industry commonly use some determinant factors such as satisfaction, involvement, perceived value, switching cost, quality and the other related factors [5, 18]. In line with researchers that commonly use bivariate analysis approach, most of the studies measure customer loyalty using conative loyalty [2, 19]. Those studies conclude that creating loyal customer can be developed through providing a high service quality and a high perceived service value that can satisfy customer need. Back [20] is the only identified researcher who uses three elements of attitude, there are a cognitive aspect, affective aspect, and conative aspect, to measure customer loyalty in the star-rating hotel.

A comprehensive study that tests the association between behavioral intention, quality of service, satisfaction, image and value of service in Taiwan hotel industry has been conducted by Clemens and colleagues [21]. Their study aims to get an exhaustive consideration of "loyalty battery". The results of their research emphasizes the important of satisfaction toward the service, service value, quality of service, and image for deciding customer intention to being loyal in the future. Other scholars [21] also report how service quality hierarchically influences customer intention to behave either directly or indirectly toward the accommodation service provider. Those studies specifically discuss behavioral intention (conative loyalty), therefore, in essence, they only focus on conative aspect of attitudinal loyalty. The other research on loyalty in the hotel industry is conducted by other scholars in star-rating hotels. Even though those researchers be able to elaborate all of the determinant factors for loyalty behavior, the effect of those factors cannot be generalized in budget accommodation. Thus, it is important that this current study focuses on satisfaction and loyalty as a sustainable strategy for non-star rating hotel.

3 Research method

The measurement of variable constructs applied is referred to the recent literature. Attitudinal loyalty is defined as customer dispositional commitment levels towards a hotel. Four item that adopted from past studies [20][16] used for measuring this construct. While service quality is defined as a complete evaluation of all the service that received by hotel customer during their staying. This variable is measured by 11 items adapted from past studies [16]. Those items adopted as they are consistent with SERVQUAL dimensions; there are tangible, reliability, assurance, empathy, and responsiveness. Customer satisfaction is measured by two factors that adapted other studies [12, 22]. The last factor, customer perceived service value is gauged by two indicators [12]. The population of this research is consisted of all guests who stayed in accommodation service providers (non-star hotel) in five main tourism destinations in West Java; there are Cianjur, Cirebon, Bandung, Sukabumi, and Ciamis. As it is almost impossible to apply random sampling, this study uses convenience sampling to select the respondents. This research is intended to assess satisfaction and loyalty as the basis for developing sustainable in non-star hotel service industry. The skewness and kurtosis tests show that the data distribution is not normal. Therefore, the estimation model of the relationship between variables was conducted by employing logistics regression method. Prior to testing the relationships, the loyalty data was transformed into binary data, loyal and not loyal.

4 Data analysis and result

The respondents are 298 domestic tourists collected from a various region in West Java. As there are four variables used (loyalty, satisfaction, quality, and value), the number of the data is adequate for using multiple regression to analyze the data [22]. The result of descriptive statistics test towards four tested variables is shown in Table 1. Refer to this analysis result, this finding notes that tourist perception of quality, value, level of satisfaction and loyalty toward the non-star hotels service is relatively low. This result is indicated by the average score in all variables less than 4. This finding suggests that domestic tourists towards non-hotel or budget-accommodation service provider are not loyal. In other words, the domestic tourists do not consider that the accommodation they stay as their first choice when coming again to the same destination.

To identify factors that lead customers loyal toward non-star hotel services, the analysis was conducted using logistics regression analysis. The result shows that the impact of service value and service quality individually towards customer satisfaction is shown in Table 2. Wald-Test is used for this testing, showing that service quality significantly affects customer satisfaction ($p<0.01$). In addition, service value does not significantly affect customer satisfaction since its significance level at $p>0.05$. The odd ratio (Exp B) value of service quality is very high, 100,755. This suggests that service quality will 100 times affect customer satisfaction.

Wald-Test result is depicted in Table 3 informs the effect of the quality of service and satisfaction individually on perceived service value. This test shows that quality significantly influences service value with significance level at $p<0.05$. While customer satisfaction does not significantly affect service value because its significance level is at $p>0.05$. The next Wald-Test was conducted to explain the impact of quality, satisfaction, value individually towards customer loyalty. The result of this assessment is depicted in Table 4. Based on the test result, all of those variables significantly affect customer loyalty with significance level at 0.009, 0.000, and 0.003 respectively. Further, service quality has the highest value of odd ration (Exp B) compared to other variables.

The result of correlation test shows that quality satisfaction, value, and customer loyalty are correlated and affected each other. This finding is supported by the result of logistics regression. Logistics regression shows that the higher the service quality, the higher the level of customer satisfaction. In the logistics regression test for customer satisfaction variable, the significance of service quality is at $p<0.01$, meaning that service quality is a key element to improve satisfaction. This result indicates that in case of non-star hotels, the quality of service, perceived service value, and satisfaction are imperative factors in determining customer loyalty behavior.

Table 1. Variable analysis based on the place where data collected.

City	Service Quality		Service Value		Customer Satisfaction		Customer Loyalty	
	Mean	Std. Dev	Mean	Std. Dev	Mean	Std. Dev	Mean	Std. Dev
Cianjur	4,363	0,5940	4,180	0,8850	4,160	0,8597	4,423	0,6812
Cirebon	3,209	0,4658	3,170	0,8841	3,112	0,6635	3,155	0,7871
Bandung	3,448	0,6521	3,393	0,7432	3,186	0,9135	3,152	0,8217
Sukabumi	3,652	0,5892	3,625	0,7033	3,385	0,6937	3,516	0,6602
Ciamis	3,655	0,6462	3,500	0,7354	3,570	0,8452	3,447	0,7013

Table 2. Variables in the equation customer satisfaction.

		β	S.E.	Wald	df	Sig.	Exp(β)
Step 1[a]							100,755
	Value	0,308	0,347	0,785	1	,376	1,360
	Constant	-15,512	2,645	34,391	1	,000	,000
a. Variable(s) entered on step 1: Servqual, Value.							

Table 3. Variables in the equation service value.

		β	S.E.	Wald	df	Sig.	Exp(β)
Step 1[a]							11,237
	Satisfaction	-,191	,393	,238	1	,626	,826
	Constant	-6,050	1,364	19,683	1	,000	,002
a. Variable(s) entered on step 1: Servqual, Satisfaction.							

Table 4. Variables in the equation customer loyalty.

		β	S.E.	Wald	df	Sig.	Exp(β)
Step 1[a]							9,008
	Satisfaction	1,166	,397	8,620	1	,003	3,209
	Value	,935	,356	6,911	1	,009	2,548
	Constant	-13,225	2,086	40,197	1	,000	,000
a. Variable(s) entered on step 1: Servqual, Satisfaction, Value.							

5 Conclusion, implication, and future research

The contribution of this research is to extend our understanding on customer loyalty model especially in non-star hotel accommodation service and identifying the driving factors of customer loyalty with its three determinant factors: the quality of the service, service value, and satisfaction level. As both satisfaction and loyalty are the basic requirements of sustaining any business organization, including in accommodation service industry, this research contributes in highlighting the pivotal role of delivering a high quality of service and creating a high perceived service value as a strategy to sustain any non-star hotel service provider. From a theoretical perspective, this study contributes to extending our understanding that delivering a quality service, creating a high service value that can satisfy customer are imperative factors in creating guest loyalty. As limited studies have been conducted in non-star hotel or budget accommodation context, thus, this paper not only confirms the result of previous studies in star hotels but it also offers a new empirical evidence in the context of budget accommodation services.

From a managerial perspective, this research provides non-star hotel service managers with the basis for developing, flourishing, and sustaining their accommodation business, especially in this Industry 4.0 era. This research reveals that the domestic tourist did not consider that accommodation service provider where they stay as their first choice when revisiting to the destination. The hotel service that they received does not make them want to choose the same accommodation or recommend it to their friends or relatives. To develop guest loyalty (revisit and recommend the hotel), non-star hotel managers should be aware the importance of delivering a high service quality and providing a high service value. In particular, this research recommends that the accommodation services should be improved with stressing on employee service skill and professionalism that can make guests trust to the hotel. Besides employee service skill and appearance, the hotel appearance including neatness, interior and exterior design, and hotel facilities that meet customer needs have to be the attention of hotel managers.

This study has several drawbacks, although it does contribute in terms of theoretical and managerial application. The first drawback related to the data gathered from a few tourist destinations in West Java province. The limited collection of the data from a few tourist destinations causes that the finding of this study cannot be generalized to other tourist destinations. The future study could replicate this research in other tourist location as the different location has different guest behavior due to the cultural differences. The second drawback is related to the model tested with including quality, value, and satisfaction as the loyalty drivers. Besides these factors, many more variables that potentially influence guest loyalty such as brand image, destination image, and customer motivation [2, 5]. Including these variables will increase the predictability of loyalty model in non-star hotel industry.

References

1. C. Jesus and M. Franco, "Cooperation networks in tourism: A study of hotels and rural tourism establishments in an inland region of Portugal," *Journal of Hospitality and Tourism Management,* vol. 29, pp. 165-175, 12 (2016)
2. J. Kandampully, T. Zhang, and E. Jaakkola, "Customer experience management in hospitality: A literature synthesis, new understanding and research agenda," *International Journal of Contemporary Hospitality Management,* vol. 30, pp. 21-56, (2018)
3. C. F. Lin and Y. Y. Lin, "Internal and external marketing for exotic restaurant," *Journal of Foodservice Business Research,* vol. 13, pp. 193-216, (2010)
4. D. Suhartanto, D. Dean, R. Nansuri, and N. N. Triyuni, "The link between tourism involvement and service performance: Evidence from frontline retail employees," *Journal of Business Research,* vol. 83, pp. 130-137, (2018)
5. A. S. Hussein, R. D. V. Hapsari, and I. Yulianti, "Experience quality and hotel boutique customer loyalty: Mediating role of hotel image and perceived value," *Journal of Quality Assurance in Hospitality & Tourism,* pp. 1-18, (2018).
6. BPS. Jumlah tamu asing pada hotel non-bintang menurut provinsi tahun 2003-2010 [Online]. Available: http://www.bps.go.id/tab_sub/view.php?tabel=1&daftar=1&id_subyek=16&notab=6
7. R. Hapsari, M. D. Clemes, and D. Dean, "The impact of service quality, customer engagement and selected marketing constructs on airline passenger loyalty," *International Journal of Quality and Service Sciences,* vol. 9, pp. 21-40, (2017)
8. R. L. Oliver, "Whence consumer loyalty?," *Journal of Marketing,* vol. 63, pp. 33-44, (1999)
9. D. Suhartanto, "The role of store coopetition and attractiveness on the performance of tourism destination and its retail stores," *International Journal of Tourism Policy,* vol. 7, pp. 151-165, (2017)
10. A. H. Eagly and S. Chaiken, "The advantages of an inclusive definition of attitude," *Social Cognition,* vol. 25, pp. 582-589, (2007).
11. D. Suhartanto, Ruhadi, and N. Triyuni, "Tourist loyalty towards shopping destination: The role of shopping satisfaction and destination image," *European Journal of Tourism Research* vol. 13, pp. 84-102, (2016).
12. B. Chitty, S. Ward, and C. Chua, "An application of the ECSI model as a predictor of satisfaction and loyalty for backpacker hostels," *Marketing Intelligence & Planning,* vol. 25, pp. 563-580, (2007).
13. C. Lovelock and J. Wirtz, *Service Marketing - people, technology, strategy*, 6 ed. Upper Saddle River: Pearson Prentice Hall, (2007).
14. V. A. Zeithaml, L. L. Berry, and A. Parasuraman, "The bahavioral consequences of service quality," *Journal of Marketing,* vol. 60, pp. 31-46, (1996)
15. J. J. Cronin, M. K. Brady, and T. M. Hult, "Assessing the effects of quality, value, and customer satisfaction on consumer behavioral intentions in service environments," *Journal of Retailing,* vol. 76, pp. 193-218, (2000)
16. X. Han, R. J. Kwortnik, and C. Wang, "Service loyalty: An integrative model and examination across service contexts," *Journal of Service Research,* vol. 11, pp. 22-42, (2008).

17. D. Suhartanto, "Tourist satisfaction with souvenir shopping: evidence from Indonesian domestic tourists," *Current Issues in Tourism,* vol. 21, pp. 663-679, (2018)
18. A. S. Lo, H. H. Im, Y. Chen, and H. Qu, "Building brand relationship quality among hotel loyalty program members," *International Journal of Contemporary Hospitality Management,* vol. 29, pp. 458-488, (2017).
19. Y. Ekinci, G. R. Massey, and P. L. Dawes, "An extended model of the antecedents and consequences of consumer satisfaction for hospitality services," *European Journal of Marketing,* vol. 42, pp. 35-68, (2008)
20. K. Back, "The effect of image congruence on customers brand loyalty in the upper middle-classs hotel industry," *Journal of Hospitality & Tourism Research,* vol. 29, pp. 448-467, (2005)
21. M. D. Clemes, J. H. Wu, B. D. Hu, and C. Gan, "An empirical study of behavioral intentions in the Taiwan hotel industry," *Innovative Marketing,* vol. 5, pp. 364-390, (2009)
22. D. Suhartanto, D. Dean, A. Sosianika, and T. Suhaeni, "Food souvenirs and their influence on tourist satisfaction and behavioural intentions," *European Journal of Tourism Research,* vol. 18, pp. 133-145, (2018)

The effects of drying time during manufacturing process on partial discharge of 83.3 mva 275/160 kv power transformer

Siswanto[1], *Agus* Indarto[1], *Radin* Rahmatullah[1], *Chairul* Hudaya[1,2*]

[1]Electric Power and Energy Studies, Department of Electrical Engineering, Faculty of Engineering University Indonesia, Kampus Baru UI Depok 16424, Republic of Indonesia
[2]Energy System Engineering, Faculty of Engineering University Indonesia, Kampus Baru UI Depok 16424, Republic of Indonesia

Abstract. Power transformer (PT), essential equipment in electrical power system proceeds complicated manufacturing process to meet high standard design. During manufacturing process of PT, the drying stage is of importance to remove excessive moisture contaminant on the solid insulation. This study investigated the effects of drying duration on partial discharge test of 83.3 MVA 275/160 kV PT. The partial discharge test measurements were carried out using IEC 60270 standards comprising of the capacitive network model based on the main quantity apparent charge concept. We found that the lowest partial discharge of 21 pC was achieved at 68 hours drying duration, while the highest value of 60 pC was reached at 81 hours drying time. The longer drying time may lead to initial degradation of cellulose insulation due to excessive heating. The amount of water extracted during drying effect to partial discharge also investigated. We conclude that the drying time at a higher and lower than 68 hours significantly contributes to the increased partial discharge values of PT.

1 Introduction

Power transformer (PT) constitutes a major portion of the capital equipment of the power grid and has significant role on the reliability of the power supply which heavily depends on the fault free-operation of transformer. The transformer failure rates have been found to follow "bathtub" curve composed of early infant mortality as the first stage. Infant mortality stage contributes to the extreme decreasing failure rate arising from imperfect material or lack of control process during manufacturing [1,2]. On the other hand, the statistical survey has shown that substantial portion of the transformer failures (~ 48%) are due to the failures on winding insulation system [2,3]. Thus, the transformer manufacturing process significantly determines the quality of the winding insulation and imposes the power transformer life.

Large power transformers (PT) proceed the complex design and take considerable time of complicated manufacturing process to meet tight standard design [4,5]. During this

* Corresponding author: c.hudaya@eng.ui.ac.id

process, special attention must be given to the drying process to remove the excessive moisture contaminant from solid insulation. The solid insulation of PT is generally made of cellulose (C6H10O5)n which is naturally hygroscopic [5]. It tends to excessively absorb additional moisture through exposure to humid environment especially on the tropical regions. The solid insulation shall be dried before it is impregnated with oil liquid insulation. In some cases, the excessive residual moisture is left on the insulation during drying process due to defect material, imperfect design or control process [5,6]. This residual moisture even in small quantity is extremely damaging for PT insulation, as it will reduce the dielectric withstand strength and accelerate the ageing process through chemical reaction which produces more moisture contents [6,7]. This phenomenon is also considered as the initial cause that lead to the partial discharge (PD) occurrence. The PD activity is classified as serious failure mode leading to the insulation breakdown failure [8]. It is therefore very important to keep the residual moisture to be very low concentration through the optimized drying duration during the manufacturing process.

In recent years, the increasing awareness on the transformer ageing have led the customers demanded the PT specifications with PD testing lower than the acceptance criteria of the internationally applicable codes and standards. To the best of our knowledge, there are no reports observing the effects of the drying duration on partial discharge of 83,3 MVA 275/160 kV power transformer. For this reason, the optimum drying time process to achieve lowest partial discharge testing was investigated. This paper will address the following research questions: (1) how many hours required for the drying time to achieve the lowest partial discharge testing value? (2) how is the relationship between the extracted water content, residual moisture and the partial discharge? (3) how to optimize the drying time of PT to achieve the efficient manufacturing process and fulfil the customer requirements on PD testing. These critical questions are utmost important for customer to keep the PD as the initial ageing prominent symptoms to a very minimum value and optimize the manufacturing process of PT, making the manufacturer easier to meet the expected delivery date.

2 Methodology

In this study, six (6) identical and newly developed PTs with specification shown on the table 1 were investigated and tested at the factory, in Indonesia. We useed the study cases emphasizing the drying process and partial discharge testing during the manufacturing process.

Table 1. Power transformer specification

Description	value	units
Nominal rating	83.3	MVA
Voltage rating	275/160	kV
Group Vector	Dyn11	-
Number of phase	1	-
Frequency	50	Hz
Impedance	14.5	%
Solid insulation weight	~3500	kg

2.1 The drying process during manufacture

In our experimental work, the vapor phase drying (VPD) method was employed to remove the moisture contaminant within the cellulose insulation, as shown in Figure 1 [9]. Briefly, the method combines heat and vacuum-based apparatus simultaneously to achieve faster and more efficient drying process. A special vapor of organic liquid with elevated temperature of 110°C was used as heat precursors. When the hot vapor circulated over the winding assemblies, it would transfer heat to the solid insulation parts until the moisture of winding insulation was removed by mass diffusion from inner layer to the outside surface of insulation, evaporated as water vapor. The water vapor and condensed organic liquid taken out by vacuum pumping system was then condensed and the separated organic liquid was recirculated on the next drying cycle. By lowering the pressure of the VPD chamber and maintain constant heating temperature, it allowed water to evolve from liquid to vapor state reducing the required thermal energy. The vacuum pressure was set to 10^{-3} bar.

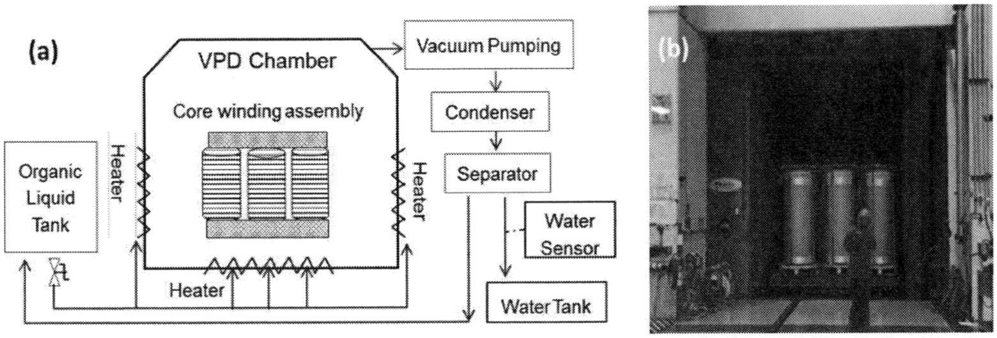

Fig. 1. (a) VPD chamber for drying process during manufacturing of the power transformer. Figure is redrawn from [9] with some adjustment according to PT manufacturer VPD facility, (b) Photo of real VPD chamber for drying the power transformer.

2.2 Methods for partial discharge testing

The PD testing and measurements were carried out in compliance with IEC 60270 standard [10]. The PD was modelled as a capacitive network based on the apparent charge concept, known as ABC model.

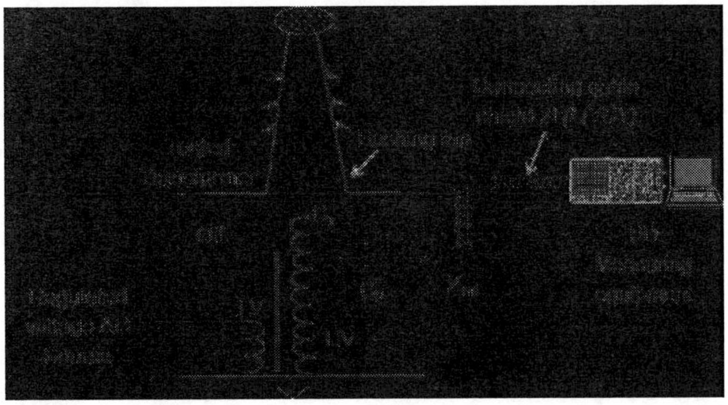

Fig. 2. PD testing circuit with bushing tap coupling method, C1 = bushing capacitance, C2 = capacitance between bushing with grounding, Zmi = measuring input impedance. Figure is redrawn from [11] with some adjustment.

Bushing tap coupling method was used to connect the measuring impedance between earthed transformer tank and bushing tap terminal. The commercial PD measuring equipment Omicron MPD 600 is connected to the terminal through the shielded coaxial cables (LiYCY) to measure and record PD. A gradually raise of AC voltage up to ~325 kV was applied across the low voltage (LV) winding, following the test voltage procedure stipulated on the IEC-60076-3 [11]. The circuit configuration test set-up for an induced voltage test of PT using bushing tap coupling method for PD measurement is shown in Figure 2.

3 Results and discussion

3.1 Effects of drying period on the partial discharge

The drying period represents the accumulation of thermal energy involved in the heat transfer mechanism absorbed by solid insulation. It must remove the excessive water contaminant without degrading the cellulose properties due to excessive heating. The moisture diffusion inside cellulose structure was a very slow process which can be modelled by Fick's law

$$\frac{\partial C}{\partial t} = D(C,T)\frac{\partial^2 C}{\partial x^2} \tag{1}$$

where $D(C,T)$ is the moisture diffusion coefficient in solid insulation (m^2/s), C is the local moisture concentration (mol/m^3)and T is the temperature (°C). The PTs have different drying period ranging from 59-82 hours depending on the water extraction rates. Figure 3 shows the relationship between maximum partial discharge apparent charge magnitude and different drying period. The PD testing results at drying duration both shorter and longer than 68 hours exhibited an increase trend. Polynomial regression analysis is applied to find the optimum drying time as follows:

$$\frac{\partial y}{\partial x} = \frac{\partial(0.1657x^2 - 22.339x + 782.15)}{\partial x} = 0 \tag{2}$$

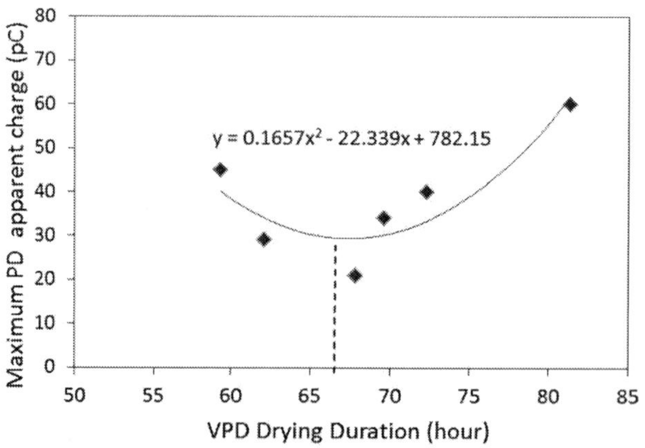

According Lawson *et al.*, the excessive thermal energy lead to initial ageing processes which cause physico-chemical structure changing on the cellulose insulation material [12].

Fig. 3. PD apparent charge test result with different drying period

The excessive drying thermal energy accumulations absorbed by cellulose solid insulation have adverse effect on cellulose structure as it will break the polymer chain and decrease the degree of polymerization value. The shortened polymer chain will reduce the physical properties of the cellulose insulation then reduce the PD inception voltage. Thus, the PD apparent charge tests increase as initial ageing symptom at the drying period longer than the optimum time. While the drying time less than 68 hours does not have enough thermal energy to remove the excessive residual in the solid insulation, which leads to the increasing PD apparent charge result.

3.2 Effects of water extraction on partial discharge

In principle, the water moisture is conductive material which has low resistivity so that easily conduct the electric charge. We found that the highest PD test result achieved at the water extraction of 146 gram, while the lowest PD test result reached at the water extraction of 2,232 gram. The amounts of water extracted during drying process determine the insulation residual moisture condition and have inverse proportional with PD test result as the residual moisture will increase the apparent charge conductivity, as reported by Suleiman and colleagues [13]. They found that the conductivity of kraft paper cellulose insulation increase proportionally with the moisture content values, depending on the dielectric response polarization dan depolarization current measured. Similar finding was also observed in [14], in which the surface partial discharge on the oil impregnated paper increased in linear with the moisture content during the experimental works. Figure 4 shows the relation between amounts extracted water content with PD apparent charge test results.

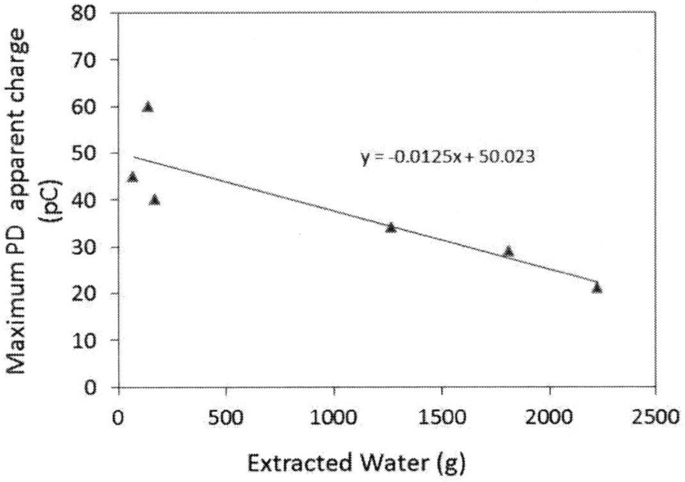

Fig. 4. PD apparent charge test result with different water extraction.

4 Conclusions

The effects of drying time and extracted water content during manufacturing process on partial discharge of 83.3 MVA 275/160 kV power transformers have been investigated. We found that the two process parameters significantly affected to the PD results. We observed that the lowest partial discharge of 21 pC was achieved at 68 hours of drying duration, while the highest value of 60 pC was reached at 81 hours of drying time within the acceptance criteria. Using polynomial regression, it was concluded that the optimum time was 67.4 hours. The longer drying time might lead to the cellulose initial degradation due to accumulated excessive heating, meanwhile the shorter drying period than optimum one might not effectively remove the excessive residual moisture content which tend to increase the PD values. The amounts of water extracted during drying process determine the insulation residual moisture condition and have inverse relationship with PD results, since the residual moisture content would increase the apparent charge conductivity. We conclude that the drying time at a higher and lower than 67.4 hours significantly contributes to the increased partial discharge values of PT. The future research could be focused on the effect of the other parameters i.e. impregnation pressure, solid insulation press-board forming to investigate the dominant parameter affecting the partial discharge testing.

Acknowledgments

The authors would like to thank the financial support provided by Universitas Indonesia through the 2018 PITTA funding scheme managed by Directorate Research and Community Engagement (DRPM) Universitas Indonesia.

References

1. V.I. Kogan et al., "Failure analysis of EHV transformers", *IEEE Transaction on Power Delivery*, **3,** 672 (1988)
2. S. Tenbohlen, J. Jagers, F. Vahidi, "Standardized survey of transformer reliability", *Conference Preceeding of ISEIM*, 593 (2017)
3. Daniel Martin, Judith Marks, Tapan Saha, "Survey of Australian PT failures and retirements", *IEEE Electr. Insulation*, **33**, 16 (2017)
4. D.J. Allan, A. White, "Transformer design for high reliability", *IEE Power Engineeering Jurnal*, **5** ,66 (1995)
5. James Harlow, *Power Transformer Engineering* (CRC Press, Boca Raton, 2012)
6. J. Gielniak et al., "Moisture in cellulose insulation of power transformers – statistics", *IEEE Trans. on Dielectric and Elect. Insulation*, **20**, 982 (2013)
7. CIGRE Working Group A2.30, "Moisture equilibrium and moisture migration within transformer insulation systems", Technical Brochure no 349 (2008)
8. Peter H.F Morshuis, "Degradation of solid dielectric due to internal PD: some thought on progress made and where to go now", *IEEE Trans. on Dielectric and Elect. Insulation*, **12**, 905 (2005)
9. LE Feather, "Drying and oil impregnation of PT insulation", *IEEE Elect. Insulation Conference*, no. 32C3-8 (1965)
10. IEC 60270, *HV test techniques - Partial discharge measurements*, (2000)
11. IEC 60076-3, *P.T.- 3: Insul. levels, dielectric tests and ext. clearances in air*, (2013)
12. W.G. Lawson, M.A. Simmons, P.S. Gale, "Thermal ageing of cellulose paper insulation", *IEEE Tran. on Elect. Insulation,* **12**,61 (1977)
13. A.A. Suleiman et al, "Moisture effect on conductivity of kraft paper immersed in PT vegetable based insulation oils", IET Generation, Transmission & Distribution, **11**, 2269 (2017)
14. Lianwei Bao et al., "Influences of moisture content on surface PD behaviors of OIP insulation under pulsating DC voltage", *IEEE HV Eng. And Application*, (2016)

Repair of The Bracket and Clutch Mechanism On Centrifugal Pump For Firefighter Motorcycle

Triyono[1,] , *Himawan* Hadi Sutrisno [1,*]
[1]Fire Safety Engineering Department, Universitas Negeri Jakarta, Indonesia

Abstract. An addition of centrifugal pump on the engine of automatic motorcycles is expected to become one of the solutions for fire incidents in densely populated areas in big cities. Many cases of fire-related incidents have not been handled properly due to limitations of access for fire engines such as narrow alleys. The effectiveness of the centrifugal pump for firefighter motorcycles is determined by the optimum function of the pump performance according to the desired specification. The repair of the brackets and the clutch mechanism for the pump in the engine of automatic motorcycles may assist to the full effectiveness of the pump. The repair is based on the stability of the number of cycle on the centrifugal pump during the cycle of the motorcycle engine without any load. There is no disruption to the motorcycle cruise control or an increase of load due to the addition of centrifugal pump in the motorcycle's engine. By using the magnetic clutch which uses the electrical energy in the connection and disconnection of the joint, this model will be useful since it does not disrupt the motorcycle cruise performance and does not add an extra load to the motorcycle torso.

1 Introduction

Densely populated areas that are commonly found in big cities such as Jakarta are the areas in which the incidents of fire are the highest. The most common causes of these fire incidents are short circuits, followed by gas cylinder explosions. Based on the data released by the fire department office in Jakarta, the incidents of fire from January to July 2017 can be seen in the graph as shown in Fugure 1.

From the reference above, that 28-78 fire incidents happened in Jakarta last year and 90% occurred in houses located in densely populated areas. One difficulty that the fire department has to deal with in the cases of fire occurred in densely populated areas is that the fire trucks have limited accesses to the fire location. By adding a centrifugal pump to the motorcycle engine, the firefighter motorcycle is then created and hoped to become one of the applicable solutions to the fire in urban areas [1]. Although the capacity of the centrifugal pump in a motorcycle is still far below the capacity of the centrifugal pump of a fire truck, the dimension of the motorcycle becomes an advantage in the city's narrow alleys. Motorcycles are also easy to ride and maneuver.

* Corresponding author :himawansutrisno@yahoo.com

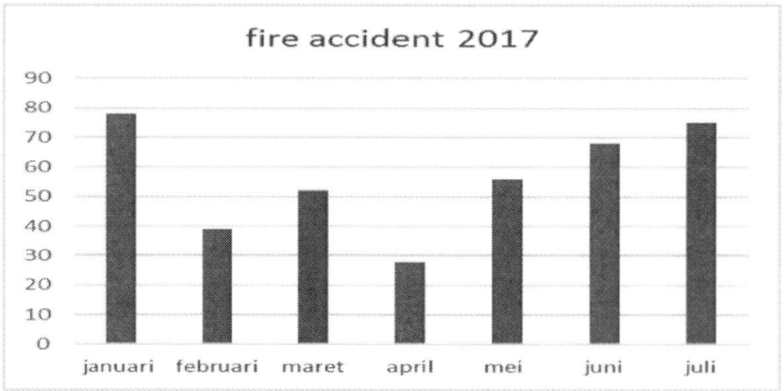

Fig. 1. The Number of Fire Incidents over the period of January – July in Jakarta

The previous research on firefighter motorcycles discusses the installation of the centrifugal pump in the motorcycle by using the manual clutch connected to the axis in the continuously variable transmission (CVT) component of an automatic motorcycle. Therefore, since the cycle of the centrifugal pump follows the CVT cycle in the engine, it is difficult to maintain stability[2]. This burdens the engine when the motorcycle is ridden to the fire location. The use of a magnetic clutch and the improvement in the mechanism of the centrifugal pump are hoped to improve the firefighter motorcycle's performance. The repair of brackets and magnetic clutch mechanism are explained in the sub-chapters below.

2 Related Works

The increase in the number of houses in urban areas has contributed to the development of densely populated areas in certain areas [3]. The growth of population is proportionate to the increase of energy consumption [4] which results in the increase of fire incidents in the urban areas [5]. Although one solution to this issue which is vertical housing such as apartments and flats has been provided by the government, without a comprehensive solution to the fire incidents, the vertical housing remains high for fire risks [6]. The efforts to prevent fire in the residential areas are influenced by the social conditions of the community and the index living condition (ILC). This was mentioned by Niklas Guldaker [7] for the Malmo Sweden study case. Lopes and Hao [8, 9] state that the mitigation of fire is a joint effort between the community and the government as the decision-maker for public policies.

To prevent the loss of lives and materials due to fire in densely populated areas, city planning must be done well [10, 11]. This may be done by adding evacuation routes or maps in case of fire as explained by Dian who discusses the fire in densely populated areas in Surabaya. This also applies to public areas such as campuses where students must be aware of the correct locations of the fire equipment and the evacuation routes during fire.

Aside from the social and cultural conditions of the community, the equipment also become a determining factor in the success of the prevention and mitigation of fire [12]. Efforts have been made for the past several years to develop an early detection of fire [13] which allows a quick response for the community during fire. Other means such as the development of fire prevention equipment [14], fire resistant materials [15], until the evolution of the fire trucks [16] have also been developed by researchers. However, all these means are not applicable to certain areas such as densely populated areas since they have not solved the issue of limited access to the areas. One example of problems that need to be solved is the narrow alleys that are not accessible by fire trucks. Therefore, a fire-fighting equipment that can be used in narrow alleys needs to be specially designed.

3 Research Method

In the beginning, the firefighter motorcycles that have been found in the market use the motorcycle's fuel to move the centrifugal pump. Since this kind of setting requires the motorcycle to provide extra spaces for the centrifugal pump, fuel, and other firefighting equipment, the dimension of the motorcycle becomes larger. In its development, the centrifugal pump in the firefighter motorcycle may be installed in the CVT of the motorcycle using the bracket as the clutch for the centrifugal pump. This, however, creates a complication in the operation. The number of cycle in the centrifugal pump and the volume of water sprayed remains unstable. Therefore, to increase the effectiveness of the centrifugal pump in the firefighter motorcycle, research was conducted, and the flow chart of the research can be seen below:

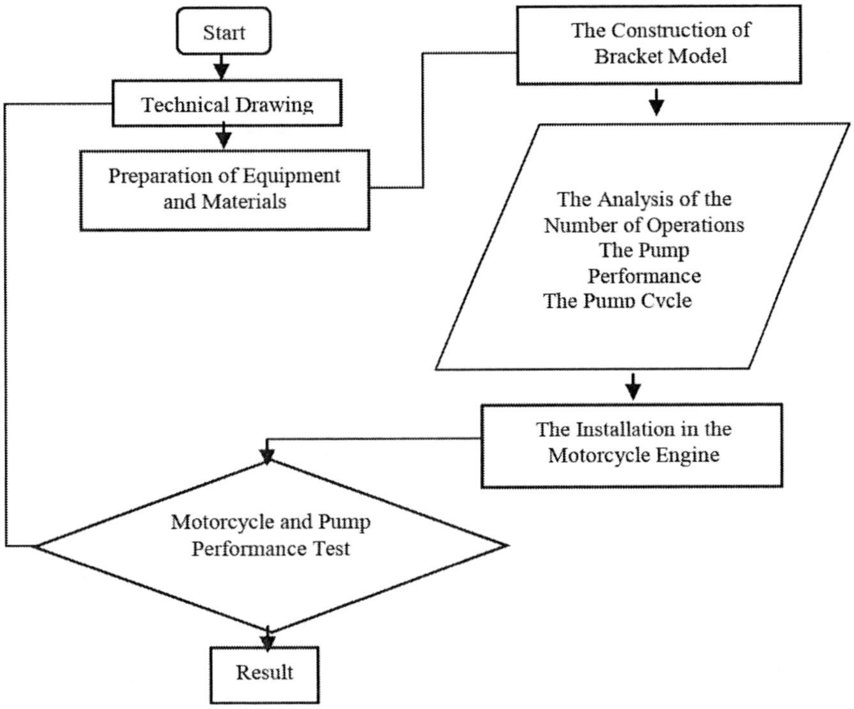

Fig. 2. Flow chart of the Research Procedure

The flow chart shows that the research started from assessing the installment of bracket from the previous research, the installment of the bracket, the tightening of the centrifugal pump, and the calculation of the new dimension after the installment of the pump in the motorcycle. The analysis of the bracket repair was based on the performance of the centrifugal pump in the motorcycle. The result of this research was a model which had passed the test for the performance of the centrifugal pump in the firefighter motorcycle.

4 Repair for Bracket and Clutch Mechanism in the Centrifugal Pump

The equipment used in the firefighter motorcycle can be seen in Table 1 below. The main component of the equipment is the centrifugal pump to spray the water. The volume of the water and the pressure produced by the centrifugal pump must be taken into account carefully in order to avoid adding an extra load to the motorcycle.

Table 1. The Firefighter motorcycle components[1]

No	Description	Specification
1	Centrifugal pump	2.2 KW
2	Bracket Pump	Custom, hand made
3	Rubber hose bracket	Custom, hand made
4	Rubber hose outlet	2 inch, length 20 m
5	Suction hose	2 inch, length 20 m
6	Store box	Capacity 20 liters
7	Siren lamp	

Also, the bracket must be designed according to the specification of the centrifugal pump used. Table 2 shows the specification of the centrifugal pump used:

Table 2. The Specification for the Centrifugal Pump[1]

no	Description	Specification
1	Power	2.2 KW
2	Dimension	H240 x W198 x D132mm
3	Weigh	4.8 kg
4	Body material	Cast iron

In the installation of the centrifugal pump, the bracket is used to support the body of the pump and to attach it to the engine which it uses as the main power. For the clutch component, the magnetic model is chosen since its electrical power is used to connect the magnet component. This model is used in many centrifugal pumps and has been patented by Kiker, et al. For the firefighter motorcycle, the Magnetic clutch Type g is used which can be seen below:

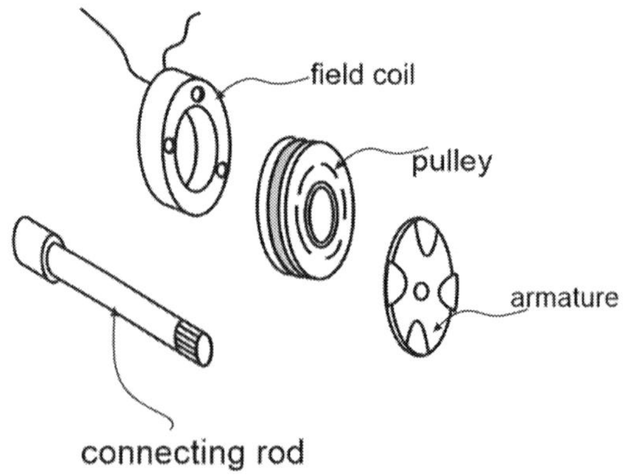

Fig. 3. Magnetic Clutch Type g

From the shape of the clutch seen above, it can be seen that the bracket for the pump is adjusted so that it could be placed next to the clutch and connected with a belt with a pulley placed on the centrifugal pump. The pulley in the centrifugal pump and the pulley in the magnetic clutch have a ratio of 1: 2 or this means that 1 cycle of the pulley in the magnetic clutch creates 2 cycles in the pump. The ratio refers to the cycle of the centrifugal pump used which is the 1450 rpm. Meanwhile, the cycle of the automatic motorcycle engine at a stationary position is 700 rpm. The construction of the bracket for the centrifugal pump is as follows:

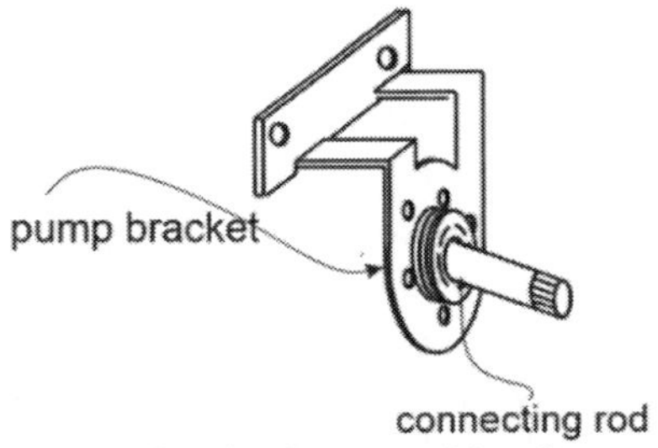

Fig. 4. The Construction of the Bracket in the Centrifugal Pump

Meanwhile, the installment of the pump with the pulley for the firefighter motorcycle is as follows:

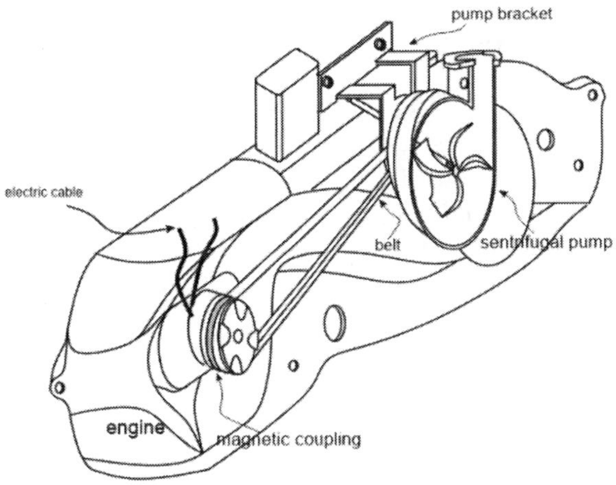

Fig. 5. The Installation of Centrifugal Pump in the Firefighter Motorcycle.

The repair and the improvement that are done can be seen in Table 3 below:

Table 3. The comparison old design and new design of bracket firefighter motorcycle

No	Items of Repair	Old Design	New Design
1	The number of cycles	depending on engine rotation	700 rpm on the engine equal with 1400 rpm on the centrifugal pump
2	The stability of the cycles	Not stabil	stabil
3	The total width of the installed pump	750mm	820mm
4	Construction of centrifugal pump	Coupled with CVT	Pulley and belt ratio 2:1
5	The number of spare parts	5 part	8 part

The following is the illustration of the water volume test in the firefighter motorcycle after the repair and the improvement.

Fig. 6. The Illustration for the Centrifugal Pump Test

5 Discussion and Conclusion

Although the repair of the brackets and the improvement of the clutch mechanism resulted in few modifications such as an addition to the motorcycle part, which is the belt component, and the shift of the centrifugal pump location due to the reduction of cycle used to reach the 1400 cycles during the normal cycle of the engine, there was no increase to the motorcycle load. This is because the magnetic clutch is not needed during cruise time. In the cycle of the centrifugal pump, when the engine is in a low cycle position (700rpm), the pulley and belt mechanism may be used with a 1:2 ratio. The following is the result of the initial trial using the repair of the brackets and the magnetic clutch mechanism in the firefighter motorcycle.

Table 4 the comparison of preliminary reseach of improve firefighter motor cycle bracket

No	Item test	Old design	New design
1	The stability of water outlet	depending on engine rotation	Stabil in the low and hight rotation
2	The water pressure	1.69 bar in the 2000 rpm	1.39 Bar in the 700 rpm (low rotation)
3	The maximum height of the water spray	5 meter	4.1. meter

The table above shows that the repair of brackets and the improvement of the clutch mechanism increased the effectiveness of the firefighter motorcycle. The repair and the improvement resulted in an increase effective mechanism of centrifugal pump coupling. With this model, the firefighter motorcycles will no longer have an additional burden at cruise time.

References

1. H. H. Sutrisno and Triyono, "Designing a Firefighter Motorcycle as an Effort to Provide an Early Response to Fire Disaster," *Journal of Mechanical Engineering* vol. SI 4, p. 13, 2017.

2. S. Himawan Hadi, "The selection of flying roller as an effort to increase the power of scooter-matic as the main power of centrifugal pump for fire fighter motor cycle," *IOP Conference Series: Materials Science and Engineering,* vol. 324, p. 012055, 2018.
3. S. Ahmed, "Assessment of urban heat island and impact of climate change on socioeconomic over Suez Governorate using remote sensing and GIS techniques," *The Egyptian Journal of Remote Sensing and Space Science,* vol. 21, p. 11, 2018.
4. E. Resch, R. A. Bohne, T. Kvamsdal, and J. Lohne, "Impact of urban density and building height on energy use in cities," *Energy Procedia,* vol. 96, p. 15, 2016.
5. A. Jonsson, C. Bonander, F. Nilson, and F. Huss, "The State of the residential fire fatality problem in sweden; Epidemiology, risk factor, and event typologies," *Journal of Safety Research,* vol. 52, p. 12, 2017.
6. J. LIU and W. K. CHOW, "Determination of fire load and heat release rate for high-rise residential buildings," *Procedia Engineering,* vol. 84, p. 7, 2014.
7. N. Guldaker and P.-o. Halim, "Spatio-temporal patterns of intentional fires, social stress and socio-economic determinants: A case study of Malmo, Sweden," *Fire Safety Journal,* vol. 70, p. 10, 2014.
8. Y. Hao-wei, "Development of Fire Science and Fire Protection Engineering in China," *Procedia Engineering,* vol. 52, p. 2, 2013.
9. S. Lopez-Aparicio, M. Vogt, P. Schneider, M. Kahila-Tani, and A. Broberg, "Public participation GIS for improving wood burning emission from residential haeting and urban environmental management," *Journal of Environmental Management,* vol. 191, p. 10, 2017.
10. E. K. Addai, S. K. Tulashie, J.-S. Annan, and I. Yeboah, "Trend of Fire Outbreaks in Ghana and Ways to Prevent These Incidents," *Safety and Health at Work,* vol. 7, p. 9, 2016.
11. A. R. Harper, S. H. Doerr, C. Santin, C. A. Froyd, and P. Sinnadurai, "Prescribed fire and its impacts on ecosystem service in the UK," *Science of the Total Environment,* vol. 624, p. 13, 2018.
12. R. Kuffel, P. Forsyth, and C. Peters, "The Role and Importance of Real Time Digital Simulation in the Development and Testing of Power System Control and Protection Equipment," *IFAC Paper Online,* vol. 49, p. 5, 2016.
13. L. Giglio, W. Schroeder, and C. O. Jusctice, "The collection 6 MODIS active fire detection algorithm and fire products," *Remote Sensing of Environment,* vol. 178, p. 11, 2016.
14. C.-p. WANG and B.-j. SHIH, "Research on the Integration of Fire Water Supply," *Procedia Engineering,* vol. 211, p. 10, 2018.
15. K. Mroz, I. Hager, and K. Korniejenko, "Material solution for passive fire protection of building and structures and their performances testing," *Procedia Engineering,* vol. 151, p. 8, 2016.
16. W. Ge, J. Yong-xing, and S. Yao-zong, "The 3rd Generation Fire Truck and its Spraying Technique," *procedia Engineering,* vol. 11, p. 7, 2011.

Knowledge management in multi-generation workforce: investigation on critical success factor and strategy

I Komang Artha Winadi[1,*], *Amalia* Suzianti[2], and *Romadhani* Ardi[3]

[1]Industrial Engineering, Universitas Indonesia, Depok, Indonesia

Abstract. This paper investigates critical success factors and strategy of knowledge management within organization that made of multi generation workforce. It is widely understood that tensions and conflicts are inevitable if employees from every generational cohort fail to understand and accept the unique characteristics of each group. As the race to gain competitive advantage between businesses growing intensely from time to time, there is a need emerging for a strong guidance in developing a correct knowledge management strategy in respect to multi-generation work force phenomenon. By incorporating knowledge management critical success factor and strategy and also generational differences obtained from literature review, we design models that can be used to choose appropriate knowledge management strategy in a multi generational workforce organization. Using these models, which were validated using Structural Equation Modelling (SEM), it is found that the appropriate knowledge management strategy for multi generation work force organization is external personalization.

1. Introduction

The rate of technology development has increased rapidly during the last decade that directly affect customer behavior. The advancement of technology creates new opportunity and subsequently new competition. To keep up with this challenge, organization should possess and continuously increase their competitive advantage, where innovation is a key factor [1]. It is widely accepted that one of the common approach to support innovation in an organization is knowledge management.

Knowledge management can be defined as the process of creating, sharing, using and managing the knowledge and information of an organization [2]. Knowledge management has been regarded as a major part of organizational success due to the increasing recognition towards knowledge as a valuable asset within the organization [3] ; [4].

[1] Corresponding author: komangwinadi89@gmail.com

Copious academic research has been conducted in the area of knowledge management, especially in terms of knowledge management implementation. Eventhough it is safe to say that there is a sufficient theoretical foundation available for practitioners to implement knowledge management strategy, there are circumstances that force organization to continuously improve their innovative power [5], such as a disruptive business environment; customer loyalty and switching provider behavior; and also first-mover advantages. These condition entitled organization to enhance their knowledge management strategy in order to gain the competitive advantage.

Judging from current researches, the aspect that is rather overlooked is the phenomenon of multi-generation work force within an organization. Multiple generations had worked in the same organization for quite a while, however they were usually separated by job descriptions and system hierarchy [6]. Right now it is very common to see younger people supervising older employee with their own generational differences. These differences could potentially ignite work tensions and conflicts if employees from each generation fail to compromise [7].

Although this condition is already being observed by several knowledge management researcher, such as [4] ; [8] ; [9], they mainly emphasize on the differences that follows generational cohorts and does not provide any insight on what kind of strategy that should be used to cope with that differences. As the race to gain competitive advantage growing intensely from time to time, there is a need emerging for a strong guidance in developing a correct knowledge management strategy in respect to multi-generation work force phenomenon [10]. Therefore the objective of this paper is to propose a model that can be used to choose knowledge management strategy in a multi generational workforce organization.

2. Literature Review

There are several factors that proved to be critical towards the success of knowledge management implementation or also known as the critical success factors of knowledge management. There are numerous researches that proposed their own respective ideas about these factors such as [11], [12] and also [13]. In terms of generalization on these factors, [14] managed to find a common ground and define a classification of knowledge management critical success factor, which consist of Culture, Structure and Procedures, Human and Financial Resources, Technology and Infrastructures, Strategy and Leadership and KM Processes.

By using Theory of Planned Behavior (TPB) perspective, [15] introduce a conceptual model that predict knowledge management success through two variables namely Performance Ability and Performance Willingness. Performance ability defined as the organization capability to implement knowledge management that is taken from TPB's perceived behavioral control and Performance Willingness is defined as the organization willingness or motivation to implement knowledge management which is taken from TPB's intention factor.

Knowledge management strategy can be defined as the basic reference that can be used in order to manage knowledge within an organization [16]. According to [17], knowledge management strategy is classified into two kinds, which are codification and personalization. Based on this information, [18] elaborate these two strategies into four strategies that incorporate type and origin of knowledge, which are Internal Personalization, Internal Codification, External Personalization and External Codification. It is suggested by [18] that organization should choose the strategy that is best suited with their IS Maturity and Knowledge Intensity. IS maturity is defined as organization's IS capability to utilize internal and external information to support knowledge management process while knowledge intensity is defined as organization's intensity to find and exploit internal and external knowledge.

In the history of workforce, it can be said that it is the first time where employees from various generations not only working side by side but also directly both with people who are as young as their children and as old as their parents [19]. These differences should be handled carefully in order to prevent potential conflict and tension between employees from each generation [7]. To get a better understanding of each generation, the summary of notable characteristic for each generation obtained from literature review will be shown in Table 2.

Table 1.Generational differences (Summarized from [6]; [7]; [20]; [21])

Aspect	Gen X	Gen Y
Philosopy	Favors order dan responsibility	Favors risk taking and freedom
	Work to live	Live first,then work
Top Developmental Areas	• Leadership • Skill training in my areas of expertise • Team Building	• Leadership • Problem solving, decision making • Skill training in my areas of expertise
Prefered Leadership Attributes	• Credible • Trusted • Farsighted	• Listen wells • Dependable • Dedicated
Individual Characteristic	Individualist	The more the merrier
	Technology savvy	Technology savvy

3. Methodology

By incorporating knowledge management critical success factor and generational differences, we design models that enable their user to choose an appropriate knowledge management strategy based on their organization characteristic. In order to validate the model, a data collection process is conducted.

A questionnaire is then designed based on the literature review and modification from the field study. The instrument items are measured on a five point (1-5) Likert scales, of which 1 indicates that the respondent strongly disagree with the statement and 5 indicates that the respondent strongly agree with the statement respectively. The questionnaire is pre-tested by three knowledge workers, pilot tested among 30 employees to ensure content validity. This study undertakes the main survey to 253 subjects among the organization that consist of multi generation workforce and located in Indonesia. Using the data collected by questionnaire, a Confirmatory Factor Analysis (CFA) is performed to specify the structure between observed indicators and latent constructs, and test the validity of measurement model. Subsequently, structural equations among latent constructs are examined to test the conceptual structural equation model (SEM). The CFA and SEM procedures are conducted using AMOS software. The result of calculation using SEM subsequently deployed into the model to choose the appropriate knowledge management strategy

4. Model and hypothesis development

4.1 Knowledge Management Implementation Model

By incorporating insight from two previous researches on knowledge management critical success factor [14]; [15], we managed to generate a model that enable organization to assess their workforce characteristic. The classification of critical success factor into KM Ability (KMA) and KM Willingness (KMW) construct is supported by the previous research conducted by [15] and [22]. The model is shown in Figure 1.

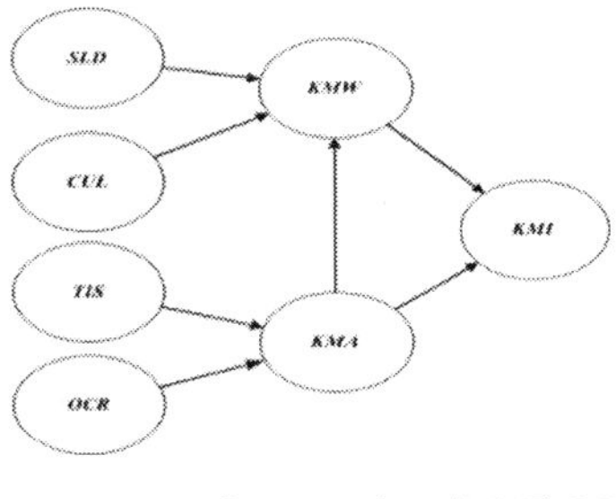

Fig.1. KM Implementation (KMI) Model

The description of each variable included in the model is as follows:

- **OCR** stands for Organizational Characteristic; this variable represents the activities and procedures that support knowledge management implementation and also the policies on people management that is the heart of organizational knowledge.
- **TIS** stand for Technology and Infrastructure, this variable represents the organization capability on information technologies and communication systems.
- **CUL** stands for Culture; this variable represents values and beliefs that are invented, shared and developed by group of people in the organization.
- **SLD** stands for Strategy and Leadership, this variable represents strategic planning and leadership issue in terms of knowledge management implementation.
- **KMA** stands for KM Ability; this variable represents organization capability to implement knowledge management.
- **KMW** stands for KM Willingness; this variable represents organization willingness or motivation to implement knowledge management.
- **KMI** stands for KM Implementation; this variable represents the magnitude of knowledge management implementation towards organizations business process.

4.2 Knowledge Management Strategy Model

Adopting the model from [18], we propose a slight adjustment which is to replace IS Maturity with KM Ability and Knowledge Intensity with KM Willingness. This adjustment is made due to the comparable definitions between them. By adjusting this model, the result of the quantitative analysis using SEM from KM Implementation model can be used to choose the right knowledge management strategy. The parameter that we proposed to be used in determining strategy position is the Standardized Regression Weight of KMA → KMI and KMW → KMI with the center point of the quadrant should be 0.5, which means a beta (β) score less than 0.5 will be considered as low and vice versa. The calculation using these models should be conducted for the entire organization and also for each generation in order to identify the right strategy for the whole organization and each cohort.

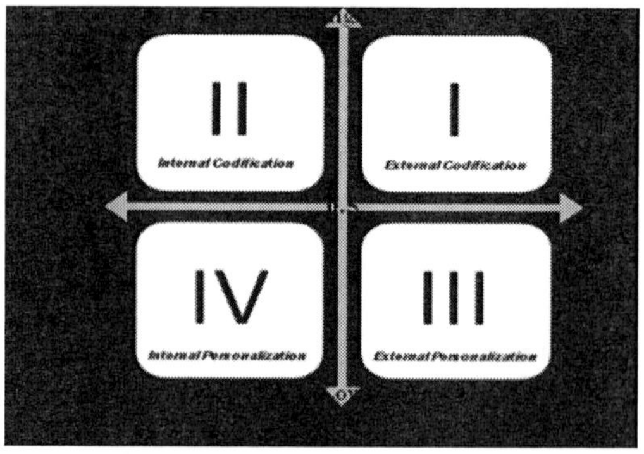

Fig.2. KM Strategy (KMS) Model

Table 2. Research hypotheses

Hypothesis	Code		
OCR has a positive and significant relation towards KMA	H1a	H1b	H1c
TIS has a positive and significant relation towards KMA	H2a	H2b	H2c
CUL has a positive and significant relation towards KMA	H3a	H3b	H3c
SLD has a positive and significant relation towards KMA	H4a	H4b	H4c
KMA has a positive and significant relation towards KMW	H5a	H5b	H5c
KMA has a positive and significant relation towards KMI	H6a	H6b	H6c
KMW has a positive and significant relation towards KMI	H7a	H7b	H7c
There is a significant differences of OCR between Gen X and Gen Y	H8		
There is a significant differences of TIS between Gen X and Gen Y	H9		
There is a significant differences of CUL between Gen X and Gen Y	H10		
There is a significant differences of SLD between Gen X and Gen Y	H11		
Gen X would prefer Internal Codification Strategy	H12		
Gen Y would prefer External Codification Strategy	H13		
Multi Generation Work Force Organization would prefer External Codification Strategy	H14		

Note: (a) Gen X; (b) Gen Y; (c) Multi Generation Work Force Organization

5. Data analysis and result

The number of respondent involved in the survey are 283, of which consist of 35% Gen X (100 respondent) and 65% Gen Y (183 respondent). Subject that has more than 10 years of work experience are 44%, while 41% and 15% have worked 2-5 years and 5-10 years, respectively. More than half of the subject has the role of staff (70.7%), while 20.8% has the role of supervisor, and the rest of them (8.5%) has the role of manager or higher.

This study first undertakes the CFA to confirm the validity of the constructs and to assess the model fit. Convergent validity of CFA results that consist of construct reliability and average variance extracted are shown by table 3. The model adequacy was assessed by the fit indices suggested by [23]; [24] and shown by table 4. The results indicate that the measurement model has good convergent validity. Therefore, the proposed measurement model is reliable and meaningful to test the structural relationships among the constructs.

Table 3. Convergent Validity

Item	OCR	TIS	CUL	SLD	KMA	KMW	KMI
CR	0.92	0.94	0.84	0.89	0.90	0.87	0.93
AVE	0.55	0.70	0.57	0.62	0.74	0.69	0.76

Table 4.Fit Indices for Measurement and Structural Model

Indices	Standard and Reference		Measurement Model	Structural Model
X^2 / df	<3.00	[23]	2.109	2.125
CFI	>0.90	[24]	0.933	0.901
SRMR	<0.08	[24]	0.036	0.051
RMSEA	<0.07	[24]	0.063	0.045

Table 5. Path Analysis Result

Path	Standardized Regression Weight			Critical Ratio (bold is significant in p < 0.10)		
	A	B	C	A	B	C
OCR→KMA	0.465	0.362	0.407	**1.84**	**2.74**	**3.53**
TIS → KMA	0.446	0.579	0.529	**1.76**	**4.26**	**4.51**
CUL→KMW	-0.350	0.374	0.180	-1.18	**2.89**	**1.91**
SLD→KMW	0.865	0.353	0.519	**2.60**	**1.72**	**3.51**
KMA→KMW	0.453	0.290	0.296	**3.11**	**2.02**	**2.83**
KMA→KMI	0.293	0.150	0.272	1.61	0.99	**2.34**
KMW→KMI	0.703	0.780	0.679	**3.78**	**4.97**	**5.73**

Note: (A) Gen X; (B) Gen Y; (C) Multi Generation Work Force Organization

Table 6. Comparison between Generations

Variable	Gen X	Gen Y	Comparison z-Score	Conclusion
OCR	0.465	0.362	-0.218	No significant differences
TIS	0.446	0.579	0.595	No significant differences
CUL	-0.35	0.374	2.252	There is a significant differences
SLD	0.865	0.353	-1.395	No significant differences

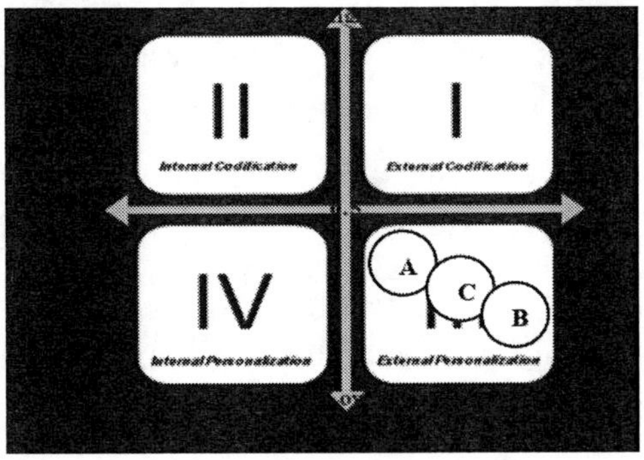

Fig.3. KM Strategy Deployment for Gen X (A), Gen Y (B) and Multi Generation Work Force Organization (C)

6. Discussion

6.1 Knowledge Management Implementation Model

Based on the SEM calculation, it is found that for Generation X, OCR and TIS has a positive and significant effect on KMA, while SLD has a positive and significant effect on KMW. These findings confirm the result of the research conducted by [11]. It is also found that CUL has a negative albeit not significant effect on KMW, where this finding somewhat reflect the individualist characteristic of Generation X [6]; [7]; [20]; [21].

For Generation Y, OCR and TIS has a positive and significant effect on KMA, while CUL and SLD also have a positive and significant effect on KMW. It can be seen that for both Generation X and Y, KMA does not have a significant effect towards KMI but have a significant effect on KMW, thus it can be said that KMA became the catalyst for KMW to enforce KMI. These findings justify other research result that stated there is a significant effect of ability to implement knowledge management towards intention to implement them [22].

Further analysis between the critical success factor of Generation X and Y discover that Culture (CUL) has a significant difference. Since culture is highly supported by collaboration, and collaboration is affected by individualism, this finding emphasizes a different characteristic between Gen X as the individualist or lone wolf and Generation Y as their polar opposite which hold the philosophy of the more the merrier [6];[7].

6.2 Knowledge Management Strategy Model

Based on the value of standardized regression weight of KMA and KMW towards KMI, it was found that External Personalization Strategy is the suitable knowledge management strategy that should be used for Gen X, Gen Y and also the whole organization. Although this result is different from the hypothesis proposed by this research, it is supported by other research that highlights the importance of motivation and willingness towards knowledge management implementation [25].

6.3 Potential Contribution towards Industry 4.0

Alongside a wide array of opportunities that comes with the automation and digitization of Industry 4.0, they also possess some challenges of their own. As stated by [26], one of the emerging challenges in Industry 4.0 is the changes in workplace demographic. The admission of younger generation in the workplace presents a new challenge on the ability to transfer knowledge, mainly because their characteristic differences with the older generation already settled in the workplace. If not treated correctly, this condition could incapacitate organization's capability to retain and at the same time expand their knowledge base.

By incorporating knowledge management critical success factor, strategy and also generational differences, this research manage to design models that could be used in deciding suitable knowledge management strategy for multi generation work force organization. With knowledge management strategy that is suitable with their respective workplace demographic, organization could then overcome the demographic challenges of Industry 4.0 and reap the maximum benefit from the opportunities that they offer.

7. Future research

As with any research, the specific context of this study limits the interpretation of the findings. Some adjustments must be made to apply these results, such as generalization in a broader region such as Asia or in the international setting. Another aspect that is interesting to be examined for future research is the introduction of Generation Z into the workforce. With their unique characteristic, the inclusion of Gen Z in the model could possibly generate a different result compared to this research.

Thanks are due to the Directorate of Research and Community Service (DRPM) Universitas Indonesia, which has supported and facilitated this research through PITTA Grant (2374/UN2.R3.1/HKP.05.00/2018).

References

1. J. M. Trejo, J. S. Gutiérrez and G. M. Guzman, The customer knowledge management and innovation, Contaduría y Administración **61**, pp. 456–477 (2016)
2. J. Girard and J. Girard, Defining knowledge management: Toward an applied compendium, Online Journal of Applied Knowledge Management, pp. 1-20 (2015)
3. R. Filippini, W. H. Güttel and A. Nosella, Ambidexterity and the evolution of knowledge management initiatives, Journal of Business Research **65**, pp. 317–324 (2012)
4. J. Ž. Brčić and K. K. Mihelič, Knowledge sharing between different generations of employees an example from Slovenia, Economic Research-Ekonomska Istraživanja Vol. **28**, pp. 853–867 (2015)
5. R. Nowacki and K. Bachnik, Innovations within knowledge management, Journal of Business Research **69**, pp. 1577–1581 (2016)
6. D. Gursoy, T. A. Maier and C. G. Chi, Generational differences An examination of work values and generational gaps in the hospitality workforce, International Journal of Hospitality Management **27**, pp. 448–458 (2008)
7. T. Angeline, Managing generational diversity at the workplace: expectations and perceptions of different generations of employee, African Journal of Business Management Vol **5**, pp. 249-255 (2011)
8. W. K. McHenry and S. R. Ash, Knowledge Management and Collaboration: Generation X vs. Generation Y, International Journal of Business and Social Science Vol.**4**, pp. 78-87 (2013)
9. K. Chidambaranathan and S. B.S., Knowledge Management as a Predictor of Organizational Effectiveness:, The Journal of Academic Librarianship **41**, pp. 758–763 (2015)
10. V.-H. Lee, A. T.-L. Foo, L.-Y. Leong and K.-B. Ooi, Can competitive advantage be achieved through knowledge management? A case study on SMEs, Expert Systems With Applications **65**, pp. 136–151 (2016)
11. L.-S. Huang and C.-P. Lai, An investigation on critical success factors for knowledge management using structural equation modeling, in *International Conference on Asia Pacific Business Innovation and Technology* (2012)
12. A. Samad, A. K. Kazi and M. Raheem, Critical Success Factors of Knowledge Management Systems Implementation, KASBIT Business Journal (KBJ), pp. 64-78 (2014)

13. L. A. Y. Al Hakim and S. Hassan, Critical Success Factors of Knowledge Management,Innovation and Organisational Performance: An Empirical Study of the Iraqi Mobile Telecommunication Sector, British Journal of Economics, Finance and Management Sciences Vol. **4**, pp. 31-49 (2012)
14. M. Sedighi and F. Zand, Knowledge Management: Review of the Critical Success Factors and Development of a Conceptual Classification Model, in *10th International Conference on ICT and Knowledge Engineering*, Bangkok (2012)
15. F. Lehner and N. Haas, Knowledge Management Success Factors – Proposal of an Empirical Research, Electronic Journal of Knowledge Management, pp. 79-90 (2010)
16. L. Mladkova, Knowledge strategy: key player or relict of the past?, in *Procedia - Social and Behavioral Sciences* (2014)
17. M. T. Hansen, N. Nohria and T. Tierney, Whats *Your Strategy for Managing Knowledge?*, Harvard Business Review (1999)
18. T. H. Kim, J. N. Lee, J. U. Chun and I. Benbasat, Understanding the effect of knowledge management strategies on knowledge management performance: A contingency perspective, Information & Management, pp. 398-416 (2014)
19. R. Zemke, C. Raines and B. Filipczak, Generations at work: *Managing the clash of Veterans, Boomers, Xers and Nexters in your workplace (2nd Ed)*, New York, NY: American Management Association (2000)
20. A. Tolbize, *Generational differences in the workplace* (2008)
21. V. Srinivasan, Multi generations in the workforce: Building collaboration, IIMB Management Review, pp. 48-66 (2012)
22. T. Khalil, K. Atieh, A. U. Mohammad and F. Bagdadlian, Examining the Social and Technical Factors Influencing School Teachers Knowledge Sharing Intentions in a Teachers Online Professional Community, The Electronic Journal of Knowledge Management Volume **12** Issue 3, pp. 157-165 (2014)
23. D. Hooper, J. Coughlan and M. R. Mullen, Structural Equation Modelling: Guidelines for Determining Model Fit, Electronic Journal of Business Research Methods Volume **6** Issue 1, pp. 53-60 (2008)
24. J. Hair, *Multivariate Data Analysis Fifth Edition*, New Jersey: Pearson Higher Education (1998)
25. A. Mathew, G. Nair and L. Rodrigues, Systems Approach to Determine the Significance of the Critical Success Factors of a Knowledge Management System, Research Journal of Management Sciences Vol **1**, pp. 28-36 (2012)
26. F. Hecklau, M. Galeitzke, S. Flachs and H. Kohl, Holistic approach for human resource management in Industry 4.0, in *6th CLF - 6th CIRP Conference on Learning Factories* (2016)

Aluminum based Composites by Severe Plastic Deformation Process as New Methods of Manufacturing Technology

Agus Pramono[1,a], Alry Mochtar Jamil[1] and Anistasia Milandia[1]
[1,] *Department Metallurgy Engineering, University of Sultan Ageng Tirtayasa. Cilegon, 42435 Banten, Indonesia*

Abstract. Composites a material was developed to replace metal and alloys, because of the properties such as light weight and unique mechanical properties. Processing of aluminum-based composites has been developing by new manufacturing technology, namely severe plastic deformation (SPD), to produce unique of mechanical properties. Some of the methods used are; equal channel angular pressing (ECAP), accumulative roll bonding (ARB) and multi-axial forging (MAF). The results of some of these methods were compared with the latest method of new SPD, namely: repetitive press roll forming (RPRF). Based on grain morphology and mechanical properties, the result of RPRF has superior to another method. The properties produced by SPD technology was varies, the highest of hardness produced by RPRF process was 88 HV10, ECAP produced 65 HV10, MAF was 46 HV10 and ARB reached 50 HV10. While the highest of tensile strength produced by MAF was 237 MPa while the RPRF process just only around 147 MPa, but the ultrafine grains just only produced by RPRF method which is 0.9 μm, compared to other methods: MAF 1.2 μm, ECAP 5.7 μm and ARB is not so far with MAF that is equal to 1.4 μm. The RPRF process can be recommended for the interest of the aluminum-based composite materials processing industry. Because currently some component product by industries have been substituted from metal alloy materials to metal-based composites.

1 Introduction

The new technology of manufacturing process by severe plastic deformation (SPD) has been developing rapidly as a processing of metals and alloys. By high pressure there will get ultrafine grained (UFG) in the material which will lead to the increase of mechanical properties. The main emerging technologies after forging and extrusion are equal channel angular pressing (ECAP), multi-axial forging (MAF) and accumulative roll bonding (ARB) [1]. Some of SPD technologies, such as ECAP, MAF and ARB, have been used for the development of composite materials. Therefore, grain miniaturization can be coupled by reinforcement from particulates for mechanical property enhancement [2-3]. To produce high mechanical properties on SPD required a long pressure cycles. Some of research on

[a] Corresponding author: agus.pramono@untirta.ac.id

SPD works such as; ECAP performed by Langdon [4] to achieve 1.9 μm grain size with tensile strength more than 100 MPa required 11 cycles' as for the ARB process conducted by Argentero [5] in his research, requires 10 cycles of rolling compression to produce 113 HV with tensile strength > 100 MPa. Improved of mechanical properties on composites has been performed by Amirkhanlou [6] using new method accumulative press bonding (APB), to produce fine grains with a ultimate tensile strength of 180 MPa from 88 MPa base material, it takes 14 cycles. The APB process was developed based on the principle of accumulative roll bonding (ARB) process, but can be readily installed in both a laboratory and industrial environment. The APB process can be applied to sheet materials and thick billets of relatively large-scale dimensions, while the ARB process can only be applied to sheet materials [7-9]. This phenomenon does not change the long cycles to produce high strength by UFG. The development of the latest methods that refine the ARB and APB processes is repetitive press roll bonding (RPRB). RPRB is a process of SPD that combines repetitive pressing forces with the rolling forces, where the compression procedure is was done continuously to meet the cycles [10]. Compression on the RPRB process has two important roles in the processing of composite materials; First pressing force gives effect to the embedded of the reinforcement, either powder or fibres into the surface of the sample plate, so that there will be a face bond between the matrix and the coupled reinforcement, latter of the roll force providing contact on the surface which will result in the distribution of the reinforces dispersed into the composite contact plane [11]. This study aims to distinguish several of SPD methods that have been applied for composite materials, among the methods applied are ECAP, MAF, and ARB. In all, the SPD method is compared with the RPRB method, in application of application in the Industry, bonding on the material will change its name to forming because if it has been applied for certain component application then bonding phase will become repetitive press roll forming (RPRF).

2 Materials and Experimental Methods

The matrix material used for ECAP process were Aluminum series 1070 (AA1070)/Al_2O_3 nano-fibers (ANF) as reinforced. For MAF and ARB used Aluminum series AA1100/Silicon Carbide (SiC) as reinforcement. As a comparison method the material matrix used on RPRF process is AA1070/ANF and in other hand AA1100 as matrix material with SiC as a reinforcement. The use of composite materials processed in SPD is presented in Table 1 to clarify the use of materials to the SPD process.

Table 1. Use of Materials for the SPD process.

SPD Process	Matrix	Reinforcement	Form
ECAP	AA1070	Al_2O_3	Powder/Fibers/Bars
MAF	AA1100	SiC	Plate/Powder
ARB	AA1100	B_4C/SiC	Plate/Powder
RPRF	AA1070/AA1100	Al_2O_3/SiC	Plate/Powder/Fibers

Characterization using Microscope Optical were used to determine the grain morphology and precipitate distribution, imaga-J software to know grain size, while the mechanical properties traced by hardness (HV10) testing to determine the ability of loading to plastic deformation and tensile strength to know the relationship of stress diagram to strain.

2.1 Equal channel angular pressing (ECAP)

Materials of composite by ECAP consist of AA1070 as matrix and ANF as reinforcement. AA1070 and ANF mixture by mechanically ball milling. The volume fractions used were 97:3%. Milling process used 60 rpm for 24 hours. After the process was complete the next stage preparation of compaction by wrapped on composite materials (AA1070/ANF) in copper tube and heated in a furnace with temperature 400^0C during holding times one hour. Pressing process of ECAP in the single pass, shown in Fig. 1.a

2.2 Multi-axial forging (MAF)

MAF process stages were cutting AA1100 size 135 mm x 25 mm x 3 mm with thickness 3 mm. The sample was brushed using a wire brush in the same direction. Before the grinding, process was done to smooth the surface; the material is first cleaned by acetone solution. Giving acetone is expected so that the material does not have dirt and no oxidation occurs. Then make a hole on each edge of the sample. The next step, done grinding and adding SiC powder, fraction of weight 0.2% wt, 0.4% wt, 0.6% wt SiC. MAF principle that is after forging, material rotated 90°, the forging up to several times pass on each side. The process scheme is presented in the figure 1.b

2.3 Accumulative roll bonding (ARB)

The material used AA6061 as a matrix and SiC for reinforced. The sample was cutting by grinding to be: 2 x 100 x 10 mm. Then the surface of the sticking sample is cleaned with a wire brush and acetone and marketed by grinding, both plates are neatly stacked and each surface is sprinkled with a powdered SiC and each edge of the plate is clamped with a rivet. The prepared sample was preheated using a muffle furnace with a temperature of 400^0C and a holding time of 60 minutes. The heated sample is directly rolled using a rolling machine. A scheme of the ARB process of composites is presented in Fig. 1.c

2.4 Repetitive press roll forming (RPRF)

Type of materials used AA1100 and SiC, preparation begins by cutting a sample of AA100 sheets measuring 2 x 1 cm^2 into 2 x 5 cm^2. SiC as reinforce was used percent by weight 0.4%. The next stage of surface preparation is done by cleaning the aluminum surface with a wire brush followed by surface grinding with a grinding machine. Then stacked with other samples that have been prepared with the prepared part of the surface, at the end of the sample in the friction weld, for the previous experiment used copper wire clamp, copper wire clamp still allows the plate to shift so that with a very strong clamping weld clamping and avoid shift plate. Heating sample 350°C for 1 hour of pressure begins with maximum load capacity of 10 tons, then rolling process with percent reduction of 50%. Procedures of the RPRF Experiment shown in Fig. 1.d

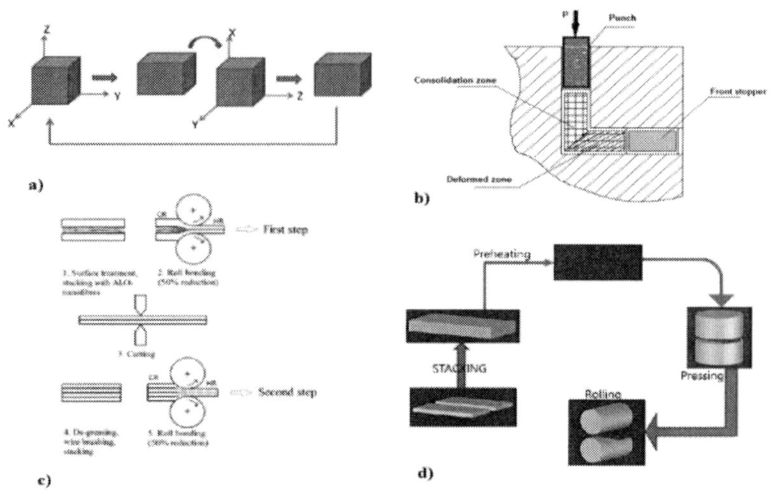

Fig. 1. The technology process of SPD on Aluminum based composite: a) Multiaxial forging (MAF) [13]. b). Equal channel angular pressing (ECAP) [12]. c) Accumulative roll bonding (ARB) [14]. d). Repetitive press roll forming (RPRF) [10].

3 Result and Discussion

Investigation of composite by SPD was conducted to determine the characteristic of mechanical and physical. For microstructures behaviour used microscope optic, was traced to determine the grain morphology and deployment precipitates on grain. Image-J software was used for quantitative microstructure analysis calculation of the grain size. Each SPD method is traced to its mechanical properties, such as the mechanical properties of ECAP, MAF and ARB processes, compared to the new of RPRF method. Investigation of grain size on microstructure to know the relation of mechanical properties.

3.1 Microstructural analysis

The microstructure of grain morphology on ECAP process of composite was observed, seen of Al_2O_3 shaped fibres, white, spread evenly among the aluminum matrix shown in fig. 2 a. Aluminum as matrix and Al_2O_3 as a reinforcement is a suitable composite material because at the time of ECAP compression, tensile and compressive stresses gave an effect on the twinning movement of dislocation in the dies that makes the diffusion work effectively, so as to form strong grain bonds [15]. In The micro-structures of ECAP process using 50-100X magnification, it appears that the alumina flakes are spread evenly on each side of the aluminum matrix, proving that aluminum with alumina has good grain bonds due to the combination high pressing and tensile forces on ECAP processes. Figure 2 b. shown the result microstructure of aluminum-based composite by MAF process, with SiC as an reinforcement. Visible grains of equaxed shape, the compressive stress will tend to cause the grain change to be equilateral [16], this essential grain change will improve mechanical properties such as strength and hardness.

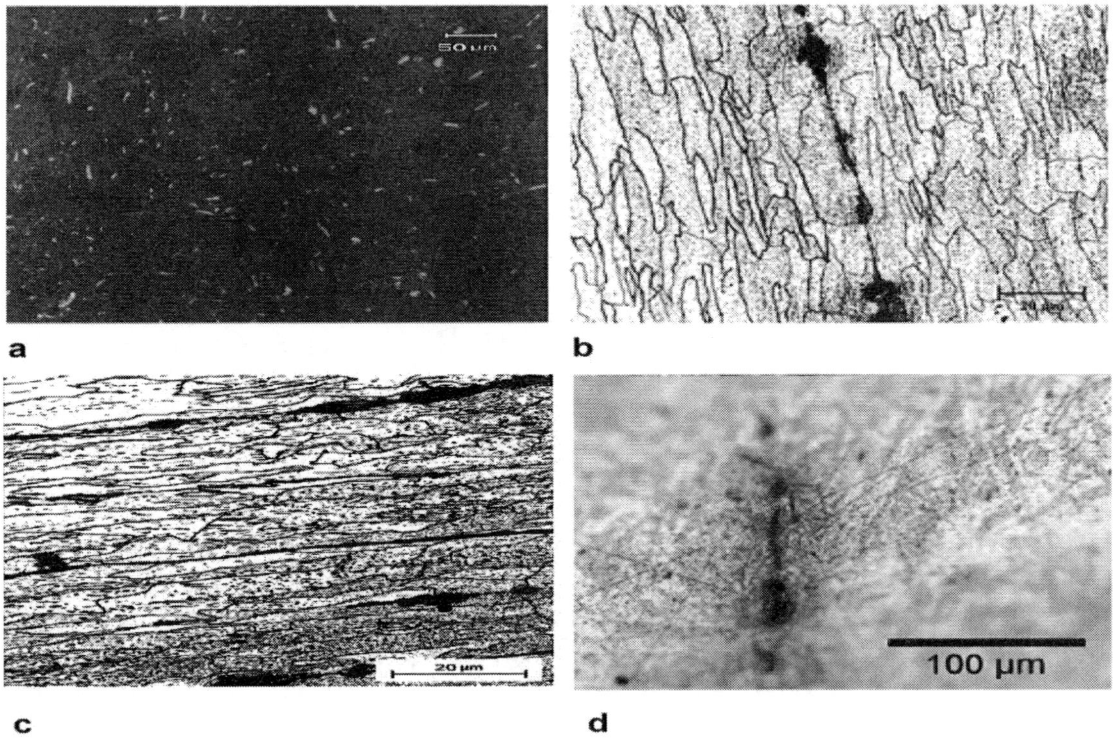

Fig. 2. Result of microstructural Optical a) Composite Al/Al_2O_3 by ECAP b) Composite Al/SiC by MAF c) Composite Al/SiC by ARB d) Composite Al/SiC by RPRF

For Al/SiC composite by ARB process it is seen that a lot of black porosity dominates the matrix area, fig 2 c. this is the result of poorly diffusion between the Aluminum as a matrix to SiC as reinforced, although at the farthest part of the grain morphology interface it turns to equaxial but the porosity dominance does not increase enough its mechanical properties. Figure 2 d. shows the microstructure of Al/SiC processed by RPRF. Seen the formation of a closed interface line across the boundary by the missing line between the interfaces, based on [14] the missing line is the result of a grain delamination process that results in a particle shift that covers the interface boundary line, it indicates the diffusion occurs perfectly.

3.2 Mechanical Properties

Hardness was compared between each method, the highest hardness was found in Al/SiC composite by RPRF process that was 88 VH10, when compared with Composite Al/Al_2O_3 by RPRF process that was 49 HV10, this is influenced by the bonding between the matrix to the reinforcement is good formed. Bonding of composite microstructure is marked by the presence of a line on the boundary bond [14], as shown in Figure 3 of the microstructure. Among the comparable processes, Composite of Al/SiC by RPRF has a missing line that dominates along the interface line. Based on grain boundaries, the RPRF method is able to produce UFG of 0.9 compared to other processes. However, grain refinement with increased hardness accompanied by decreased elongation to 0.5%.

Table 1. Mechanical properties of Composite based aluminum by SPD.

SPD Process	Hardness (VH10)	Grains Size (μm)	Ultimate Tensile Strength (MPa)	Elongation (%)
ECAP	65	5.7	124	3
MAF	46	1.2	237	8
ARB	50	1.4	102	7
RPRF	88	0.9	147	0.5

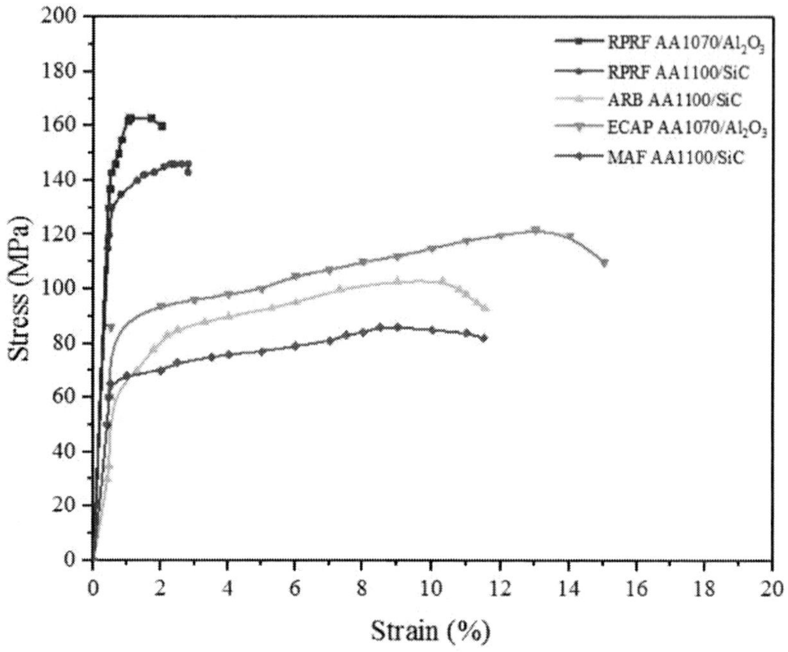

Fig 3. Stress Strains of Aluminum based composite by SPD technology

In other hand MAF method is capable of producing the highest tensile strength of 237 MPa, the increase is accompanied by significant elongation reaching 8%, the highest compared to other processes, but the grain gain reaches only 1.2 μm. The ECAP and ARB processes result in relatively identical mechanical properties, ie 65 VH10 ECAP results and 50 results from ARB, tensile strength for ECAP 124 MPa and on ARB of 102 MPa. Improvements in the SPD ECAP and ARB processes require long cycles to improve mechanical properties and fine grain performance. In ECAP, to produce fine grains with high mechanical properties required 4-6 times cycles of pressing [15], whereas ARB requires 8-12 cycles' compression of rolling for significant mechanical properties improvement with fine grain achievement [5].

3.3 Application Area of Severe Plastic Deformation (SPD) Technology

The markets of materials produced by the SPD technology being exist in the sectors where highest of properties, such as; specific strength or increased fatigue life, are needed. Hardness and tensile strength is important properties for component application. Formal market analysis specializing in innovation metals production was conducted by company of Metallicum. It has found more than 100 specific markets for production of component in areas on aerospace, transportation, medical equipment, chemical processing, electronics and defence [1]. Some of the applications that use SPD materials are aluminum based composites with high yield strength and superplastic properties, higher mechanical properties in cold and cryogenic temperatures and high-strength lightweight alloys with shape-memory effect for space, medical and many other applications [17-18]

Fig 4. The application of Severe Plastic Deformation: a) Bolt nut connection; b) plate implants made of nanostructure metal; c) engine block machine fabricated from AA1420 by ECAP; d) body of military helicopter from high strength materials; e) Super-plasticity sheet materials for Models of hollow blades from Ti-6Al-4V. f) military device - amphibious assault vehicle, courtesy of BAE Systems by ARB.

4 Conclusion

The comparation research of SPD Technology showed the applicability of kind of SPD: ECAP, MAF, ARB and new methods of RPRF for producing composite based aluminium with SiC and Al_2O_3 as reinforcement. The Technology of SPD has a potential for manufacturing aluminum based composites, to produce the ideal tensile strength MAF is the right process for its application, Producing ideal tensile strength MAF is an appropriate process for its application, rather than an ECAP or ARB process, the MAF process is more appropriate for processing Aluminum-based composites in producing high mechanical properties. Development of the latest SPD technology RPRF is present as specific for the processing of aluminum-based composite materials, in the process procedure have 2 stages, namely repeated pressure to immerse the reinforcement particles and rollers to distribute the reinforcement to evenly.

Acknowledgements:

The principal author is thankful to Faculty Engineering – University of Sultan Ageng Tirtayasa, for supported research grant "Hibah Penelitian Dosen Madya".

References

1. Agus Pramono, Investigation of Severe Plastic Deformation Processes for Aluminum Based Composites. Ph.D. Dissertation, Department Mechanical Engineering. Tallinn University of Technology., Tallinn-Estonia, (2016).
2. Senkov, O.N.; Miracle, D.B.; Scott, J.M.; Senkova, S.V., Equal channel angular extrusion compaction of semi-amorphous Al85Ni10Y2.5La2.5 alloy powder, Journal of Alloys and Compounds, 1,**365**, 126-133 (2004).
3. Robertson, J.; Im, J.T.; Karaman, I.; Hartwig, K.T.; Anderson, I.E., Consolidation of amorphous copper based powder by equal channel angular extrusion, Journal of Non-Crystalline Solids, 1-2, 144-151, **317** (2003).
4. Megumi Kawasaki, Yi Huang, Cheng Xu, Minoru Furukawa, Zenji Horita, Terence G. Langdon. Quantitative study of cavity development in the tensile testing of an aluminum metal matrix composite processed by equal-channel angular pressing. Materials Science and Engineering A. 402–407, **410** (2005).
5. Argentero, S., Accumulative roll bonding technology of aluminum alloys, Proceedings of Strategic Management Factors of MNC's Subsidiaries–Comparative Analysis of Metal Manufacturing and Other Industries in the Czech Republic, 1-6 (2012).
6. Sajjad Amirkhanlou, Mostafa Ketabchi, Nader Parvin, Shohreh Khorsand, Reza Bahrami.Accumulative press bonding; a novel manufacturing process of nanostructured metal matrix composites. Materials and Design, 367–374, **51,** (2013).
7. Sajjad Amirkhanlou, Mostafa Ketabchi, Nader Parvin, Shohreh Khorsand and Fernando Carreño. Manufacturing of nanostructured Al/WCp metalmatrix composites by accumulative press bonding. IOP Conf. Series: Materials Science and Engineering **63** (2014) (6th International Conference on Nanomaterials by Severe Plastic Deformation IOP Publishing).
8. Sajjad Amirkhanlou, Masoomeh Askarian, Mostafa Ketabchi, Navvid Azimi, Nader Parvin, Fernando Carreño. Gradual formation of nano/ultrafine structure under accumulative press bonding (APB) process. Materials Characterization, 57–65, **109** (2015).
9. Sajjad Amirkhanlou, MostafaKetabchi, NaderParvin, Masoomeh Askarian, Fernando Carreño. Achieving ultrafine grained and homogeneous AA1050/ZnO nano composite with well-developed high angle grain boundaries through accumulative press bonding. Materials Science & Engineering A, 374–380, **627** (2015).
10. Agus Pramono, Lauri Kollo and Renno Veinthal, Aluminum based composite by novel process: repetitive press roll bonding (RPRB), Procedia Chemistry, 473-479, **16** (2015).
11. Agus Pramono, Merging method of aluminum fiber composite materials with alternating pressure and rolling process using Repetitive Press Roll Forming technology. Application number: 001/P/ HKI/2000. University of Sultan Ageng Tirtayasa Banten. Patent, **P00201709889** (2017).
12. Agus Pramono, Lauri Kollo, Kaspar Kallip, Renno Veinthal and Jaana-Kateriina Gomon. Heat Treatment of Ultrafine Grained High-Strength Aluminum Alloy. Key Engineering Materials Vol. **604** pp 273-276 (2014).
13. Zrink J, Dobatkn V, Mamuzic I. Processing Of Metals By Severe Plastic Deformation (SPD) – Structure And Mechanical Properties Respond. Metalurgija. **47** (3) 211-216 (2008).

14. Agus Pramono, Lauri Kollo and Renno Veinthal. Hot and cold regions during accumulative roll bonding of Al/Al_2O_3 nanofibre composites Proceedings of the Estonian Academy of Sciences. **65**, 2, 132–137 (2016)
15. Agus Pramono. Lembit Kommel. Lauri Kollo and RennoVeinthal, Hot and Cold of Pressing Effect on ECAP-Parallel Channel Composite Based on Al/ANF Material, Advanced Materials Research. **1123**: 343–347 (2015).
16. Agus Pramono, Lauri Kollo, Kaspar Kallip, Renno Veinthal and Jaana-Kateriina Gomon. Heat Treatment of Ultrafine Grained AA6061 Consolidation by Equal Channel Angular Pressing. Applied Mechanics and Materials Vol **771,** pp 252-256 (2015).
17. Agus Pramono, Anistasia Milandia, Kurnia Nugraha and Moch Fawaid. Aluminum Alloys by ECAP Consolidation For Industrial Application. VANOS Journal of Mechanical Engineering Education. Vol 2. **2 .** 117-125 (2017).
18. Agus Pramono. The new of technology Severe plastic deformation (SPD) For Military Device Applications. Journal DEFENDONESIA. Vol 3, **1:** 10-18 (2017).

Augmented Reality Chemical Reaction with User-Centered Design

Finsa Nurpandi[1*], *and Agung* Gumelar[1]

[1]Informatics Engineering Program, Faculty of Engineering, Suryakancana University, Indonesia.

Abstract. One of chemistry is the chemical element that is represented by the symbol on the periodic table. The low level of activity, interest, and the result of chemistry learning in school is caused by the students generally having difficulty in solving problems related to chemical reactions. In addition, most of the chemical concepts are abstract so it is difficult to imagine the structure of molecules clearly. Augmented Reality can integrate digital elements with the real world in real time and follow the circumstances surrounding environment. Augmented Reality can provide a new more interactive concept in the learning process because users can directly interact naturally. By using Augmented Reality, the atoms in the periodic table will be scanned using a camera from an Android-based smartphone that has installed this app. The scan results are then compared with existing data and will show the molecular structure in three-dimensional form. Users can also observe reactions between atoms by combining multiple markers simultaneously. Augmented Reality application is built using the concept of user-centered design and Unity with personal license as development tools. By using this app, studying chemical reactions no longer requires a variety of chemicals that could be harmful to users.

1 Introduction

Augmented Reality (AR) is a technology where there is integration between digital elements with the real world in real time and follow the circumstances surrounding environment. Azuma [1] proposes augmented reality as the incorporation of real and virtual objects in real environment that runs interactively in real time, where virtual objects are integrated in the real world. The basic concept of Augmented Reality is to take objects or elements in the real world as a base (marker) using the base image, then add a virtual element to be accepted by the user. This virtual element can be text, audio, visual, or any other form. Many of the benefits of the development of Augmented Reality that can be applied to various human activities, especially in the field of education. This is because Augmented Reality technology can provide a new way more interactive in the learning process because users can directly interact naturally.

One of the materials in chemistry is the chemical elements depicted using symbols in the periodic table. The use of this symbol is due to the chemical elements can't be observed

* Corresponding author: finsa@unsur.ac.id

using the senses. Based on research by Sunyono [2] in 2005, low activity, interest, and low student's chemistry result is caused by the general difficulties in solving problems related to chemical reaction and chemical count. This shows that chemistry is difficult to learn, one of which is because most of the chemical concepts themselves are abstract and learners can't imagine the molecular structure clearly. Chemical practice conducted in schools can't be done optimally, because it requires a long time of preparation in a limited time, so that student learning time can't be utilized optimally. Expenditure of funds large enough for each lab also became one of the reasons lab chemistry in school is not optimal. Some of the substances that are harmful to students also become one of the problems that arise if the chemistry lab is not supervised by experts.

By using AR, the atoms in the periodic table will be scanned using the device in this case is the camera, then will be compared with the existing data in the application and bring up the molecular structure in 3D form. Users can also observe the reaction between atoms by combining multiple markers at once. Making this application itself using Unity with personal license and Vuforia Software Development Kit (SDK). Using this app, to learn about chemical reactions does not require expensive and dangerous equipment and lab materials, ordinary people can learn chemical reactions only with the help of android-based smartphones and cards that serve as markers.

2 Related Work

According to Banu [3] and Purnama [4] in their work that made AR based Geometry learning system. Banu explained that the system he developed would produce 3D geometry objects derived from 2D geometry handwritten. While Purnama made an AR system specifically intended for elementary school as a learning process in understanding geometry.

Martinez [5] and Lee [6] in their work made a lesson to improve of using vocabulary and grammatical structures for English using AR technology.

The AR system developed by Nanthanasit [7] is a system for learning art. The system will explain how light reacts to objects such as sphere, cylinder, and cube. While based on what was done by Chen [8], is an AR system that is used to display a graphical representation of engineering structure.

3 Methodology

This application was developed using Multimedia Development Life Cycle (MDLC) model. As shown in Fig 1, Binanto [9] describes a software development model derived from Luther [10] consisting of six stages: concept, design, collecting material, assembly, testing, and distributing. This software development model is intended for multimedia development.

The concept stage is the initial stage to determine the type of application and purpose for which the application is made. The next stage, design, make specifications about the program architecture, style, appearance, and material needs. Material collecting is the stage by which the collection of materials matches the need. This stage can be done parallel to the assembly stage, the stage where all objects or multimedia materials are created. In some cases, the material collecting stage and assembly stage will be worked out in a linear fashion not parallel. The testing phase is also referred to as the alpha test stage (alpha test) where the test is performed by the manufacturer or the manufacturer's own environment. This last stage will store multimedia applications in a storage medium and distribute it.

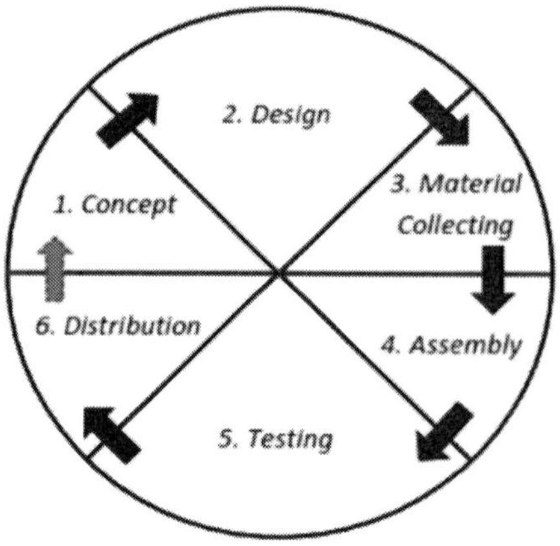

Fig. 1. MDLC Process

4 Design System

4.1 Use Case

In this system, there are two actors namely User and Camera. Users are represented as people who operate the system, and the camera as a device that interacts with markers. As shown in Fig 2, there are four Use Cases in this system. Users can scan markers and view a list of existing chemical reactions and view application info, and the camera can identify the markers that were previously directed by the user.

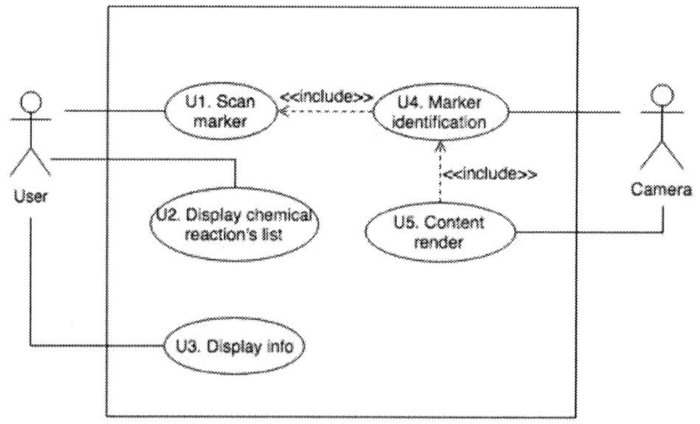

Fig. 2. Use Case Diagram

4.2 Activity Diagram

The Activity diagram as shown in Fig 3, describes the entire process of the Augmented Reality Chemical Reaction system. In the process User is given three menu options, namely menu Scan, Minipedia, and Info. Scan menu is the main process of this system. The system will scan the marker that has been given. When identifying the marker, the system will check whether the marker is in the library or not. The library marker is stored in the Vuforia SDK, which previously the marker image must be uploaded to Vuforia SDK. If yes, then the system will display 3D objects and their chemical reactions.

The Minipedia and Info menu displays only available chemical reaction information and information about the application developer. Chemical reactions contained in Minipedia is a chemical reaction that can be obtained from the scanned Marker. In a sense, the number of chemical reactions in Minipedia is equal to the number of markers that can be scanned.

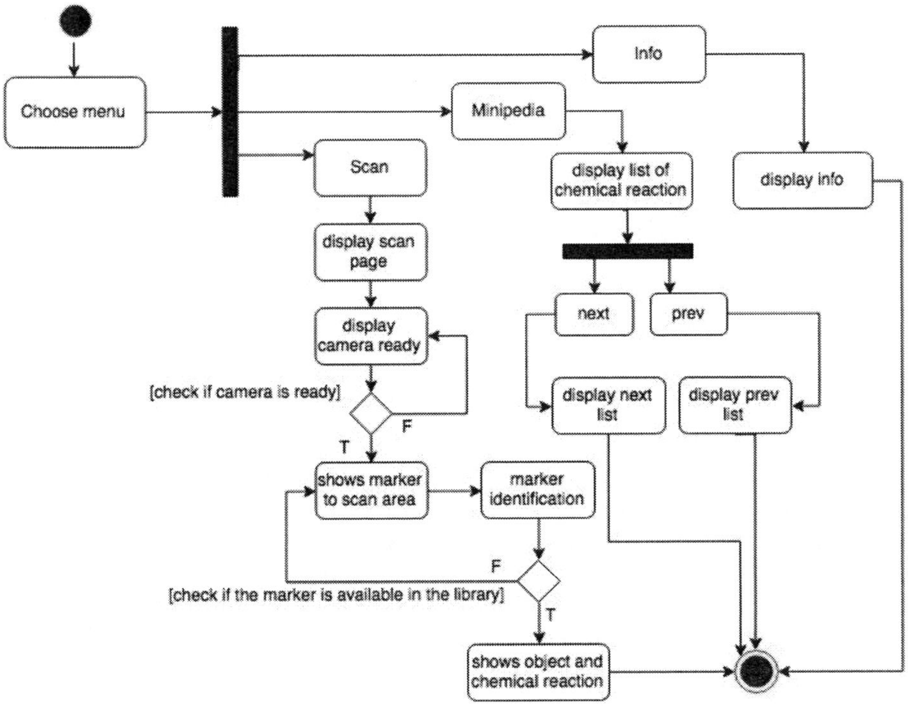

Fig. 3. Activity Diagram

4.3 Class Diagram

In this stage, classes will be created in the structure of the program, in which there are inputs, processes, outputs and data. Each class will connect with other classes or with attributes of other classes. In this stage, it takes a Class Diagram to be used as a blueprint or design for the structure of the program to be created.

Based on Fig 4, Class Diagram Augmented Reality Chemical Reaction application can be known relation of all existing classes in the system, Content class has a relationship composition with Scan class, this is because the class has a dependence on the Scan class. Scan class has associate relationship with Menu class. Menu class associated with the Minipedia class.

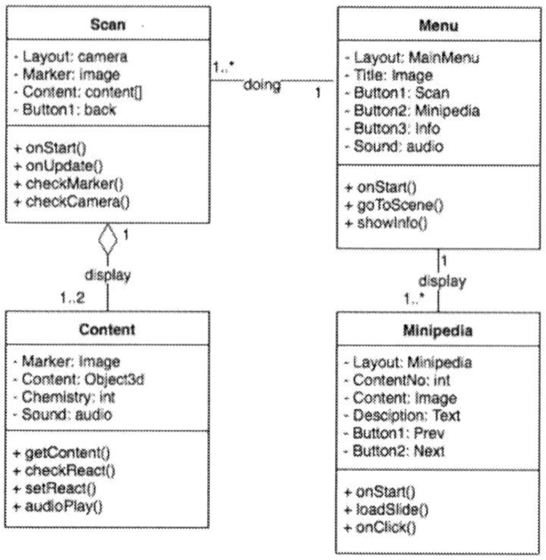

Fig. 4 Class Diagram

5 Result

In the selection of the initial interface, the use of less-bright colors and easily readable fonts is generally suitable for the user, the selection of instrumental music can also be used to avoid boredom. For the interface design is obtained as follows:

a. Use blue as the background colour of the app, because blue represents innovation, knowledge.
b. Use bright colors like red and yellow to attract user attention or warning.
c. Use a small amount of color so that users do not feel confused with the number of colors.
d. Use Bebas Neue fonts, because in this chemical simulation media application is required is the ease of users in reading and understanding writing, not the beauty of shapes and colors.

The interface of the AR Chemical Reaction application can be viewed as in Fig 5.

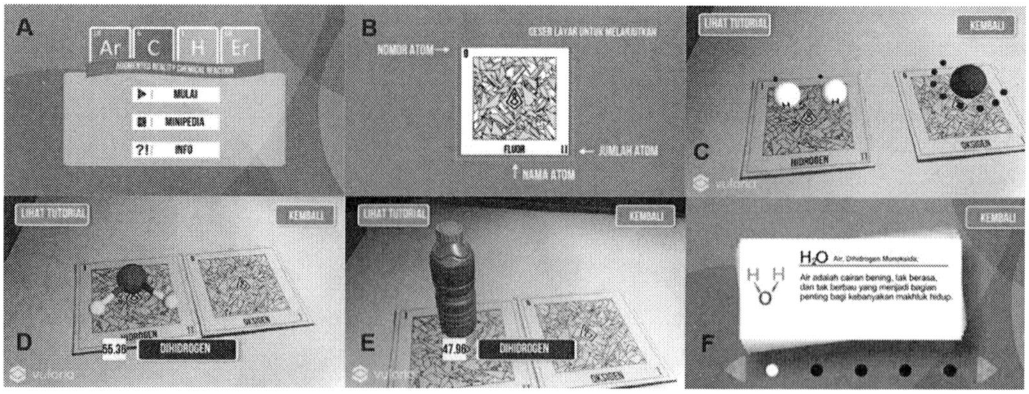

Fig. 5. Interface of AR Chemical Reaction, (A) Main Menu, (B) Tutorial scan marker, (C) Scan marker, (D) Scan with combine marker, (E) Scan combine marker with object < 50 from camera, (F) Minipedia page

6 Conclusion

AR Chemical Reaction application serves as a simulation media of chemical reactions that can be used by students, teacher, and the general public. Inside the user can see a list of chemical reactions and can perform simulations of basic chemical reactions. For its development, some features can be added to this app. Such as adding a quiz feature to guess and calculate the chemical reaction of a given problem.

References

1. R. T. Azuma, "A Survey of Augmented Reality Navigation," *Presence Teleoperators Virtual Environ.*, vol. 6, no. 4, pp. 355–385, 1997.
2. Sunyono, *Optimalisasi Pembelajaran Kimia Kelas IX Semester I SMA*. 2005.
3. S. M. Banu, "Augmented reality system based on sketches for geometry education," *2012 Int. Conf. E-Learning E-Technologies Educ. ICEEE 2012*, pp. 166–170, 2012.
4. J. Purnama, D. Andrew, and M. Galinium, "Geometry Learning Tool for Elementary School using Augmented Reality," *Iaict*, no. August, pp. 145–148, 2014.
5. A. A. Martinez, J. R. L. Benito, E. A. Gonzalez, and E. B. Ajuria, "An experience of the application of Augmented Reality to learn English in Infant Education," *2017 Int. Symp. Comput. Educ. SIIE 2017*, vol. 2018–January, pp. 1–6, 2018.
6. L.-K. Lee, C.-H. Chau, C.-H. Chau, and C.-T. Ng, "Using Augmented Reality to Teach Kindergarten Students English Vocabulary," *2017 Int. Symp. Educ. Technol.*, pp. 53–57, 2017.
7. A. Nanthanasit and N. Wongta, "Approach augmented reality real-time rendering for understanding light and shade in art education," *3rd Int. Conf. Digit. Arts, Media Technol. ICDAMT 2018*, pp. 71–74, 2018.
8. H. Chen, K. Feng, C. Mo, S. Cheng, Z. Guo, and Y. Huang, "Application of augmented reality in engineering graphics education," *ITME 2011 - Proc. 2011 IEEE Int. Symp. IT Med. Educ.*, vol. 2, pp. 362–365, 2011.
9. I. Binanto, *Multimedia Digital Dasar Teori dan Pengembangannya*. Yogyakarta: Andi, 2010.
10. L. A. C, *Authoring Interactive Multimedia*. Massachusettes: Academic Press, Inc, 1994.

Quality Control of the Steel Wire Rod Product by Integration Lean Six Sigma and Taguchi Method

Asep Ridwan[1,*] , *Ratna* Ekawati[1], and *Ayu* Novitasari[1]
[1]Department of Industrial Engineering, Sultan Ageng Tirtayasa University, Banten, Indonesia

Abstract. RMW Inc. is the global company in Indonesia. One of its product is the steel wire rod. This research aims to control the production process and improve the product quality. This research is using the integration between lean six sigma and taguchi method. The first step is to reduce the waste using lean six sigma method while the second step is to reduce the defect using taguchi method. Based on the identification of seven waste, the most dominant waste is the defect that is caused by there is no setting of the roll entry guide. The product has 11 CTQ (critical to quality) with the focus on laps defect, underfill defect, and overfill defect. Then, the cause of failure is identified using FMEA (failure mode effect analysis). The level of sigma capability is achieved at 4.54 with DPMO (defect per million opportunities) is 4006.55. The first improvement can reduce the time until 8.9 minutes. The second improvement sets the roll entry guide. The optimum factors for setting of the roll entry guide is A factor with 150 tonnages of wear level, C factor by 5 bar of cooling and D factor by 6 ml of lubrication.

1. Introduction

1.1 Background

RMW Inc. is a steel company consisting of several factories. One of the factory is the wire rod steel. Over the past few years, RMW Inc. often perform production processes resulting in waste and declining product quality. SWRM 8 is the product that has the largest number of defects compared to other grades. The total defect of SWRM 8 is 404 coils of total production of 100959 coils. Improving the quality and control of production processes needs to be done continuously by eliminating waste to produce optimal production process and reduce the defect of steel wire rod products SWRM 8.

Many researchers have developed lean six sigma in manufacturing companies [13,14,2]. At the same time, some researchers have investigated taguchi method in manufacturing companies [17,16,18,15]. The integration of lean six sigma methodology and taguchi method has not yet considered by previous studies.

Based on the problems contained in the wire rod steel factory of RMW Inc. and previous studies, this research proposes to build a model of quality control that integrates a lean six sigma methodology with taguchi method to control and improve the production process and the product quality in manufacturing companies.

[1] Corresponding author: asep.ridwan@untirta.ac.id

1.2 Research Objectives

Based on the formulation of the problems, the purpose of this study are as follows:

1. Identify the waste contained in the production process of SWRM 8 in the wire rod factory of RMW Inc.
2. Identify factors causing defects in SWRM 8 products.
3. Determining the level of sigma capability in the production process of SWRM 8 products.
4. Designing proposed improvements to the production process of SWRM 8 products.
5. Determining the optimum factors to minimize the potential failure of the product.

2. Literature Review

2.1 Lean

The concept sourced from the National Institute of Standards and Technology (NIST) in the United States states that, lean is an approach used to identify and eliminate activities that do not add value (waste) continuously by distributing products only to increase consumer needs [4].

2.2 Seven Waste

According to [9], waste in Toyota's production system has seven types consists of over production, waiting, transportation, unnecessary inventory, unnecessary motions, and defects. Waste consists of three different types of activities that occur in the production system operations performed:

1. Value added
   All activities that produces products or services that provide added value for consumers, then consumers are willing to pay for the activity.
2. Necessary non value added
   Any activity that produces products or services that do not add value to the consumer but is required and under current operating procedures, unless there is a change in the existing process.
3. Non value added
   All activities that produces products or services that do not provide added value for consumers.

2.3 Value Stream Mapping (VSM)

Value stream mapping (VSM) is a mapping method that shapes the flow of the value of the production process from the initial arrival of the raw material (supplier) to the output and customer. The VSM is made in detail to identify and eliminate waste, and also to improve the flow of production process [2].

2.4 Six Sigma

Six Sigma Methodology is a vision of quality improvement towards the target of 3.4 failure per million opportunities (DPMO) for each product transaction, goods or services [11]. According to [10], there are five steps for implementation of quality improvement with six sigma method using DMAIC (define, measure, analyze, improve, and control). Control stage as continuous process control to improve process capability toward six sigma target [1]. Lean six sigma is a combination of lean concept and six sigma methodology [11].

2.5 Failure Mode and Effect Analysis (FMEA)

FMEA is a methodology used to evaluate failures occurring in a system, design, process, or service [3]. Failure mode and effect analysis (FMEA) is a procedure to identify and prevent the failure of a product so that the output of the production can meet company standard [5]

2.6 Taguchi Method

Taguchi method is a methodology in the engineering field that aims to improve the quality of products and processes and can reduce costs and resources to a minimum. The target of taguchi method is to make the product robust to noise, because it is often referred as robust design [8].

3. Research Methods

This research using integration of lean six sigma and taguchi method. The lean six sigma is required to reduce the waste and the Taguchi method to reduce the defect, one of the seven waste based on Toyota's production System.

The flow of research is shown in the figure 1 as follow:

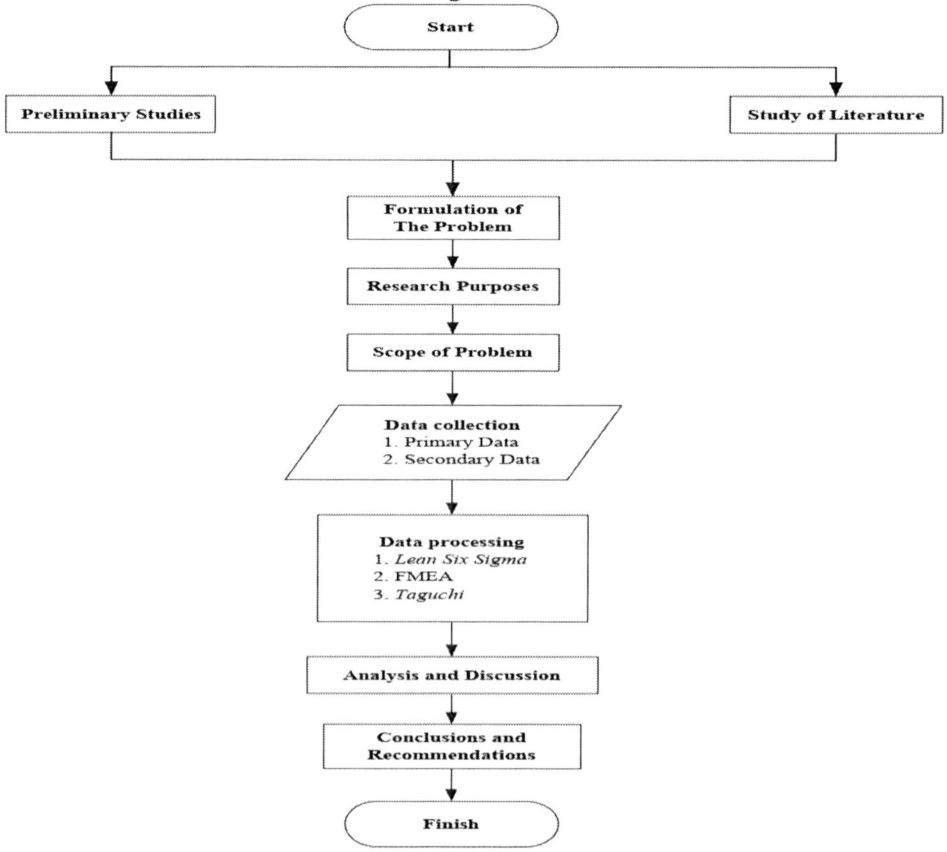

Fig.1. Flow Chart Research

4. Result And Discussion

4.1 Critical Identification of Quality

CTQ is an attribute that must be realized because it relates to customer needs and satisfaction. In SWRM 8 product there are 11 quality characteristics. The first process improvement key is to identify the main area (the main problem) and focus on the main problem. The pareto principle states that about 80% of the problems are caused by 20% of the causes [12]. Type of defect that must be overcome based on the rules of the pareto diagram are laps defects, underfill defects, and overfill defects.

4.2 Calculation of DPMO and Sigma Levels

The result of the Cp value is 1.514. It shows that the production process of SWRM 8 product still needs to be improved continuously till the target of 6-sigma.

4.3 Current State Value Stream Mapping SWRM 8

Current state value stream mapping of production process SWRM 8 is used to know the percentage of efficiency of production process flow of SWRM 8. The calculation of this efficiency is 61.13%.

4.4 Waste Identification

In the waste identification stage using questionnaires. Questionnaires were filled by 6 respondents, including four supervisors, one senior engineering, and one production cadre. The results obtained for the most dominant waste is waste defect, then brainstorming with one senior engineering to identify the cause of the Five why method that is currently used to identify the causes of defective product using the 5 why method. The next stage creates mapping tools using VALSAT (value stream mapping tools). Based on mapping tools, the tools used in this research is PAM (process activity mapping).

4.5 Identification of Production Process Activities

In this research, the whole time of value added activities are 59.83 minutes with a percentage of time is 61.13%. All non value added activities are 38.04 minutes with a time percentage is 38.87% and no activity is included in non value added activities. The total time is 97.87 minutes. The proposed improvement is done with the design of action based on the highest RPN value of FMEA and optimum factors of setting the roll entry guide using taguchi method.

4.6 Taguchi Method

The following are steps of applying taguchi method in this research:

1. The quality characteristics determination for SWRM 8 products is smaller the better so that the quality characteristics observed in SWRM 8 products are defective or not defective in SWRM 8 with the purpose to minimize the defect categories.
2. Application of orthogonal arrays. The degree of freedom in this research is 8 then the orthogonal matrix chosen is $L_9(3^4)$, which shows 3 levels, 4 factors, and 9 experiments [6].
3. Determining the number of levels and level values.

Table 1. Determining Amount And Value Factor Level

Code	Influential Factors	Level Fator		
		1	2	3
A	Wear level	150 tonnages	200 tonnages	300 tonnages
B	Assembly unit	<1 mm	±1 mm	>1 mm
C	Cooling	4 bar	5 bar	6 bar
D	Lubrication	3 ml	6 ml	9 ml

4. The Taguchi experimental results data shown that 3 replications were performed for each combination of factor level. The data can be seen in table 2.

Table 2. Taguchi Experimental Results

Control Factor				Results			Average
A	B	C	D	1	2	3	
1	1	1	1	12	12	12	12,000
1	2	2	2	10	10	10	10,000
1	3	3	3	12	12	11	11,667
2	1	2	3	12	13	12	12,333
2	2	3	1	13	11	13	12,333
2	3	1	2	13	13	13	13,000
3	1	3	2	10	11	11	10,667
3	2	1	3	12	12	12	12,000
3	3	2	1	12	12	11	11,667

Table 3. Average ANOVA

Source	SS	D_f	MS	F_{ratio}	SS'	Ratio%
A	9.185	2	4.593	15.500	8.593	34.118
B	2.074	2	1.037	3.500	1.481	5.882
C	4.963	2	2.481	8.375	4.370	17.353
D	3.630	2	1.815	6.125	3.037	12.059
Error	5.333	18	0.296	1	7.704	30.588
SSt	25.185	26			25.185	100
Mean	3721.815	1				
SStotal	3747	27				

The results obtained for optimum level combinations are A level 1 factor (level of tire wear roll entry guide 150 tonnages), C level 2 factor (cooling 5 bar), and level D factor 2 (6 ml lubrication).

The next step is calculating the confirmation experiment. Confirmation experiment results can be accepted with confidence interval because the confirmation experiment is still within the optimum yield interval. This means that the results of the taguchi experiments can be applied and the optimal level setting can be used as a reference in the production process of SWRM 8 products to reduce product defects. According to [7], decisions of optimal conditions can be accepted or not, with compare the average estimation value and the average confirmation experiment results with each trust interval.

4.7 Future State Value Stream Mapping Production SWRM 8

The following suggestions for improvement of NNVA activity include giving directions of vehicle entrance and exit reduced transport time from billet check to raw material warehouse by 3.2 minutes, cleaning magnet crane 401 from billet reduced the time of billet transfer by 1.1 minute and billet transportation from the raw material warehouse to the billet hit number checking reduced the time by 2.02 minutes, reducing the checking time of the billet hit number by positioning the billet hit number facing the pulpit so that the time is reduced by 2.6 minutes.

Based on the proposed future state value stream mapping product SWRM 8, produces stream flow production efficiency of wire rod SWRM 8 67,26%. The efficiency of the production process flow increase by 6.13 %.

5. Conclusion

The following is a conclusion obtained on the research as follow :

1. Waste in the production process of SWRM 8 in the wire rod factory of RMW Inc. are overproduction, transportation, waiting, excess process, inventory, motion and defect. Waste defect is a priority for improvement because it has the largest average value.
2. Factors that cause the dominant defect in SWRM 8 is no standard setting of roll entry guide.
3. The level of sigma capability in the production process of SWRM 8 is 4.54 with DPMO (defect per million opportunities) is 4006.55.
4. Proposed improvements to the production process of SWRM 8 in the wire rod factory of RMW Inc. are:
   a. Reducing the billet checking time to the raw material warehouse by providing road signs and directions for vehicle directions so the time can be reduced by 3.2 minutes.
   b. Reducing the billet loading time by cleaning the magnet crane 401 so can be decreased by 1.1 minutes.
   c. Reducing the billet transportation time from the raw material warehouse to the billet hit number checking by clearing magnetic crane 401 so the time can be reduced by 2.02 minutes.
   d. Reducing the checking time of the billet hit number by positioning the billet hit number facing the pulpit so can be decreased by 2.6 minutes.
5. Designed improvements to minimize the cause of potential failures on SWRM 8 products is to determine the standard setting of the roll entry guide. The optimum factors obtained include A factor with level 1 (150 tonnages of wear level), C factor with level 2 (5 bar of cooling) and D factor with level 2 (6 ml of lubrication).

Reference

1. A. Ridwan, Kulsum, and S. Murni, Supply Chain Performance Measurement with Lean Six Sigma Supply Chain Management Approach, *J. Indust. Serv.* **3**, 1a (2017).
2. A. Ridwan, P.F. Ferdinant, and R. Aliandru. Improvement Design of Deformed Production Process 16 Using the Lean Six Sigma Method Through Value Stream Mapping and Design of Experiment, *J. Indust. Serv.* **3**, 2(2018).
3. D. H. Stamatis, *Failure Mode and Effect Analysis: FMEA from Theory to Execution*, (1995).
4. D. Sarkar, *Lean for Service Organizations and Offices : A Holistic Approach for Achieving Operational Excellence and Improvements* (2008).
5. D.F. Mayangsari, H. Adianto, and Y. Yuniati. Product Quality Control of Isolator with Failure Mode and Effect Analysis (FMEA) Method and Fault Tree Analysis (FTA), *J.Ind. Eng. Itenas* **3**, 2 (2015).
6. I. Soejanto, *Experiment Design with Taguchi Method* (2009).
7. N. Belavendram, *Quality by Design : Taguchi Techniques for Industrial Experimentation* (1995).
8. N. Fitria, Analysis of the Taguchi Experimental Design Method in Quality Characteristics Optimization, *Thesis* (2009).
9. P. Hines and D. Taylor, *Going Lean* (2000).
10. P. Pande and L. Holpp, *What is Six Sigma ?* (2002).
11. V. Gaspersz, *Lean Six Sigma for Manufacturing and Service Industries* (2007).
12. V. Gaspersz, *Total Quality Management* (2005).
13. W.M. Goriwondo and N. Maunga. Lean Six Sigma Application for Sustainable Production: A Case Study for Margarine Production in Zimbabwe. *Int. J. of ITEE* **1**, 5(2012).
14. S.Gultom, T.S.Sinaga, and S.Sinulingga, Study of Quality Control Using Lean Six Sigma Approach at XYZ Inc., *J. Ind. Eng. FT USU* **3**, 2 (2013).

15. D.M. Kastanja, N.Widha S., and R.Yanuar E., Application of the Taguchi Method to Reduce the Number of Defective Products of Standard Candles HAN 17, *J. Eng.Ind. Syst.Mgt* **3**, 3. (2015).
16. A.Telaumbanua, K. Siregar, and T.S. Sinaga, Quality Control Analysis with Taguchi Method Approach at Asahan Crumb Rubber Inc., *J. Ind. Eng. FT USU* **3**, 5 (2013).
17. P. Sidi, and M. T. Wahyudi, Application of Taguchi Method to Know the Roundness of CNC Lathe Process. *J. Mach. Eng.* **4**, 2 (2013).
18. E.A. Sigalingging, I.P.Tama, and L. P. Riawati, Application of Lean Manufacturing to Reduce Waste in Production of Cigarettes Filter Using WAM and Taguchi Method, *J. Eng. Ind. Syst Mgt* **2,** 3 (2014).

Development System for Reducing Fuel Consumption with Logarithmic Approach for Aperture Valve System on Throttle Body

Muhamad Farhan[1*], *Mohamad* Ramdhani[1], and *Estananto*[1]
[1]Electrical Engineering, Electrical Department, 40257 Telkom University, Indonesia

Abstract. Decreasing of petroleum reserve and increasing of fossil fuel consumed lead to an energy crisis. To anticipate that case, it is required for the development of modified motor vehicles to optimize fuel consumption. A modification toward the valve on the throttle body regulating the mixed amount of the incoming gasoline and air into the combustion chamber can be an alternative solution of the case. By changing the valve aperture system which has initially been linier become logarithmic with the gas handle, make the mixture of incoming gasoline and air into the combustion chamber more efficient. In this research Motorcycle Carburettor 110 cc is used. The result of the research shows that the fuel consumption can be optimized to 37% lower consumption compared to normal mode.

1. Introduction

Recently, there are many motorized vehicles operating in our daily life either as public transportation, delivery of goods, or just as a personal transportation [1]. A huge hike on increasing number of vehicles will have a great impact on fuel consumption, and because fossil fuel is still dominant, it will cause big problem due to limited petroleum resources [2]. More petroleum is being consumed, however in the other hand the reserve is scarcer. This might be one of the causes of rising price fuel, and it can make operational cost of vehicle rising [2]. Therefore, it is necessary to develop a system that can save the use of fuel to optimize the daily operational costs as well as to improve environmental impact of fossil fuel consumption.

In the automotive industry some efforts have been widely developed on fuel efficiency of automobile, such as throttle by wire (TBW) system that can focus on fuel efficiency and comforting while driving [3], In this research, a new approach is developed that focuses on fuel efficiency of two-wheeled vehicle with small cubical centimeter (cc), By modifying the valve aperture system which has initially been linier to logarithmic one with the gas handle, the mixing process of incoming gasoline and air into the combustion chamber can

[1] Corresponding author: m.farhanmlt@gmail.com

be more efficient, which in turn can optimize daily operational costs and improving environment quality.

2. Performance Parameters of The Machine

Specific Fuel Consumption (SFC), also known as Thrust-Specific Fuel Consumption (TSFC), is the fuel efficiency of engine to spend energy for an hour.

$$m_f = \frac{\text{sgf}.V_f.3600}{1000.t_f}\ (\frac{Kg}{Hour}) \qquad (1)$$

where m_f is fuel consumption (kg/hour), sgf is specific gravitation (gr/ml) which has a value 0.715, Vf is volume of tested fuel (ml), tf is time to deplete tested fuel (seconds).

3. Design and Application

Figure 1 describes the overall system block diagram. In this diagram, it can be seen that the accelerator uses a resistance changing sensor to provide an input signal to the microcontroller and it controls the motor servo to open the valve of the throttle body. The opening valve in the throttle body will affect the least amount of air and gasoline mixture into the combustion chamber. The process is shown in Fig. 1 and Fig. 2.

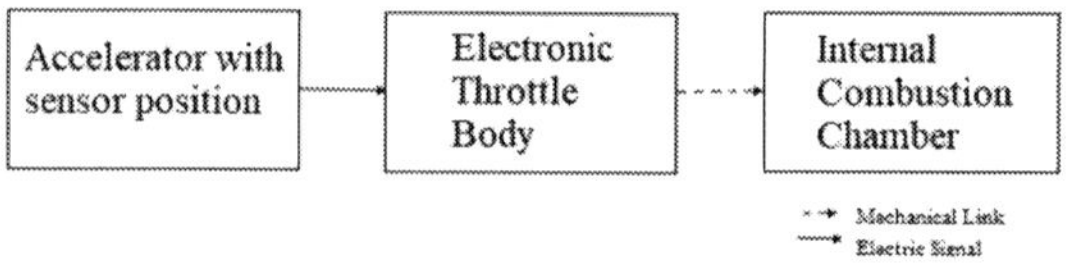

Fig. 1 Electronic Throttle System

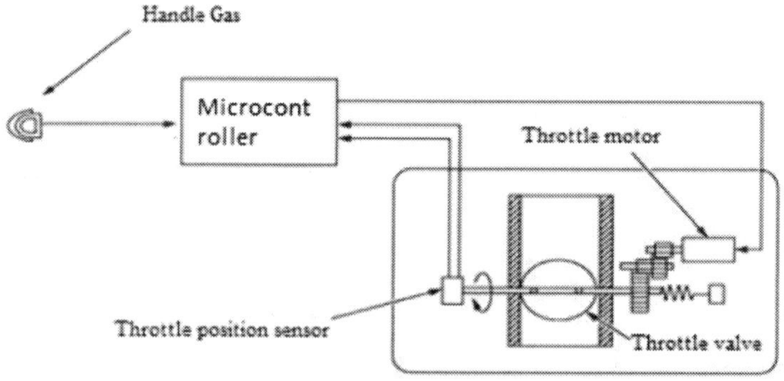

Fig. 2 Design System

By modifying the position on the accelerator, it will change the resistances from 0 - 1000 Ohm, then the value is read by the microcontroller to be processed, and output from microcontroller makes the motor servo working in two modes: linear or logarithmic, depending on which mode has been chosen for opening valve on the throttle body. The two modes are depicted in figure 3.

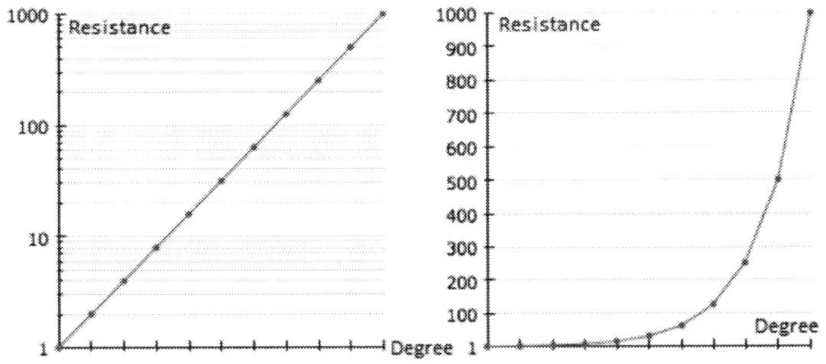

Fig. 3 Curve Changing Degree of Valve Opening System

Fig. 4 Installation of System

4. Test Results and Discussion

Before the system is installed, the motorbike under test is tested and the test data is taken. Under normal circumstances the motorbike is able to travel surrounding the field forming 1100-meter circumference circle. The motorbike's specification is 110 cc carburettor engine type and the fuel consumed is 1.05 litter Pertamina's Pertamax 92-octane gasoline. Rider's weight about 68 kg. The engine temperature after the vehicle is turned on for 5 minutes.

One day after, the test of the engine with the new proposed system with logarithmic opening valve with the same vehicle and same condition is conducted. At that time, experiment with efficient mode shows significant improved result, with no negative impact during the test. The engine works well and the system runs stable. Data is displayed in Figure 7-10.

The difference in average speeds obtained after the collection of test data between the two systems found that the speed traveled per round ranges from 30 to 40 km / h, but at the time of the round to 19 vehicles that use normal mode cannot run again due to exhausted gasoline , while vehicles that use efficient mode can still drive up to 31 rounds.

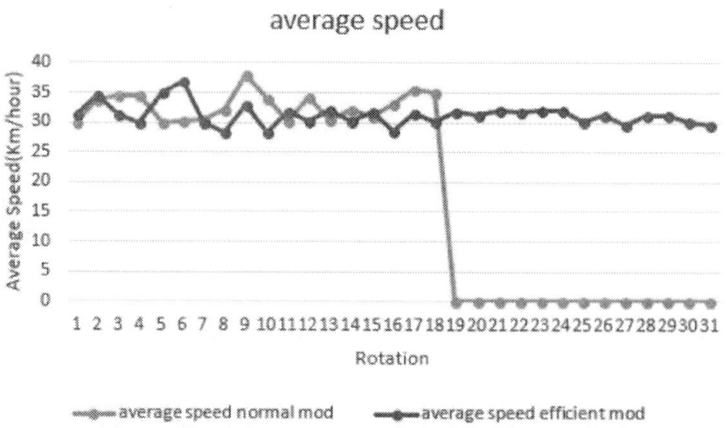

Fig. 5 Comparing average speed between two modes

Distance used when sampling ranges is between 1080 meters until 1100 meters, with the same Distance of vehicles with normal mode can only travel as much as 18 rounds.

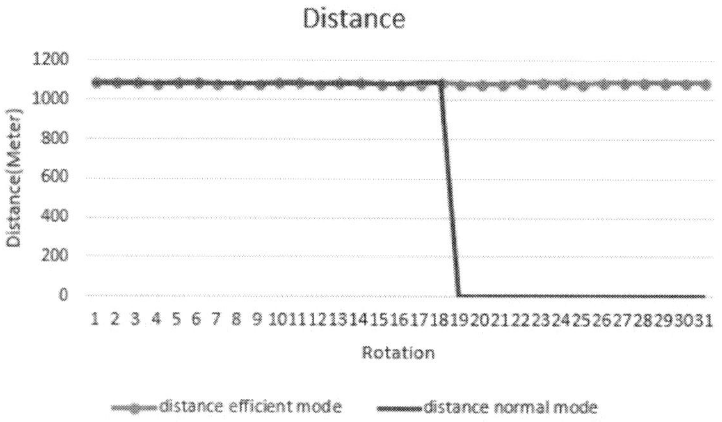

Fig. 6 Comparing Mileage between two modes

With an average speed of 30km / h -40 km / h, the distance to round the track as far as 1100 m is ranging between 120 - 140 seconds. Vehicles with normal mode can only travel as much as 18 rounds while vehicles using the efficient mode can still go up to 31 rounds.

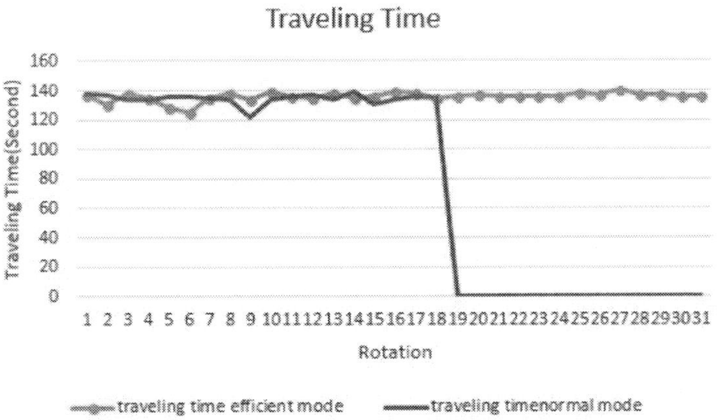

Fig. 7 Comparing Traveling time needed between two modes

5. Conclusion

According to the results obtained from the efficient mode installation it can be seen that, by using efficient mode, a vehicle with 110 cc carburettor type can save fuel usage up to 31 rounds with a distance of 33 km, in comparison in normal mode it can only take 18 rounds or 19 km, both with fuel consumption of 1 litter of Pertamax. If we calculate the figures based on fuel consumption per hour (m_f) then we can know the fuel consumption while using efficient mode is only 0.86 liter/hour, in comparison with 1.36 liter/hour in normal mode. This is 37% saving of fuel consumption in comparable same condition.

6. Future Works

Further improvement is needed to implement the proposed system in various type of motorcycles in larger scale. This concept can save up to millions of litres of fuel consumption, billion rupiah in operating costs, and can reduce much air pollution.

References

1. Giulio Panzani, Matteo Corno, Sergio M. Savaresi, "Design of an Adaptive Throttle-by-Wire Control System for a Sport Motorbike," International Federation of Automatic Control (IFAC), Sept (2011)
2. Xiong Fang, Wenwu Mao, Xisheng Shen, Jianyu LI, Wenchao Huang and Zhixin Chen, "The Quantum Energy Saver design and Fuel-saving application," IOP Conf. Series: Materials Science and Engineering 157, Aug (2016)
3. Pratik Parsania, Ketan Saradava, "Drive-By-Wire Systems In Automobiles," Conference: Journal of systematic computing, (VVP Engineering, 2012)
4. [4] Abhishek Kumar, Abhijeet Kumar, Ujjwal Ashish, Ashok B (2016). Design of Electronically Controlled Fuel Injection System For Carburetor Based Engine. IJRET(International Journal of Research in Engineering and Technology) ,Volume: 05 ,Special Issue: 03, SYNERGY-2016, Mar (2016)
5. T.Balasubramani, S.Jeyapaul, K.Karthick, M.Karthick, G.Manikandan (2016). Fuel Efficiency in A Petrol Engine by Using Water Injection. IJLTET(International Journal of Latest Trends In Engineering And Technology), Volume 6, Issue :04

Incremental Innovation towards Business Performance: Data Management Challenges in Healthcare Industry in Indonesia

Muharman Lubis[1,a], *Arif Ridho* Lubis[2], *Bastian* Lubis[3] *and Asmin* Lubis[3]

[1]School of Industrial Engineerin, Telkom University, Bandung, Indonesia
[2]Department of Computer Engineering and Informatics, Politeknik Negeri Medan, Indonesia
[3]Department of Anesthesiology and Reanimation, Universitas Sumatera Utara, Indonesia

Abstract. The Industry 4.0 indicate the increases in competitive pressures, the margins reduction, the availability of new technology and the marketing development techniques suggest making more complex decisions to make and sustain the success. Data management is essential for organizations because the administrative process, which data is acquired, validated, stored, protected and processed through its accessibility, reliability and timeliness ensures that the needs of data are met. It is also a process of developing data architectures, practices and procedures that address the data and then implement these aspects on a regular basis. All these process will ease and smoothen the business flow. Therefore, there are number of data management challenges to maintain sheer volume of data, taking reactive approach, lack of process and data handling, fragmented data ownership and driving a data culture. This study investigate the current issues in health industry to overcome through recognizing both the importance of quality data and having more sophisticated approach to manage data as the organization begin shifting to be more data centric model.

1 Introduction

The definitions of innovation for medical products may not be sufficient to evaluate medical devices and system. The characteristics that drive innovation are largely different, which the current standards and thresholds for capturing the added benefit of innovative molecules may not be representative of the marginal utility produced by other medical technologies [1]. According to data from the Ministry of Health, there are a total of 2,457 hospitals in Indonesia currently. The increase in private hospitals, both profit and non-profit has grown faster and now outnumber the number of public hospitals with 61% market share [2]. Meanwhile, at around 70% of hospitals are governmental hospitals, which hospital care services accounted for 60% of total healthcare expenditure [3]. Nowadays, organized data, systematic data retrieval and storing loads of data are crucial subject to every hospital, even to every organization, especially for large enterprises that deal with large number of data and clients' information like hospital, government body and administration department. An

[a] Corresponding author: muharmanlubis@telkomuniveristy.ac.id

innovation should bring agility on every autonomous implemented idea which further clubbed together based on symmetric behavioral pattern and business needs customization to formalize a radical innovation thus continuing the ideation and implementation cycle. A prelude to evaluate the quality of the data and the context of use is largely unknown, because the ability to assess has been offered in the generic manner [4].

Indonesia's healthcare expenditure at 2.9% of GDP is one of the lowest in the region and below ASEAN countries' average of 4%, despite growing at 14% for the last five years [2]. In the process of doing business in Indusrty 4.0, the organization will involve certain procedures, policies, regulations and problems based on the sector involved, with aligning to the customer expectations satisfaction. However, data management is one of the important solutions that require great attention from the organization to support the commercial process in terms of analysis and development. Therefore, the hospital collects many interrelated data, such as when the patient received treatment, medication, theraphy, medical condition, dose of medicine, patient diagnosis and information, medical examination, check-ups, etc. They need to be done immediately and accurately where data collection and storage process should be integrated from paper-based into digital-based record for comprehensive information retrieval and worst incident avoidance to develop database center or even data warehouse [5]. Furthermore, the changes in regulations such as negative investment lists, patents, price ceilings and anti-trusts should impact the pharmaceutical industry as a whole [2], so the health standard always developed reactively and take long time to be implemented. Therefore, innovation is required to overcome issues and challenges in the health industry, so the understanding of their characteristics can provide more insight to improve the service delivery and data management such as several criterion below.

Table 1. Innovation criterion for differences

No	Innovation Matrix	Incremental	Responsive	Disruptive	Radical
1.	Organizational Strategy	Bottom-up	Top-Down	Indirect	Direct
2.	Market Type	Low	Closed	Open	High
3.	Learning Curves	Slow	Trial	Quick	Paradox
4.	Value Creation	Followers	Collaborative	Partnership	Pioneer
5.	Product/Service Improvement	Minor	Major	Significant	Trivial
6.	Purposes	Simplification	Expansion	Penetration	Replacement
7.	Discovery Approach	Modification	Reflective	Insights	Metamorphosis
8.	Enhancement	Existing Platform	End Result	Acknowledgement	Component Delivery
9.	Research Source	Consolidation	Performance	Brainstorming	Re-conceptualization
10.	Business Process	Refinement	Exploration	Transformation	Elimination
11.	Changes	Constant	Phases	Rapid	Risky
12.	Achievement Goals	Productivity	Representational	Exponential	Opportunities
13.	Delivery Objective	Retention	Loyalty	Satisfaction	Potential
14.	General Operation	Packaging	Distinguished Features	Price Wars	Product/Service Inception
15.	Measurement	User Comforts	Added Value	Substantial Value	Competitive Advantages
16.	Evaluation Instrument	Financial Resources	Human Resources	Access Resources	Demand Resources

2 Literature review

2.1 Data Accuracy

Identification of five hospitals in Indonesia that has been using Hospital Information System (HIS) to handle their health record management, presented numerous issues that could lead to inaccurate data about patient information. For example many data in the form of physical copy, the lack of Internet infrastructure in each health division, high computer illiteracy in technical staff to operate and troubleshoot certain features in HIS, less interest to adopt digital approach, little socialization and training to relevant staff and inappropriate utility in term of electricity, backup, recovery, compression and antivirus [6]. At certain extent, the inaccuracy of data can hinder the doctors to diagnose the type of disease from patient, prevent the cross-validate data to produce fast decision making and suggest unspecific medicine that no longer as the treatment process. It estimates approximately the range of 0.5-30% errors or tiny to small differences in the databases could compromise the decision making process that might endanger the patient long term condition [7]. After registration in particular Policlinic and Posyandu, commonly patients are directed to the respected poly according to their illness or disease. Prior, they are registered into the register book in each poly, which data integrity has been less considered because every staff or other unknown people can have opportunity to change or even manipulate data indirectly or directly without notification over its entire life cycle [8].

In other context, there are also some internal management activities that do not obey the rules such as the regulation to govern hazardous waste management that has been enacted since 1999, but most of healthcare facilities have not yet complied. Although, in May 2008, the Indonesian government enacted a new Act of Solid Waste Management with the Act 18/2008, which is followed by more technical regulations and policies on how to implement waste generators. Indeed, there is also other regulation to guide the process like the Ministerial Health Decree No 1204/2004 concerning Environmental Health Standards, still the disobedience have been found regularly due to the lack of the enforcement [9]. Unfortunately, out of 75 hospitals to be investigated, only 45 hospitals (60.0%) have a sanitation unit, which can be assumed that missing information exist in the hospital in the managerial aspect especially in regard to procedure to throw away the garbage or waste after treatment service has been done [9]. Furthermore, the problem occur as there is no record at all in regard these medical waste due to the lack of documentation and platform. Actually, the information sharing mechanism between relevant unit can increases the productivity but has been done manually which tend to be forgotten during the process. Thus, the health industry should concern to eliminate the missing data and reducing overlapping of data that might also occur in term of consent and clarity to be given to the patients as there are lack of operational standard [10]. Meanwhile, basis data is prepared to overcome data inconsistency, data redundancy, data isolation, data duplication, data dependency and data accessibility by developing the uniqueness of data format and data type, while in particular hospital still use notes or excel to record the patient data, the list of operation schedule and the responsibility of the doctors [11]. Hence, 90% of raw materials for medicines in Indonesia are imported from faraway countries, which between the end of 2012 and now, the IDR has depreciated by a staggering 40% increase the price of medicine at higher cost [2]. The changes and the dependency of the imported medicine should be recorded accurately to identify the proper strategy to be made in solving the issue of the expensive price in hospital and drugstore.

2.2 Data Entry

In general, data collection of outpatient still used manual records, which often officers or staff make mistake and error in entry the data such as name and address, to some extent they even forget to input the whole data due to number of patient for intensive care due to emergency [12]. The complex cases of a new variable system of third-party actions have become other obstacles in the health information system. Historical data shows that the hospital still conserves large quantities of manual stocks or physical documents [13].

Surprisingly, employers' associations, business halls and even a number of government officials rejected the mandatory component of integrating national identity into the HIS as the lack of coordination between national hospital delay the medical treatment [14]. Multiple and mutual factors lead to invalid and inaccuracies data in the entry process, for instance person's attitude and behavior, lacking the technical skill and understanding, bad intention to hide some information, the improper interface, computer performance, personal mood, etc. After the adoption of the drug for sale, companies usually collect information about their products to ensure safety and quality, which is generally considered as the last phase of the innovation cycle as post-marketing surveillance [15] but the problem can arises as there is no monitoring procedure to keep track the patient location as the climate, eating and sleep order can influence the results against the success clinical trial in the lab. Normally, changes in patient or employee data must be determined first before the system development process begins. The marital status, the treatment method, the identification number, the patient's address, the passport number, the telephone number and the primary email are the kind of value that can change over time. Individual may be once in a single status but got married next time but the data value in the system still stores the old one. The data is valid, but the changes might affect the decision making or tracking process. On the other hand, there is also no mechanism to track and locate the origin of the data services organized by hospitals include inpatient, outpatient, emergency, referral and support after long period of time. Study to 28 hospital pharmacists found that there was much dissatisfaction with current pharmaceutical policy and participants recommended the need for improvement in the relevant regulations, which government should strictly regulate medicine prices and distribution in the market [3].

In some cases, a drug compound may not have regulatory approval to use for certain indicator but not the others. Actually, the failures in research of pharmaceutical compounds that do not meet the standards and unsuccessful to become commercialized products can provide useful information for future research. In fact, the failure rates of small molecule and biotherapeutic products are approximately 86% and 96% respectively. As the drug classes utilize new physiological pathways, future research is expected to reveal new uses for existing compounds [13-16]. The long process of exploration requires the uniqueness of data to prevent data redundancy with similar attribute such as in the model simulation and therapy process. Another risk to highlight to meet the requirements of industry 4.0 is related to the impact of Act no. 33/2014 that is effective since 17 Oct 2014 on Halal Product assurance. It requires all pharmaceutical products to be Halal certified by 2019. Naturally, there is around 90% of raw materials are imported, which is virtually extremely difficult for pharmaceutical companies to obtain the certification [2]. Meanwhile, the operational challenges has been identified in regard to the project scalability, which are small coverage in area and short term goal [7]. On the other hand, three main problem related the inventory management practice has been identified such as overstock, unjustified forecasting technique and lack of the IT support [13]. Commonly, it is reported that lacks of disabled parking areas, the absence of wheelchair accessible ramps and lack of elevators become the external factors including poor design, inadequate examination and improper equipment that make healthcare services inaccessible for people with limited mobility and hearing disabilities [14], in the end raise complex problem in data management.

3 Research methodology

This study use grounded theory to investigate the existing problem face by healthcare industry through methodic gathering and data analysis. It is started through simple questions of the current implementation of data management in healthcare industry. Then, the repeated ideas, concepts or elements are tagged with codes to be reviewed to extract new information. Meanwhile, it also use prior literature review in journals, conference proceedings, theses and articles of expert author. For data collection, health websites and public repositories are examined as well as the respected database including IEEE, Springer, ACM and Science Direct are explored and research paper within 10 years are studied to expand the scope of understanding in term of historical attempts. Across the great variety of illnesses, patients and their families are often confronted with similar types of problems. The site targets people who know about the daily troubles of rare disease patients whether that is the patient himself or other people connected to patients.

4 Fixing data problem

The lack of a common understanding of how data quality is quantified lead become prohibitive factor to exploit data accuracy lead to isolated process and partial measures taken over a period of time within the open data community [4]. Different system integration is required to provide the automatic cooperation and coordination between each organization and related institution such as regulator, health department, CRO (contract research organization), hospital, drug laboratory, clinics and university [7]. On the other hand, the cost of outsourcing can be estimated by setting the size of data storage space (TB) or number of data or images by using the clearance system to verify and validate the completeness and correctness of data. It is necessary to keep the updated information remain clear and accurate although be handled by different unit or division. The system is embedded in the HIS that accept the data sent by relevant staff periodically should be through proper check mechanism and strict authorization structure. Thus, the security procedure need to be matched with the readiness of system involving the user, activities, structure and technology in the observed environment [17]. Furthermore, there are number of regulation with specific issues have been enacted by legislative over decades that might have different principles, attributes and consideration lead to different interpretation in the court, which complicate the citizen even certain legal experts to understand the regulation [18]. However, intensive regulation that codifies an industry's past can pose a significant barrier to incremental innovator who are trying to improve on entrenched the traditional or conventional business [19].

In general, low metadata quality will affect the satisfaction about the service introduced by digital libraries. Automated metadata tools can be implemented in HIS to produce some alerts for the low quality record. The provision of earlier treatment can reduce the health costs and can eliminate the risk of chronic diseases [20]. Therefore, incremental innovation is the attempt to heighten the existing therapeutic classes by improving complex molecular structures, reformulating medicines to enchance patient administration, adding the number of available dose options, discovering new physiological interactions of known medicines, exploring new uses for existing medicines and replenishing other properties of existing medicines. It is often based on the experiences of healthcare providers and patients' needs. Prevention is always better than cure, which the advances of Big Data analytics, it is easy to capture, analyze and compare patient symptoms earlier to offer a preventive care in a better way. Clear strategy that aims and visions communicated vigorously, bottom-up innovations facilitated and funded at a considerable scale and an innovation center orchestrating, fueling, and reflecting the processes of change, these are the main ingredients to produce constant changes for incremental innovation strategy [15]. Studies of gender in the health sciences may show potential for development when viewed against the medical model of health and disability, which are unavoidably connected with poverty, inaccessible education, inadequate healthcare services and inability to access health information and employment due to discrimination on the grounds of gender and disability [9, 21]. Big data can more accurately pinpoint where education and prevention is needed to produce healthier populations at lower costs [20].

5 Conclusion

Medical data must be kept as a reference for up to 20 to 30 years, which consumed a lot money, time and effort due to its difficulty to maintain and highly risk for long-term digital data management. In the process of data storage and monitoring, the total cost account for data management in the long term, with all the tasks necessary to maintain minimum levels of service. It is also taking into account the additional limitations, such as data security, privacy protection, retention and the seamless connection to allow fluid communication within the current system. Incremental innovations can provide advances drugs through the expansion of therapeutic classes, increasing the number of available dose options, discovering new physiological reactions and improving other properties of the existing drugs. Once the new drug is introduced, the important knowledge of health care providers during the "post-marketing monitoring" phase helps researchers to identify the needs of unattended patients. This kind of innovation ranges from design modifications to the complex molecular structures. However, there may be the complexity in the improvements process but the overall goal of further innovation remains the same, which is related to the increases the number of treatment options available to patients.

References

1. C. Oriana, A. Patrizio, B.P. Roberta, C. Marianna, J. Claudio, T. Rosanna. "De innovation: The concept of innovation for medical technologies and its implicaitons for healthcare policy-making". Health Policy and Technology, **5**: 47-64, (2016).
2. E.A. Tanuwijaya. "Indonesia Industry Focus: Indonesia Healthcare Sector." DBS Group Research, (26 Nov 2015).
3. V. Gones. "Hospital pharmacy management: Australian and Indonesian perspectives." Thesis, University of Sydney, (2016).
4. H. Setiaji, F. Wahid. "Masalah implementasi sistem informasi rumah sakit: pelajaran dari beberapa proyek." Seminar Nasional Informatika Medis, p. 97, (2015).
5. A. Harsono. "Analisis implementasi sistem informasi manajemen rumah sakit umum daerah (SIM-RSUD) terintegrasi di provinsi Kalimantan Barat." Eksplora Informatika, **5(1)**, 11-22, (2015).
6. S. Sadiq and M. Indulska. "Open data: quality over quantity." Int. J. of Information Management, **37**, 150-154, (2017).
7. M. Lubis and M. Kartiwi. "Data management challenges in pharmaceutical industry." IIUM Press. Kuala Lumpur. ISBN 978-967- 4180-84-3 in M. Kartiwi and M.Z.M Akram, eds. Data management: issues, challenges and opportunities, (2011).
8. Iswahyudi, V. Sahfitri, S.D. Purnamasari. "Analisis dan perancangan basis data pasien rawat jalan pada rumah sakit Dr. Ibnu Sutowo Baturaja." Skripsi, Bina Darma, 2013.E.N.
9. S. Irianti. "Current status and future challenges of healthcare waste management in Indonesia." Media Litbangkes, **23(2)**, 73-81, (2013).
10. C.P.D. Yunita, C.T. Punami, Y. Dharmawan. "Evaluasi kegiatan manajemen sensus harian rawat inap rumah sakit umum daerah Banyumas tahun 2011." Jurnal Kesehatan Masyarakat, **1(2)**, 143-153, (2012).

11. Huda. "Desain basis data administrasi rawat inap pada rumah sakit bhayangkara akademi kepolisian Semarang." Skripsi, UDINUS, (2013).
12. Sucipto and I.U. Wardati. "Sistem informasi pendataan pasien pada rumah bersalin dan balai pengobatan pelayanan kesejahteraan umat (PKU) Muhammadiyah Batuwarno." Indonesian J. on Networking and Security, vol. **4(2)**, 27-32, (2015)
13. I.N. Rachmania and M.H. Basri. "Pharmaceutical inventory management issues in hospital supply chains." Management**, 3(1)**, 1-5, (2013).
14. H. Thabrany. "Politics of national health insurance of Indonesia: a new era of Universal coverage." European Conference on Health Economics, NHI Indonesia, (2008).
15. E. Mohr. "Incremental innovations: changing the culture of teaching and learning." Educause Review, (September/October 2007).
16. Supeno, D. Novita and F. Prihatini. "Sistem informasi rekam medik unit kebidanan dan kandungan pad rumah sakit ibu dan anak Widiyanti Palembang." Skripsi, STMIK GI MDP, (2014).
17. A.R. Ahlan, M. Lubis and A.R. Lubis. "Information security awareness at the knowledge-based institution: its antecedents and measures." Procedia Computer Science 72, pp. 361-373, (2015).
18. M. Lubis, M. Kartiwi and S. Zulhuda."Current state of personal data protection in electronic voting: criteria and indicator for effective implementation." Telkomnika, 16(1), 290-301.
19. E. Kregor. "Food trucks, incremental innovation and regulatory ruts." 82 University of Chicago Law Review Dialogue 1, (2015).
20. S. Patel and A. Patel. "A big data revolution in healthcare sector: opportunities, challenges and technological advancements." Int. J. of Information Sciences and Techniques (IJIST), vol. **6**, no. ½, (2016).
21. S. Kamal. "Pengolahan data rekam medis rumah sakit menggunakan pemograman visual foxpro-9." Majalah Ilmiah, **24(1)**, 114-120, (2017).

Enterprise application integration for high school students using blended learning system

Muharman Lubis[1a], *Rahmat* Fauzi[1] and *Arif Ridho* Lubis[2]

[1]School of Industrial Engineerin, Telkom University, Bandung, Indonesia
[2]Department of Computer Engineering and Informatics, Politeknik Negeri Medan, Indonesia

Abstract. Education has purposes to improve quality of human resources and to develop readiness to look at future. It also play critical role to reduce ignorance, recognize the potency and shape the behavior in the community. There are two types well-know of education in Indonesia, which are formal (kindergarten, elementary, junior high, senior high and university) and non-formal (computer courses, English lessons, tutoring session, vocational training and professional certification). Of course, people will go to formal education to obtain graduation degree for the necessity of finding job and position in the market. But, some people tend to seek alternative to provide adequate education for their future. It hinted certain problems have been resided in the current educational model, which focus on face to face approach. Along with the rising of IT in the era of Industry 4.0 in every aspect of daily life, people learn mostly every subject through Internet without considering much on its credibility. Thus, this paper explore the possibility of using hybrid method, which is known for blended learning to be implemented in nonformal education to help facilitate the public on information availability and develop mastery-based presentation to improve the understanding of student efficiently.

1 Introduction

The current educational system, which adopted by many countries has shortcomings such as failing to meet the individual needs for sufficient ratio between students and teachers in teaching learning system, difficult to provide active engagement to subject material in the class, concern only to intellectual quotient from student by excluding any sort assessment to others quotients (emotional, spiritual, ecological, etc.) and serving the students' differences in term of resources, performances and capabilities [1, 2]. On the other hand, eLearning approach come to the picture as if to be the solution to solve the problem, while in reality, there are others problems caused by the implementation such as limited bandwidth and quota from student to access the system, evaluation and assessment control for grading, the high rate of drop out per subject and there is lack of interactivity between student and teachers [3-5]. This new environment approach combines the positive

[a] Corresponding author: muharmanlubis@telkomuniveristy.ac.id

approach of e-Learning and classical learning to overwhelm drawbacks of current mechanism of teaching learning.

At the end of the transformation process to be the leaders that dominate the market, successful industrial companies should engage with creative way in innovate the breakthrough, with physical products in essence, add digital interfaces and innovative services based on data. Collaborative digital activities will become the mainstream process whereby the clients and the suppliers create digital ecosystems of Industry 4.0. An effective BL environment is critical to ensure the pedagogical approaches with the collaborative use of the teaching and learning, in which online learning only focus on superficial technological adoption rather than conceptual change process while offline only focus on quantity over quality [26]. By developing BL application for private and non-formal education is always challenging, especially to convince providers to put their trust and teach them on mechanism to align with their learning curves and outcomes. Even though, it is challenges the status quo, it balance both approach as the resources used will be optimized and minimalized. Actually, there are several types of BL methods proposed to use in the environment such as Rotation, Competency, Flex, A la Carte, Skill-Driven, Attitude-Driven and Enriched Virtual Model [6]. In short, parallel with the growing use of IT and the changes of student behavior to be distracted, BL offers an active learning environment with flexibility in using resources and provide interferences to increase student participation and performance in learning outcome based. Thus, this study investigates the possibility to implement the application to maximize the benefits of BL in non-formal education as it has non-strict regulated industry compare to formal education.

2 Literature review

Blended Learning can be defined as learning in an educational context characterized by a deliberate collaboration between Internet-based and classroom-based interventions to support learning process [7]. Therefore, learning process is mostly influenced by various aspect such as learning techniques, technology, individual differences, students' background, perception, content delivery, course structure, ethical, resources and management [8]. However, academicians have broad concept of the motivation as a process, not as a product, which can be distinguished by the student's belief and behavior such as choice of activities and quality of engagement [9]. In addition, comprehensive research program has defined a learning strategy as the special thought that help student to understand, learn and maintain new information through meta-cognitive, cognitive, social mediation, self-regulatory and resource management strategies [10, 11]. Despite the advantages of BL, it has been shown that most students required more training on how to use application in teaching learning environment where some consider studying online to be difficult, annoying and boring [12]. The implementation of BL requires comprehensive strategy plan, good teacher, suitable tutor training and ongoing evaluation such as WEBLEI, HELAM, TAM, ELearning Framework or Rubric-based [13].

The disappointment of BL often related to the multimedia use, self-assessment test, access to digital literature and collaborative activities but course management, material used, teacher engagement and timely feedback have been perceived well [14]. Interestingly, BL has become increasingly popular for several reasons namely cost effective, raising awareness among students, suits different learning styles, offers flexibility and reduces student anxiety [15]. In addition, future studies should revolve around how to use class time most effectively, what types of active learning exercise promote effective learning rather than performance and the attempts to move beyond using lecture as a control [16, 17]. In fact, the developmental process should be started in the early childhood, which is shaped by values, culture and experiences of families and communities, but driven by mentality and customs [18]. Therefore, it is necessary to increase the learning outcome of BL by applying constructivism and conversation theories such as discussing the relevant experience on particular topic [19]. When understanding the visualization, it is also necessary to provide some ideas and opinions for better enhancement, either through suggestion or recommendation in evaluating the system use in the environment [11, 20]. However, the extent to which the enjoyment in the online environment was largely dependent on whether students found that the material have clarity in line with the applied courses [21].

3 Research methodology

This study uses action-based and case-based research method that consist of qualitative approach include interviews, focus groups and observations in Indonesia. The researchers asked a group of students to develop the application for non-formal education market using classic lifecycle to make it easier implementation and evaluation. To analyze the requirement, the team uses Unified Model Language (UML), which involves Use Case diagrams, Class diagrams, Activity diagrams, Sequence diagrams and Entity Relationship diagrams to represent the relationship between entities and characteristics. At the development stage, it uses PHP to allow integration of service with HTML for website and MySQL for DBMS. However, the developed innovation in the culture and climate of each institution must adapt to the dynamic conversation and interaction in the institution. There is no universal approach that ensures success, nor does success come rapidly, but rather is obtained through accumulated effort over years. Meanwhile, according to survey, there are primary advantages of BL selected by students related to the flexibility to be able to complete the task anywhere and any time, the convenience of not necessarily coming to campus as often and the benefits of the online components when job responsibilities and commitments make it difficult to attend classes [22].

4 System development

4.1 Institutional goals alignment

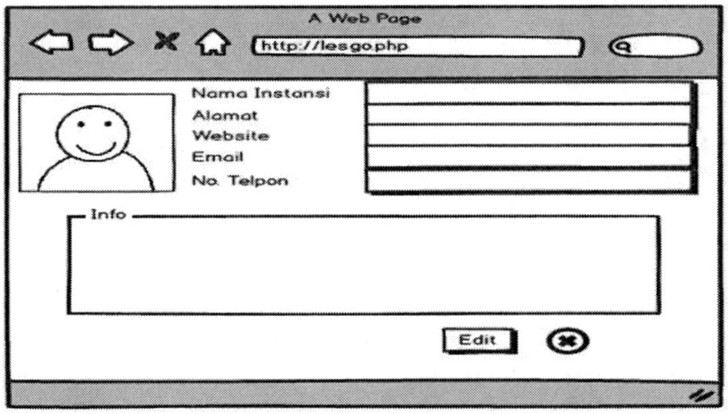

Fig. 1. Prototype of User Profile in Indonesian

Success is the achievement of specific goal in which an institution seeking to achieve it in the BL implementation. Thus, it should have the sense of purposes and the outcomes that is describe in the activity process such as division, unit, teacher or student goals. Thus, the decision makers should align the policy with the institution-centered goals to deliver the efficient use of classroom resources or expanding the reach of campus. The purpose of oriented goals can include better process through curriculum planning development and adoption of student centered of education practices such as greater convenience and flexibility, greater access, increased student academic success, or expanded information literacy. Some preliminary question also should be aligned with right strategy, such as why the institution engage in BL, what kind of advantages they want to achieve in short and long term, what type of course or program that will be offered use BL mechanism, how the infrastructure support the learning process and what level of investment should be prepared to gain return or profit [27]. The senior executives in the institution must play a critical role in formulating objectives for BL, together with the list of responsibilities to each relevant parties in teaching-learning process. However, the goals cannot be exclusive or even mandatory by the competence of superior leadership in topdown but also bottom-up approach. Over decades, several factors have been identified as discouraging academic staff from teaching use BL techniques such as inadequate support and training, not sufficient time for creating online materials, big fears of failure and lack of beliefs in technological value in education [23].

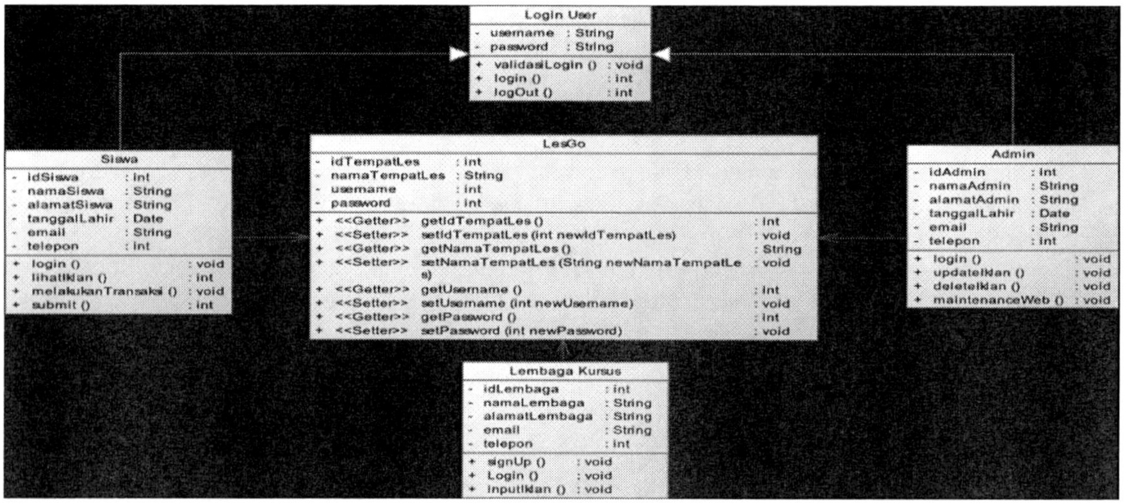

Fig. 2. Class diagram of LesGO application

4.2 Proactive Policy Development

Support for BL approach can take many forms such as direct phone support, voice mail, email, instant messaging, websites with documentation, video tutorial or drop-in centers. Alternatively, institutions can outsource support doamin to particular company to provide the service. In particular, BL requires the elements that are the same as other network services such as servers, load balancing, traffic management, network bandwidth and remote access. Those requirements are more stringent in terms of reliability and consistency compare to common network services, in which technical side should be well designed and properly monitored. The study also clearly shows that the mere application of technology and removing the physical classroom cannot immediately deliver the most ideal learning environment for foreign language learners or any subjects [24].

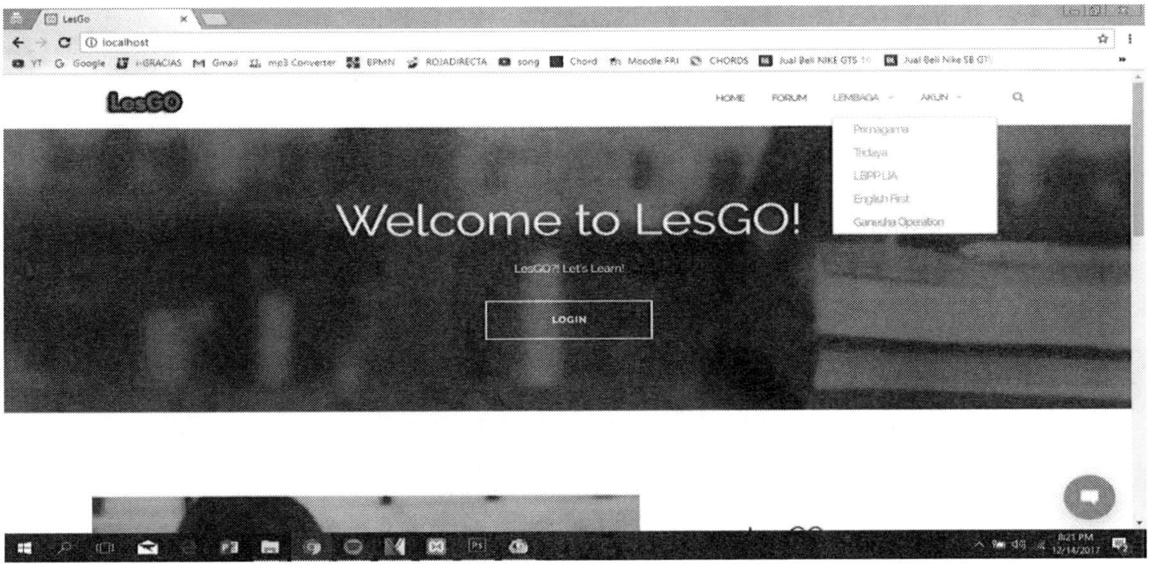

Fig. 3. Home page of LesGO application

Producing rich data sets over time brings an institution to pinpoint the effect of BL on students, staff and the teaching-learning process. Initially, the data gathering answer the questions such as does it run well or is it good method but later, the accumulated data will inform the institution to initiate continuous quality assessment to reach maturity model. Meanwhile, the changes to the existing course system, concentrate on BL approach, involving the improved topography of new generic skills or new cutting edge of technology like laser scanning or even enhanced skill-development activities, ubiquitous learning and balanced distribution of content among units, typically have been used to further alleviate the uniqueness of learning components over the others [25]. On the other hand, the ubiquity of the technology improved the efficiency and timeliness of content delivery in BL environment [26]. Organizations must understand how related products or customer services can serve as a vital basis for companies to identify customer opinions and the social, demographic and psychological factors that influence their decision making regarding the use of the connected product [27]. On the other hand, the interface is a communication mechanism between users and the system that can receive and provide information to users to support their activities until a solution is found, which presented in the figure 3.

4 Conclusion

As BL traverses almost all sector of the learning teaching environment, it requires careful software development and policy implementation. Relevant executives should consider the modalities due to their impact on infrastructure requirements, course development and strategic planning. Therefore, employees should take the opportunity to raise their potential to accommodate the learning process to be more efficient. In addition, students should reconsider their concerns about the way of the educational system work and what is requires for competency. The modality of BL should be exercised in the consistent approach with its context and objective to maintain the organizational capacity. It also requires advanced support at all stages within organizational structure and student learning support mechanisms. Of course, they must be developed in acceptable condition of the institutional culture through adequate investment of resources.

Acknowledgement

Thank you for the list of programmers from our students namely Bagus Aria Nugraha, Ramona Dwi Utari, Teguh Piganta Putra, Islamia Annida Sulistyo and Fitri Muthia Syifa.

References

1. K.L. Lalima, Dangwal. "Blended learning: an innovative approach." Universal Journal of Educational Research 5(1): 129-136. HRPUB (2017).
2. J. Kenney. "Adopting a blended learning approach: challenges encountered and lessons learned in an action research study." Journal of Asynchronous Learning Networks 15(1), 45-57, 2011. Online Learning Consortium.
3. P. Arinto. "Issues and challenges in open and distance e-Learning: perspective from the Philippines". International Review of Research in Open and Distributed Learning (17) 162-180. (2016).
4. H. Rana, Rajiv, M. Lal. "E-Learning: issues and challenges". International Journal of Computer Application 97(5), 20-24, (2014).
5. T. Capra. "Online education: promise and problems." Journal of Online Learning and Teaching 7(2), 288-293. Merlot (2011).
6. H. Staker, M.B. Horn. "Classifying K-12 blended learning". 2012. Innosight Institute.
7. R. Boelens, S. Van Laer, B. De Wever, J. Elen. "Blended learning in adult education: towards a definition of blended learning". (2015).
8. P. Ramakrisnan, Yuraidza bt Yahya., M.N.H. Hasrol, A.A. Aziz. "Blended learning: a suitable framework for e-Learning in higher education." In Procedia Social and Behavioral Sciences 67, 513-526, 2012. Science Direct.
9. W. Liu, H. Yu, "Effectiveness study of English learning in blended learning environment." Theory and practice in language studies, 2(3), 524-530. Academy Publisher. (2012)
10. J.M. O'Malley, A.U. Chamot. "Learning strategies in second language acquisition". Cambridge University Press. (1990)
11. P.R. Pintrick, E. De Groot. "Motivational and self-regulated learning components of classroom academic performance." Journal of Educational Psychology, 82, 33-40, (1990).
12. G. Torrisi-Steele, S. Drew. "The literature landscape of blended learning in higher education: the need to better understanding of academic blended practice." International Journal for Academic Development 18(4), 371-383. Taylor & Francis. (2013)
13. J. Bowyer, L. Chambers. "Evaluating blended learning: bringing the elements together." Research Matters 23. UCLES. (2017)
14. M. Zuvic-Butorac, N. Roncevic, D. Nemcanin, Z. Nebic. "Blended elearning in higher education: research on students' perspective." Issues in Informing Sciencce and Information Technology, 8, 409-429, (2011).
15. Y.M. Luna, S.A.Winters. "Why did you blend my learning? A comparison of student success in lecture and blended learning introduction to sociology courses". In Teaching Sociology, 45(2), 116130, SAGE (2017).
16. M. Lubis, M. Kartiwi and S. Zulhuda. "Privacy and personal data protection in electronic voting: factors and measures." Telkomnika 15(1), pp. 512-521, (2017).

17. M. Delucchi. "Measuring student learning in social statistics: a pretestposttest study of knowledge gain". Teaching Sociology 42(3), 231-239. SAGE. (2014)
18. S. Patrick, C. Sturgis. "Maximizing competency education and blended learning: insights from experts". CompetencyWorks Issue Brief. (2015)
19. A.M. Al-Huneidi, J. Schreurs. "Contructivism based blended learning in higher education." IJET 7(1), (2012)
20. M. Lubis, M. Kartiwi and S. Zulhuda."Current state of personal data protection in electronic voting: criteria and indicator for effective implementation." Telkomnika, 16(1), 290-301.
21. B. Kistow. "Blended learning in higher education: a study of a graduate school of business, Trinidad and Tobago." Carribean Teaching Scholar 1(2), 115-128, (2011).
22. M.J. Kintu, C. Zhu and E. Kagambe. "Blended learning effectiveness: the relationship between student characteristics, design features and outcomes." Int. J. of Educational Tech. in H. Education 14(7), (2017).
23. P. Moskal, C. Dziuban, J. Hartman. "Blended learning: a dangerous idea?" Internet and Higher Education 18, 15-23, (2013).
24. L.M. Jeffrey, A. Higgins. "Blended learning: how teachers balance the blend of online and classroom components." J. of IT Education: Research, 13, 121-140, (2014).
25. A. El-Mowafy, M. Kuhn and T. Snow. "Blended learning in higher education: current and future challenges in surveying education." Issues in Educational Research, 23(2), 132-150, (2013).
26. F.Z. Azizan. "Blended learning in higher education institution in Malaysia." Proc. Of Regional Conf. on Knowledge Integration in ICT. 454-466. (2010)
27. V. Roblek, M. Mesko and A. Krapez. "A Complex view of industry 4.0" Sage Open, pp. 1-11. (April-June 2016)

Six sigma implementation model based on critical success factors (CSFs) for indonesian small and medium industries

Teuku Yuri M. Zagloel[1,a], *Romadhani* Ardi[1,b], and *Wahyu* Poncotoyo[1,c]

[1]Department of Industrial Engineering, Faculty of Engineering, Universitas Indonesia

Abstract. This study aims to build a model of critical success factors in the application of Six Sigma for Indonesian small and medium industries and their relation to the industry performance. The critical success factors, successful Six Sigma implementation and the impact on industry performance are validated and analyzed. The method used in this study is Structural Equation Modeling (SEM) based on Partial Least Square (PLS) by using SmartPLS software. This study uses 54 data from Indonesian small and medium industries that have implemented Six Sigma. The results of this study indicate that a critical success factors that has a positive impact on the successful Six Sigma implementation is the involvement and commitment top management, training and education, cultural change and industrial infrastructure. It is surprising that teamwork has a negative impact on the successful Six Sigma implementation. In addition, the successful Six Sigma implementation has a positive impact on the performance of Indonesian small and medium industries. Then, the model is valid to shows the relationship between critical success factors in the successful Six Sigma implementation on the performance of small and medium industries in Indonesia.

1 Introduction

In Southeast Asia (ASEAN), Indonesia becomes the country with the highest Gross Domestic Product. In the world, Indonesia's ranking is 16th with the number of Gross Domestic Product at $932,259 and then followed by Thailand at $407,026 (25th), Philippines at $304,905 (35th), Singapore at $296,975 (36th), Malaysia at $296,535 (37th), Vietnam at $205,279 (46th), Myanmar at $63,225 (70th), Cambodia at $20,016 (106th), Laos at $15,805 (113th), and the last one is Brunei Darussalam at $11,400 (125th) [1]. Gross Domestic Product is a summary of the number of products produced in a country within a year [2] and is an important variable in analyzing a country's economic growth [3]. As much as 20.51% of Gross Domestic Product in Indonesia is produced by manufacturing industry [4].

For modern economic today, small and medium industries play an important role in sustaining the economy of a country and even the world economy. "Small and medium

[c] Corresponding Author: wahyu.poncotoyo@ui.ac.id

industries are the source of modern economic life" [5]. In order to keeping their contribution to the modern economic growth, small and medium industries are required to improve production quality to meet the increasingly high market demands and expects high precision levels. On the other hand, a study shows that the success of small and medium industries has a direct impact on economic development in developed and developing countries [6]. They have the ability to produce products with minimum cost, pioneers in the field of innovation and have a high flexibility that enables them to meet customer needs [7]. In 2015 there are 3,385,851 units of small and medium industries in Indonesia [4]. Small, medium and also large industries can be classified on the basis of profits, units of electricity consumed, the amount of labor and considered as machines for economic growth in Europe [8]. In Indonesia the definition of industrial scale is regulated in the regulation of the Minister of Industry number 64 of 2016 [9].

However, when compared to the level of global competitiveness for the year 2016-2017, Indonesia's ranking is 41th [10]. In terms of competitiveness in Southeast Asia (ASEAN), Indonesia should be left behind by Singapore, Malaysia, and Thailand respectively ranked 2nd, 25th, and 34th. To improve competitiveness, many ways can be implemented. Increasing productivity is a way to improve competitiveness. In increasing productivity, one way is to do continuous improvement program. With continuous improvement the industry will improve their performance to achieve its goals to meet the consumer needs. As a business strategy, Six Sigma focuses on improving understanding of customer needs, increasing productivity levels, and industry performance [11]. A survey of 2,870 respondents undertaken to examine ongoing improvement programs that provide the best outcomes for industries that implement them. Survey results showed that 53.6% of respondents chose Six Sigma as the most important program, 26.3% respondents chose Lean, and 10.3% of respondents chose Total Quality Management (TQM) [12]. This activity is done by the industry as a form of effort in improving their performance, so that if achieved will have an impact on customer satisfaction. Efficiency, profitability, responsiveness and customer satisfaction have been a major concern of the company [13].

The success rate of implementing Six Sigma for the industry is influenced by several critical success factors. A critical success factor is an important ingredient required for the success of a Six Sigma project within an industry [14]. According to Rungasamy, Antony, and Ghosh (2002), critical success factors are factors needed to ensure success of any initiative or program to be implemented, in other words if the critical factor of success in the implementation of Six Sigma is not achieved, six sigma will fail in the implementation [15]. To implement Six Sigma, top industry management must have an understanding of quality management that builds Six Sigma methodology, in addition to understanding top management should also be involved directly so that the implementation of Six Sigma does not lose the goal because top management understands and participates in the application of Six Sigma [16]. By increasingly demonstrating that successful implementation of Six Sigma methods will affect industry performance [17], this study assumes that industry performance can be an indicator of successful implementation of Six Sigma. So bring up some interesting questions, namely:

1. What is the critical success factors that have a positive impact on the successful implementation of Six Sigma in small and medium industries in Indonesia?
2. Does successful implementation of Six Sigma have a positive impact on the performance of small and medium industries in Indonesia?
3. How is the involvement and commitment of top management as the main foundation of critical success factors in successful implementation of Six Sigma in small and medium industries, influencing other critical success factors?

2 Literature review

According to Tack-Wei Leong and Pee-Lee Tea (2012) in his study concluded that after a thorough review of the critical success factors theories that were important for this study were measured by five factors: involvement and commitment top management, training and education, teamwork, cultural change, and industrial infrastructure [19, 25]. The initial conceptual model in this study builds on a conceptual model of critical success factors in the application of Six Sigma in the OEM or original equipment manufacturing industry in Malaysia [18]. The use of this conceptual model is based on the fact that the model is tested in the industrial sector in developing countries. This model consists of five critical success factors that have also been used in previous research, namely:

1. Involvement and commitment top management. According to many experienced experts, Six Sigma recognizes that involvement and commitment top management are critical success factors [19]. Many previous studies that use this critical success factor, use similar terms, as one of the success factors in the application of Six Sigma. These include top management and leadership commitment [20]; management commitment [21] top management commitment [22], executive engagement [23].
2. Training and Education. Samsung believe that training and education is an important step that companies must take in preparing for the implementation of Six Sigma, enabling companies to establish and create new goals as employees are asked to think and act in different ways [24]. Providing training and education is important in the application of Six Sigma [25]. It is indispensable to plan the application of Six Sigma to continue to grow and not run in place due to limited knowledge [26]. Some studies have used different terms such as: Training [22], education and training [27].
3. Teamwork. This is related to the sharing of knowledge and experience [28]. Teamwork can be defined as the extent to which people in the group actively support and assist one another with their work [29]. There are several terms used and have a single meaning with teamwork that has been used in previous studies including: effective communication [30], team communication [8], as well as team involvement and commitment [31].
4. Cultural Changes. The success of an industry in controlling market share is highly dependent on the industrial working culture [32]. Successful introduction and implementation of Six Sigma requires adjustment to organizational culture and changes in the attitudes of its employees. Employees must be motivated and accept responsibility for the quality of their own work [22]. In the previous studies that use several terms of cultural change include: organizational culture change [33], cultural change [34].
5. Industrial Infrastructure. The importance of this critical success factor has been conveyed by several studies [19, 21, 25, 36]. Critical success factors in the application of Six Sigma are related to Brun's research (2010), Antony and Banuelas (2002), and Tlapa et al, (2016). Some studies use different terms including: organizational infrastructure [35], infrastructure and organizational culture [36].

A research model reveals that Lean Six Sigma has become the most influential quality method for the performance of the automotive industry, especially car manufacturers in their efforts to improve competitiveness, in addition to making those industries able to improve innovation, technology development, quality, production process and customer satisfaction [37]. Good Six Sigma project performance affects industry performance because Six Sigma projects must have some characteristics related to business priorities and are critical to the industry [17]. This is because if the industry has succeeded in implementing Six Sigma there will be defect reduction, cycle time reduction, lot size reduction, and inventory reduction, which will lead to reduced production costs and increased customer satisfaction. Measuring industrial performance is generally done in any industry, whether applied to measuring aspects of finance, non-financial aspects, or both. Based on this theory, Kaplan and Norton (1992) proposed several performance measures using a balanced scorecard approach. This comprehensive measure of performance is based on four perspectives: finance, customers, processes or industrial operations, and innovation. Industry performance indicators consist of finance, consumer, internal business, and innovation [38]. The models of Tack-Wei Leong and Pee-Lee Tea (2012) and Habidin and Yusof (2013) were modified to examine relations empirically and obtain models that fit the conditions of small and medium industries in Indonesia. Here are the considerations:

1. Any successful quality method such as Six Sigma requires top management involvement as the provision of appropriate resources and training [39]. The involvement and commitment of top management is recognized as a critical success factor in the successful implementation of Six Sigma by those who have implemented Six Sigma [19]. Jack Welch, General Electric's chief, is heavily involved in restructuring business organizations and changing employee attitudes toward the application of Six Sigma [19]. Therefore, in this study "the involvement and commitment of top management" is the critical success factor that will influence other critical success factors.
2. Semi-structured interviews with three Six Sigma experts are conducted to get a direct picture of the relationship between critical success factors in the application of Six Sigma. Based on the interviews it is found that in this research critical factor of success is the involvement and commitment of top management has an indication of positive relationship to other critical success factors that are: training and education, teamwork, cultural change, and industrial infrastructure. In addition, according to Six Sigma experts, states that by doing training and education about Six Sigma an industry will be more receptive to cultural changes that come in conjunction with the implementation of Six Sigma. Teamwork will be easy to establish and build when employees are trained and educated and openness to new cultures.

Based on these two considerations, the hypothesis on the conceptual model proposed in this study is as follows:

H1. The involvement and commitment of top management has a positive impact on training and education, cultural change, teamwork, industry infrastructure and the successful implementation of Six Sigma.

H2. Training and education have a positive impact on teamwork, cultural change and the successful implementation of Six Sigma.

H3. Teamwork has a positive impact on the successful implementation of Six Sigma.

H4. Cultural changes have a positive impact on teamwork and successful implementation of Six Sigma.

H5. Industrial infrastructure has a positive influence on the successful implementation of Six Sigma.

H6. Successful implementation of Six Sigma has a positive influence on industry performance.

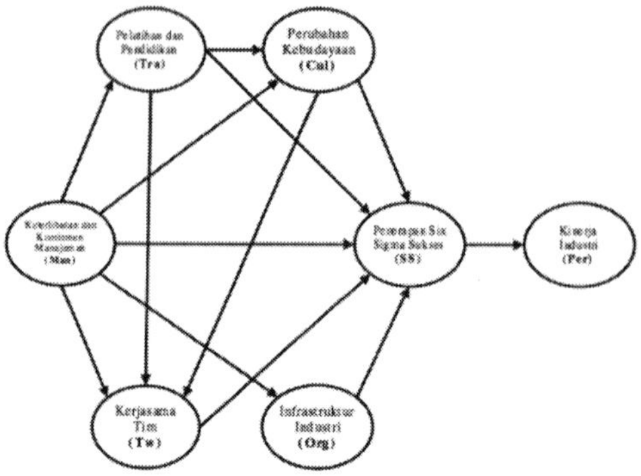

Fig. 1. The Conceptual model

3 Research methodology

In this study, there are 24 measuring variables obtained through theoretical reviews and interviews with Six Sigma experts. Pilot testing was conducted on ten academics aimed at knowing the clarity of terms used in the questionnaire instrument. Likert scale is worth 1 (strongly disagree) up to 5 (critical) is used to know the respondent's assessment of the questionnaire instrument. This survey was conducted among small and medium-sized industrial companies that have implemented Six Sigma in Indonesia. A total of 60 questionnaires were sent to the respondents and a total of 54 questionnaires were returned. A total of 54 questionnaires are considered valid. Based on the respondents involved in the research, 90 percent are manufacturing industry, 4 percent is construction industry, 2 percent is industry in finance, 2 percent is mining industry, and 2 percent is industry service area. The proportion of industrial scale is 87 percent is medium industry. To estimate the SEM model parameters, this study uses the Partial Least Square (PLS) procedure. PLS-SEM provides many advantages for researchers working with structural equation models, such as using non-distributed data, small sample size and research models with formative indicators [40]. According to Barclay et al. (1995), states that the minimum sample size for the PLS model should be equal to greater than: ten times the number of indicators most used to measure a latent variable, or ten times the number of model paths (arrows) directed to a latent variable in inner model [41]. In this study, the minimum number of samples to be achieved is as many as 40 samples. Thus, the sample size in this study is appropriate for use with PLS procedures.

4 Data analysis

4.1 The testing of measurement model

Manurut Ghozali (2008), in the using PLS-SEM requires two stages to assess the Fit Model of a study. These stages include the stage of analysis or evaluation of measurement models and structural model analysis. There are three criteria in using data analysis techniques with SmartPLS to assess the validity and reliability of the outer model of convergent validity, discriminant validity, and composite reliability.

4.1.1 Convergent validity

Convergent validity aims to ensure indicators of latent variables are highly correlated with these variables. Convergent validity can be seen from the weighting factor and Average Variance Extracted (AVE). The measuring variable that has a weighting factor of > 0.5 is considered very significant [42]. Then for Average Variance Extracted (AVE) must also have value > 0.5 [43]. Thus, a value of 0.5 is the cutting point used to ensure that measure variables can represent latent variables. In this study, all measure variables had loading factors value > 0.5 ranging from 0.574 to 0.944 . Thus it can be said that all measure variables can represent latent variables. And then, for Average Variance Extracted (AVE) values is MAN = 0.696, TRA = 0.740, CUL = 0.729, TW = 0.771, ORG = 0.659, SS = 0.811, and PER = 0.725.

4.1.2 Discriminant Validity

Discriminant validity ensures that indicators of different latent variables should not be highly correlated [43]. This can be seen from the value of cross loading for each latent variable > 0.5 [43]. Discriminant validity also sees the square root of AVE (√AVE), for each latent variable and compared with the correlation value between latent variables [43]. The rule of thumb of the square root AVE (√AVE), is the square root AVE (√AVE), must be greater than the correlation value between the latent variables [43]. Based on the calculation then the overall value of cross loading and square root of AVE (√AVE) in accordance with the criteria (see table I).

4.1.3 Composite Reliability

Reliability test or internal consistency reliability in PLS can use two parameters that is cronbach's alpha and composite reliability [43].

Table 1. Measure of *Discriminant validity* (Source: Data Processing Using SmartPLS 3.2.7)

Laten		Nilai Korelasi dengan Variabel						
		Man	**Tra**	**Cul**	**Tw**	**Org**	**SS**	**Per**
MAN	Man1	**0,868**	0,778	0,753	0,852	0,815	0,837	0,818
	Man2	**0,895**	0,881	0,755	0,781	0,805	0,836	0,807
	Man3	**0,838**	0,773	0,677	0,699	0,596	0,75	0,733
	Man4	**0,73**	0,695	0,629	0,588	0,613	0,729	0,571
TRA	Tra1	0,819	**0,843**	0,684	0,732	0,723	0,804	0,842
	Tra2	0,86	**0,927**	0,761	0,725	0,739	0,809	0,764
	Tra3	0,741	**0,808**	0,679	0,718	0,707	0,791	0,585
CUL	Cul1	0,788	0,722	**0,89**	0,786	0,739	0,805	0,82
	Cul2	0,813	0,832	**0,884**	0,832	0,745	0,83	0,749
	Cul3	0,526	0,518	**0,783**	0,518	0,632	0,633	0,439
TW	Tw1	0,803	0,763	0,75	**0,89**	0,766	0,819	0,83
	Tw2	0,704	0,7	0,723	**0,862**	0,736	0,73	0,665
	Tw3	0,81	0,755	0,787	**0,882**	0,861	0,839	0,701
ORG	Org1	0,559	0,528	0,502	0,598	**0,766**	0,542	0,586
	Org2	0,626	0,647	0,671	0,716	**0,828**	0,699	0,548
	Org3	0,855	0,829	0,805	0,848	**0,908**	0,907	0,822
SS	Ss1	0,875	0,867	0,785	0,818	0,816	**0,92**	0,919
	Ss2	0,786	0,834	0,768	0,751	0,782	**0,866**	0,735
	Ss3	0,828	0,794	0,81	0,808	0,832	**0,906**	0,791
	Ss4	0,914	0,858	0,856	0,891	0,83	**0,915**	0,818
PER	Per1	0,846	0,856	0,756	0,796	0,78	0,858	**0,944**
	Per2	0,408	0,374	0,498	0,456	0,51	0,533	**0,574**
	Per3	0,804	0,72	0,711	0,771	0,71	0,821	**0,922**
	Per4	0,871	0,859	0,743	0,767	0,752	0,838	**0,912**

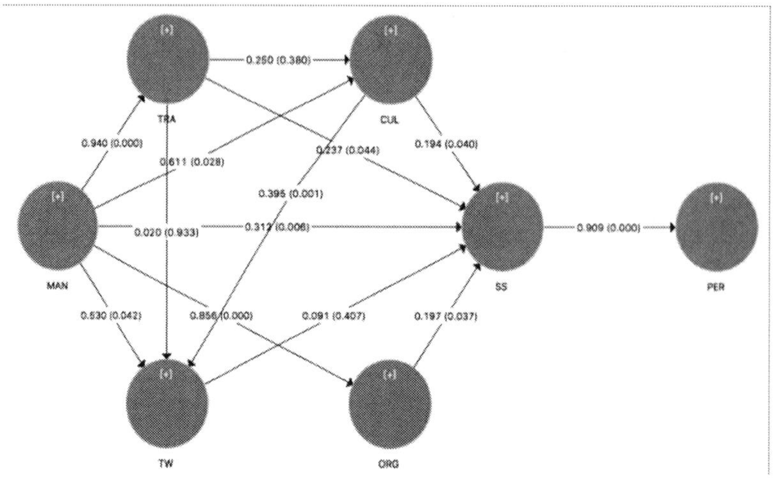

Fig. 2. The Result of Path Coefficient
Note: The value without parentheses is the value of the Path Coefficient and the value inside parentheses is the P-Value.

Cronbach's alpha assumes that all indicators are reliable and composite reliability assumes that indicators have different loading values [44]. Hair et al. (2010) explains that the cut off point for estimating the reliability of a latent variable is 0.7. Therefore, any latent variant that has a score of > 0.7 can be said to have good reliability. In this study, all latent variables had values > 0.7 ranging from 0.739 to 0.922. Thus it can be said that all latent variables have a good level of reliability so that it can be passed on the next tes

4.2 The test of structural model

The evaluation of structural model or inner model aims to predict the relationship between latent variables [43]. the structural model is evaluated by looking at the percentage of variance by looking at R-Square values for endogenous variables, predictive relevance test, and average variance extracted for prediction values using resampling procedures such as jackknifing and bootstrapping in order to stabilize model measurement [43]. There are four criteria to evaluate the structural model, namely coefficient determination (R-Square), path coefficient, Cohen (f2), and predictive relecance (Q2) [44].

4.2.1 Coefficient determination (R-Square)

R-Square value for coefficient determination can show influence value of exogenous variable to endogenous variable. The value of R-Square 0.75 denotes a stronger model, 0.50 indicates a moderate model, and 0.25 denotes a weak model. R-Square is only possessed by endogenous variables, that is Training and Education (TRA) with 0.833, Cultural Change (CUL) with 0.723, Teamwork (TW) with 0.823, Industrial Infrastructure (ORG) with 0.732, Six Sigma Success (SS) with 0.950 and Industrial Performance (PER) with 0.826.

4.2.2 Path coefficient

The coefficient path can be used to see if the path or relationship between latent variables and other latent variables has an effect on the model. Path or relationship between latent variables must be seen significant level (P-Value) of the influence of the path. The path coefficient should be > 0.100 with a significance value of 0.05 for the path to be considered to have an effect on the model. The value of Path Coefficient (β) and significant value (P-Value) can be seen in Figure 2. The results show that the path for TW➔SS, TRA➔CUL, and TRA➔TW have negative magnitude so that according to the path coefficient test, it is considered no effect on the model.

4.2.3 Effect size cohen (f2)

Cohen Effect Size Testing (f^2) is used to evaluate the effect of exogenous variables on endogenous variables. The rule of thumb of the value of f^2 is 0.02 is considered to have a weak influence, the value of f^2 is 0.15 is considered to have medium effect, and the value of f^2 is 0.35 is considered to have a strong influence. The following table of values of f^2 is related to the relationship between exogenous variables and endogenous variables for critical success factors research model in the application of Six Sigma in Small and Medium Industry in Indonesia (see table III). The results are MAN ➔ SS = 0.167 "sedang", TRA ➔ SS = 0.123 "sedang", CUL ➔ SS = 0.163 "sedang", TW ➔ SS = 0.021 "lemah", ORG ➔ SS = 0.125 "sedang", SS ➔ PER = 4.759 "kuat", MAN ➔ TRA = 7.541 "kuat", MAN ➔ CUL = 0.158 "sedang", MAN ➔ TW = 0.161 "sedang", MAN ➔ ORG = 2.731 "kuat", TRA ➔ CUL = 0.026 "lemah", TRA ➔ TW = 0.000 "lemah", and CUL ➔ TW = 0.244 "sedang"

4.2.4 Predictive relecance (Q2)

Predictive relevance (Q^2) represents integration of cross validation and fitting functions with predictions of latent variables and estimates of latent variable indicators. Predictive relevance evaluates whether interrelated latent variables are predicted to have relationships among latent variables. Based on the rule of thumb, Q^2 with value 0.02 indicates a weak predictive relevance level, 0.15 indicates moderate predictive relevance, and 0.35 indicates strong predictive relevance. Based on SmartPLS software, the value of TRA is 0.881, CUL is 0.712, TW is 0.813, ORG is 0.727, SS is 0.944, and PER is 0.823.

5 Results and discussion

Based on the result of measurement model and structural model, the path that is formed is: MAN ➔ SS ➔ PER means the involvement and commitment of top management (MAN) has a positive impact on the successful implementation of Six Sigma (SS), and the successful implementation of Six Sigma (SS) has a positive impact on industry performance (PER). MAN ➔TRA ➔SS ➔PER means the involvement and commitment of top management (MAN) has a positive impact on training and education (TRA), then training and education (TRA) has a positive impact on the successful implementation of Six Sigma (SS), and the successful implementation of Six Sigma (SS) has a positive impact on industry performance (PER). MAN ➔ CUL ➔ SS ➔ PER means the involvement and commitment of top management (MAN) has a positive impact on cultural change (CUL), then cultural change (CUL) has a positive impact on the successful implementation of Six Sigma (SS), and the successful implementation of Six Sigma (SS) has a positive impact industry performance (PER). MAN ➔ ORG ➔ SS ➔ PER means the involvement and commitment of top management (MAN) have a positive impact on industrial infrastructure (ORG), and industrial infrastructure (ORG) has a positive impact on the successful implementation of Six Sigma (SS), and the successful implementation of Six Sigma (SS) has a positive impact on industry performance (PER).

6 Conclusion

The purpose of this research is to build a model of Six Sigma implementation based on critical success factors, to study the relationship between critical success factors for successful implementation of Six Sigma in small and medium industries in Indonesia, to study the relationship between successful implementation of Six Sigma on the performance of small and medium industries in Indonesia, confirmed the model developed for small and medium industries in Indonesia. The Six Sigma implementation model in this study has 5 latent variables of critical success factors that make Six Sigma successful for implementation, that is top management involvement and commitment, training and education, teamwork, cultural change and industrial infrastructure. There are 4 measuring variables, 3 measuring variables, 3 measuring variables, 3 measuring variables, 3 valid measuring variables representing each critical success factor. While the other two latent variables are the successful application of Six Sigma and latent variables of industry performance are represented by each of the four measuring variables. A total of 13 hypotheses were successfully developed in this study, but only 10 hypotheses were expressed to have a positive impact. The results revealed that four latent variables of critical success factors, namely: involvement and commitment of top management, training and education, cultural change and industrial infrastructure have a positive relationship to the successful implementation of Six Sigma in small and medium industries in Indonesia. And also the successful implementation of Six Sigma has a positive impact to the performance of small and medium industries in Indonesia. In addition, the relationship between critical success factors is also known. It can thus be deduced that if small and medium industries in Indonesia want to achieve success in implementing Six Sigma they must ensure that they have met these critical factors. So the resources they spend in order to implement Six Sigma are not wasted. The limitation of this research is to focus on small and medium industries and the most of the respondents are in manufacturing sector, so the next research is interesting to be done on big industry. In addition, future research can also consider the methods of Analytical Hierarchy Process (AHP), Neural Network, Analytical Network Process (ANP) and Topsis method to be used as a step to follow up the model that has been developed from this research so as to obtain a measurement matrix of success in application of Six Sigma.

Acknowledgement

This research received funding from the PITTA Grant at Universitas Indonesia.

References

1. World Bank. (2016). *The World Bank*. Retrieved Oktober 26, (2017)
2. Bureau of Economic Analysis. (2007). *Measuring the Economy: A primer on GDP and the National Income and Product Accounts*. Bureau of Economic Analysis, United States Department of Commerce, Economics and Statistics Administration., United States of America.
3. Henderson, J. V., Storeygard, A., & Weil, D. N. Measuring Economic Growth from Outer Space. *American Economi Review , 102* (2), 994-1028. (2012).
4. Central Bereau of Indonesia Statistics. *Badan Pusat Statistik*. Retrieved September 16, (2017)
5. Antony, J., Kumar, M., & Madu, C. N. Six sigma in small- and medium-sized UK manufacturing enterprises: Some empirical observations. *International Journal of Quality & Reliability Management , 22* (8), 860-874. (2005)

6. Demirbag, M., Tatoglu, E., Tekinkus, M., & Zaim, S. An analysis of the relationship between TQM implementation and organizational performance:Evidence from Turkish SMEs. *Journal of Manufacturing Technology Management , 17* (6), 829-847. (2006)
7. Brock, W., & Evans, D. The Economics of Small Business: Their Roles and Regulations in US Economy. (1986).
8. Kumar, M., Antony, J., Madu, C. N., Montgomery, D. C., & Park, S. H. Common myths of Six Sigma demystified. *International Journal of Quality & Reliability Management , 25* (8), 878-895. (2008).
9. Ministry of Industry of the Republic of Indonesia. *Peraturan Menteri Perindustrian Republik Indonesia Nomor 64 Tahun 2016.* (2016).
10. World Economic Forum. *World Economic Forum*. Retrieved Oktober 26, 2017, (2016-2017).
11. Kwak, Y. H., & Anbari, F. T. Benefits, obstacles, and future of six sigma approach. *Technovation , 26,* 708-715. (2006)
12. Dusharme, D. *Six Sigma Survey: Big Success...But What About the other 98 percent?* Retrieved January 11, 2018, (2004)
13. Green, K. W., Zelbst, P. J., Meacham, J., & Bhadauria, V. S. Green supply chain management practices: impact on performance. *Supply Chain Management: An International Journal , 17* (3), 290-305. (2012).
14. Coronado, R. B., & Antony, J. Critical success factors for the successful implementation of six sigma projects in organisations. *The TQM Magazine , 14* (2), 92-99. (2002)
15. Rungasamy, Antony, & Ghosh. Critical success factors for SPC implementation. *TQM Magazine , 14* (4), 217 - 224. (2002)
16. Pande, P. S., Neuman, R. P., & Cavanagh, R. R. How GE, Motorola, and Other Top Companies Are Honing Their Performance. The McGraw-Hill Companies, Inc. (2000)
17. Snee, R. D., & Jr., W. F. The Project Selection Process. Quality Progress , 35 (9), 78-80. (2002)
18. Leong, Tack-Wei., & Tea, Pei-Lee. Critical Success Factors of Six Sigma in Original Equipment Manufacturer Company in Malaysia. *International Journal of Synergy and Research, Vol. 1, No. 1,* 7-21. (2012)
19. Henderson, K., & Evans, J. Successful implementation of Six Sigma: benchmarking general electric company . *Benchmarking and International Journal , 7* (4), 260-281. (2000)
20. Sakthivel, P. B., Rajendran, G., & Raju, R. TQM implementation and students satisfaction of academic performance. The TQM Magazine, 17(6), 573-589. doi: 10.1108/09544780510627660. (2005)
21. Anbari, F.T. and Kwak, Y.H. "Success factors in managing Six Sigma projects", paper presented at the Project Management Institute Research Conference, London, 11-14 July. (2004)
22. Antony, J., & Banuelas, R. Key ingredients for the effective implementation of Six Sigma program. Measuring Business Excellence , 6 (4), 20-27 (2002)
23. Hayes, B.J, "Six Sigma critical success factors", iSixSigma, (2002)
24. Szeto, A. and Tsang, A. 'Antecedents to successful implementation of Six Sigma', International Journal of Six Sigma and Competitive Advantage, Vol. **1**, No. 3, pp.307–322 (2005)
25. Coronado, R. B., & Antony, J. Critical success factors for the successful implementation of six sigma projects in organisations. The TQM Magazine , 14 (2), 92-99 (2002)

26. Heckl, D., Moormann, J. and Rosemann, M., "Uptake and success factors of Six Sigma in the financial services industry", Business Process Management Journal, Vol. 16, No. 3, pp. 436-472 (2010)
27. Brun, A. Critical success factors of Six Sigma implementations in Italian companies. *Int. Journal Production Economics , 131*, 158-164 (2010)
28. Rus, I., and Lindvall M. "Knowledge Management in Software Engineering," IEEE Software (19:3), 2002, pp. 26-38.
29. Hurley RF, Hult GTM. Innovation, market orientation, and organiza- tional learning: an integration and empirical examination. J Mark 1998;62:42 – 54.
30. Carvalho, M.M., Ho, L.L. and Pinto, S.H.B., "The Six Sigma program: an empirical study of Brazilian companies", Journal of Manufacturing Technology Management, Vol. 25 No. 5, pp. 602-630 (2014)
31. Brady, J.E. and Allen, T.T., "Six Sigma literature: a review and agenda for future research", Quality and Reliability Engineering International, Vol. 22 No. 3, pp. 335-367 (2006)
32. Sohal, A. "Assessing manufacturing or quality culture and practices in Asian companies", International Journal of Quality & Reliability Management, Vol. 15 No. 8/9, pp. 920-30 (1998)
33. Kundi, O.H.K., "A study of Six Sigma implementation and critical success factors", paper presented at the 9th Pakistan International Convention on Quality Improvement, 14-15 November, Karachi (2005)
34. Chakrabarty, A., & Tan, K. C. The current state of six sigma application in services . *Managing Service Quality: An International Journal , 17* (2), 194-208 (2007).
35. Tlapa, D., Limon, J., García-Alcaraz, J. L., Baez, Y., & Sánchez, C. Six Sigma enablers in Mexican manufacturing companies: A proposed model. Industrial Management & Data Systems, 116, 926–959. doi:10.1108/IMDS-06-2015-0265 (2016).
36. Zu, X., Robbins, T. and Lawrence, F. "Mapping the critical links between organizational culture and TQM/Six Sigma practices", International Journal of Production Economics, Vol. 123 No. 1, pp. 86-106 (2009)
37. Habidin, N. F., & Yusof, S. M. Critical success factors of Lean Six Sigma for the Malaysian automotive industry . *International Journal of Lea n Six Sigma , 4* (1), 60-82 (2013).
38. Kaplan, R., & Norton, D. The Balanced Scorecard: Measures That Drive Performance. *Harvard Business Review , 70* (1), 71-79 (1992)
39. Halliday, S. "So what exactly is Six Sigma?"', Works Management, Vol. **54** No. 1, p. 15 (2001)
40. Hair, J., Black, W., Babin, B. and Anderson, R. Multivariate Data Analysis, 7th ed. (electronic edition), Pearson Education Limited, London (2014)
41. Barclay, D.W., Higgins, C.A. and Thompson, R. "The partial least squares approach to causal modeling: personal computer adoption and use as illustration", Technology Studies, Vol. 2 No. 2, pp. 285-309 (1995)
42. Ghozali, I. Structural equation modeling; alternative method with partial least square. Badan Penerbit Universitas Dipenogoro. Semarang (2006)
43. Latan, H dan I. Ghozali. Partial least squares: The concept, technique, and SmartPLS 2.0 M3 application. Badan Penerbit Universitas Dipenogoro. Semarang (2012)
44. Urbach, N., & Ahlemann, F. Structural equation modeling in information systems research using partial least squares. Journal of Information Technology Theory and Application, 11(2), 5–40 (2010)

Identify eight aspects of ergonomics to determine the improvement of human-machine interaction work (case studies in manufacturing industry)

Wahyu Susihono[1*],*Tania* Anggi Saputri[2]

[1],[2]Industrial Engineering Department, Engineering Faculty, University of Sultan Ageng Tirtayasa, Banten, Indonesia

Abstract. Manufacturing Industry is one of the industrial activities in Indonesia, manufacturing industry is an industry with main activities is to change raw materials, components, or other parts into goods which is according to company specifications standards. In the production floor, activity in the manufacturing industry, the workers have different job specifications with each other. Some works consist of human-machine interaction is found by the activity between workers and lathe machine, welding maching, milling machine, frais machine, and others. The manufacturing industry will increase its ability to serve a variety of better quality products caused by the desire or encouragement of the customers. In general, according to the increase of corporate targets, its also need improvement from the aspect of work performance. To obtain a description of the proposed improvement based on human performance, it is necessary to identify the eight aspects of ergonomics include the consumption of nutritionalfor workers (energy), muscle power, body posture, environment, time conditions, social conditions, information conditions, and human machine interaction. This research use cross sectional method approach that is research done at one time, no follw up, to find the correlation between independent variable (risk factor) with dependent variable (effect). The conclusion of this research is needed nutrition intake or nutrition to recover the workers, it is necessary to design facilities such as manufacturing aids to reduce the use of excess muscle or appropriate technology (TTG). After the application of TTG (Appropriate Technology) to reduce the excessive use of muscle to the workers, the company should provide the nutritional intake accordance with workload of employees in the manufacturing industry

1 Introduction

Ergonomics is a component of activities which is within a broad scope, among others, include the reciprocal harmonixation of work on labor for efficiency and comfort of work

* Corresponding author: pmy_wahyu@yahoo.co.id

[1,2]. The target of ergonomics is the entire workforce of both the modern sector as well as in the traditional and informal sectors. Basically all activities that use man in it will always be related to ergonomics because of its relation to the correct working procedure.

The main problem in activity on the production floor of manufacturing industry, is still very complex. The need for applying the ideal conditions of healthy manufacturing industry still needs to be supported by several factors that need to be improved. One of the factors, begin from the division of labor which is unwell organized, until the adoption of technology that still needs attention. Safety and health issues of workers is still far from ideal conditions. As example in production floor activities to produce goods, until nowadays required evaluation on worker's activities start from applying of good working methods, work nutrition, and the use of work tools. Other condition of concern are musculoskeletal complaints by workers during work on this production plant, because workers bend their body during work, and also hot on production plant can make workers do not feel comfortable while working.

From the initial observation there is still a lot of work done without considering the aspects of ergonomics either from the method of work or the facilities used. The ergonomics approach offers an approach that can evaluate 8 aspects of ergonomics within the framework of improvement of work methods. Eight aspects of ergonomics in the settlement of ergonomic poblems include: energy or nutritional status (nutrition), musculosceletal energy use, work posture, time, social conditions, information conditions, and machine interface [3,4,5]. Problems are identified together to gain success in improving performance of work that can be shared.

Manufacturing activities that exist in this manufactuing industry are: manufacture of spare parts to fulfill orders from other companies, receiving machine service, and fixing the construction objects. There are 40 workers in it including operators and administrative staff with working time of 8 hours of work and 1 hour breaktime for 6 work days.

In this manufacturing industry, there are still workers' muscle complaints that become one of the main factors causing the decline of work performance in this industrial sector. Based on this problem, it is necessary to conduct an evaluation study of 8 aspects of ergonomics, so as to provide alternatives in the form of: (1) the availability of appropriate technology (TTG) on production plant, and (2) the evaluation of work nutrition as work recovery

2 Methods

This research was conducted on manufacturing industry studies on the operator of the process of turning, welding, fraising, and milling. This research was conducted by cross sectional, on estudy and one time, no follow-up to find the correlation between independent variable (risk factor), and dependent factor (effect).

The stages of this research are preliminary study, identification of 8 aspects of ergonomics, identification of appropriate technology (TTG), operational nutrition assessment of operator, suggesting for improvement, giving thr conclution and suggection as the final.

3 Results And Discussion

Based on the results of preliminary study and observations that have been done on the manufacturing industry on the identification of 8 aspects of ergonomics is known as follows.

3.1 Nutrition

On the production floor of manufacturing industry has a hot environment so the consumption of carbohydrates and minerals is dominant [3]. The fulfillment of nutrients is just eating and drinking that has no rules regarding the criteria of nutritional intake. The purpose of providing nutrition at both active and lunch breaks is for energu recovery after completion of work. So the energy released by the body is proportional to the nutritional intake that should be received by the body. Supplementary nutriion or nutrition to workers (average age 37.73 and weight 64 kg) can be calculated to find the value of the adequacy of nutrition work. The amount of additional nutrition work is calculated according to the needs of calories by the workers as known as follows: a) the average worker;s wight, b) the value of the Basal Metabolism Rate (BMR), c) the energy adequacy rate based on the level of physical activity, d) Specific Dynamic Action or 10% intake. Intake for additional nutrition or nutrition at the time of active rest is calculated by 10% of the total calorific value. The calculation of the overall worker calorie requirement is as follows.

a. Basal Metabolisme Rate (BMR) is 11,6 weight + 876 = 1618.4 calories.
b. The total energy requirement or energy adequacy based on the level of physical activity is 1.88 x 1618,4 = 3042.56 calories.
c. Specific Dynamic Action or intake (10%) is 3346.85 calories.

In order to meet the nutritional need of workers, the overall worker calorie requirement of 3346,85 caloriea is a reference for the division of nutritional adequacy and the food menu from provided in the morning and afternoon. If the food intake is lowered, the worker will experience malnutrition and potentially adverse effect on the health condition of the employee, however, the exess of food intake such as provide high calorie also effect the incidence of excessive body fat, and if not balanced with the activity which can burn the body's calories.

3.2 Muscle power

Muscle expenditure on some tasks performed by operator included in activities that require great muscle power. For example in material transport activities in station. In addition, other activities such as welding, and turning do not require large muscle power except when the activity set up raw materials to the machine.

3.3 Body posture

Some tasks performed by operators are known not ergonomic, for example, the operators who use the lathe machine while doing set up of the machine, with a bent posture can cause injury to the backbone. Here are the pictures of worker's posture while working.

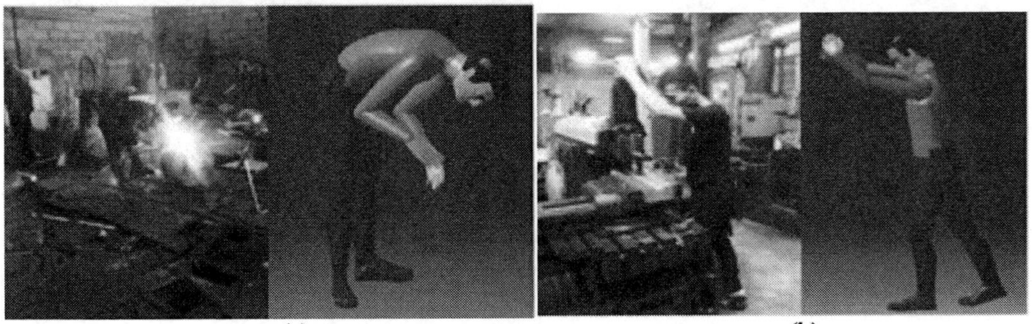

(a) (b)

Fig. 1. (a) Body posture at welding, **(b)** Body posture at frais

Based on figure 1 (a), it shows the posture when workers welding the objects is bending. This can cause back muscle injury. Based on the figure 1(b), shows that the work posture on the frais stastion, there is a bad poture such as lifting of the shoulders and elbows upward when doing the process of frais on the workpiece. This can cause arm muscle injury.

(a) (b)

Fig. 2. (a) Body posture at hand milling station, (b) Body posture at milling station

Based on the figure 2 (a), shows the body posture of the operator while doing the milling process is slightly bent and shoulder lifted. This can cause neck and spine injuries. Based on the figure 2(b), the posture of milling operator with lifted leg. This isn't belongs to the healthy posture (the body working in a natural position or straight leg).

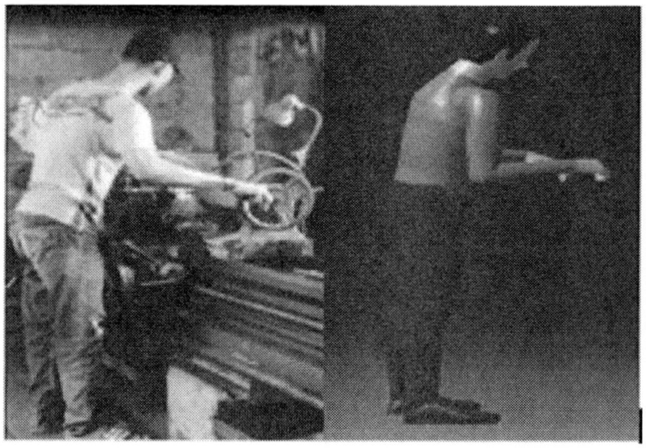

Fig. 3. Postur kerja stasiun bubut

Based on figure 3, the posture of the operator at the process of lathing is with raised shoulders. This does not correspind to the supposed working position of the body working with normal shoulder position.

3.4 Environment

3.4.1 Welding Station

Based onthe results of physical environmental data at welding stations, it is known that the average of humidity is 64%, if this compared with the threshold value according to the desicion of the Minister of Health RI 1405/MENKES/SK/XI/2002 is 40-60%. Excessively high humidity accompanied by high environmental temperatures can increase heart rate and accelerate the evaporation of sweat ehich affects worker will feel prolonged thirst [6,7]. The average value of the temperature of this manufacturing industry is 32.9°C, and when compared with the thershold value according to the desicion of Minister of Helath RI No. 1405/MENKES/SK/XI/2002 can be said this condition have exceeded the limit of thershold value for medium type of work is 26.7 °C and the heavy work load is 25°C. The average noise rating of the manufacturing industry is 43.42dB, the noise thershold value with 8 hours of exposure time in one day is 85dB, so it can be categorized as safe.

3.4.2 Frais Station

Based on the results of the temperature at the frais station, the average lighting on this station is 258.2 lux, if this results is compared with lighting thershold value, then it can be categorized as safe. The average of humidity of this manufacturing industry is 58%, this result if compared with thershold value at frais station can be categorized as safe. The average of temperature value of this station is 33.5 °C, when this results is compared with the optimal working temperature of thershold value 26.7 °C for moderate work, and 25 °C for heavy work, so this average result can be categorized has exceeded the safe limit. The average noise output at frais station is 78.76 dB, when compared with thershold value that is 85dB with exposure time is 8 hours, so it can be categorized as safe.

3.4.3 Milling Station

Based on the average result of the lighting on the production floor of the milling station in manufacturing industry is known with the result is 199,2lux, if compared with the lighting thershold value on Minister of Health RI No. 1405/MENKES/SK/XI/2002 is 100 lux, so the average of these illustrations can be categorized as safe. The average of humidity in manufacturing industry is 60,8%, when compared with thershold value is 40-60%, then this result can be said to have exceeded the limit. The average temperature at the milling station is 33.5 °C, while the thershold value for moderate work is 26.7 °C, and for heavy work is 25 °C, the average of temperature can be categorized to have exceeded the thershold limit. The average noise rating is 81.06dB, this results if compared to thershold value of noise is 85dB, so this average value can be categorized as safe.

3.4.4 Turning Station

Based on the average results of lighting on the production floor of turning station in manufacturing industry is known with the result of 115,4lux, if it compared with lighting thershold value in Minister of Health RI No. 1405/MENKES/SK/XI/2002 is 100 lux, so the average of these results can be categorized as safe. The average of humidity in manufacturing industry is 58%, if compared with humidity thershold value is 40-60% then this results can be categorized as safe. The average temperature at the milling station is 33.3°C, while the thershold value for moderate work is 26.7 °C, and the thershold for heavy work is 25°C, the average of this temperature can be categorized as exceeded the thershold limit. The average noise is 81,66dB, if this results is compared to noise thershold value is 85dB, so this average can be categorized as safe.

3.5 Time Conditions

Effort to restore power certainly require rest time to release tired and reduce the workload [8,9]. Working in hours at manufacturing industry is accordance with the rules applied by government, the optimal working is eight hours with an hour of break time during the day.

3.6 Social

The social life at work can be the interaction between workers, family, and company [9,10,11]. In the manufacturing industry has a habbit of briefing among workers in the morning before work begins. This is done because the type of work is changing accorrding to demand so the communication between operational managers and machine operators is must. Unfortunately, in this manufacturing industry is still lacking awareness of workers to use the Safety Equipment on each job so that the accidents is probably happen.

3.7 Information

The information condition in the production floor such as: applying the display in the work area enable the distribution of information to avoid mistakes of work's procedure, knowing the location of the retievl and the re-laying of goods, as well minimize the questioning to the operational manager.

3.8 Man-Machine Inteface

Human-machine interaction in the manufacturing industry, shows by how the workers using the equipment or work system that focusing on man to adjust their work (fitting mn to the job). For examples, at the turning station, operator uses beam as footrest so that operator can do their set up carefully. Then, for lifting the heavy workpiece, the company has provided manual crane as a tool for material transportation. Based on the result of posture analysis using CATIA, the score of each operator of machining process as follows.

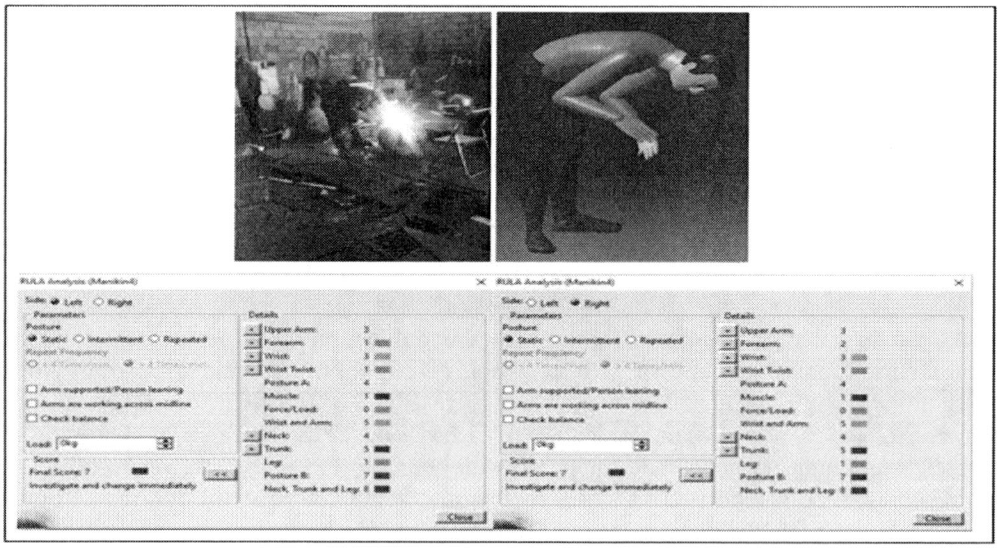

Fig. 4. RULA Score of Welding Operator

Based on the posture analyze onthe welding process, the RULA Score is seven as shown in figure 4, which is categorized as high risk level. In thiis condition, needed the repair action at work system immediately, because this situation can causes the injury of worker. In the frais process, the RULA Score of operator is 5, as shown in figure 2(a). This score can be categorized the medium level, in this condition required improvement of work system in the future. Than at the lathe process the RULA score is as shown in figure 2(b). This score shows the medium level risk in this condition so required repair of work system in the future. Based on the results of the RULA score, it is known that highest and most dangerous scores are in the welding operation where the posture is bent, so that's why can cause injury at work if not repair well. Therefore, the the appropriate technoogy for this manufacturing industry is located in welding station. This kind of aprropriate technology is adjustable welding table with hydraulic, so that the height of object could be adjusted by the height of operator. This following is a drawing of the hydraulic welding table for welding station.

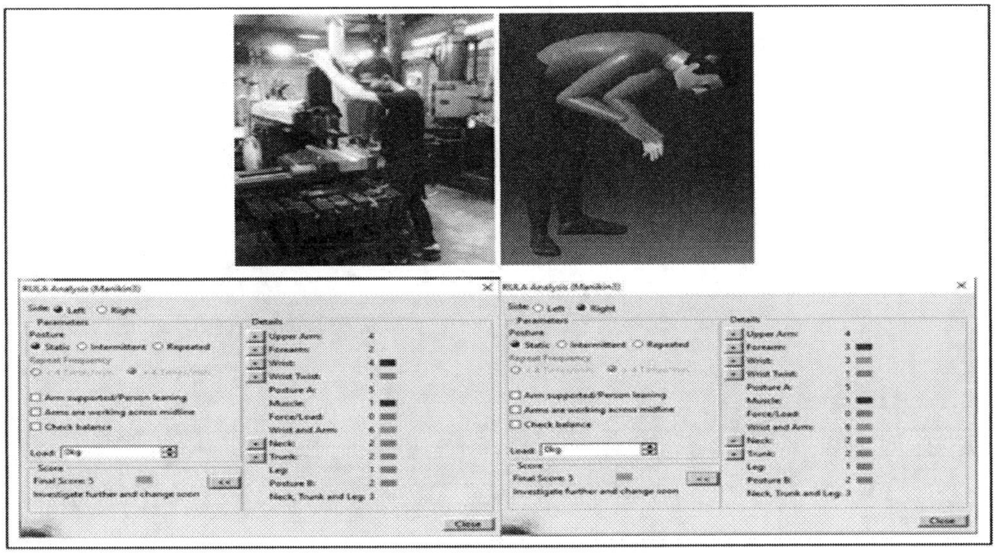

Fig. 5. RULA Score of frais operator

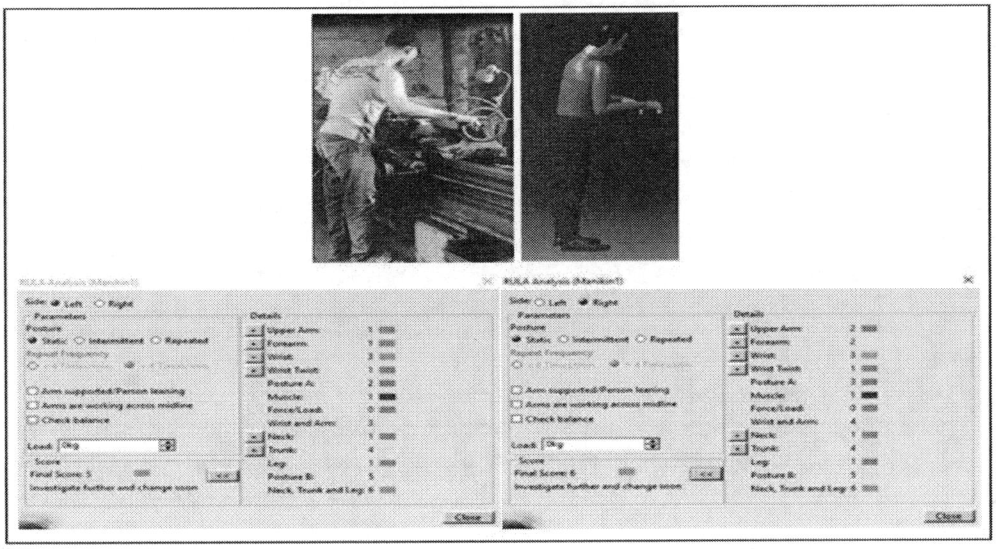

Fig. 6. RULA Score of turning operator

This technology has a hydraulic working principle with a lever rotation control that function is to adjust the height of table with the height of operator and the height of the workpiece. This hydraulic table has a thickness 8 cm, width 1 m, and length 2 m. This hydraulic table is designed with refractory material

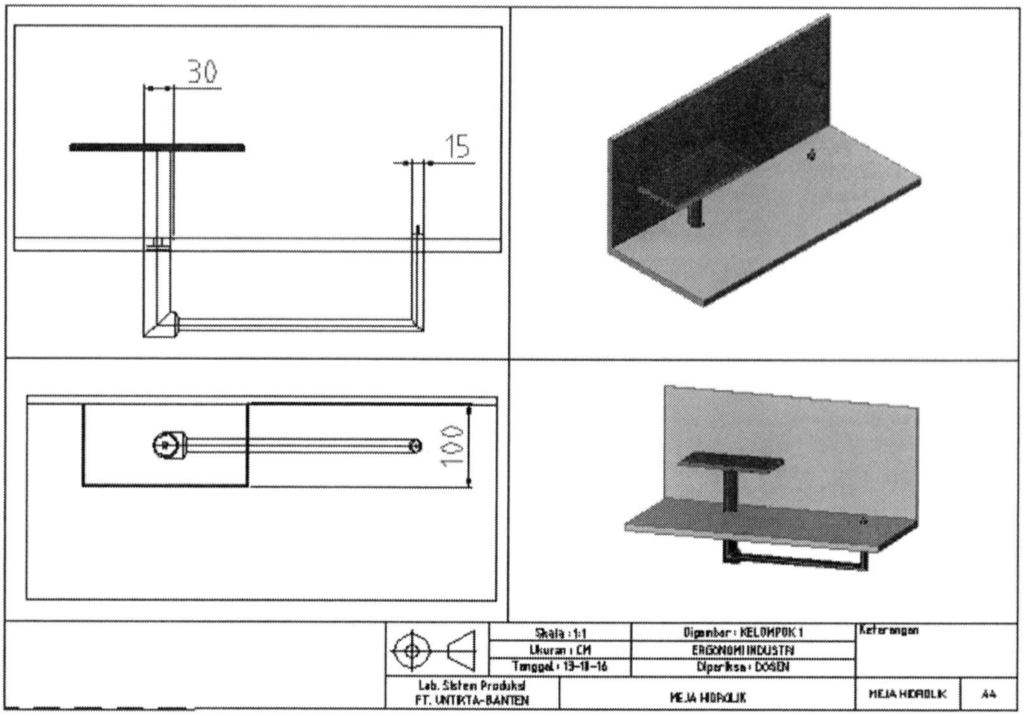

Fig. 7. Hydraulic welding table.

This hydraulic welding table provides technical benefits that facilitate the operator in performing the process on the workpiece with varying heights, then the benefits in terms of ergonomics is to improve the health posture of operator thereby reducing complaints musculosceletal disorders, then economically can increase the productivity of the company thus increasing revenue of the company, in addition to the economic terms of this tool also does not reqiure expensive investment because of its simple working principle and do not use advanced technology.

4 Conclusion

The following is the conclusion and suggestion of the discussion and analysis that has been done:

1. Based on 8 aspects od ergonomics, the aspects that need to be evaluated are: the oprator's posture tends was harmful to the back muscles and the worker's arms so that it is necessary to adjust the anthropometry of the worker. In the work area need a well-organized layout and does not exceed the thershold value. The evaluation on the habbit of this manufacturing industry is by applying the personal protective equipment (PPE) while working as a safety and health management system.
2. Basal Metabolisme Rate (BMR) is 1618.4 calories, while total energy adequacy value based on physical activity is 3402,56 calories. So the obtained Spesific Dynamic Action or intake (10%) is 3346.85 calories.
3. Appropriate technology results are used in welding stations (the most dangerous by RULA Analysis score) this technology is using a hydraulic welding table that can adjusted the height so that can fix the bad postural while operator welding the objects.

Acknowledgements

Thanks to all those who have supproted this research, both in the manufacturing industry of this research and all the operators of the company's production floor. Laboratory of Work System and Ergonomics Major Industrial Engineering UNTIRTA.

References

1. Suma'mur. Company Hygiene and Occupational Health. Jakarta : PT Toko Gunung Agung. (1989)
2. Manuaba, A. Implementation of ergonomics to improve the quality of human resources and productivity; The material of the K3 Seminar with the theme Through culture of K3 improve the quality of human resources. Ergonomics potpourri. Vol 1. Denpasar. Departemen of Physiology. Udayana University. (1992)
3. Grandjean, E. Fitiing The Task to The Man. 4th edition. London : Taylor & Francis. (1993)
4. Susihono, Wahyu. Automatic Semi Cutting Machine with Attention of Ergonomic Working Aspects. Proceeding National Seminar K3 and Ergonomic Program Application at Work. University of USU. Medan. (2009)
5. Manuaba, A. The effort of ergonomic culture in PTP XXI-XXXII; Seminar Material Cultivate Ergonomics at Sugar Factory PTP XXI-XXII Surabaya. Ergonomics potpourri. Vol 1. Denpasar. Departemen of Physiology. Udayana University. (1992)
6. Susihono, W. Ergonomics in Metal Casting Industry. Serang : Untirta Press. (2016)
7. Susihono, W. Ergonomics Approach Analysis As Basis for Identification and Repairing Working Conditions in Industrial Metal Kitchen Casting System Induction; Case Study at PT. X Ceper Klaten. (2014)
8. Mulyadi. Batan, I Made Londen. Testing Prototype e Multipurpose Wheelchair. Department of Mechanical Engineering University of Muhammadiyah Sidoarjo and ITS Surabaya. East Java. (2016)
9. Proceeding National Seminar of Industrial Engineering Conference. Department of Industrial Engineering, University of Sebelas Maret. Surakarta. Hal 284-290. ISBN 978-602-70259-2-9.
10. Anizar, Suriadi Joko. Posture Analysis of Operators in Boiler Section with Ovako Working Posture Analysis System Method at PTPN V SEI Rokan Riau. Paper in the National Seminar on Industrial Engineering and Congress BKSTI V. Makassar, 16-17 Juli (2008)
11. Chapanis, A. National and Cultural Variables in Ergonomics. Ergonomics-17. P 153-175. (1974)

Design of standard operating procedure (SOP) based at ergonomic working attitude through musculoskeletal disorders (Msd's) complaints

Wahyu Susihono[1*], *Ariesca*[1], *Suryanawati*[1], *Mirajiani*[2], *Gugun* Gunawan[2]

[1]Industrial Engineering Department, Engineering Faculty, University of Sultan Ageng Tirtayasa, Banten, Indonesia
[2] Agribusiness Department, Agriculture Faculty, University of Sultan Ageng Tirtayasa, Banten, Indonesia

Abstract. PT PJC is a company synthetic resin. Synthetic resin is that produced by PT PJC is sold to the market in three types of packing, one of then is the drum packing type that weigh 208 kg. One of the activities that use the manual handling product (manual material handling) is the activity of filling, the drum-packed product will be transfered with drum packaging to the finishing goods area which done by the man force. Manual material handling concept has several advantages, one of which is easy to do by humans. However, manual handling material handling becomes an issue if the material load exceeds the maximum lifting of the worker because of the potential risk of injury to the human body, especially if it is carried out simultaneously for a relatively long period of time. One of the risks of human injury that can be experienced is Musculoskeletal Disorders (MSDs). This research will investigate the musculoskeletal complaints on Filling workers which will form the basic of the ergonomically-based Standard Operating Procedure (SOP) design basic. Data is taken by a Nordic Body Map (NBM) questionnaire. The result shows that most of the dominant body pain felt by the worker as the impact of previous day's work was the complaints of pain on the left and right wrists, left and right palms, left and right knees, left and right calves, left and right ankles, while the pains felt by the workers after the activity are pain in the back, waist, left and right elbow, left and right thighs, left and right calvs, left and right ankles. The pain work will disappear after the night rest, but there is no guarantee that the worker's body will be fit in the morning, because of the remnants of body pain that have not been fully fit, thus affecting the readiness of physical conditions at work in the morning. These body pain can be reduced by designing the right standard operating procedure (SOP) based on an ergonomic work attitude. Musculoskeletal pain before the activity were 39.3 ± 10.6 and after activity was 76.5 ± 14.9 or an increase in musculoskeletal pain of 94.66%.

1 Introduction

* Corresponding author: pmy_wahyu@yahoo.co.id

PT PJC is a company that engage in the production of synthetic resin. Synthetic resin is produced by PT PJC is sold to the market in three types of packing, namely packing of cans, bottles, and drums. Especially the drum packaging has the greatest weight among the other which weighs 208 kg. In the production activities, PT PJC still using human as part of the work completion process. Although the use of machines is categorized as high technology, but various supporting activities can not be separated from the use of human force.

Human have limited abilities including muscle ability to carry load and material loads, because these problems will have implications for the performance of the company in general. As to be seen from the some perspective of ergonomic glass panels, various aspects of human beings can be seen from the factors of energy use or the necessary nutrients, the strength of muscle power, the condition of body posture at work, working environment condition, time consumed for work and rest, socio-cultural circumstances that exist in the community and internal enterprise, the condition of information available within the company and others and, human-machine interaction. From these eight aspects it can be concluded that the main problem of the workers is the aspect of muscle strength or energy.

Until now, the work pattern has been using the correct rules, which is the existence of standard operating procedures that issued by the management company to operate the machine and other manual material handling, but as time passed the process and attitude of human work, there still found various problems that occur in the production such as workers bent below 900, the round of bodies that cause pain in certain body part and generally occurs to L5 / S1. There are complaints from several body parts. The work that done repeatedly and for a long time will cause premature aging or musculoskeletal disolders. The effort taken by the company's management for the recovery of each personnel specifically on the fulfillment of energy or nutritional status which provids in the form of mineral water, coffee, and the sweets. Meanwhile, at the break time the operators are given lunch that contains carbohydrates, proteins, and vitamins.

Furthermore, it is found on the production floor that the operator's work attitude is relatively unnatural on material transfer activities. Manual Material Handling (MMH) is all load-lifting work that includ rotating, bending, grabbing, damping, pushing, and pulling, activities performed by the workers in order to move the load from one ponit to a particular destination location (Nurmianto, 1998). Manual Material Handling (MMH) is still widely used by the company because it has several advantages, such as more flexible movement of material transfer in an irregular or untidy work location, cheaper, and easier for light load (Santiasih, 2013).

Based on this phenomenon, it is found that manual material handling is still needed in order to complete the production process so it is still important to note. MMH can facilitate the activity of the material displacement with heavy load or relatively light material. In the company heavy PJC manual drum manually packaged material (manual material handling) is 208 kg. So with the weight, manual handling material becomes unnatural and can cause the risk of injury to the limb due to over-exertion-lifting and carrying damage body tissue caused by excessive weight lifting. The injury will be more severe if the activity is done continuously for long periods with improper work posture. One type of injury that can arise from over-exertion lifting and carrying is Musculoskeletal Disorders (MSDs). Musculoskeletal Disorders is a pain on the skeletal muscle sections felt by a person ranging from very mild to very painful complaints (Nugraha, 2013).

When judged from the aspect of the working environment, the physical environment of the work varied form of lighting, noise, temperature, humidity, and wind speed slightly above the critical point of the threshold value, so that the physical environment can affect the operator's work activities. This is caused by the physiology of the human body which sensitive to the state of the physical environment.

For the aspect of information, as seen from the flow of coordination and communication of the superiors and the subordinates are very good and in accordance to the structure that has been set at the company, in addition to the delivery of information related to job, cleanliness of the work area, findings on the work area, and activities in each day at a briefing before the work begun. For information in the form of hazard info, warning, regulations, and safety induction had been done in the form of picture display, as well as voice display from company radio. PT PJC applies the culture of wearing personal protective equipment in the form of helmet, mask, cloth gloves, and safety shoes. So that in the interaction between man and machine hopefully could minimize the work accident. In addition, the work instruction manual has been provided in the form of standard operational procedure (SOP) based on work process, but in SOP there is only work instruction in the form of each process stage, not in giving instruction in the form of work attitude on each process, so that in between one operator to another operators, the varying body movements are still exist. This condition caused the occurrence of musculoskeletal pain that varies between one operator to another operator. Based on the eight aspects of ergonomic identified has the dominant problem is musculoskeletal pain that caused by the variation of work attitude of the operator in manual material handling activity and the absence of work-based SOP.

To overcome the afformentioned problems, the company has made improvements in the form of procurement of material handling tools in the form of hand forklifts, but in the application, the use of such tools slows down the production process because of its operation which is considered to be complicated and heavy by the operator. So that the operator re-do material handling activities manually without tools. The Company also made SOP improvements for each period, but SOP has not paid attention to the operational attitude aspect of the operator so there is no significant decrease in risk of musculoskeletal disorders between before and after. Msukuloskeletal pain are also experienced by physical workers in general (Choobineh, et al., 2007; Choi, et al., 2009; Errico, et al., 2010; (Bernards, et al., 2011; Nonnenmann, et al., 2010)

In order to reduce the risk of musculoskeletal disorders in the operation, it is important to have a standard operational procedure (SOP) based on good working attitude and correctness in the manual activity of drum manual handling material. This intervention is an alternative to decrease musculoskeletal pain through the ergonomic approach improvement. Improvements with ergonomic approaches could reduce the impact of musculoskeletal complaints of workers (Das, et al., 2013; Kirkhorn, et al., 2010). So it is necessary to conduct a research in the form of identification of musculoskeletal filling operator's complaints as the basis for improvement of Standard Operational Procedure (SOP). To find out the type and magnitude of pain complaints experienced by operators in the filling section, this study used a questionnaire Nordic Body Map (NBM). Nordic Body Map is one of the subjective measurement methods to measure workers' muscle pain (Wilson, 1995). The Nordic Body Map is the most commonly used questionnaire for discomfort to workers, and it is most commonly used because it is standardized and well-organized (Kroemer, 2001).

2 Methods

The design in this research is cross sectional. Data were analyzed descriptively and quantitative. Descriptive research is a research that aims to make a description of a social phenomenon or natural in a systematic, factual, and accurate (Wardiyanta, 2006). While quantitative research is a study by obtaining data in the form of numbers or qualitative data that is suspected (Sugiyono, 2003). Cross sectional is an approach that is momentary or in and not followed in a certain time (Hadjar, 1996). Subjects in this study were all (population) operator of filling section at PT PJC (N = 10 people). Data were collected using a Nordic Body Map (NBM) questionnaire with Likert scale. Criteria subject in this study is the operator with age between 23-48 years, weight 50-68 kg, at least have worked for 1 month.

The stages of this research are preliminary study, research problem determination, data collection using Nordic Body Map (NBM) questionnaire, processed questionnaire data using graph comparison data between all population of operator before pre-test and after activity manual material handling (Post-Test)

3 Results and discussion

Based on the measurement of the physical environment in the work area, it is found that the physical environment measurement value obtained in the lighting aspect is 837.53 Lux, if the limitations of Lighting Threshold Level (NAB) is 100-200 Lux. Then for the noise aspect, the physical environment measurement value is 86.92 dB, if the NAB noise limit is 85 dB. Furthermore, for temperature measurement results show the results of 28.93 oC, if viewed NAB temperature is equal to 25.9 oC. Then the results of humidity measurements show the results of 63.84 oC, if viewed NAB moisture is equal to 40-60 oC. Then wind speed measurement results show the results of 0.48%, if viewed NAB moisture is equal to 0.25%. So it can be seen that the value of the physical environment when data collection Pre-Test and Post-Test there is a tendency above the NAV according to the Regulation of the Minister of Manpower and Transmigration RI No Per. 13 / MN / X / 2011. Work environment conditions affect the results of work done (Keun and Park, 2007).

Distribution of Nordic Body Map (NBM) questionnaires to all filling operators in the morning at 09.00 before the operator starts the activity (Pre-Test), and in the afternoon at 16.00 after the activity is completed (Post-Test). Based on the data of the questionnaire distributed, then the difference test is done on both data that is Pre-Test and Post-Test complaints data to know whether there is difference between questionnaire distribution result during Pre-Test and Post-Test. Hypotheses in this study are as follows:
H0: There is a difference between the results of the Pre-Test questionnaire and the Post-Test.
H1: There is no difference between the results of the Pre-Test questionnaire and the Post-Test.

Different test results using Wilcoxon Signed Rank Test on SPSS obtained results in Figure 1.

	POSTTEST - PRETEST
Z	-2.236[b]
Asymp. Sig. (2-tailed)	.025

Fig. 1. Test results of different wilcoxon signed rank test

Based on Figure 1, the results can conclude that H0 is accepted, meaning there is a difference between Pre-Test data with Post-Test data. After the differences between Pre-Test and Post-Test, the average calculation of musculoskeletal complaints of the operator before the start of the activity (Pre-Test) was obtained by 39.3 ± 10.6. Of the 28 items of the statement of pain complaints, obtained the top ten highest pain complaints felt by the filling operator at the time before the start of the activity (Pre-Test) presented in the graph below:

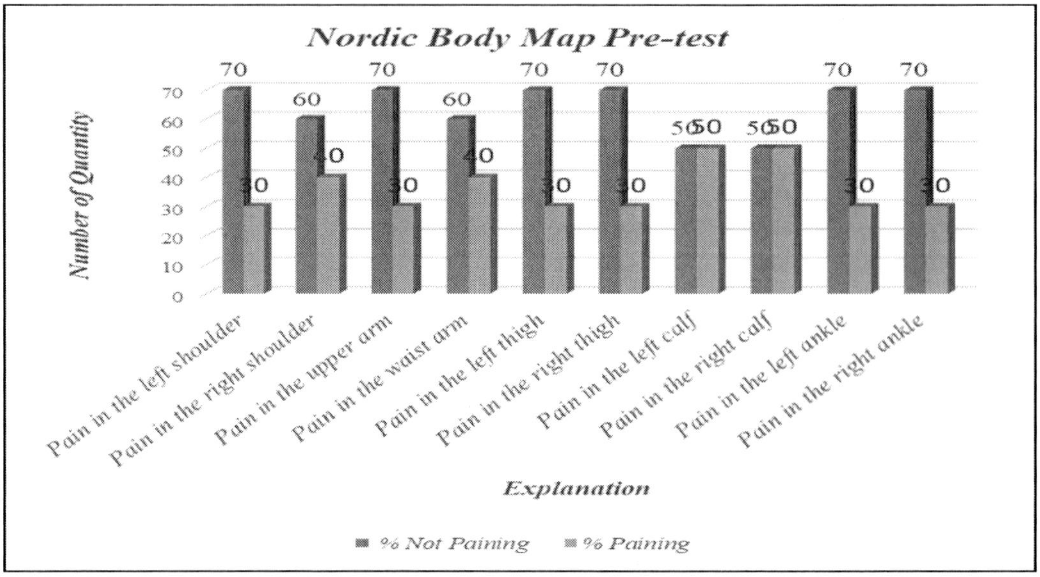

Fig. 2. Graph of nordic body map (nbm) pre-test

Based on Figure 2. Nordic Body Map Map Pre-Test, it can be seen that the highest pain complaints felt by filling operators before starting activity or work are complaints of pain in the left shoulder, pain in the right shoulder, pain in the right upper arm, waist, aching on left thigh, pain in right thigh, pain in left calf, pain in right calf, pain in left ankle, pain in right ankle. These painful causes of complaints in various dimensions of the body (Yokoyama, et al., 2006), if not immediately looked for solutions may cause stress symptoms (Lee, et al., 2005)

While based on the average calculation of operator pain after work activity (Post-Test) got value equal to 76,5 ± 14,9. Of the 28 items of the statement of pain complaints, obtained the top ten highest pain complaints felt by filling operators after the work activities (Post-Test) are presented in the graph below:

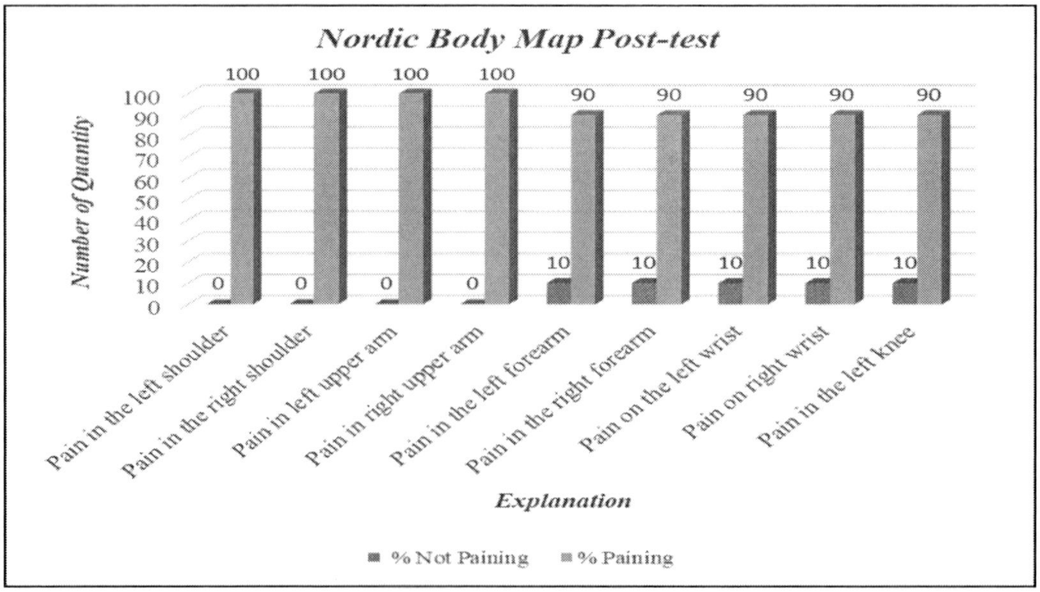

Fig. 3. Graph of nordic body map (NBM) post-test

According to Figure 3, the Nordic Body Map Post-Test graph, it can be seen that the highest pain complaints felt by operators of filling after work activity are complaints of pain in the left shoulder, pain in the right shoulder, pain in the upper left arm, pain in the upper right arm, pain in the left forearm, pain in the right forearm, pain in the left wrist, pain in the right wrist, pain in the left knee, and pain in the right knee.

Complaints in various parts of the body will cause a delay in work activities. Musculoskeletal complaints are closely related to subjective complaints (Jungsun, et al., 2006). This complaint begins with muscle fatigue. An immediate effort is to recover muscle tension (Gawke, et al., 2012). Increased fatigue causes the decline on the concentration and endurance . Decreased fatigue is caused by decreased occupational health and concentration (Nagai, et al., 2011). Ergonomic's SOP re capable of providing work improvements characterized by changes in musculoskeletal complaints. SOP-based ergonomic design among others is :

1. Use of work activities as a basis for designing SOPs
2. Consider the human factor as a major part of the work activity
3. Apply the job to the man fitting, before fitting the man to the job
4. Measuring ability, skill and human limitations
5. There is no forced body stance, it works naturally
6. Human machine interaction puts the human factor first

With the SOP based on ergonomy, workers achieve job satisfaction. Feasibility can be achieved if the worker considers the performance of the personnel (Hakansson, et al., 2011). Increased musculoskeletal complaints are consistent with length of work and the burden of work. Based on the average of musculoskeletal complaints of filling operators, the percentage of grievance change between before and after work was 95%. This condition is likely to change larger, if the work time continues to be added, such as soft work and additional work after the operator returned from the company. There needs to be a holistic evaluation of employee activity, and it is not advisable to do work on the worker's abilities.

4 Conclusion

After the discussion of results, then the conclusion is the musculoskeletal complaints of the filling section before the work activity is equal to 39.3 ± 10.6. Musculoskeletal operators of the filling section at the time after work activity is equal to 76.5 ± 14.9. The percentage change in musculoskeletal complaints between before and after work was 95%.

Acknowledgements

Thanks to the management and operators PT PJC who has been willing to become subjects in this study, as well as Laboratory Engineering System Work and Ergonomics FT. Untirta who has provided facilities in the form of data retrieval tool

References

1. Bernards, C. Courouve, L, Bouee, S. Adjemian, A. Chretien, J.C. Niedhammer, I. Biomechanical and Psychosocial Work Exposures and Musculoskeletal Symptoms among Vineyard Workers. Journal Occupational Health (53). p: 297-311 (2011)
2. Choi, W.J. Kang, Y.J. Kim, J.Y. Han, S.H. Symptom Prevalence of Musculoskeletal Disorders and the Effects of Prior Acute among Aging Male Steelworkers. Journal Occupational Health (51). p: 273-282(2009)

3. Choobineh, A. Tabatabaei, S.H. Mokhtarzadeh, A. Salehi, M. Musculoskeletal Problems among Workers of an Iranian Rubber Factory. Journal Occupational Health (49). p: 418-423 (2007)
4. Das, B. Ghosh, T. Gangopadhyay, S. Child work in agricultural in waste Bengal, India; Assessment of Musculoskeletal disorders and occupational health problems. Journal Occupational Health (55). p: 244-258 (2013)
5. Errico, A.D. Caputro, P. Falcone, U. Fubini, L. Gilardi, L. Mamo, C. Migliardi, A. Quarta, D. Coffano, E. Risk Factors for Upper Extremity Musculoskeletal Symptoms among Call Center Employees. Journal Occupational Health (52). p: 115-124 (2010)
6. Gawke, J.C. Gorgievski, M.J. Linden, D.V.D. Office Work and Complaint of the Arm, Neck and Shoulders; The Role of Job Characteristics, Muscular Tension and Need for Recovery. Journal Occupational Health (54). p: 323-330 (2012)
7. Hadjar, Ibnu. The Basics of Quantitative Research Methodology in Education (Written in Bahasa Indonesia). Jakarta: PT Raja Grafindo Persada (1996).
8. Hakansson, C. Bjorkelund, C. Eklund, M. Associations between women's subjective perceptions of daily occupations and life satisfaction, and the role of perceived control. Australian Occupational Therapy journal (58). p: 397-404 (2011)
9. Jungsun, P. Mina, H. Yunjeong, Y.I. Yangho, K.I.M. Subjective Fatique and Stress Hormone Levels in Urine According to Duration of Shiftwork. Journal Occupational Health (48). p: 446-450 (2006)
10. Keun, Y.K dan Park, H.S. Workers Perception of the Changes of Work Environment and its Relation to the Occurrence of Work Related Musculoskeletal Disorders. Journal Occupational Health (49). p: 152-154 (2007)
11. Kirkhorn, S.R. Richardson, G.E. Banks, R.J. Ergonomic Risks and Musculoskeletal Disorders in Production Agriculture: Recommendations for Effective Research to Practice. Journal of Agromedicine (15). p: 280-299 (2010)
12. Kroemer, K. Ergonomics How to Design for Ease and Efficiency. New Jersey: Prentice Hall International (2001)
13. Lee, H.Y. Yeh, W.Y. Chin, C.W. Wang, J.D. Prevalence and Psychosocial risk Factors of Upper Extremity Musculoskeletal Pain in Industries of Taiwan: A Nationwide Study. Journal Occupational Health (47). p: 311-318 (2005)
14. Nagai, M. Morikawa, Y. Kitaoka, K. Nakamura, K. Sakurai, M. Nishijo, M. Hamazaki, Y. Maruzen, S. Nakagawa, H. Effects of Fatigue on Immune Function in Nurses Performing Shift Work. Journal Occupational Health (53). p: 312-319 (2011)
15. Nonnenmann, M.W. Hussain, A. Shirley, M. Shepherd, S. Levin, J.L. Risk Factors for Musculoskeletal Symptoms Among Crawfish Farmers in Louisiana; A Pilot Study. Journal of Agromedicine (15). p: 386-363 (2010)
16. Nugraha, Harvian. Repair Analysis of Operator Posture Using RULA Method To Reduce Musculoskeletal Disorders Risk (Case Study on Bad Stock Warehouse PT X Surabaya) (Written in Bahasa Indonesia). Malang:Brawijaya University (2013)
17. Nurmianto, Eko. Basic Concepts and Application (Written in Bahasa Indonesia). Editio I. Printed II. Oktober. Jakarta: Guna Widya (1998)
18. Nurmianto, Eko. Basic Concepts Ergonomics and Application (Written in Bahasa Indonesia). Surabaya: Guna Widya (2003).
19. Santiasih, Indri. Manual Material Handling Review Against Low Back Pain Incidents On Textile Workers (Written in Bahasa Indonesia). Surabaya: Politeknik Perkapalan Negeri Surabaya (2013)

20. Santoso, Gempur. Ergonomics, People, Equipment and the Environment (Written in Bahasa Indonesia). Prestasi Pustaka, Jakarta (2004)
21. Sugiyono. Business Research Methods (Written in Bahasa Indonesia). Edisi 1. Bandung: Alfabeta(2003)
22. Wardiyanta. Research Methods of Tourism (Written in Bahasa Indonesia). Yogyakarta: (ANDI 2006).
23. Wilson, J.R., Corlett, E.N. *Evaluation of Human Work: A Practical Ergonomics Methodology*. 2nd and Revised Edition. London: Taylor and Francis (1995)
24. Yokoyama, Wakui, T. Harada, N. Prevalence of Cervical Spondylosis and Musculoskeletal Symptoms among Coolies in a City of Bangladesh. *Journal Occupational Health* (48). p: 69-73 (2006)

Assessing bank performance measurement in islamic banking industry

Mokhamad Ikhsan Ramdhoni[1*]

[1]Applied Master of Islamic Banking and Finance, Politeknik Negeri Bandung, Indonesia

Abstract. The Indonesian Islamic banking industry, attaining the age of more than 25 years tend to adopt conventional approach to measure its bank performance. Although this method could well inform the performance of the bank financial, this measurement fails to inform both the characteristics of Islamic banking. This study is designed to evaluate three Islamic bank performance measurements: Sharia Maqashid Index, Sharia Conformity and Profitability (SCnP), and CAMELS. Sharia Maqashid Index approach has three variables: educating individual, establishing justice, and public interest. SCnP approach has two variables: Sharia conformity and profitability. While CAMELS approach has five variables: Capital, Asset Quality, Management, Earning, and Liquidity. This study uses data from Islamic Commercial Banks registered on Bank Indonesia from 2012-2016. The data analytical technique is quantitative method using descriptive approach. The results of the study using Sharia Maqashid Index reveal that there is a difference in Islamic Commercial Banks performance. The used of SCnP reveals that none of Islamic Commercial Banks on upper right quadrant. The used of CAMELS reveals that all of Islamic Commercial Banks in Indonesia have Fairly Healthy predicate. These findings suggest that, to provide a comprehensive picture of its performance, the Islamic banks need to adopt both Sharia-based as well as financial based measurement.

1 Introduction

Currently, the world is entering the era of Industrial Revolution 4.0. This also applies to the banking industry where they are competing to utilize technology to provide excellent service to its customers. This is in line with Senge et al. [1] which states the current challenge is to develop sustainable businesses that are compatible with current economic realities. The financial services sector was an early adopter of key technologies associated with business transformation and it is currently one of the most intensive users of information and communication technologies [2] .

Indonesia's Islamic banking is over 25 years old, but in its age of more than a quarter century it has not had its own performance measurement tools and is still adopting conventional measuring instruments and limited only to financial ratio measurement. The

* Corresponding author: ikhsanramdhoni@gmail.com

use of conventional measuring instruments as the only measure of performance in Islamic banks is considered less suitable.

First, scholars [3] state that Islamic banks and conventional banks are characteristically difficult to distinguish. Secondly, several studies have shown that the measurement of performance of Islamic banks with conventional banks is not the same, since the views of core functions and operational characteristics are clearly different [4]. Thirdly, most Islamic banks rely on financial benchmarks to measure their performance results, therefore, many stakeholders cannot clearly see the difference between Islamic and conventional banking. Furthermore, by using conventional benchmarks to measure the performance of Islamic banking there is a discrepancy between conventional banking performance indicators and the broader objectives of Islamic banking. [5]

This is confirmed by researchers [6] which states that the characteristic difference between Islamic banking and conventional banking is something very basic. So the characteristics of services produced will be different, and evaluation of both types of banking will also be different [7]. Therefore, efforts are needed to develop performance measurement tools that will complement the financial objectives of Islamic banking [5]. The development of measuring instruments for performance of Islamic banks is needed to answer one of the research results that the Islamic finance industry is fast growing and innovating but slow in establishing its credibility [8].

In fact, research related to measuring instruments that have been aligned with the characteristics of Islamic banking has been done. Past researchers have attempted to improve performance measurements to align with the characteristics and objectives of Islamic banking [5] with the Sharia Maqashid Index approach, and research with the Sharia Conformity and Profitability (SCnP) method [9]. The results show that the model can be a strategic alternative approach that can illustrate how well the performance of Islamic banking is more universal and can be implemented in the form of a comprehensive policy strategy [10]. The difference of this study with previous studies is in this study the author use several approaches in assessing the performance of Islamic banking. Considering one of the functions of banking is as financial intermediary, the bank is required to always maintain the level of business performance and business health level in order to gain the trust of stakeholders [7]. On the other hand, scholars [10] maintain that because Islamic banks have different characteristics from conventional banks in both concept and practice, Islamic banking should consider performance that not only focus on financial ratios, but also pay attention to Non-financial ratio.

Based on the above background, this study is intended to assess three models of bank performance measurement, namely the measurement of Islamic banking performance using Sharia Maqashid Index model, Sharia Conformity and Profitability model as well as CAMELS model. Such a study is important to assist the Islamic banking industry to evaluate their business performance and subsequently developing appropriate competitive strategy.

2 Literature review

In the highly competitive business environment, as as in Islamic bank industry, having a high performance is imperative [11]. Performance appraisal is a management tool for knowing how far the company's objectives have been achieved, evaluating business performance, managers, divisions and individuals within the company, as well as for predicting future company expectations [12]. The information used as a performance measurement or assessment tool is grouped into two categories namely financial performance and Non-financial information [13]. The performance of a bank can be identified by using a measuring tool known as CAMELS (Capital, Asset Quality, Management, Earning, Liquidity, and Sensitivity of market risk) analysis. Bank health measurement and rating model through CAMELS analysis has been determined by Bank Indonesia [14]. Non-financial information can increase confidence in the quality management supervision process. A comprehensive performance measurement technique developed by several organizations and companies is a balanced scorecard which includes four measured aspects: financial perspective, customer satisfaction, internal process efficiency, and learning and growth [13].

Mohammed [5] uses a method to define the concept of Maqashid Sharia into a measuring instrument for measuring the performance of Islamic banking. This method describes the Concept (C) into observable behavior and is called Dimension (D). The dimension is then translated into several measurable behaviors called Elements (E). This measurement of performance that have been developed from the framework of Maqashid Sharia and then weighted so that the concept can be used to measure the performance of Islamic banking by weighting each Objective (O) and Element (E) [5].

In the Sharia Conformity and Profitability (SCnP) model there is a merger between profitability indicators to assess conventional financial performance with an index of conformity to sharia values to assess the socio-economic obligations of Islamic banks. The sharia compliance variable consists of the ratio of sharia investment, sharia income, and profit sharing ratio. While the profitability variable consists of ROA, ROE, and profit margin [9]. From each sharia compliance ratio and profitability ratios are averaged and the result is a graph of four quadrants.

One of the bank's performance appraisal tool is CAMELS method, as stated in Bank Indonesia Regulation dated January 24, 2007 [14] regarding Rating System of Commercial Banks Based on Sharia Principles. The Bank's Health Level is the result of qualitative assessment of various aspects affecting the condition or performance of an Islamic commercial bank or Islamic business unit through; quantitative assessment and qualitative assessment of capital factors, asset quality, profitability, liquidity, sensitivity to market risk; and qualitative assessment of management factors. The CAMELS approach in this study do not include sensitivity to market risk factor due to the limited data obtained, especially related to the data of excess capital and the potential loss of exchange rate which become the formula/ratio to calculate the sensitivity factor.

3 Method

This research useses descriptive quantitative approach. In this case, the researchers describe the quantitative data that have been obtained. After the statistical process, the data is interpreted. The reason for choosing this research method is because all data is in the form of financial ratios and can be directly and easily explained. The population of this study are all Islamic Commercial Banks registered in Bank Indonesia from 2012-2016 as many as 13 Banks [15]. While the sample for this research consists of four Islamic Foreign Exchange Banks.

4 Result

Table 4.1. Islamic Commercial Bank rating based on Sharia Maqashid Index

BANK	SHARIA MAQASHID INDEX				
	PI of 1st O	PI of 2nd O	PI of 3rd O	SMI	RANKED
Bank Syariah Mandiri	0.00235	0.18596	0.10431	0.29262	2
BNI Syariah	0.00663	0.17580	0.09888	0.28131	3

Bank Muamalat	0.00391	0.21483	0.11298	0.33172	1
Bank Mega Syariah	0.00071	0.15546	0.10621	0.26237	4

Based on Table 4.1 the total calculation of Sharia Maqashid Index of Bank Muamalat was ranked first followed by Bank Syariah Mandiri at the second level, while BNI Syariah and Bank Mega Syariah ranked third and fourth. These results are similar to the results of Prasetyowati, et. al [16] study which reveals that Bank Muamalat is the best performing bank based on Sharia Maqashid Index measurement compared to other Islamic banks.

Table 4.2. Performance of Islamic Commercial Banks based on SCnP

BANK	2012			2013			2014			2015			2016		
	SC (%)	P (%)	Q	SC (%)	P (%)	Q	SC (%)	P (%)	Q	SC (%)	P (%)	Q	SC (%)	P (%)	Q
Bank Syariah Mandiri	74	12	LRQ	74	8	LRQ	73	2	LLQ	75	4	LRQ	76	4	LRQ
BNI Syariah	71	8	LLQ	72	8	LLQ	72	8	ULQ	73	7	ULQ	73	7	ULQ
Bank Muamalat	81	12	LRQ	83	6	LRQ	82	2	LRQ	83	2	LRQ	84	2	LRQ
Bank Mega Syariah	67	25	ULQ	67	13	ULQ	66	4	ULQ	66	4	LLQ	67	7	ULQ

Table 4.2 shows the performance measurements using the SCnP approach. It indicates that for five years all Islamic Commercial Bank in Indonesia spread in upper left quadrant, lower right quadrant, and lower left quadrant. These results are fairly similar to the finding of Ratnaputri [17] research. This result suggests that there is no Islamic Commercial Bank in Indonesia is located in upper right quadrant, indicating a high sharia compliance while having high profitability.

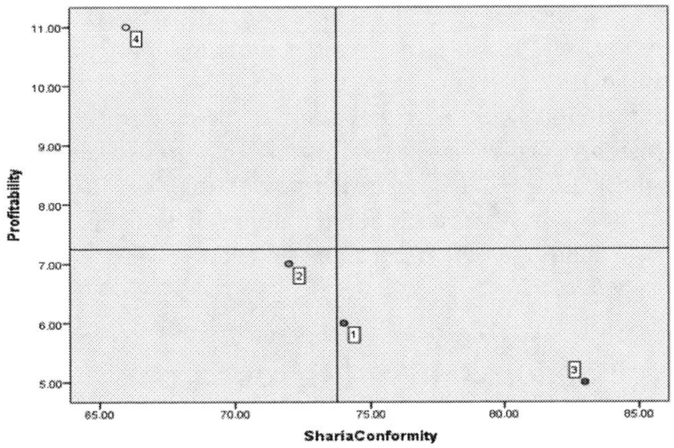

Fig. 4. SCnP Quadrant

In general, Islamic commercial bank performance based on SCnP approach in the period of 2012-2016 is depicted in Figure 4.1

Note:

1. Bank Syariah Mandiri (1) and Bank Muamalat (3) are in lower right quadrant, meaning have a high level of sharia compliance but having low profitability level.
2. BNI Syariah (2) is in lower left quadrant, indicating of a low level of sharia compliance and profitability level.

3. Bank Mega Syariah (4) is in upper left quadrant, means having a low level of sharia compliance but has a high level of profitability.

Table 4.3 Performance of Islamic Commercial Banks based on CAMELS

Years	2012		2013		2014		2015		2016	
Bank	Result	Predicate	Result	Predicate	Result	Predicate	Result	Predicate	Result	Predicate
Bank Syariah Mandiri	76.75	Fairly Healthy	76.81	Fairly Healthy	66.28	Fairly Healthy	70.17	Fairly Healthy	70.47	Fairly Healthy
BNI Syariah	70.62	Fairly Healthy	76.51	Fairly Healthy	75.50	Fairly Healthy	76.60	Fairly Healthy	76.68	Fairly Healthy
Bank Muamalat	76.16	Fairly Healthy	69.49	Fairly Healthy	66.97	Fairly Healthy	67.36	Fairly Healthy	67.27	Fairly Healthy
Bank Mega Syariah	78.49	Fairly Healthy	77.67	Fairly Healthy	69.02	Fairly Healthy	69.34	Fairly Healthy	76.89	Fairly Healthy

Table 4.4 Comprehensive Measurement Performance

Bank	Ranked	Quadrant	Predicate
Bank Syariah Mandiri	2	LRQ	Fairly Healthy
BNI Syariah	3	LLQ	Fairly Healthy
Bank Muamalat	1	LRQ	Fairly Healthy
Bank Mega Syariah	4	ULQ	Fairly Healthy

Table 4.4 shows that two Islamic banks (Bank Syariah Mandiri and Bank Muamalat) which have the highest index maqashid value and have a fairly healthy predicate not necessarily that the banks have a high level of sharia suitability and a high profitability. This finding also shows that there is also an Islamic bank which is in the lowest maqashid index (Bank Mega Syariah) and has a fairly healthy predicate but has a high profitability level (in upper left quadrant).

However, for Islamic banks with first and second ranking with the index maqashid method, the result consistently shows the possibility of being in two quadrants, i.e. in upper right quadrat or low right quadrant. This means that an Islamic bank with a first and second ranking maqashid index does not necessarily have a high level of profitability. Further, the result indicates that every Islamic bank with a first and second ranking maqashid index has a high sharia conformity value.

5 Conclusion, implication, and future research

The measurement of Islamic Commercial Banks performance in 2012-2016 using maqashid index method shows the varied results of Islamic Commercial Banks in Indonesia. The best performance was achieved by Bank Muamalat who consistently maintained the performance that was ranked first. Performance measurements using the SCnP approach indicate that for five years all Islamic Commercial Banks in Indonesia are spread out in upper left quadrant, lower right quadrant, lower left quadrant, and that means no Islamic Commercial Banks in Indonesia in the upper right quadrant which has a high sharia suitability and has a high profitability. The measurement of performance using CAMEL approach shows that for five years all banks has a fairly healthy performance rating. These findings suggest that, to provide a comprehensive picture of their performance, Islamic banks need to adopt both Sharia-based and financial-based performance measures concurrently. The inclusion of these approaches will enable the Islamic bank growing fast and building the public confidence. This study bears a drawback related to the sample used, which contains of four banks. Considering the important of this study's finding; thus, it is recommended that this performance measurement assessment need to be tested in other Islamic banks and in other financial institution.

References

1. P.M. Senge, G. Carstedt, and P.L. Porter, "Innovating Our Way to the Next Industrial Revolution," *MIT Sloan Management Review*, ***42*(2)**, pp. 24-38 (2001)
2. S.V. Scott, J.V. Reenen, and M. Zachariadis, "The Long-Term Effect of Digital Innovation on Bank Performance: An Empirical Study of SWIFT Adoption in Financial Services," *Centre for Economic Performance*, ***992***, pp. 1-49 (2017)
3. R.M. Zaman, H. Movassaghi, "Interest-Free Islamic Banking: Ideals and Reality," *International Journal of Finance*, ***14*(4)**, pp. 2428-2442 (2002)
4. M.H. Bedoui, "Sharia-based Ethical Performance Measurement Framework," *Chairs for Ethics and Financial Norms*, pp. 1-12 (2012)
5. M.O. Mohammed, D.A Razak, "The Performance Measures of Islamic Banking Based on the Maqashid Framework," *Paper of IIUM International Accounting Conference (INTAC IV) held at Putra Jaya Marroitt*, pp. 1-17 (2008)
6. U. Chapra, Sistem Moneter Islam, Jakarta: Gema Insasi Press & Tazkia Press (2000)
7. D. Suhartanto, et al., "Loyalty Intention towards Islamic Bank: The Role of Religiosity, Image, and Trust," *International Journal of Economics and Management*, **12(1)**, pp. 119-132 (2018)
8. M. Ayub, C. Paldi, "Dichotomy between Syariah Compliance and The Economic Goals of Islamic Financial Institutions," *Journal of Islamic Business and Management*, ***5*(2)**, pp. 69-98 (2015)
9. M. Kuppusamy, A.S. Saleha, A. Samudhram, "Measurement of Islamic Banks Performance Using a Shariah Conformity and Profitablity Model," *Review of Islamic Economics*, ***13*(2)**, pp. 35-48 (2010)
10. M.S. Antonio, Y.D. Sanrego, M. Taufiq, "An Analysis of Islamic Banking Performance: Maqashid Index Implementation in Indonesia and Jordani,". *Journal of Islamic Finance*, ***1*(1)**: p. 012-029 (2012)
11. D. Suhartanto, et al., "The link between tourism involvement and service performance: Evidence from frontline retail employees," *Journal of Business Research*, **83**, pp. 130-137 (2018)
12. S. Yuwono, E. Sukarno, M. Ichsan, "Petunjuk Praktis Penyusunan Balanced Scorecard: Menuju Organisasi Yang Berfokus Pada Strategi," Jakarta: Gramedia Pustaka Utama, (2002)
13. Ulum, "Akuntansi Sektor Publik Sebuah Pengantar," Malang: Universitas Muhammadiyah Malang, (2005)
14. B. Indonesia, "Sistem Penilaian Tingkat Kesehatan Bank Umum Berdasarkan Prinsip Syariah", in *9/24/DPbS*, (2007)
15. O.J. Keuangan, "Statistik Perbankan Syariah," Jakarta, (2018)
16. L.A. Prasetyowati, L.H. Handoko, "Pengukuran Kinerja Bank Umum Syariah dengan Maqashid Index dan Sharia Conformity and Profitability (SCnP)," *Jurnal Akuntansi dan Keuangan Islam*, ***4*(2)**: pp. 107-130 (2016)
17. W. Ratnaputri, "The Analysis of Islamic Bank Financial Performance by Using CAMEL, Shariah Conformity and Profitability (SCnP)," *Jurnal Dinamika Manajemen*, ***4*(2)**: pp. 220-232, (2013)

Design of a small and medium industry cluster in cilegon city with supply chain system approach

Dyah Lintang Trenggonowati[1,*], *Asep* Ridwan[2], and *Zunika* Dwi Anggraini[3]

[1,2,3]Engineering Department Faculty of Engineering, University of Sultan Ageng Tirtayasa, Banten

Abstract. Future industrial development strategies, derived from current ideas, that is industrial development through a cluster approach. Cilegon city is a city with many petrochemical industries and other, it should be a potential that can give a positive impact for the welfare of Cilegon city community, the researchers want to design the cluster of the petrochemical derivative of IKM with supply chain system approach. The objective of the research is to design the development of a petrochemical derivative cluster of IKM in Cilegon, to determine the factors that influence the development of SMEs clusters, and to design improvement scenarios in the development of SMEs clusters with dynamic system simulations. Based on the results of SMEs cluster design, the stakeholders in the design are government, big companies, and business actors. Types of products to be produced are plastic packaging, plastic bottles, and plastic bags. The location of the right cluster of IKM development is located in Grogol sub-district. Raw materials used by petrochemical derivatives can be obtained from factories producing polyethylene in Grogol sub-district such as PT Lotte Chemical Titan. Factors that have the significant effect on SMEs cluster development are the factors of government support and local business power factor. The simulation result shows that the second alternative proposal is considered reliable enough to face the fluctuating market demand. From the simulation results obtained the average production of 228,513,6914 psc, the average production stock 821,742.4 and the average sales of 200,000 psc.□

1 Preliminary

The era of globalization today has brought about a very rapid renewal and a major impact on the economy both at home and abroad. The impact is most felt is increasingly fierce competition globally in the industrial sector. The industrial sector is a major sector in national economic development. The industrial sector in Indonesia in of global competition cannot be released from the strategy implemented. The strategy of the industrial development in the future, adapting to the ideas developed at the moment, namely the development of the industry through the cluster approach. According to Sa`id in Nusantoro (2011) it clear that clusters are the relationship between one type of economic activity, start

* Corresponding author: dyahlintang@untirta.ac.id

from the activities of the production of primary, collectors, processing products and a half or the finished product. According to Chopra (2007), The benefits that can be obtained in general is to promote the specialization of production in accordance with the competencies of the core and encourage the transformation of a comparative advantage to be a competitive advantage. The competitive cannot be separated from the supply chain the manufacture of products from upstream to downstream. Ridwan (2017) said that the supply chain is a system in the organization that transfers goods production and services to its customers. Cilegon city which is in the western tip of Java island is a city that is taken by the main trade of goods and services from Java to Sumatra Island and otherwise. The location of Cilegon, which is very strategic and supported by the existence of those industries with large capital intensive which is in the Cilegon city, and surrounding areas, should be the potential for being, and it's got to be able to provide a positive impact for the welfare of the city of Cilegon. Based on the condition it is in this study researchers want to do research on the establishment and development of clusters IKM in the city of Cilegon, with the supply chain approach, especially in the petrochemical industry as well as analyze what factors affect the development of clusters of industry and downstream petrochemicals with that of the Partial Least Square. The research also aims to provide a recommendation that can optimize the development of clusters with dynamic systems simulation method.

2 Methodology

The purpose of the use of PLS is to do the relationship between construct. In contrast to PLS which aims to make predictions, the use of SEM is intended as a method to confirm the theory, so that it can be said that PLS is more oriented to the prediction and SEM to be on the theory. The excess of PLS, according to Abdillah and Jogiyanto (2009) is:

1. Being able to model a lot of variables dependent and variable independent.
2. Being able to manage the problem of multicollinearity between variables are independent.

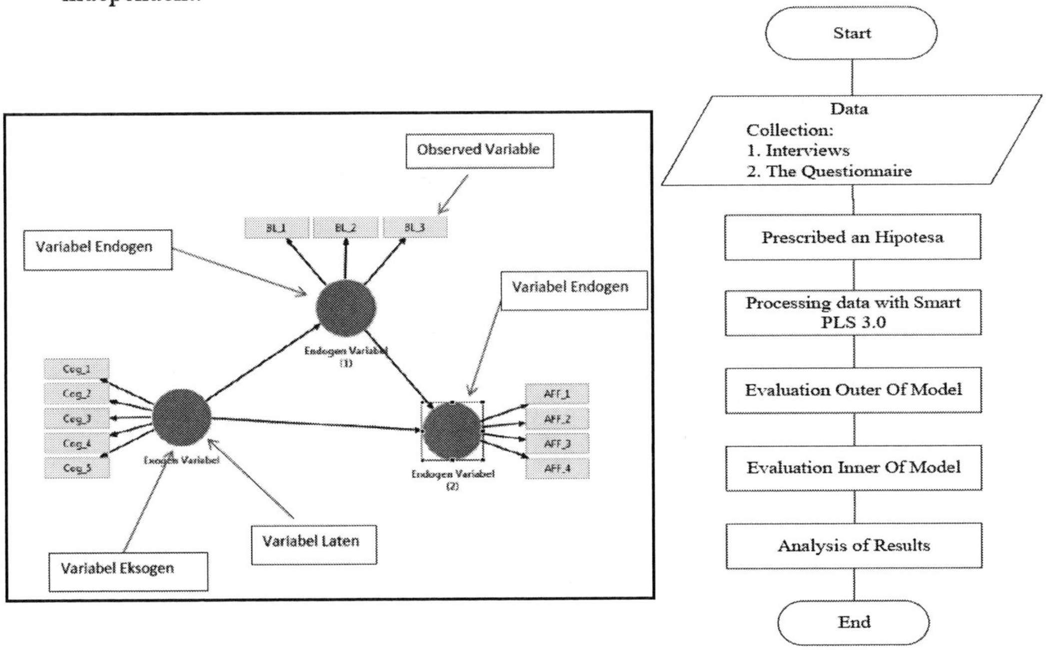

Fig. 1. Conceptual Model of PLS Method and Flow Chart Analysis Of Factors Using Methods PLS.

3. The result remains strong even though there is the data that is not normal and goes.
4. Making a variable is latent independent directly based on cross-product that involves variables latent dependent as the predictions.
5. Can be used in constructed reflective and formative.
6. Can be used in the tiny sample.
7. It requires data distribution.
8. Can be used in the data to the type of scale.

From Fig. 1. above can be seen that there are several reasons that cause used PLS in a research. In this study the reasons, namely: first, PLS is a method of analysis of data based on the samples don't have to be big, which is the number of samples less than 100 can do the analysis. Second, PLS can be used to analyze the theory still said to be weak, because PLS can be used for predictions. Third, PLS allow the algorithms by using analysis of the series Ordinary Least Square (OLS) to achieve the efficiency of the logarithm.Fourth, on the approach PLS, it is assumed that all variance can be used to explain. Here is a flowchart analysis of factors using the method of PLS.

3 Model development

3.1 Design Of Clusters IKM a Descent Petrochemical of Cilegon City

Here is the design of cluster IKM a descent petrochemical of Cilegon city :

1. *Stakeholder*
   The development of clusters IKM needs cooperation between the actors involved. Stakeholder a role in the development of clusters IKM is the government, the company, and business leaders.
2. Type of Product
   Products that can be produced by clusters of IKM the petrochemical Cilegon, among others, are plastic packaging, plastic bottles, and a plastic bag.
3. The Location Cluster IKM of Plastic of Cilegon
   Seen from a lot of companies petrochemicals that are around the district Grogol.
4. The Need For Market
   The need for plastic in Indonesia increasingly, it can be seen from a statement from the Minister of Industry (Menperin) who said that "The plastic products to 4, 6 million tonnes and increased by five percent in the last five years".
5. The availability of Raw Material
   The source of raw materials is the most important factor in the selection of the clusters IKM, especially on cluster IKM the raw materials. The raw material of plastic is polyethylene. The raw material can be obtained from plants that produce polyethylene in the district Grogol like PT Lotte Chemical Titan, and could also be from the company around like PT Chandra Asri.
6. The Transportation Facilities
   The Klaster IKM of plastic, they would build in the district Grogol, Cilegon, close to the factory that produces raw materials for polyethylene., Grogol also close to the facilities and infrastructure of transportation such as toll roads Jakarta-Merak, the airport in, means of transport by train, as well as the port. This will provide facilities for the delivery of the product, operational and administrative management of the management cluster IKM.

7. The Government Policy
   In accordance with the policy of the development of industry, The government has decided the area of Cilegon, as the industry that is open to foreign investors. the Government as facilitator has provided facilities for licensing, taxes and other things related to the technical implementation of the creation of a factory.

3.2 Testing conceptual model

3.2.1 *Convergent Validity*

Convergent validity aims to measure compatibility between indicators of measurement variables and the theory that explains existence as an indicator of the test of these variables. The convergent validity can be evaluated in two stages is to look at the outer loadings and a height of variance extracted (AVE).

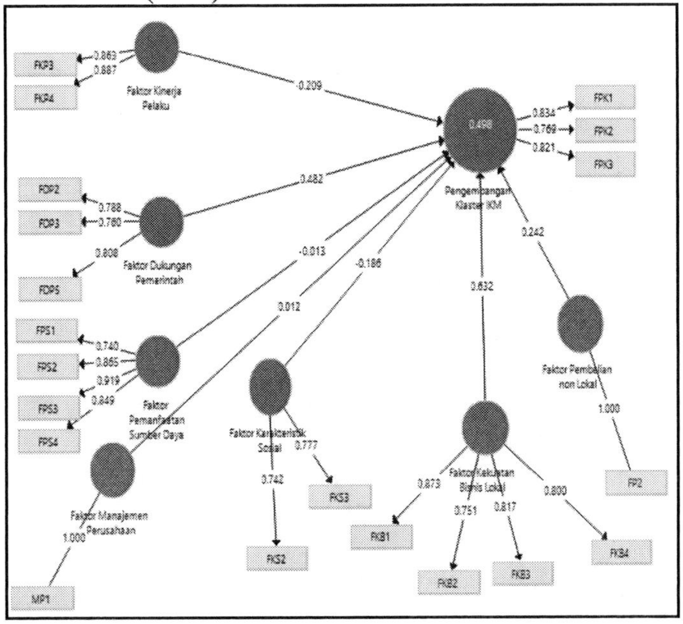

Fig. 2. Diagram of re-estimate line

From the Fig. 2. Can be seen that the outer loading must be greater than 0, 7 it is said to be valid. From Fig.2 the results of calculations Algorithm Report Smart PLS still there is the outer loading under 0.7 For it's done re-estimate to eliminate indicators, which is under 0.7.

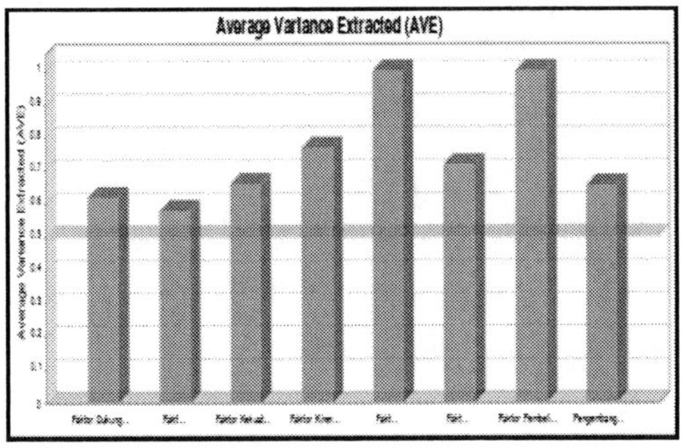

Fig. 3. Graph of AVE

The next step is to look at the AVE. The AVE aims to measure the level of variance a construct component to figures compiled from the indicators by adjusting the error rate. The AVE a minimum recommended is 0, 50. Output AVE can be seen in the picture below. From the Fig. 3. above can be seen that output AVE the respective variables latent is greater than 0, 5 it shows that each of these variables latent was said to be valid.

3.2.2 *Discriminant Validity*

Discriminant validity is the level of the difference an indicator of measuring constructs the instrument. The assessment discriminant validity is generated by Smart PLS and discussed in this study the Fornell-Lacker Criterion and cross the loading. The following is the result of the cross-loadings of PLS Algorithm Report Smart PLS. From Cross Loadings process is known that each construct indicator correlated are higher with constructing the association compared with constructed so it can be said to have discriminant validity. The next step in discriminant validity is to compare the correlation between variables with √ AVE. The measurements have discriminant validity is a good thing if √ AVE every variable is greater than the correlation between the variables. The √ AVE was that for judging from the output Fornell-Lacker Criterion Smart PLS are presented in Table 1. below.

Table 1. Output fornell-lacker criterion

	FDP	**FKS**	**FKB**	**FKP**	**MP**	**FPS**	**FP**	**FPK**
FDP	0,786*							
FKS	0,608	0,759*						
FKB	0,326	0,610	0,811*					
FKP	0,263	0,343	0,600	0,875*				
MP	0,136	0,194	0,325	0,201	1,000*			
FPS	0,209	0,367	0,694	0,628	0,113	0,846*		
FP	-0,220	-0,075	0,175	0,345	0,167	0,161	1,000*	
FPK	0,465	0,400	0,586	0,310	0,243	0,366	0,188	0,809*

From the results of each variable shows √ AVE more than a correlation between variables, so that the terms discriminant validity with √ AVE are met

4 Result and Discussion

There are significant differences between the system simulation existing a simulated improvement repairs 1 and improvement repairs, the after ANOVA with the use of the Least of all the Significant Differences (LSD) to know the average of the condition or treatment which different. Here is the result of the test after ANOVA using software SPSS.

Based on Table 2. the test results after ANOVA, known to the lower bound and the upper bound in every comparison of outputs that have not passed through, then rejected H0 which means there are significant differences between the average output system simulation existing with simulated improvement repairs.

Based on Table 3. the best condition is a model of the improvement of the 2 with an average of the product as much as 228513.6914 psc, the next is improvement repairs one of the product as much as 185577.6629 psc and the last one is existing the average product as much as 125025.1 psc.

Table 2. The results of the test after ANOVA with software SPSS.

Multiple Comparisons

Dependent Variable: Simulasi

LSD

(I) Kondisi	(J) Kondisi	Mean Difference (I-J)	Std. Error	Sig.	95% Confidence Interval	
					Lower Bound	Upper Bound
1.00	2.00	-60552.57000*	15286.84713	.001	-92669.0441	-28436.0959
	3.00	-103488.59857*	15286.84713	.000	-135605.0726	-71372.1245
2.00	1.00	60552.57000*	15286.84713	.001	28436.0959	92669.0441
	3.00	-42936.02857*	15286.84713	.012	-75052.5026	-10819.5545
3.00	1.00	103488.59857*	15286.84713	.000	71372.1245	135605.0726
	2.00	42936.02857*	15286.84713	.012	10819.5545	75052.5026

*. The mean difference is significant at the 0.05 level.

Table 3. The sequence of the best

No	Condition	Average of *Output* (Psc)
1	Improvement Scenario 2	228513.6914
2	Improvement Scenario 1	185577.6629
3	Existing	125025.1

5 Conclusion

Based on the results of research then obtained by the following conclusion :

1. Based on the clusters IKM the industry downstream petrochemical and can be seen that stakeholders in designing are the government, big companies, and business leaders. This type of products that are produced is plastic packaging, plastic bottles, and plastic bags. The development of clusters IKM the derivatives industry downstream petrochemicals the right is to be in the district Grogol. The raw material used in making products IKM the chemical can be obtained from the factory that produces polyethylene in the district Grogol like PT Lotte Chemical Titan. The prevailing means of transportation available, which is the toll Jakarta-Merak, the airport in the train, as well as the port.
2. The factor that has a significant effect on the development of clusters IKM is a factor of government support and the power of local businesses.
3. Based on the results of the simulation found that the proposal the second alternative is considered to be big enough to deal with market demand had been fluctuating in the proposed improvements in the changes in the structure of the decision is made of stock that is not static, but follows the average of sales. From the simulation then earned an average of 228.513.6914 psc, the average stock production 821.742, 4 and the average sales of 200,000 psc.

References

1. Abdillah, W. dan Jogiyanto, H. M. Konsep Dan Aplikasi PLS (*Partial Least Square*) Untuk Penelitian Empiris. Badan Penerbit Fakultas Ekonomi Dan Bisnis UGM, Yogyakarta (2009)

2. Chopra, S., & Meindl, P, *Supply Chain Management.* New Jersey: Pearson Prentice Hall (2007)
3. Nusantoro, J. Model Pengembangan Produk Unggulan Daerah Melalui Pendekatan Klaster Di Provinsi Lampung (2011)
4. Ridwan A., Kulsum, Murni S., Pengukuran *Dasar Kinerja Aliran Rantai Pasok dengan Pendekatan Metode Lean Six Sigma Supply Chain Management,* Jurnal Teknik Industri Universitas Sultan Ageng Tirtayasa, Cilegon. Vol. **3** No. 1a (2017)
5. Ridwan A., Trenggonowati D.L., Haryanto H., Kustiningsih I., Sulaiman F., Masterplan Pengembangan Sentra Industri Kecil dan Menengah Kota Cilegon. Kerjasama COE dan BAPPEDA, Cilegon (2017)

Integration house of risk and analytical network process for supply chain risk mitigation of cassava opak chips industry

Nurul Ummi[1,a], *Putro Ferro* Ferdinant , *Ade* Irman M.S and *Akbar* Gunawan

[1]Majoring in Industrial Engineering Universitas Sultan Ageng Tirtayasa, Jendral Sudirman Km. 3 Cilegon, 42435, Indonesia

Abstract. Supply chain risk is difficult to predict and cause losses that will affect business growth. Supply chain management is a method used for risk control by being used in the cassava industry supply chain. This objective is to determine the risks that will occur and potentially disrupt the supply chain, to determine the risk agent sequence from the Aggregate Risk Potential (ARP) score, and to determine priority mitigation actions. This study uses integration methods. (HOR), and Analytic Network Process (ANP) with pareto diagrams and method-based ratings. The SCOR method is used to map supply chain activities and classify supply chain activities. Meanwhile, the HOR method is used in identifying risk events and risk agents, as well as assessments based on the assessment of the FMEA method. The output of the HOR method is the ARP score used for the priority of the risk agent. Pareto diagrams are used in the priority of risk agents that must be immediately addressed based on the ARP score. The ANP method is used to determine the priority of mitigation actions. Based on this study, researchers obtained 21 risk events and 30 risk agents with 17 risk agents including priority risk agents and 13 non-priority risk agents, and 10 mitigation measures.

1 Introduction

Opak Singkong is a traditional snack similar to chips which is very popular, made from thin circular boiled cassava with crisp texture. Opak Singkong has been a source of livelihood for the people of refined cassava craftsmen. This business is classified as a type of home industry.

The demand for Opak Singkong chips is quite high, but the processing is still conventional and the risk of processing failure causes the demand for Opak Singkong chips has not all been fulfilled. To meet the high and fluctuating demand for Opak Singkong chips, adequate supply chain management is needed.

---

[a] Corresponding author : Nurul.ummi@untirta.ac.id

The supply chain includes all activities related with the effort to meet consumer needs, both directly and indirectly [1]. A good supply chain management will affect the process chain in creating value of products both in terms of cost efficiency and speed in meeting consumer needs (responsiveness Effective supply chain management is characterized by high demand fulfillment rates, always on time deliveries and efficient operating and sales expenses. The supply chain is a network of companies that work together to create and deliver a product to the endusers [2].

The development of supply chain management concept spawned a special concern of the risk impacts of a supply chain within the company. Every supply chain activity has the potential to face risks. This risk is a factor that hinders operations in the supply chain, where the risk in the supply chain can occur from the upstream suppliers, factories, distributors, consumers.

In the context of the supply chain, increased risk is part of the network complexity as a result of outside sources of the company. A study from [3] states that networks between organizations can increase company risk factors, especially when dealing with small and medium business partners. While according to supply chain risk is classified into operating risk and risk of interference [4]. The level of dependence and complexity of the supply chain network today makes the supply chain as a whole more vulnerable to disruption. Any interference that occurs in one of the supply chain players can affect the supply chain network as a whole, such as the cessation of information and resources from upstream to downstream in the supply chain. This can cause an imbalance between supply and demand. Therefore, risks in the supply chain can be defined as disruption of information and resource flows in supply chain networks due to uncertain terminations and variations [5]. Concerning the issue of potential risk deviations that occur in the entire series of production processes and mitigation management such as supply management, demand management, production management, information.

Management and safety management [6]. Supply chain risk management focuses on how to understand and overcome chain effects when small or large risks occur in supply chain networks. Furthermore, ensuring that when the risk occurs, supply chain players have the cability to return to normal conditions and continue their business [7]. Supply chain risk management consists of risk identification, risk analysis, risk evaluation and risk mitigation. Risk identification is a fundamental step in the risk management process [8] [9] SCOR is a reference model of supply chain operations. The SCOR model is basically a process-based model [10]. This model integrates three main elements in management, namely business process reengineering, benchmarking, and process measurement into the cross-functional framework in the supply chain. The SCOR model divides supply chain processes into five core processes, namely plan, source, make, deliver, and return [2].

The House of Risk (HOR) method is a supply chain risk management model using the concept of House of Quality and Failure Modes and Effects Analysis (FMEA) to develop a framework for managing supply chain risk [11]. HOR is a method for managing risk proactively, where the risk agent identified as the cause of risk event can be managed by determining priorities based on the magnitude of the impact that may be caused.

There exist many studies in Risk supply chain i.e. Hadiguna [12] Risk Assessment Model Based on Performance for Sustainable Palm Oil Supply Chain in Indonesia with Non Numeric Multi Criteria Decision Making Approach. Ulfah [13] Refined Sugar Supply Chain Risk Management With a House Of Risk Approach. Pujawan [11] Supply chain house of risk : a model risk management for proactive supply chain. Ummi et al [14] Risk identification Supply Chain in Banten Typical Foods With House Of Risk Method

The analytical network process (ANP), as an extensive and complementary method of the AHP, was introduced and further developed by Saaty [15,16,17,18,19,20,21,22]. The ANP method can be used to make decisions that cannot be arranged hierarchically and do not have independent assumptions. Since its introduction, the ANP method has been applied to various fields. For example, Lee and Kim [23] suggest an information system project (IS) method that uses ANP in a modeling model to solve IS project selection problems. Hafeez et al. [24] provides a structured framework to determine company capabilities using ANP. Karsak et al. [25] use ANP to evaluate the relationship between customer expectations and product technical requirements (PTR) to determine the level of quality at home PTR (HOQ).

Niemira and Saaty [26] developed a model of problem mismatch to estimate the possibility of financial problems based on the ANP framework. Chung et al. [27] use ANP to select and evaluate different products in electronic assemblers. Kahraman et al. [28] uses ANP to get the same objective function and propose a fuzzy optimization design for quality use planning (QFD) using ANP. Bayazit and Karpak [29] developed an ANP-based framework to assess total quality management (TQM). Aktar and Ustun [30] conduct integrated planning from Archimede and Goal Programming (AGP) and Analytic Network Process (ANP) to assess suppliers and determine their futures delivery locations with a number of appropriate and inappropriate criteria. Caballero-Luque et al. [31] presents ANP-based models to help managers work to clarify whether their site content is appropriate to meet the goals they have run. In addition to the above fields, ANP has also been widely used in risk evaluation and assessment analysis. Dagdeviren et al. [32] uses ANP to determine the significant risk of wrong behavior (FBR) in surviving work In ANP, having similarities with AHP, two priority issues need to be solved for the reciprocal pairwise comparison matrix (RPCM here after): test consistency and identification and similarity of inconsistent elements. However, this problem is more difficult in ANP than in AHP because there are many comparison matrices in ANP. There are many studies in consistency tests [33,34,35,36,37,38,39,40]. The most well-known consistency index for the pairwise comparison matrix in AHP / ANP is the ratio of consistency by Saaty. That is, $CR = CI / RI < 0.1$, where the consistency index $CI = (\lambda_{max} - n) / (n-1)$, RI is an average random index based on the size of the matrix n, and λ_{max} is the maximum eigenvalue of the matrix A [36]. If CR is less than 0.1, the comparison matrix calculates consistency tests. Otherwise, entries in the comparison matrix for which the same value must be revised. For element identification methods that are less consistent, this method does not have a generally accepted method for AHP / ANP. In the AHP Expert Choice software, the error matrix $\varepsilon_{ij} = a_{ij} (w_j / w_i)$ is built to assess the elements that are least in line with expectations [18]

When ANP is used to evaluate and analyze risk factors and possible risks, as well as the impact of the decision results for events that arise, constant comparison matrix and elements that are not constant must be known and adjusted. Risk identification and decision analysis of the circumstances that arise are important time-critical information services that are highly dependent on time and information

2 Experimental programs

This paper proposes the integration of HOR into ANP method for Supply Chain Risk Mitigation Cassava Opak Chips. HOR by using Failure Modes and Effect Analysis (FMEA), used to determine the severity of risk events and the emergence probability of risk causes, and the ANP Method is used to determine risk mitigation actions based on a priority scale. Based on the order of priorities can be determined proactive and effective steps to reduce the possibility of risks. The following are steps for integrating HOR and ANP:

1. Mapping of supply chain activities based on the SCOR method
2. Identification of supply chain risks based on SCOR classification

This stage includes identifying risks that might occur and potentially occur in supply chain activities. One important aspect that will be carried out in identifying risks is to list the risks that may occur as much as possible by means of field surveys, interviews and questionnaires (Ulfah, 2016).

3. Risk Measurement
Assess the impact (severity) of such risk event (if happened). We use a 1-10 scale where 10 represents extremely severe or catastrophic impact (see Shahin (2004) for a detailed verbal description about the scale). Identify risk agents and assess the likelihood of occurrence of each risk agent. Here, a scale of 1-10 is also applied where 1 means almost never occurred and a value of 10means almost certain to happen. Develop a relationship matrix, i.e. relationship between each risk agent and each risk event, Rij {0, 1, 3, 9} where 0 represents no correlation and 1, 3, and 9 represent, respectively, low, moderate, and high correlations.
4. Calculating Aggregate Risk Potential (ARPj) is determined by the possibility of a risk agent and aggregate the impact of the risk event generated.

$$ARP_j = O_j \sum S_j R_{ij} \qquad (1)$$

If Oj is the probability of occurrence of risk agent j, Si is the severity of impact if risk event i occurred, and Rij is the correlation between risk agent j and risk event I, than ARPj Aggregate Risk Potential of Risk agent j.

5. Creating a risk agent priority based on the aggregate risk potential (ARP)
6. Risk Mitigation Stage
7. This stage uses the ANP method to mitigate risks and reduce the consequences of risks and prioritize risk control follow-up. The stages that must be carried out in the ANP method are as follows:
   a. Create a decision network hierarchy that shows the relationship between decision factors
   b. Form a pairwise comparison matrix between factors that influence decisions. This pairwise comparison matrix is needed to calculate the impact on alternatives compared to the 1-9 measurement ratio scale developed by Saaty.
   c. Calculate relative importance weight vectors
   The level of inconsistency in the response is called the inconsistency ratio (CR). The steps in calculating CR are as follows:
      a) Pairwise comparison value multiplied by the weighting matrix (eigen) resulting in a value of the results. The results values will then be divided by the eigenvalues of each row to get an average value. Then calculate the value of λ max.
      b) The value of λ max is obtained from the number of results divided by the number of indicators compared

$$\lambda \max = \frac{\text{number of yield values}}{\text{number of indicators}} \qquad (2)$$

      c) Consistency index (CI) is derived from the calculation of the value of λ max minus the number of indicators, then divided the number of indicators minus 1.
      d) Consistency Ratio (CR)

$$CR = \frac{CI}{Random\ Index} \qquad (3)$$

Saaty (1994) proposed the maximum allowable inconsistency ratio is 0.10 or 10%.

d. Supermatrix establishment and analysis. Supermatrix is a matrix composed of relative importance weight vectors. Then normalize the supermatrix so that the numbers in each column in the matrix have a value of 1 (one).
e. Calculation of the final weight to determine the importance of each subcriteria.

3 Results and discussions

Based on the results of clarification with the company, we found total of 21 risk events with their representative degree of severity, and total of 30 risk agents with their respective degree of occurrence. The next step we can calculate the aggregate risk potentials of each risk agent. The relationships between each risk agent and each risk event is shown in HOR1 in Table 2.

As can be seen from Table 2, the calculated values range from 35 to 441. The Pareto diagram of the aggregate risk potentials for all 21 risk events is shown in Figure 1. From the Pareto analysis shows that the first seventeen risk agents contribute to about 81,57 percent of the total commulative ARP values and it should be prioritize for reduction those with high-aggregate risk potentials, and 13 risk agents contribute to 18,43 percent of the total commulative ARP value are not prioritize for reduction.

Table 1. Some of risk events and risk agent, with Severity and Occurrence Degree

Business Process	code	Risk Event	Severity	code	Risk Agent	average
Plan	E1	Product demand is not met in large quantities	6	A1	limited number of grinding machines	5.5
				A2	weather environmental factors	4
				A3	unpreparedness to face large orders	5
	E2	Incompatibility between supply chains and financial planning	5	A4	still mixes between personal finance and business finance	7.5
Source	E3	Damage to raw materials during storage	5	A5	the absence of a standard warehouse for raw materials	5
	E4	harvesting age of raw materials varies	5	A6	different planting times	5
	E5	different types of cassava raw materials	5	A7	there are two different types of cassava raw materials	5
Make	E6	old cassava stripping time	4	A8	Stripping is still manual	7
	E7	boiling process is still manual (not standardized)	5	A9	the absence of a cassava peeler machine	7
	E8	Uneven frying	6	A10	frying techniques are not standard	4.5
				A11	the amount of fried opaque exceeds the capacity of the pan	5.5
				A12	Fluctuating fireplace (inconsistent)	5.5
	E9	Uneven seasoning	5	A13	the absence of a standard composition of ingredients and the amount of	6
	E10	Less varied taste	5	A14	taste innovation has not yet been created	5.5
	E11	there is still oil at the time of opaque draining	7	A15	the draining process is still manual	7
				A16	the absence of a slicing machine	6
	E12	there are still products that are not perfect	6	A17	the absence of an opaque manufacturing process standard	6.5
	E13	packaging is not glue	7	A18	the adhesive is not good	4.5
	E14	There is plenty of air in the package	6	A19	the sealer is less glue perfectly	4.5
				A20	Lack of inspection at the time of gluing	5.5
	E15	lack of label inventory	5	A21	lack of label inventory control	4.5
	E16	difficulty getting the packaging tube	6	A22	financial condition	6
				A23	distance of packaging packaging is too far	6.5
				A24	The late process of making and sending labels	5
	E17	sealer temperature settings that are still not suitable	7	A25	Lack Of understanding about proceduress for using sealers and plastic	5
	E18	plastic packaging has not been standardized	5	A26	Depend on the availability of packaging on the market	5
Deliver	E19	the product is easily damaged when shipping	8	A27	Packaging that does not protect the product when shipping	5.5
				A28	Shipping negligence	4.5
	E20	error type of product specifications ordered	6	A29	error in recording orders	4.5
Return	E21	product returns	5	A30	the product is easily rancid	5.5

Table 2. House Of Risk (HOR)

SCOR	*Risk Event*	*Risk Agent*																														*Severity*
		A1	A2	A3	A4	A5	A6	A7	A8	A9	A10	A11	A12	A13	A14	A15	A16	A17	A18	A19	A20	A21	A22	A23	A24	A25	A26	A27	A28	A29	A30	
Plan	E1	9	3	9																												6
	E2				3																											5
Source	E3					9																										5
	E4						9																									5
	E5							3																								5
Make	E6								9																							4
	E7									9																						5
	E8										9	1	3																			6
	E9													9																		5
	E10														3																	5
	E11															9	9															7
	E12																	3														6
	E13																		1													7
	E14																			3	3											6
	E15																					9										5
	E16																						3	9	3							5
	E17																									9						6
	E18																										3					5
Deliver	E19																											3	9			8
	E20																													9		7
Return	E21																														9	5
	Occurrence	6	4	5	8	5	5	5	7	7	5	6	6	6	6	7	6	7	5	5	6	5	6	7	5	5	5	6	5	5	6	
	ARP	324	72	270	120	225	225	75	252	315	270	36	108	270	90	441	378	126	35	90	108	225	90	315	75	270	75	144	360	315	270	
	Priority Rank	4	28	8	19	14	14	25	13	5	8	29	20	8	22	1	2	18	30	22	20	14	22	5	25	8	25	17	3	5	8	

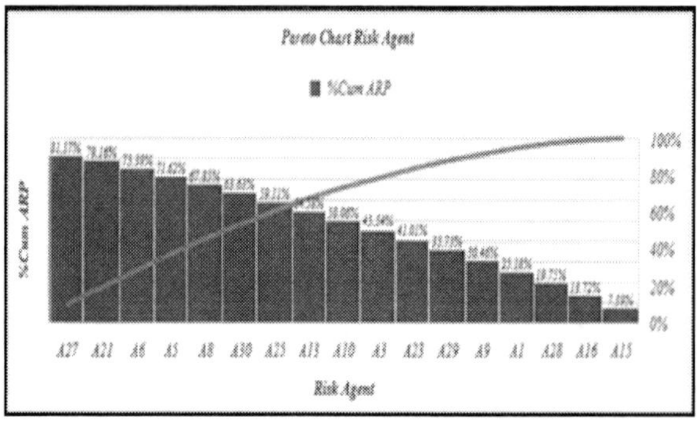

Fig. 1. Pareto risk agent diagram

Table 3 Supply Chain Risk Mitigation

Criteria	Proactive Action	Code
Plan	*Re-layout of Warehouse According to Standards*	P1
	Production Planning Techniques	P2
Source	Adding Grinding Machine	S1
	Spinner Vacuum Machine	S2
	Cassava Peeler	S3
	Additional QC Staff	S4
Make	*Safety Stock and Reorder Point*	M1
	Membuat Standard Operating Procedure (SOP)	M2
	Making Product Batch According to Product Type	M3
Deliver	Make a cooperation agreement with the Shipping service that contains a guarantee of product protection guarantee upon delivery	D1
Return		

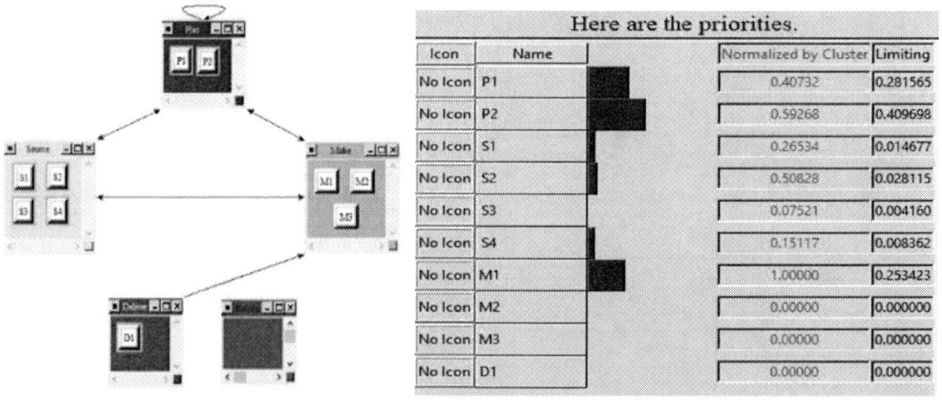

Fig. 2 ANP model selection of mitigation action on super decision software

From ANP Model Selection of Mitigation Action by using Super Decision Software show that the value results of Normalized by All Element, a priority rating is obtained on the proposed mitigation actions : Production Planning Techniques (P2), *Re-layout of Warehouse According to Standards (P1), Safety Stock and Reorder Point (M1),* Spinner Vacuum Machine (S2), Adding Grinding Machine (S1), Additional QC Staff *(S4),* Cassava Peeler Machine (s3), Make a Standard Operating Procedure (SOP) (M2), Making Product Batch According to Product Type (M3), Make a cooperation agreement with the Shipping service that contains a guarantee of product protection guarantee upon delivery (D1)

4 Conclusion

1. Integration HOR and ANP, can be applied to assess the key factors of the risk and the potential risk, determine risk level and risk consequences, and analyze for Supply Chain Risk Mitigation**.** This integration model HOR and ANP can assign the risk agents base on large aggregate risk potentials, i.e. those with high probability of occurring and causing many risk events with severe impacts and can be used to mitigate supply chain risks and reduce the consequences of risks based on priorities.
2. In the case of the Cassava Opak Chips industry, 21 risk events were identified and had an effect on supply chain risk and there were 17 supply chain risks that needed to be mitigated prioritized.
3. Obtained 10 proposals for supply chain risk mitigation actions in the Cassava Opak chips industry. From the value results of Normalized by All Element, a priority rating is obtained on the proposed mitigation actions namely Production Planning Techniques (P2), *Re-layout of Warehouse According to Standards (P1), Safety Stock and Reorder Point (M1),* Spinner Vacuum Machine (S2), Adding Grinding Machine (S1), Additional QC Staff *(S4),* Cassava Peeler Machine (s3), Makea Standard Operating Procedure (SOP) (M2), Making Product Batch According to Product Type (M3), Make a cooperation agreement with the Shipping service that contains a guarantee of product protection guarantee upon delivery (D1).

In the future, it is expected that the processing of opak chips cassava can support the government's roadmap towards Indonesia 4.0 with the fourth industrial revolution (4IR) with one of its initiatives, there is empowering small and medium industries

References

1. Chopra, Sunil, Meindl, Peter. *Supply Chain Management : Strategy, Planning, and Operations, 2nd edition*. New Jersey : Prentice-Hall (2004)
2. Pujawan, I.N., *Supply Chain Management.* First Edition. Institut Teknologi Sepuluh Nopember : Surabaya (2005)
3. Finch, P., Supply Chain Management: An International Journal, Vol. **9** No. 2, pp. 183 96 (2004)
4. Tang CS, Tomlin B. *International Journal Production Economic*. (116):12-17 (2008)
5. Juttner, U., International Journal of Logistics Management, Vol. **16** No. 1, pp. 120-41 (2005).
6. Zsidisin. *International Journal of Supply Chain Management*. 5(4):87-197 (2000)
7. Chapman P, Christopher M, Juttner U, Peck,Wilding R. *Logistics and Transportation Focus*. 4(4):59-63 (2002)
8. Hallikas, Virolainen VM. *International Journal of Production Economics*. 90(1):47-58 (2004)
9. Norrman A, Lindroth R., Supply chain risk: ashgate Publishing Limited (2004)
10. [SCOR] Supply Chain Operation Reference. Version 10.0 *Overview, Supply Chain Council,* Pittsburg
11. Pujawan, Geraldin LH. *Business Process Management Journal*. 15(6):953-67 (2009)
12. Hadiguna. Jurnal Teknik Industri, Vol. **14,** No. 1, Juni 2012, 13-24 ISSN 1411-2485 print / ISSN 2087-7439 online (2012)
13. Ulfah, Maria.. "International Journal of Engineering Technology and Scientific Innovation Framework Of Risk Mitigation Of Management Of Refined Sugar Supply

Chain With The House Of Risk Model." (4): 400–414 (1851)

14. Ummi, Nurul et al. Jurnal Industrial Servicess 3 (1): 342-50 (2017)
15. Saaty TL. Decision making with dependence and feedback: the analytic network process. Pittsburgh, Pennsylvania: RWS Publications; 0-9620317-9-8.
16. Saaty TL. Fundamentals of the analytic network process. In: ISAHP 1999, Kobe, Japan, (August 12–14; 1999).
17. Saaty, TL. Decision making with the analytic network process (ANP) and its ''Super Decisions'' software: The National Missile Defense (NMD) Example. In: ISAHP 2001 Proceedings, Bern, Switzerland, (August 2–4; 2001)
18. Saaty TL. Theory of the analytic hierarchy and analytic network process—examples part 2.2. The International Journal of Systems Research and Information. Technologies (2003:2)
19. Saaty TL. Fundamentals of the analytic network process: dependence and feedback in decision-making with a single network. Journal of Systems Science and Systems Engineering 2004;13(2):129–57. (published at Tsinghua University, Beijing)
20. Saaty TL. Theory and applications of the analytic network process: decision making with benefits, opportunities, costs and risks. Pittsburgh, Pennsylvania: (RWS Publications; 1-888603-06-2)
21. Saaty TL. Rank from comparisons and from ratings in the analytic hierarchy/network processes. European Journal of Operational Research, 168:557–70 (2006)
22. Saaty TL.. Iranian Journal of Operations Research, 1(1):1–27(2008)
23. Lee JW, Kim SH. Using analytic network process and goal programmming for interdependent information system project selection. Computers and Operations Research, 27(4):367–82 (2000)
24. Hafeez K, Zhang Y, Malak N. International Journal of Production Economics, 76(1):39–51 (2002)
25. Karsak EE, Sozer S, Alptekin SE. Product planning in quality function deployment using a combined analytic network process and goal programming approach. Computers and Industrial Engineering, 44(1):171–90 (2003)
26. Niemira MP, Saaty TL. An analytic network process model for financial-crisis forecasting. International Journal of Forecasting, 20(4):573–87(2004)
27. Chung SH, Lee AHI, Pearn WL. International Journal of Production Economics, 96(1):15–36 (2005)
28. Kahraman C, Ertay T, Buyukozkan G. A fuzz. European Journal of Operational Research, 171(2):390–411 (2006)
29. Bayazit O, Karpak B. International Journal of Production Economics, 105(1):79–96 (2007)
30. Aktar DE, Ustun O. Analytic network process and multi-period goal programming integration in purchasing decisions. Computers and Industrial Engineering, 56(2):677–90 (2009)
31. Caballero-Luque A, Aragone´ s-Beltra´n P, Garcı´a-Melo´n M, Dema-Pe´ rezc. International Journal of Information Technology & Decision Making, 9(3):419–36 (2010)
32. Dagdeviren M, Yuksel I, Kurt M. A fuzzy analytic network process (ANP) model to identify faulty behavior risk (FBR) in work system. Safety Science, 46(5):771–783 (2008)

33. Saaty TL. Axiomatic foundation of the analytic hierarchy process. Management Science, 32(7):841–55 (1986)
34. Saaty RW. The analytic hierarchy process—what it is and how it is used. Mathematical Modelling, 9(3-5):161–761987, 9:161–76 (1987)
35. Harker P, Vargas L. The theory of ratio scale estimation: Saaty's analytic hierarchy process. Management Science, 33(11):1383–403 (1987)
36. Saaty TL. Some mathematical concepts of the analytic hierarchy process. Behaviormetrika, No.29:1–9 (1991)
37. Liu W. A new method of rectifying judgment matrix. System Engineering—Theory & Practice, 6:30–4. (in Chinese 1999)
38. Wei CP, Zhang ZM. An algorithm to improve the consistency of a comparison matrix. System Engineering—Theory & Practice , 8:62–6. (in Chinese 2000)
39. Li H, Ma L. Detecting and adjusting ordinal and cardinal inconsistencies through a graphical and optimal approach in AHP models. Computers & Operations Research, 34(3):780–98 (2007)
40. Cao D, Leung LC, Law JS. Modifying inconsistent comparison matrix in analytic hierarchy process: a heuristic approach. Decision Support Systems, 44:944–53 (2008)

Proposed supply chain risk mitigation strategy of chicken slaughter house PT X by house of risk method

Maria Ulfah[1]*, *Dyah Lintang* Trenggonowati[1], and *Fadila Zahra* Yasmin[1]

[1]Departement of Industrial Engineering, Faculty of Engineering, University of Sultan Ageng Tirtayasa, Banten, Indonesia

Abstract. Slaughter House PT X is a company engaged in the production, trade, and distribution of chicken meat that having a role to provide the needs of raw material in a form of fresh chicken meat in the chicken meat processing industry. However, this company has the potential to experience a variety of supply chain risk that can disrupt the company's business processes. The purposes of this study are to identify risk events that have been potentially occurred, identify priority risk agent based on Aggregate Risk Potential (ARP) value, and determine the priority sequence of a risk mitigation strategy in Slaughter House PT X. This study uses the Supply Chain Operation Reference (SCOR) to mapping the company's supply chain activity. Thereafter House of Risk (HOR) method is used to identify, analyze, evaluate, and designing mitigation strategy to minimize potential occurrence of risk agent. The results of the study showed that there are 33 risk events and 32 risk agents identified. Then, based on the result of risk evaluation, there are 15 priority risk agents for mitigation action. There are 11 proposed mitigation strategies to be implemented in Slaughter House PT X.

1 Introduction

Supply chain management is not a new thing for the company nowadays. The company should have to manage and run the supply chain to fulfill customer needs [1]. The complexity of multiple stakeholder within the supply chain structure and the number of uncertainties poses challenge in supply chain management.

The emergence of risks in supply chain activity becomes unavoidable. To reduce and overcome the various risks that occurred within the supply chain, it will be required to undertake a continuous improvement of supply chain performance by overcoming and preventing various risks which potentially occurred [2]. Supply chain risk management becomes an effort to manage the risks that have been occurring along the supply chain flow.

Chicken slaughter house PT X is a company engaged in production, trade, and distribution of chicken meat to fulfill the needs of raw materials in the form of fresh

[1] Corresponding author: maria67_ulfah@yahoo.com

chicken meat for chicken meat processed industry or sold directly to customers through the retailer. The complexity within agro industry supply chain can caused some problem on its supply chain flow. The problems that occurred in chicken slaughter house PT X are based on the monthly report data of production result in 2016 and 2017, showed unstable production results for every month in a year also there are several months that cannot fulfill the production target. Furthermore, there was a decreased value of the yield percentage of the total net weight of life bird as a raw material to the total net weight of finished product in the form of chicken meat where the yield value of 2016 is 75,19% decreased to 74,37% in 2017. The causes of these problems are variable, it is not only came from the company, but also from the farm in the form of life bird as the supplier. If there is a risk in the supplier as an upstream supply chain, it will affect the customer as a downstream in the supply chain. Therefore, supply chain risk management in PT X is required to reduce or even eliminate potential risks in the company's supply chain activity.

The goals of the research are to identify risk events that potentially occurred within the business process in chicken slaughter house PT X, identify priority risk agents based on Aggregate Risk Potential (ARP) value that causes the risk to occur within a business process in chicken slaughter house PT X and determine the priority sequence of applying proposed risk mitigation based on Effectiveness to Difficulty Ratio (ETD) value on chicken slaughter house PT X.

2 Research method

House of Risk (HOR) is formed from modification of FMEA (Failure Modes and Effect Analysis) and House of Quality (HOQ) model to prioritize which risk agent that selected first to take the most effective action in order to reduce the risk potential of risk agent.

There are two phases used in House of Risk (HOR) approach that is HOR 1 which is used to determine the priority level of risk agent that supposed to give a preventive action and HOR 2 that used to determine priority in taking effective action [3].

The steps of House of Risk (HOR) first phase are [2]:

1. Identify the risk event that may occur in every business process.
2. Estimate the impact of some risk events (if they have occurred).
3. Identify the risk agent and asses the probability of risk occurrence for each risk agent.
4. Develop the linkage matrix relationship between each risk agent and risk event.
5. Calculate the Aggregate Risk Potential of agent j (ARP_j) value.

$$ARP_j = O_j \Sigma S_i R_{ij} \quad (1)$$

ARP_j= Aggregate Risk Potential of Risk agent j
O_j = Occurrence value
S_i = Severity value
R_{ij} = Correlation value between risk event i and risk agent j
$R_{ij} \in \{0,1,3,9\}$

6. Rank the risk agent based on aggregate risk potential value in decreasing order (from highest to lowest).

The steps of House of Risk (HOR) second phase are [2]:

1. Select a number of risk agents with highest priority rank using Pareto analysis of ARP_j value, declare in the second phase of House of Risk (HOR).
2. Identify relevant action that considered to prevent the risk agent or Proactive Action (PA).
3. Determine the relationship between each risk agent and proactive action, E_{jk}.

4. Calculate the Total Effectiveness value for each action as follows [2]:

$$TE_k = \Sigma_j \, ARP_j \, E_{jk} f \quad (2)$$

TEk = Total Effectiveness
ARPj = ARP value
Ejkf = correlation value of risk agent and mitigation action

5. Estimate the degree of difficulty in performing each action, Dk. Calculate the effectiveness to difficulty ratio.

$$ETD_k = \frac{TE_k}{D_k} \quad (3)$$

ETD = Effectiveness to Difficulty
Dk = degree of difficulty

6. Rank the priority of each action (Rk) which means first rank is the action with highest ETD value.

This is the figure 1 of flowchart in this research.

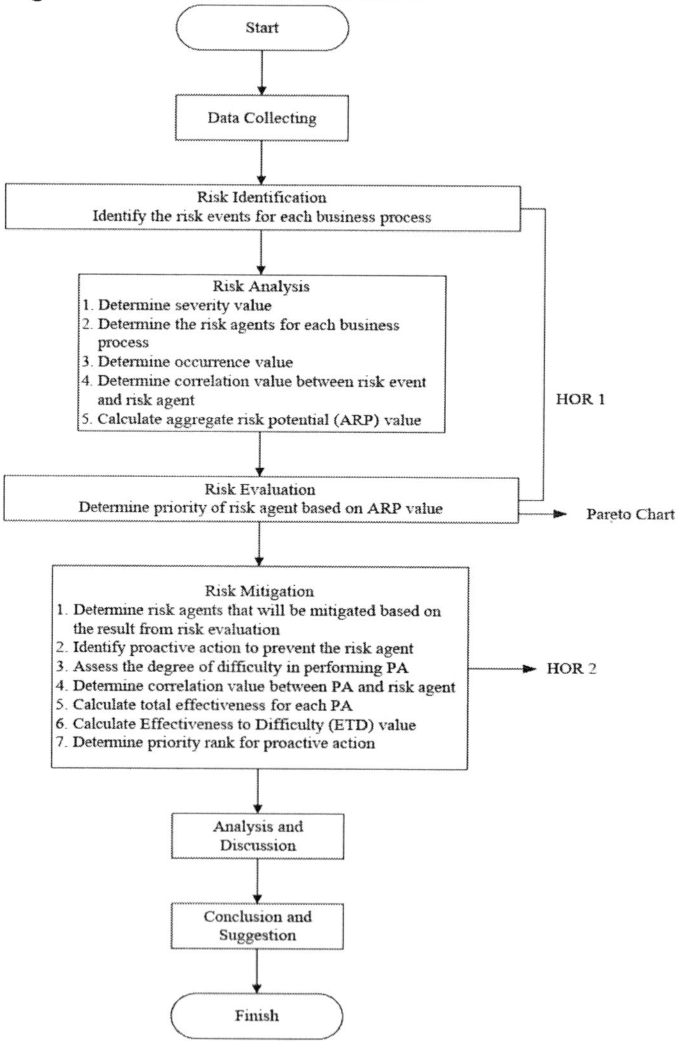

Fig 1 Research flowchart

3 Result and discussion

Data processing in this study consist of risk identification, risk analysis, and risk evaluation that belong to House of Risk first phase, and risk mitigation that belongs to House of Risk second phase.

3.1 Risk identification

Risk identification is an early stage for the study using House of Risk (HOR) method in the first phase. In this stage, the risk identification performed by the mapping of supply chain activity that has been identified previously using the Supply Chain Operation Reference (SCOR) method. The risk that has been identified either unprecedented or have a probability to occur in the supply chain activity of chicken slaughter house PT X. There are 33 risk events identified in the supply chain on chicken slaughter house PT X.

3.2 Risk analysis

The next stage is risk analysis.

Table 1 Risk agent assessment

Code	Risk Agent	Occurrence (O)
A1	Inappropriate product demand planning	5
A2	Fluctuating market condition	9
A3	Lack of raw materials needed	7
A4	Unexpectedly product demand from customer	9
A5	Inappropriate raw material planning	5
A6	The farm does not provide the required chicken size	6
A7	Insufficient ordering of live chicken's quantity that comply the specification	5
A8	External factors (currency exchange rate, government policy)	3
A9	Delay in support material procurement from packaging warehouse	6
A10	Nature disaster	6
A11	Inaccurate inspection of raw material receipts	3
A12	Variation size of live chicken weight	7
A13	Delay of raw material	7
A14	Disrupted power supply	2
A15	Machine downtime	8
A16	Lack of machine maintenance	7
A17	Inaccurate operator	7
A18	Error during packing pressing process	4
A19	Error during final weighing of finished product	4
A20	Error printing on the packaging	4
A21	Imperfect defeathering process	3
A22	Bacterial contamination	2
A23	Unhygenic production process	1
A24	Inaccurate physical checking of finished product	2
A25	Inappropriate water temperature on drum chiller process	1
A26	Lack of coordination between shipping departement and warehouse departement	5
A27	Error while order recording	3
A28	No transport truck	5
A29	Lack of communication with sales departement	5
A30	Inappropriate coolant temperature in the truck	2
A31	Lack of coordination with warehouse departement	5
A32	Delay of the transporter to take retur product	3

Based on table 1 above, the process of risk agent identification has been obtained 32 risk agent in the supply chain of chicken slaughter house PT X. Risk agent with rating 8 (high) or frequent is a machine downtime (A15), this is caused by frequent component replacement. Risk agent with rating 9 and 10 (very high) or hard to avoid, first is fluctuating market condition (A2), this is often occurring because of the increasing or decreasing the need of chicken meat in some certain period in a year. The second is unexpectedly product demand from customers (A4), this is often occurring because of the customer wants to increase their production result, so that they need to add the order beyond initial order.

3.3 Risk evaluation

Risk evaluation is the stage to determine priority sequence of risk agent that will be mitigated based on Aggregate Risk Potential (ARP) value. ARP value is used as a reference and input to determine the priority of which risk agent that should be handled first by sorting the ARP value from the largest to the smallest to analyze using Pareto chart [4].

Aggregate Risk Potential (ARP) is the result of multiplication of occurrence value with sigma of the result of multiplication of severity value with correlation value. ARP value is determinant of how much the risk affects chicken slaughter house PT X. The greater ARP value in a risk agent, the more harmful and dangerous the risk is, it means supposed to take the action immediately [5].

Based on the result of risk agent prioritization, there are 15 risk agents that have been obtained belong to priority risk agent and 17 risk agents belong to non-priority risk agent.

The risk agents belong to a priority risk agents are A4, A15, A16, A17, A2, A6, A7, A12, A9, A28, A19, A3, A13, A27, and A5, whereas the risk agents belong to non-priority risk agents are A26, A10, A29, A1, A31, A18, A20, A11, A21, A22, A32, A23, A14, A30, A8, A25, and A24. This in table 2 of the result of Aggregate Risk Potential (ARP) value.

Table 2 Aggregate Risk Potential (ARP) calculation

Code	Risk Agent	ARP	%ARP	*%Cum ARP*	*Rank*	Category
A4	Unexpectedly product demand from customer	2133	11,818	11,818	1	Priority Risk Agent (A)
A15	Machine downtime	1656	9,176	20,994	2	
A16	Lack of machine maintenance	1575	8,727	29,721	3	
A17	Inaccurate operator	1554	8,610	38,331	4	
A2	Fluctuating market condition	1458	8,078	46,410	5	
A6	The farm does not provide the required chicken size	1188	6,582	52,992	6	
A7	Insufficient ordering of life bird's quantity that comply the specification	720	3,989	56,981	7	
A12	Variation size of life bird weight	672	3,723	60,705	8	
A9	Delay in support material procurement from packaging warehouse	648	3,590	64,295	9	
A28	No transport truck	630	3,491	67,786	10	
A19	Error during final weighing of finished product	504	2,793	70,578	11	
A3	Lack of raw materials needed	441	2,443	73,022	12	
A13	Delay of raw material	441	2,443	75,465	13	
A27	Delay of raw material	432	2,394	77,859	14	
A5	Inappropriate raw material planning	360	1,995	79,854	15	

The following is fig 2 of the Pareto chart to determine the priority risk agent to be mitigated.

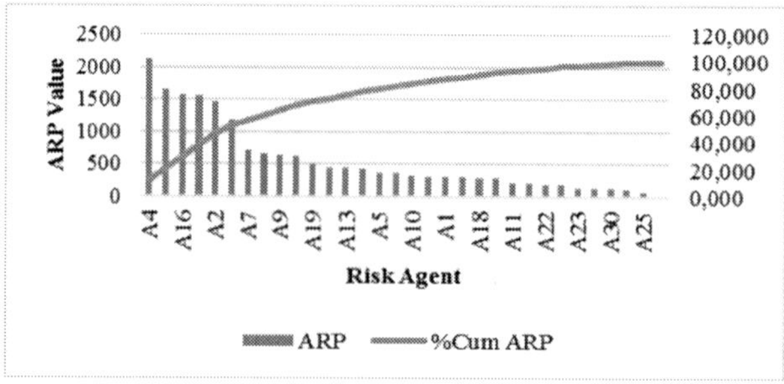

Fig 2 Pareto chart of risk agent

The priority risk agent is obtained from 79,854% (80%) total of cumulative ARP, while the non-priority risk agent is derived from remaining 20% of cumulative ARP. Intake of 80% from total cumulative ARP based on 80:20 Pareto chart principle, which is 20% of total risk agent contribute about 80% of total cumulative ARP [4].

3.4 Risk mitigation

Risk mitigation includes the activity to manage and monitor risk, create mitigation measures, reduce risk impact, and reduce the likelihood of occurrence. The proposed mitigation strategy was proposed by researcher then discussed with expert judgement in chicken slaughter house PT X. Based on 15 priority risk agent as the result of House of Risk (HOR) first phase, there are 11 mitigation strategy to prevent risk occurring in the supply chain.

Then performed the assessment for degree of difficulty to implementing each proposed mitigation strategy for the company. The next step is calculating Total Effectiveness (TEk) value for each mitigation strategy. TEk value that has calculated can not represent how proper these precautions can effectively fix some risk. Therefore, it is necessary to calculate Effectiveness to Difficulty Ratio (ETD) value for each mitigation strategy. ETD value can represent the level of effectiveness of mitigation strategy to be realized and fixed the risk. ETD value is needed to determine the priority sequence of mitigation strategy by sorting the value from the highest to the lowest [6].

Prioritizing the implementation of mitigation strategy is to determine alternative risk mitigation assessment that can be implemented by concerned the limitation of cost, human resources, and other aspects of the company [6].

Table 3 Priority of Risk Mitigation Strategy

Code	*Proactive Action*	*Rank*
PA2	Performing reliability maintenance	1
PA3	Conduct regular training for workers	2
PA8	Promotes implementation of work instruction	3
PA4	Intensive market research	4
PA11	Forecasting and collaborative planning	5
PA9	Perform evaluation supplier performance of live chicken suppliers	6
PA10	Improve communication and coordination with customers	7
PA6	Looking for alternative suppliers for live chicken	8
PA1	Set up buffer stock	9
PA5	Selective supplier of live chicken selection	10
PA7	Improve coordination with the shipping company	11

4 Conclusion

Based on the research result and analysis, it can be concluded that:

1. There are 33 risk events that potentially arise in the business process chicken slaughter house PT X among others are error in determining the needs of live chicken, required size of live chicken is unavailable, the farm cannot fulfill the order, live chicken size that has been received is inappropriate, and shackle rail plummeted.
2. There are 15 priority risk agents based on Aggregate Risk Potential (ARP) value belong to risk agent A category. Risk agent sequence based on Aggregate Risk Potential (ARP) value is unexpectedly product demand from customer (A4) with ARP value is 2133, machine downtime (A15) with ARP value is 1656, lack of machine maintenance (A16) with ARP value is 1575, inaccurate operator (A17) with ARP value is 1554, and fluctuating market condition (A2) with ARP value is 1458.
3. The sequence of priority of proposed mitigation strategy to be implemented based on Effectiveness to Difficulty Ratio (ETD) value in chicken slaughter house PT X are performing reliability maintenance (PA2), conduct regular training for workers (PA3), promotes implementation of work instruction (PA8), intensive market research (PA4), forecasting and collaborative planning (PA11), perform evaluation supplier performance of live chicken suppliers (PA9), improve communication and coordination with customers (PA10), looking for alternative suppliers for live chicken (PA6), set up buffer stock (PA1), selective supplier of live chicken selection (PA5), and improve coordination with the shipping company (PA7).

Reference

1. Utari, R. dan Baihaqi, I. *Perancangan Strategi Mitigasi Resiko Supply Chain di PT Atlas Copco Nusantara dengan Metoda House of Risk.* B-19-2. (2015)
2. Ulfah, M. Maarif, M. S. Sukardi. Raharja, S. Analisis dan Perbaikan Manajemen Risiko Rantai Pasok Gula Rafinasi dengan Pendekatan House of Risk. **26 (1)**, 87-103. (2016)
3. Pertiwi, Y. E. Susanty, A. Analisis Strategi Mitigasi Resiko pada Supply Chain V Surya CIP dengan House of Risk Model. **Vol 6 No 1**, 2. (2017)
4. Nugraheni, S. A. Yuniarti, R. Sari, R. A. The Analysis of Supply Chain Risk on Ready To Drink (RTD) Product using House of Risk Method. **Vol 5 No 1**, 49. (2017)
5. Arafiany, N. M. *Usulan Aksi Mitigasi Risiko Rantai Pasok Pabrik Baja Lembaran Dingin PT XYZ Menggunakan Metode House of Risk (HOR)*. 144. (2017)
6. Elvandra, A. R. Maarif, M. S. Sukardi. Management of Supply Chain Risk in Cattle Slice Fattening at PT Catur Mitra Taruma.. **Vol 4 No 1**, 97. (2018)

Do service activity impact on the risk of Indonesian banking industry?

Yona Friantina[1,a], *Ruhadi*[1] and *M. Edman* Syarief[1]

[1]Department of Islamic Banking and Finance, Bandung State Polytechnic, Bandung, Indonesia

Abstract. The Indonesian banking industry has developed technology innovation will provide financial services to generate income. The purpose of this paper is to compare the service activity and to explore the effect of the service activity on risk in Islamic and conventional banks. By analyzing 12 Islamic banks and 38 conventional banks in Indonesia during the period from 2015 to 2017, we use Structural Equation Modelling (SEM) with path analysis and multiple group analysis. The result shows that the service activity of Islamic banks appears to be more variable than conventional banks. The risk level of Islamic banks is more stable than conventional banks. For Islamic banks, non-financing income has a significant negative effect on risk, commission income and trading income have a significant positive effect on risk, and other non-financing income has no effect on risk. However, no significant effect has been observed between non-interest income, commission income, trading income, other non-interest income and risk level in conventional banks. This finding indicates that the combination of funding activities, lending activities, and service activities allows Islamic banks to obtain the diversification income till reduce risks. For conventional banks, diversification in activities contributes to the higher volatility of bank revenue.

1 Introduction

The development of information technology can change the behavior of everyday society in conducting financial transactions with banks. In this era, the public would require effective and efficient financial transaction services. In the banking system, financial transaction services depend on customer demand. Bankers, as in other service industry, must accommodate all customers' needs so that compete in the industry [21]. Financial transaction services open new innovations that must be supported by professional human resources and information technology that has been prepared in order to improve the quality of banking services and facilities. The innovation of banking products and services shows that the bank is increasingly diversified its business activities.

The diversification of banking business activities in terms of financial services such as transfer, collection, letter of credit, safe deposit box, etc., then the banking will generate profits called fee-based income. In banking, the income earned from this fee-based income plays a very important role as an alternative source of income, thereby reducing the over-

[a] Corresponding author : yona.friantina@gmail.com

reliance on funding activities and lending activities that are the main source of income [22]. The bank services consist of commission and provision income, revenue from foreign exchange transactions, and other operating income. Income derived from these bank services are relatively small but contain a certainty. This is because the risk of these bank services is smaller than the credit/ financing. Conventional banks are known as non-interest income but Islamic banks are expressed as non-financing income that is unrelated to the funding activities and lending activities [15].

In general, banks that diversify incomes through the providing of financial services to both conventional banks and Islamic banks are very useful to reduce risks in the banking industry. Several studies have discussed this problem, but no study has been conducted banking service activities on the risks in Indonesia context of Islamic banks and conventional banks, and also different risk measurement methods from previous studies using the parameters of the solvency ratio. Moreover, there is the difference in applying the analysis method that uses Structural Equation Modeling (SEM) with path analysis.

This study aims to compare risk and service activities in Islamic banks and conventional banks in Indonesia as well as to analyze the impact of banking service activities on risks, as an independent variable consists of non-interest income, commission income, trading income and other non-interest income and as a dependent variable measured by the debt to equity ratio between Islamic banks and conventional banks in Indonesia during the period 2015-2017. This study consist of several sections. First, it will begin with the introduction. Next, in the second part, the theoretical review will be explained. Third, the methodology used in this study. Fourth, the result. Last, the conclusion followed by the important implication and future research of this study.

2 Literature Review

According to Rivai and Ismal [20], uncertainty refers to the notion of an unexpected risk, whereas the term risk refers to the expected risks. In general, there are three sources that can cause uncertainty such as information, time, and ability/skill. In everyday life, we also often hear the term "high risk: high return". The term indicates a positive relationship between risk and profit levels. Means, the higher the risk the greater the return earned. Vice versa, the lower the risk the smaller the return.

Research on the risks and activities of banking services is not a new study. So far, a number of studies have been conducted by several researchers from international and Indonesia. The most recent research is conducted by Moudud-Ul-Huq, et al. [16] support that the diversification of income benefits the bank by improving performance and reducing risks. The study [19] shows that non-interest income as a whole is reducing bank risk in South Asia. Diversification is important because it increases bank income and also decreases the possibility of such pressure as the bank crisis [7]. Research conducted by Ekanayake and Wanamalie [4] found that risk-adjusted return on equity is positively influenced by higher non-interest income activities, indicating that a marginal increase in non-interest income increases trade-off shareholder risk. In contrast, the study [13] shows that diversification of non-traditional activities tends to increase bank risk regardless of bank size. Banks (especially dominant ones) migrated to non-traditional activities because these sources of income could be attributed to an increased risk of the bank as a whole [17].

In addition to research from abroad, researchers from Indonesia have also been researching the risks and activities of banking services. The research of [9] shows that the impact of product diversification on banks and insolvency risk is highly dependent on the size of bank assets. A study by Alfarisi [1] shows that Islamic banks earn relatively lower non-interest (traditional) income compared to conventional banks. A recent study conducted by Hardianto and Wulandari [8] found that Islamic banks have a higher fee-based service activity compared to conventional banks.

3 Method

This study uses secondary data obtained from the publication of financial statements issued by the Financial Services Authority. The data used as the sample is taken by using purposive sampling method. Thus, the researcher determines the company based on several criteria of assessment as follows: 1) Commercial banks operating in Indonesia since the second quarter of 2015 until the fourth quarter of 2017 and routinely publish financial reports reported to the Financial Services Authority. 2) Conventional banks already listed on the Indonesia Stock Exchange (IDX) no later than March 31, 2015, and recorded until the fourth quarter of 2017. Based on these criteria, from the total of 116 commercial banks, 50 banks are made up of 38 conventional banks and 12 Islamic banks. Thus, the total observation of all research samples to 550 observations.

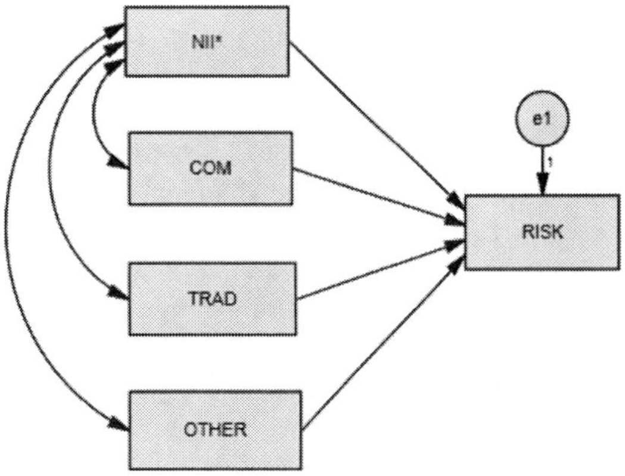

Fig. 1. Model Construction

This study provides a statistical description then the data is analyzed using Structural Equation Modeling (SEM) with data condition that has been a normal distribution. One of the advantages of SEM analysis is the ability to process models to predict the relationship between service activity and risk. SEM with path analysis is used because it can learn the direct effects (direct effect) and indirect effects (indirect effect) of the variable. There are several steps that must be passed in the path analysis. First, make a specification of the path analysis model. In making a path analysis model the relationship of one variable to another variable should be based on the existing theoretical basis. Second, make an estimate to get the path analysis coefficient. Third, conduct a path analysis significance test. Thus, the model to be used is as follows:

$$RISK = \alpha_1 NII^* + \alpha_2 COM + \alpha_3 TRAD + \alpha_4 OTHER + e \qquad (1)$$

*NFI for Islamic bank
where NII is Non-Interest Income, COM is Commission Income, TRAD is Trading Income and OTHER is Other Non-Interest Income.

Furthermore, the researchers conducted a multiple group analysis that will divide the sample based on the characteristics of conventional banks and Islamic banks, with the following hypothesis:

H0: There is no difference between two or more groups
H1: There are differences between two or more groups

We measure risk and service activity with four proxies: non-interest income, commission income, trading income and other non-interest income.

(a) Risk: Following Horne and Wachowicz [10], we use the parameters of the solvency ratio. Debt to equity ratio equals the ratio of total debt to shareholders' equity.
(b) Service activity: Following [9] and [14], we use non-interest income (NII) as the first proxy of service activity.

NII equals the ratio of non-interest income to total operating income, where non-interest income is the composition of commission income, trading income, and other non-interest income. Operating income is the summation of net interest income and non-interest income.

4 Result

The test result indicates that the goodness-of-fit index of the model is good (GFI: 0.961, AGFI: 0.805, NFI: 0.949, IFI: 0.955 and RMSEA: 0.128). We use multiple group analysis to compare Islamic and conventional that result is the probability level (0,000) has below 0.05 (H0 is rejected). This shows that there is a difference between an Islamic bank model and conventional bank model.

The risk rate of Islamic bank is more stable than a conventional bank. Islamic banks do not apply interest costs, so capital is not much needed for banking services activities. While the level of risk in conventional banks easily fluctuates due to the necessity to pay interest. However, service activity in Islamic banks is more variable than conventional banks, which is consistent with [3, 8] that Islamic banks have higher service activity compared to conventional banks. Due to the limitations of Islamic banks in channeling funds directly, Islamic banks have various service products developed as service contracts to seek other income alternatives that reduce risk and increase profitability.

Table 1. The result of hypotheses testing

Hypothesis	Path	α	C.R.	P	Note
H1	RISK<--- NFI[a]	-0,010	-2,885	0,004	Supported
	RISK<--- NII[b]	0,003	1,254	0,210	Rejected
H2	RISK<--- COM[a]	0,039	8,130	****	Supported
	RISK<--- COM[b]	-0,001	-0,218	0,827	Rejected
H3	RISK<--- TRAD[a]	0,141	2,427	0,015	Supported
	RISK<--- TRAD[b]	-0,006	-0,209	0,835	Rejected
H4	RISK<--- OTHER[a]	0,006	0,148	0,883	Rejected
	RISK<--- OTHER[b]	0,004	0,175	0,861	Rejected

a. Islamic banks
b. Conventional banks

$$RISK = -0{,}010NFI + 0{,}039COM + 0{,}141TRAD + 0{,}006\,OTHER + e \quad (1a)$$

$$RISK = 0{,}003NII - 0{,}001COM - 0{,}006TRAD + 0{,}004\,OTHER + e \quad (1b)$$

In Table 1, we can see from the results of Models (1a) that non-financing income (NFI) has a negative and significant relationship with risk (DER), which is consistent with [5, 7, 9, 16, 18]. The result shows that the increase of non-financing income will decrease the risk level so that banking service activity must be maximized. We look deeper by splitting non-financing income into commission income, trading income and other non-financing income in Model (1a). Similar to [9] we found that commission income (COM) shows a positive and highly significant relationship with risk. Trading income (TRAD) has shown a positive and significant relationship with risk, which is consistent with [4, 9, 14]. Other non-financing income shows a positive relationship with risk, which is not consistent with [4, 14, 19].

In Table 1, we can see from the results of Models (1b) that non-interest income (NII) has a positive relationship with risk (DER), which is consistent with [24]. The result shows that the increase in non-interest income has a positive impact direction while increasing diversification from non-interest income may have a negative impact on risk, not statistically significant. We look deeper by splitting non-interest income into commission income, trading income and other non-interest income in Model (1b). Similar to [4] and [14], we found that commission income (COM) has shown a negative relationship with risk. Trading income (TRAD) has shown a negative relationship with risk, which is not consistent with [4, 9, 14]. Other non-interest income shows a positive relationship with risk, which is not consistent with [4, 14, 19].

Following past studies [22, 23], this study proposes Structural Equation Modeling (SEM) with path analysis to predict the relationship between exogenous and endogenous variables in the Indonesian banking industry context. From the category of Islamic banks, we estimate overall the impact of service activities can reduce risks represented by non-financing income. To test a deeper model of service activities, we use three proxies such as commission income, trading income, and other non-financing income. Commission income shows the strongest relationship in this SEM where increasing commission income will increase risk. Likewise, an increase in trading income shows a strong relationship to increased risk. Meanwhile, other non-financing income does not show the strength of the relationship only has a positive direction to risk. By measuring risk using this debt to equity ratio, we find interesting results for conventional bank categories. Compared to other risk measurement methods commonly used in analyzing the relationship of service activity and risk in conventional bank categories, this method can clearly help researchers understand the relationship of non-interest income, commission income, trading income and other non-interest income as proxies of service activities does not show the strength of the relationship to risk in this SEM.

5 Conclusion

This study examines the impact of service activity on risk in Islamic banks and conventional banks during the period 2015-2017. Service activity in Islamic banks is more variable than conventional banks. However, the level of risk in Islamic banks is more stable than conventional banks. In Islamic banks, service activity measured by non-financing income has a significant negative effect on risk, commission income has a significant positive effect on risk, trading income has a significant positive effect on risk and other non-financing income has no effect on risk. In conventional banks, service activities measured by non-interest income, commission income, trading income and other non-interest income have no effect on risk. The result of this study indicates that the combination of funding activities, lending activities, and service activities enable Islamic banks to gain diversified benefits thereby reducing risks. For conventional banks, diversification in activities shows that contribute to higher income volatility of banks.

The result of this study provides clear evidence that service activity led to Islamic banks can reduce bank risks well but in conventional banks, these service activities have a less obvious impact on bank risk. This study has important policy implications related to bank supervision from the perspective of bank risk in the Indonesian banking industry. Thus, the financial authorities should carefully monitor the behavior of Islamic banks associated with various bank and insolvency risks, such as financing risk, liquidity risk, operational risk, compliance risk, strategic risk, reputation risk, legal risk, the rate of return risk, equity investment risk, and market risk. Islamic banks operate as universal banks offering retail financing coupled with banking services. In accordance with the research [12], banks that expand into service activity will be more stable if the bank leads to a retail-oriented income. In this case, Islamic banks are expected to implement marketing strategies that lead to retail. Retail marketing strategy is done by small-scale allocation, its market characteristics as follows: 1) Has standard financial services products such as savings, demand deposits, time deposits, micro or consumptive financing, 2) Margin or interest rate refers to market price because retail business not too sensitive to the level of margin or interest rate, 3) The promotion is relatively large because the target is the number of customers, 4) Distribution service (marketing outlet) using KC, KCP, unit, and other media such as ATM, CDM, and mobile branch, 5) Retail business market more using personal approach, 6) Service is massive and service level agreement is relatively short, 7) Condition of building, decoration, layout, and others have an important role.

We have focused on discussions on how services activities affect risks in Islamic banks and conventional banks. However, when we examine the role of service activity in conventional banks does not affect the risk. For the next researcher can add the mediation variables to show the relationship between independent and dependent variables either direct or indirect effect so that the path analysis model used will be more complex.

References

1. Alfarisi, M.F., *Impact of Financial Crisis on Non-Traditional Income: Islamic Banks Vis a Vis Conventional Banks.* Journal of Islamic Finance, 2015. **4**(1): p. 031-038.
2. Apergis, N., *The long-term role of non-traditional banking in profitability and risk profiles: Evidence from a panel of U.S. banking institutions.* Journal of International Money and Finance, 2014. **45**: p. 61-73.
3. Beck, T., A. Demirguc-Kunt, and O. Merrouche, *Islamic vs conventional banking: Business model, efficiency, and stability.* Journal of Banking & Finance, 2013. **37**: p. 433-447.
4. Ekanayake and Wanamalie, *Income Diversification and Bank Risk-Return Trade-Off: Evidence from an Emerging Economy.* Asian Economic and Financial Review, 2017. **7**(7): p. 644-655.
5. Grassa, R., *Islamic banks' income structure, and risk: evidence from GCC countries.* Accounting Research Journal, 2012. **25**(3): p. 227-241.
6. Gurbuz, A.O., S. Yanik, and Y. Ayturk, *Income Diversification, and Bank Performance: Evidence from Turkish Banking Sector.* BDDK Bankacilik ve Finansal Piyasalar, 2013. **7**(1): p. 9-29.
7. Hamdi, H., A. Hakimi, and K. Zaghdoudi, *Diversification, bank performance, and risk: have Tunisian banks adopted the new business model?* Financial Innovation, 2017. **3**(22): p. 1-25.
8. Hardianto, D.S. and P. Wulandari, *Islamic Bank Vs Conventional Bank: Intermediation, Fee Based Service Activity and Efficiency.* International Journal of Islamic and Middle Eastern Finance and Management, 2016. **9**(2): p. 1-18.
9. Hidayat, W.Y., M. Kakinaka, and H. Miyamoto, *Bank Risk and Non-Interest Income Activities in the Indonesian Banking Industry.* Economics & Management Series, 2012. **03**: p. 1-25.
10. Horne, J.C.V. and J. John M. Wachowicz, *Fundamentals of Financial Management 13th Edition.* 2008: Prentice Hall Inc.

11. Kiweu, J.M., *Income Diversification in the Banking Sector and Earnings Volatility: Evidence from Kenyan Commercial Banks.* KBA Centre for Research on Financial Markets and Policy, 2012. **02**: p. 1-32.
12. Kohler, M., *Does Non-Interest Income Make Banks More Risky? Retail-versus Investment-Oriented Banks.* Review of Financial Economics, 2013. **23**: p. 182-193.
13. Le, T., *The effect of income diversification on bank risk: Evidence from Vietnam.* SSRN Electronic Journal, 2016.
14. Meslier, C., R. Tacneng, and A. Tarazi, *Bank Diversification, Risk and Profitability in an Emerging Economy with Regulatory Asset Structure Constraints: Evidence from the Philippines.* Archives Ouvertes, 2013: p. 1-55.
15. Molyneux, P. and J. Yip, *Income diversification, and performance of Islamic banks.* Journal of Financial Management, Market, and Institutions, 2013. **1**(1): p. 47-66.
16. Moudud-Ul-Huq, S., et al., *Does Bank Diversification Heterogeneously Affect Performance and Risk-taking in ASEAN Emerging Economies?* Research in International Business and Finance, 2018: p. 1-22.
17. Nguyen, M., S. Perera, and M. Skully, *Bank market power, ownership, regional presence and revenue diversification: Evidence from Africa.* Emerging Markets Review, 2016: p. 1-27.
18. Nguyen, T.C., D.V. Vo, and V.C. Nguyen, *Risk and Income Diversification in the Vietnamese Banking System.* Journal of Applied Finance & Banking 2015. **5**(1): p. 99-115.
19. Nisar, S., et al., *The Impact of Revenue Diversification on Bank Profitability and Stability: Empirical Evidence from South Asian Countries.* International Journal of Financial Studies, 2018. **6**(40): p. 1-25.
20. Rivai, V. and R. Ismal, *Islamic Risk Management for Islamic Bank* 2013, Jakarta: PT. Gramedia Pustaka Utama.
21. Suhartanto, D., et al., *The link between tourism involvement and service performance: Evidence from frontline retail employees.* Journal of Business Research, 2018. **83**(Supplement C): p. 130-137.
22. Suhartanto, D., et al., *Loyalty Intention towards Islamic Bank: The Role of Religiosity, Image, and Trust.* International Journal of Economics and Management, 2018. **12**(1): p. 137-151.
23. Suhartanto, D., *The role of store coopetition and attractiveness on the performance of tourism destination and its retail stores.* International Journal of Tourism Policy, 2017. **7**(2): p. 151-165.
24. Trivedi, S.R., *Banking Innovations, and New Income Streams: Impact on Banks' Performance.* Munich Personal RePEc Archive, 2015: p. 1-27.

An Application of Generalized Moments Method to Examine the Management Behavior during Peak Season A Study in Islamic Micro Finance Industry

Yayat Nurhidayat[1,*]

Applied Master of Islamic Banking and Finance, Politeknik Negeri Bandung, Indonesia

Abstract. In the current highly competitive finance industry, it is important that any financial institution to develop a proper strategy to compete in the market. More specifically, the finance managers need to have an appropriate liquidity and stability strategy during the peak season where their customers have a high demand of cash. Finding such a strategy is challenging for micro finance institution due to their limited resources, in terms of finance and human resources. This study is aimed to (1) analyze the impact of liquidity risk to the stability of micro finance institutions and to (2) assess the relationship between liquidity risk and credit risk in the period of peak season by using Generalized Moments Method. The focus of this study is Islamic micro finance institutions in West Java for the period 2012-2017. Using monthly data, this study reveals that the risk of liquidity has a negative effect on the level of stability of micro finance institutions at peak season and has a positive influence on the level of stability at off peak season. Further, the study finds that the relationship between liquidity risk and credit risk is significant during the period of peak season and off peak season period. Therefore, it can be concluded that in general credit risk impacts on liquidity risk. The findings of this study provide significant contributions in terms of enlarging our understanding on the management behavior related to institution liquidity and stability during the peak and non-peak season. From managerial perspective, this study helps the Islamic micro finance institution to remain stable and competitive during the peak season.

1. Introduction

In terms of the collection of third party funds or obligations and capital sourced from low income communities, the Islamic Micro Finance Institution (IMFI) has a high risk. Therefore, the probability of supply of third party funds originating from deposits and saving will have a high risk of fluctuating income. the reason for the high risk of low-income people to third-party fund collection is illiquidity with income fluctuations and informal money-storage mechanisms [1] [2]. Such as storing cash in the house, buying and

[a] Corresponding author: ya_nur81@yahoo.com

selling farm and agricultural goods that are not durable. In particular, Zeller and Sharma [3] suggest that IMFI have difficulties in diversifying their portfolios because farmers have income that depends on the harvest period. Therefore, IMFI will have risks from two sides, namely in terms of IMFI assets and in terms of obligations and capital of IMFI. Therefore, IMFI has a vital role in maintaining its stability so that liquidity mismatches and negative balance conditions do not occur because of the collection and financing of funds to "unsecured creditor"

IMFI liquidity can be well maintained at the level of "health" if the IMFI can carry out its two main functions of creating liquidity and transforming risk [4]. These two functions are often referred to as qualitative asset transformation functions [5]. IMFI can create liquidity both on the balance sheet (on balance sheet) and off balance sheet. At the same time, IMFI creates liquidity through financing assets that are relatively less liquid with relatively liquid debt [6]. Likewise on off balance sheet, IMFI can make financing commitments and claims on liquid funds [7]. In fact, the existence of a IMFI financing commitment can provide a mechanism to optimally share risk, reduce financing rationalization, and reduce the impact of information friction between debtors and IMFI [8]. The second function, namely risk transformation, is carried out by IMFI in line with the implementation of the liquidity creation function, among its forms, namely that IMFI issues time and savings (DPK) savings products at low risk to fund risky financing. Although the relationship is not perfect, the amount of liquidity that has been created will be positively correlated to the level of risk being transformed. Therefore, the level of risk on liquidity will also affect the level of stability of the IMFI itself.

In Indonesia, the level of risk to liquidity is strongly influenced by the pattern or habit of the community in a certain period which has become a culture for many years and generations, the habit is a consumptive pattern in the Peak Season period. The term Peak Season can be foreign in the eyes of the world, but in Indonesia this phenomenon is real and has succeeded in making some financial institutions unstable. Peak season is a site that is not much different from high season. However, peak season is much more crowded and crowded than high season. In the world of Peak traveling, Season is often identified with the peak period of the holiday season. Peak Season in this case is the peak period where people flock to attract savings / savings for seasonal consumption purposes, namely the Ied-ul Fitr session and the session to attend children's school. This period is expected to cause IMFI to be vulnerable to financial distress or unstable. Based on the assumptions and results obtained from previous findings, this paper has more value by combining the previous assumptions in the Peak Season period. This is because the results found by Elgari [8] indicate the stability and resilience of IMFI in the long run if due to high growth compared to conventional financial institutions, while according to Sulung and Wahyudi [9] Liquidity risk significantly affects the IMFI stability level.

Therefore, researchers want to examine more about the creation of liquidity by IMFI in the Peak Season period and try to answer the following questions about the relationship between liquidity risk and the level of stability in the IMFI and the relationship between liquidity risk and credit risk in influencing the stability of the IMFI. Then, in Part 2 will be discussed about literature review, Section 3, will discuss the research methodology, Section 4, will discuss the results and analysis, and Section 5, will discuss the conclusions of the research questions discussed in Chapter 1.

2. Literature

The fundamental function of the IMFI is to channel funds from those who have excess funds to those who need funds. Specifically, the excess funds collected by the IMFI from depositors, and the rest are called third party funds (TPF), in the form of savings-based contract savings, both wadiah yad amanah (pure deposit) and wadiah yad dhamanah (deposit in the form of debt), as well as deposits. Futures such as deposits with the basis of the mudaraba syirkah contract. Then, the TPF that has been collected will be channeled back to the community in the form of a financing contract. In general, the financing agreements commonly used by IMFI can now be grouped into two groups, namely debt-based contracts and syirkah-based contracts. Debt agreements are divided into two groups, namely pure debt (qardhul hasan) and debt arising from buying and selling activities (buying and selling greetings and muajjal). The syirkah contract commonly used is mudharabah, musyarakah, musaqat and muzara'ah. The last two contracts are only used for the agricultural and plantation sectors.

Like financial industry products from other financial institutions, Islamic financial products will also be affected by credit and liquidity risks. Liquidity risk occurs because of the difficulty in selling assets quickly without suffering a lot of losses [10]. Liquidity risk is defined as a maturity mismatch between assets and liabilities at an unsynchronized time between cash in and cash out of business [11]. As for the sharia concept, we will also add "forms of sale and purchase contracts", and "sharia restrictions on the sale of debt" into the liquidity risk variable in the case of IMFI.

2.1 Effect of Liquidity Risk on the Stability Level of IMFI

Related to the relationship between liquidity risk and the level of stability of IMFI. Some findings from Bryant [6], Diamond and Dybvig [12], and Berger and Bouwman [4] say that liquidity risk will affect the stability of the IMFI can be measured among others by Z-score, non-performing loan, and others. Based on the assumptions and results obtained from previous findings, this paper has more value by combining the previous assumptions in the Peak Season period, therefore, our hypothesis is related to the relationship between liquidity risk and the stability level of the IMFI, namely:

H1: Liquidity risk has a negative influence on the level of stability in the IMFI S in the peak season period.

2.2 Reciprocal Relationship between Liquidity Risk and Credit Risk in Affecting the Stability of IMFI

Other factors that affect the level of stability of the IMFI and have a relationship with liquidity risk are the level of credit risk. The classical theory of microeconomics says that there is a close relationship between liquidity risk and credit. The Monti- Klein Framework, Bryant [6] or Diamond and Dybvig [12] state that the structure of assets and debt will relate, more specifically to bad credit and withdrawal of funds by customers [13]. This does not only occur in the IMFI balance sheet but also on financing and business funding especially on off-balance sheets. Based on this, the relationship between liquidity risk and credit will also affect the level of stability of the IMFI. Several papers that show both of these risks have an influence on the level of stability of IMFI, namely Goldstein and Pauzner [14], Acharya and Viswanathan [15] and Acharya and Mora [16].

Based on the assumptions and results obtained from previous findings, this paper has more value by combining the previous assumptions in the Peak Season period, therefore, our hypothesis is related to the relationship between liquidity risk and the stability level of the IMFI, namely:

H2: Credit risk has a positive influence on liquidity risk in IMFI

H3: Liquidity risk has a positive influence on credit risk in IMFI

3. Method

The sample used in the study amounted to 4 IMFI in West Java. Each sample has 5 years for the period 2013 to 2017. This study uses secondary data IMFI financial statements needed consisting of balance sheet reports, income statements, and financial ratio reports. All financial report data were obtained through the Sharia Cooperative Center (SCC). Answer. The explanation of the variables used in this study is explained in Table 1 Operationalization of Research Variables. The methodology used in this study uses the Fixed Effect Data on testing the effect of liquidity risk on the stability of the IMFI. Then, we also use the Generalized Moments Method (GMM) to test the reciprocal relationship between liquidity risk and credit in influencing the stability level of the IMFI.

Table 1. Operationalization of research variables

Type of dependent variable	Description
Z-Score	The ratio of Return on Assets (ROA) plus the equity to asset ratio (CAR) is divided by the standard deviation of ROA (SDROA)
Free	
Liquidity Risk (LR) it-1	([Immediate obligations + Wadiah savings + Other liabilities + Financing received] - [Cash] + Placement with other Banks) / Total assets
Credit Risk (CR)it-1	NPF Ratio
Control	
ROA	Profit (loss) ratio to total assets
ROE	Profit (loss) to total equity ratio
LOAN TO ASSET	Total credit ratio to total assets
PEAK SEASON	Peak period Savings collection is taking above average in one year which is indicated by number 1 if there is peak season and 0 if it does not occur

The dependent variable in this study is the risk of IMFI where the Z-score is used as the main indicator in measuring risk. The Z-score is defined as how much the standard deviation of return on assets of an IMFI must fall to make the IMFI become insolvent [16]. The higher the value of the z-score shows that the IMFI has a lower risk and tends to be stable, whereas the lower the value of the z-score indicates that the IMFI faces higher risks and tends to be unstable.

Variables for liquidity risk (LR) are calculated by subtracting all assets that can be quickly changed in cash to cover withdrawals from short-term liabilities. We do not use derivatives and securities because they do not appear in the balance sheet report of IMFI and not the main business of the IMFI. Then for other independent and reciprocal variables, credit risk is calculated from the ratio of non-performing financing (NPF) to calculate the value of credit risk.

3.1 Research Hypothesis

Some findings from Bryant [6], Diamond and Dybvig [12], Diamond and Rajan [11], and Berger and Bouwman [4] say that liquidity risk will affect the stability of IMFI. Based on this explanation, the following is the research hypothesis related to the effect of non-interest income on the z-score:

$\mathbf{H_1}$: Liquidity risk has a negative influence on the level of stability in the IMFI in the peak

season period.
Where the higher the value of z-score indicates that the IMFI has a lower risk and tends to be stable, on the contrary the lower the value of the z-score indicates that the IMFI faces a higher risk and tends to be unstable.

3.2 Research Model

Goldstein and Pauzner [14], Acharya and Viswanathan [15] and Acharya and Mora [16] show the relationship between liquidity risk and credit in influencing the stability of IMFI. Therefore, the reciprocal hypothesis which states the relationship between the two risks is as follows:

H_2: Credit risk has a positive influence on liquidity risk in influencing the level of stability of IMFI.

H_3: Liquidity risk has a positive influence on credit risk in influencing the level of stability of IMFI.

4. Result

Based on the results of the IMFI from 2012 to 2017, descriptive statistics on the Z-score showed an average score of 6.08 with a maximum value of 16.26 and a minimum value of 3.10. As for liquidity risk with an average value of 1.47 with a maximum value of 2.34 and a minimum value of 0.22.

Table 2. Descriptive Statistic

	Mean	Max	Min	SDev
Z-Score	6.08212	16.26502	3.10780	2.46218
LR	1.47635	2.34520	0.22029	0.49941
CR	0.03694	0.05326	0.03015	0.00492
ROA	0.00925	0.05739	(0.02844)	0.01497
ROE	0,01310	0,54763	(0.82985)	0,15516
Loan to Asset	0,41825	0,67738	0,23591	0,08546
Peak Season	0.30000	1.00000	-	0.46212

The value for liquidity risk is so large because in current assets this value depends only on cash and placements with other IMFI or BANKS. As for the value of credit risk, the average value of this risk is 3.69% with a maximum value of 5.32%. This value is a reasonable value.

4.2.1 Effect of Liquidity Risk on the Stability Level of IMFI

To see the influence of liquidity risk on the stability level of IMFI, this study uses panel data with a fixed effect model for each model from model 1 to 3. Each model uses the Z-score variable per month to see the changes in each period. As for the independent variables, we use existing variations, either only by using liquidity risk variables, or only credit risk variables, or interactions between the two risks. In addition, we also use several control variables including return on equity, return on assets, and loan to assets and peak season. We process data using a single risk (liquidity and credit) in model 1 while models 2 and 3 are to see a combination of all the independent variables and control variables that exist.

Table 3. Effects of Liquidity and Credit Risk on the Stability Level of IMFI

	1	2	3
LR	0,31367	-0,973973	-0,230735
CR	-0,89162	-0,291031	-0,342701
LOAN TO ASSET		-2,552251	-4,595383
ROA			6,212034
ROE			-3,422416
PEAKSESON		-0,794571	-0,620559
Adj R-Square	0,121633	-0,252466	0,218814
J-Statistic	0,00000	0,089137	0,932345

Based on Table 3, the value on liquidity risk shows a positive value in model one and shows a negative value on models two and three. According to Sulung LAK, Wahyudi, I [17] the higher the liquidity risk (LR) shows the lower z-score value, indicating that the more unstable the IMFI is, the lower the value of liquidity risk, the higher the z-score given, this indicates a high z-score tends to be more stable for financial institutions. Therefore, there is an inverse relationship between liquidity risk and the stability of the IMFI. With a confidence level of 99%, it can be said that liquidity risk in the peak season period has a significant negative influence on the level of IMFI stability measured from the z-score but not in the non-peak season period.

The negative relationship between liquidity risk and stability of IMFI needs to get more serious attention from several interested parties because the average liquidity risk value of IMFI is 147.63% with the highest value of 234.52% based on the results of descriptive statistics (see Table 2). This further strengthens the results of research conducted by Wahyudi, I (2015) where the average value of liquidity risk reached 50.52% with the highest value of 182%. Strengthening the results of this previous study is because in this study, researchers included elements of the peak season period as a control variable so that it strengthens that liquidity risk must be a focus and needs attention from certain parties, especially Islamic microfinance institutions whose notes move in low-income societies. Who are susceptible to peak season so that cash and placements with other banks cannot finance their current liabilities.

The effect of significant Liquidity and Peak Risks on the stability of the IMFI is seen consistently from model 2 and model 3. This indicates that the reliability of the results obtained from the influence of the risk of liquidity and stability of the IMFI can be accounted for.

4.2.2 Relationship between Liquidity Risk and Credit Risk

Based on Table 4, the significant effects of pool and random effects for the GMM model can be seen from the effect of credit risk on both liquidity risk for the alpha level of 1%, 5% and 10%. Then the effect of liquidity risk on credit can only be seen in the pool effect of GMM only for the 3rd lag. Therefore, it can be concluded that there is no general effect of credit risk on liquidity risk.

Table 4. Relationship between Liquidity Risk and Credit Risk

	Dependent Variable=LR			**Dependent Variable =CR**	
	Pool	Random		Pool	Random
C	-1,0266	-1,0266	C	-0,0339	-0,1361
CR(-1)	-0,1874	-0,1874	CR(-1)	0,7648	0,4718
CR(-2)	0,2851	0,2851	CR(-2)	-1,0594	-0,2676
CR(-3)	0,1466	0,1466	CR(-3)	1,0967	0,6859
LR(-1)	0,3718	0,3718	LR(-1)	-0,0574	-0,1184
LR(-2)	-0,0353	-0,0353	LR(-2)	-0,0201	0,0059
LR(-3)	0,2194	0,2194	LR(-3)	0,0020	0,1589
Adjusted R-Squared	0,6887			0,2362	0,1204
J-Statistic	54,9893			16,0749	14.3982

5. Conclusion

The results of this study have positive implications for liquidity management in Islamic microfinance institutions. Based on the results obtained, liquidity risk significantly affects the stability of an IMFI during the peak season period in all models combined [18]. This implies that the results obtained can be accounted for because of consistency. Consistency in this case is that every model added with peak season variable has an average value of IMFI liquidity risk that is not so good value. Therefore, this needs to be considered by the stakeholders because it will affect the level of stability of the IMFI which will ultimately affect the financing of low-income people [19] [20]. Then, the relationship between liquidity risk and credit shows insignificant results in the reciprocal relationship between the two because credit risk has a significant effect on liquidity risk at a certain level of significance but not the opposite on liquidity risk to affect credit risk. For future research, it is necessary to investigate further about what factors affecting the Indonesian community attract saving at IMFI during peak season period.

Reference

1. Collins D, Morduch J, Rutherford S, Ruthven O. *Portfolios of the poor: how the world's poor live on $2 a day*: Princeton University Press; (2010).
2. Dupas P, Robinson JJAER. *Why don't the poor save more? Evidence from health savings experiments*. 2013;103(4):1138-71.
3. Zeller M, Sharma M. *Rural finance and poverty alleviation. International Food Policy Research Institute (IFPRI)*, (1998).
4. Berger AN, Bouwman CHJTrofs. *Bank liquidity creation*. 2009;22(9):p.3779-837.
5. Bhattacharya S, Thakor AVJJofI. *Contemporary banking theory*. 1993;3(1):p.2-50.
6. Bryant JJJob, *finance. A model of reserves, bank runs, and deposit insurance*. 1980;4(4):335-44.
7. Holmström B, Tirole JJJoPE. *Public and private supply of liquidity*. 1998;106(1):p.1-40.
8. Elgari MA. *Credit risk in Islamic banking and finance*. (2003).
9. Wahyudi I, Sulung L. *Mengkaji Perilaku Manajemen Likuiditas Pada Bank Perkreditan Rakyat Syariah di Indonesia*. (2015).

10. Ali SS. *State of liquidity management in Islamic financial institutions*. (2013).
11. Diamond DW, Rajan RGJJopE. *Liquidity risk, liquidity creation, and financial fragility: A theory of banking*. 2001;109(2):p.287-327.
12. Diamond DW, Dybvig PHJJope. *Bank runs, deposit insurance, and liquidity*. 1983;91(3): 401-19.
13. Suhartanto D, Dean D, Nansuri R, Triyuni NJJoBR. *The link between tourism involvement and service performance*: Evidence from frontline retail employees. 2018;83:130-7.
14. Goldstein I, Pauzner AJtJoF. *Demand–deposit contracts and the probability of bank runs*. 2005;60(3):1293-327.
15. Acharya VV, Viswanathan SJTJoF. *Leverage, moral hazard, and liquidity*. 2011;66(1):99-138.
16. Acharya VV, Mora NJTjoF. *A crisis of banks as liquidity providers*. 2015;70(1):1-43.
17. Sulung LAK, Wahyudi IJJKdP. *Mengkaji Perilaku Manajemen Likuiditas Pada Bank Perkreditan Rakyat Syariah Di Indonesia*. 2017;17(2).
18. Demirgüç-Kunt A, Huizinga HJJoFE. *Bank activity and funding strategies: The impact on risk and returns*. 2010;98(3):626-50.
19. Wahyudi I, Dewi MK, Rosmanita F, Prasetyo MB, Putri NIS, Haidir BM. *Manajemen Risiko pada Perbankan Islam: Konsep dan Praktek di Indonesia*. . Jakarta: Salemba Empat; (2013).
20. Suhartanto DJIJoTP. *The role of store coopetition and attractiveness on the performance of tourism destination and its retail stores*. 2017;7(2):151-65.

Copper Electrodeposition onto Aluminum from a Copper Acid Baths In The Presence Of Poly Ethylene Glycol (PEG)

Ferry Budhi Susetyo[1,4], *Evi* M.H. Situmorang[1,3] , *Syifa* Luthfiya[2], *Bambang* Soegijono[2,*]

[1]Graduate Program, Department of Physics, Universitas Indonesia, Jakarta Indonesia
[2]Department of Physics, Universitas Indonesia, Jakarta Indonesia
[3]School of Medicine and Health Sciences, Atmajaya Chatolic University of Indonesia
[4]Mechanical Engineering, Universitas Negeri Jakarta, Jakarta Indonesia

Abstract. Many hydrophobic metal coatings onto Aluminum have been studied in recent years because of their significant potential applications. Unfortunately, most of them lost the original metallic luster due to the micro/nano binary structures. In this paper, a facile method was developed to prepare a hydrophobic and high-glossy copper coating onto aluminum substrates. The wettability and gloss could be tuned by the value of current density. With the increase of current deensity, the surface roughness raised, thus resulting in increase of contact angle and decrease of glossiness. When the current density was less than 1 mA/cm^2, the coating exhibited good luster and prefered orientation. Lowering the Current density the coating showed different corrossion resistance and good water reppelence.

1 Introduction

In semiconductor manufacturing, like using copper (Cu) technology, metal lines are defined first by lithography and etching processes, and are then filled up with metal for metallization. The electrochemical plating (ECP) method is widely used for the preparation of metallization, such as Cu, and Ni layers. Before the Cu-metal is deposited into the patterns by ECP, other method such as physical-vapor-deposition (PVD) is widely used to form a tantalum (Ta) base diffusion barrier between the dielectric and Cu films [1–4]. Plating techniques also have the advantages of low processing temperature, low tool cost, and good via/trench _filling capability. However, electroplating method needs an external current source and a seed layer of uniform thickness. Furthermore, the electrical uniformity of electrodeposited films and via/trench _filling capability depend on the current distribution for electroplating thorough out the specimens. Usually, stabilizer and surfactant were also added to the plating bath to decrease the surface tension of the solution and for increase the bath stability. Surfactant has a favorable effect on the surface roughness and

* Corresponding author : naufal@ui.ac.id

crystal orientation of electroless deposits [5-7]. These factors influence the electrical characteristic of metal films such as resistance to electro migration and conductivity.

Literature showed that the sheet stress and resistance of Cu films change as a function of time, and are closely related with the grain size growth describe as self annealing. This phenomenon depend on several factors, such as plating current density, thickness, plating method and the additives in the plating solution [8–12].

Therefore, Electrochemical Plating were used to deposit Cu on Aluminum substrate. Effects of Polyethylene Glycol (PEG) and current on the properties of copper deposit films using such as morphology, crystal structure and corrosion resistance were investigated.

2 Experimental Procedures

DC electroplating process of the Cu films was performed at room temperature on the Aluminum substrate with diameter 23.37 mm. A virgin make-up solution (VMS) consisting of 0.25-mol/L CuSO4·5H2O and 50-ppm chloride ions was used as the standard solutions. Organic additives polyethylene glycol (PEG), were added to the standard solution.

The Cu film was electroplated under galvanostatic control at room temperature with a constant plating current 0.245 A, 0.174 A and 0.124 A. The thickness of Cu film was almost 500 μm. The crystalline structure of the Cu films was identified by X-ray diffraction (Rigaku RINT 2000 with CuK_ radiation). The XRD pattern scans from 20 to 90 degrees with the resolution of 0.01 degree. The surface morphology of Cu were observed by using optical microscope. Highscore plus software were used to refined the XRD pattern to obtained crystal parameter.

3 Results and Discussion

3.1 Optical Microscope

Figure 1 show surface morphology taken using optical microscope of the three samples. Samples with different current show different morphology. Surface roughness decrease as current plating decrease. Plating with higher current density will form a higher dislocation loops and increase the total energy stored in the Cu films lead to increasing in surface roughness.

a. I = 0.245 A　　b. I = 0.174 A　　c. I = 0.124 A

Fig. 1. Surface morphology taken by optical microscope of the three samples Cu-I=0.245 A, Cu-I=0.174 A and Cu-I=0.124 A

3.2 X-ray Diffraction

X-ray diffraction were used to investigate the phase and structure of the three samples Cu-I=0.245 A, Cu-I=0.174 A and Cu-I=0.124 A. The pattern obtained were refined using Rietveld Analysis. The evolution of the crystal parameter were tabulated in Table 1.

Table 1. Crystall parameter of Cu-I=0.245 A, Cu-I=0.174 A and Cu-I=0.124 A

Parameter	Sample		
	Cu-I-0.124	Cu-I-0.174	Cu-I=0.245
a=*b* =*c*(Å)	3.6234	3.6259	3.62383
V (Å^3)	47.5741	47.6711	47.5885
ρ (g cm^{-3})	8.87	8.85	8.87
D (nm)	1501.0	929.3	987.3
Micro strain (%)	0.022	0.035	0.040
R_{wp} (%)	9.775	9.595	14.155
GOF	4.68	4.04	2.78

Higher current during plating, the unit cell volume increase due to higher number of defect. Arrangement of atom during depositions will be more perfect if the current decrease. Furthermore, microstrain differ markly when the current become higher.

Figure 2 showed the XRD pattern of the samples Cu-I=0.245 A, Cu-I=0.174 A and Cu-I=0.124 A. The intensity of the peak around 2θ =74.75 increase as the current decrease. It seem that the samples exhibit preferred orientation. From XRD pattern, crystallinity become better when the current decrease.

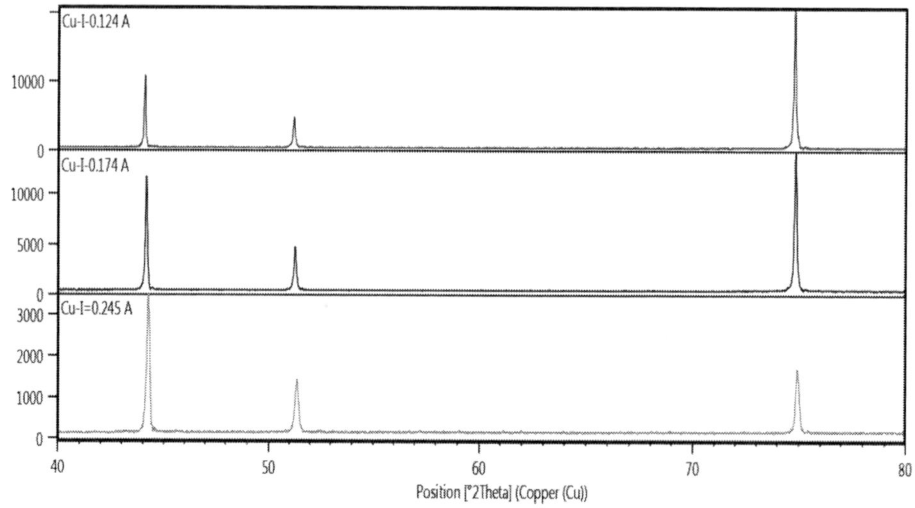

Fig. 2. XRD Pattern of the three samples Cu-I=0.245 A, Cu-I=0.174 A and Cu-I=0.124 A .

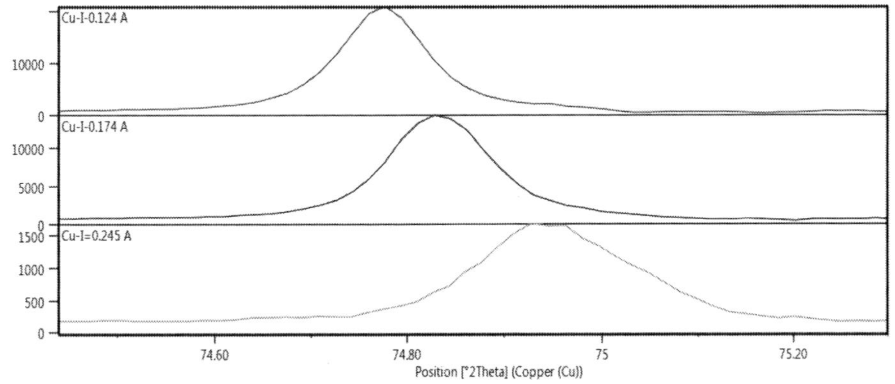

Fig. 3. Peak shifting around 2θ =74.75

Figure 3 showed peak shifting around 2θ =74.75 when the current plating decrease. The crystal become denser and less defect.

3.3 Corrosion Test

Figure 4 showed Corrosion test of the substrat Aluminum, Cu-I=0.124 A and Cu-I= 0.245 A. Alumunium is more resistance in corrosion than Copper because of alumunium oxide .
It seems that decreasing the current, the corrosion resistance decrease because the crystall become more perfect and less dislocation formed

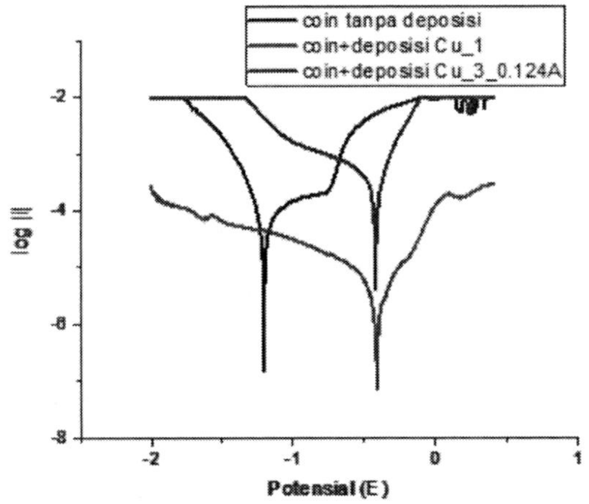

Fig. 4. Corrosion test of the substrat (Al), Cu-I=0.124 A and Cu-I=0.245 A.

4 Conclusions

Electrochemical plating of Copper on aluminum have been taken successfully. Surface roughness decrease as the plating current decrease. It exhibit highly preferred orientation when the current decrease. The single phase copper have been obtained. The unit cell parameter change as the current decrease. Decreasing current, the crystal become denser.

5 Acknowledgment

The authors acknowledge with gratitude the financial supports from the Universitas Indonesia under grant Publikasi Internasional Terindeks untuk Tugas Akhir Mahasiswa (PITTA), No. 2239/UN2.R3.1/HKPO5.00/2018

References

1. C. C. Chang, J. S. Chen, and W. S. Hsu, J. Electrochem. Soc. 151 **11**, 746 (2004).
2. Y. L. Chin, B. S. Chiou, and W. F. Wu, Jpn. J. Appl. Phys. **41**, 3057 (2002).
3. A. Istratov and E. R. Weber, J. Electrochem. Soc. 149 **1**, G21 (2002).
4. Stavrev, D. Fischer, A. Preub, C. Wenzel, and N. Mattern, Microelectr. Eng. **33**, 269(1997).
5. Valery M. Dubin and Yosi Shacham-Diamand, J. Electrochem. Soc. 144, 898 (1997).
6. H-K. Kang, J. S. H. Cho and S. S. Wong, Elec. Dev. Lett. 13, 448 (1992).
7. J. Tao, N. W. Cheung, C. Hu, H-K. Kang and S. S. Wong, Elec. Dev. Lett. 13, 175 (1992).
8. R. Junginger and G. Elsner, J. Electrochem. Soc. 135, 2304 (1988).
9. S. C. Chang, J. M. Shieh, B. T. Dai, M. S. Feng, and Y. H. Li, J. Electrochem. Soc. **149**, G535 (2002).
10 J. Koike, M. Wada, M. Sanada, and K. Maruyama, Appl. Phys. Lett. **81**, 1017 (2002).
11 C. Detavernier, S. Rossnagel, C. Noyan, S.Guha, C. Cabral, and C. Lavoie, J. Appl. Phys. **95**, 2874 (2003).
12 H. Lee, W. D. Nix, and S. S. Wong, J. Vac. Sci. Technol. B **22**, 2369 (2004).

Engineering of magnetic properties in doped bismuth ferrite materials

Hamdan Akbar Notonegoro[1,2], Bambang Soegijono[1,a], Isom Mudzakir[1]

[1]*Dept. of Physic, Universitas Indonesia, Depok 16424, Indonesia*
[2]*Dept. of Mechanical Engineering, Universitas Sultan Ageng Tirtayasa, Cilegon 42435, Indonesia*

Abstract. The engineering of magnetic behaviour of Li/Zn doped $BiFeO_3$ had been done by synthesized a polycrystalline of $BiFeO_3$, $Bi_{0.96}Li_{0.02}FeO_3$, and $Bi_{0.95}Zn_{0.05}FeO_3$. Investigation of crystallite structure and magnetic properties of the sampel had been done by X-ray diffraction and VSM analysis. At room temperature, the lithium and zinc doped bismuth ferrite has conducted a different magnetic behaviour. Within the ferromagnetic region, an increases of magnetic saturation or enlarger magnetic coercivity were identified. Doping lithium resulted in increasing magnetic saturation (M_s) and magnetic remanent (M_r), significantly. Meanwhile, doping zinc resulted in enlarger of magnetic coercivity coincide with the reveal of $Bi_{20}FeO_{40}$ as the second phase.

1 Intoduction

Bismuth ferrite or $BiFeO_3$ (BFO) become the most studied materials due its wide range of physical properties which promises for many applications [1]. Ferroelectric (T_C=830 °C) and antiferromagnetic (T_N=370 °C) ordering above room temperature, were devoted many researchers to understand and tailor these physical properties to fulfil the requirement of specific technology applications, i.e. multiferroic.

One of the vital engineerings on the multiferroic application is the enhancement of magnetic properties. Many works have been done to enlighten the role of doping towards the magnetic properties of BFO. Different types of doping can be rising the magnetic moment of BFO, while the reduction size of BFO was also increased by the magnetic moment [2]–[4]. Du *et.al.* [5] found that the La-doped BFO in Bi-site caused the effect of lattice parameter increases which resulted in increasing of magnetic moment. In common, Khomchenko *et. al.*[6] have reported that Sm-doped BFO in Bi-site was significantly enhanced for the spontaneous magnetization, while the phase transition from a rhombohedral to an orthorhombic was conducted by the composition. On the other hand, Zn or Li-doped BFO into Fe-site can significantly be enhanced the magnetic saturation of BFO by the effect of crystallinity improved or particle-size reduction, have been reported [7], [8]. Recently, Baqiah *et. al.* [9] have reported that $BiTiO_3$ phase content incorporation with the BFO matrix tends to increase the magnetic saturation [10].

[a] Corresponding author : naufal@ui.ac.id

Hence, it is interesting to engineering the magnetization of doping BFO. Different doping-site and different type of materials affect the complexity of magnetic behavior in BFO [11]. The increases in magnetic coercivity or magnetic saturation have different benefits and applications [12], [13]. In order to synthesize doped BFO, there are several preparations technique such as hydrothermal, solid-state reaction, co-precipitation, and sol-gel [5], [8], [14], [15]. Of all these processes, a sol-gel technique has been widely used because of its easy, straight-forward and its final product is highly pure [7], [16]. Moreover, this technique produces an ultrafine porous powder uniformly which can be accommodating industrial-scale production.

This paper aimed to perform engineering of magnetic behavior in low content Li/Zn doped BFO. To further understand by XRD and VSM analyze, the polycrystalline sample of $BiFeO_3$, $Bi_{0.96}Li_{0.02}FeO_3$, and $Bi_{0.95}Zn_{0.05}FeO_3$ were synthesized through a sol-gel route. The phase present and magnetic behavior of the sample is assessed.

2 Experimental procedures

2.1 Sample preparation

Polycrystalline samples of $BiFeO_3$ (BFO), $Bi_{0.96}Li_{0.02}FeO_3$ (BLFO), and $Bi_{0.95}Zn_{0.05}FeO_3$ (BFO-Z) with a low content Li/Zn were synthesized via a sol−gel method. All the nitric salt and acetic salt were purchased from Sigma Aldrich and Merck with ≥99.5% purity. Proportional wt% amounts of $Fe(NO_3)_3{\cdot}9H_2O$, $Bi(NO_3)_3{\cdot}5H_2O$, $CH_3COOLi{\cdot}2H_2O$ and Zn granule (which dissolved with 5 ml of concentrated nitric acid) were weighed and dissolved in 200 ml of de–ionized water in a glass beaker. Then, citric acid (3:1 citric acid and metal ratio) was added as a chelating agent. These solutions were mixed and heated at 80 °C with constant stirring until it becomes a very thick gel, then dried in an oven at 120 °C for 24 h to be xerogel. During the drying process, keep it from the impurity or uncontrolled combustion. After that the xerogel was collected, ground and heated at 600-750 °C for 5 h. Finally, all the samples were ground.

2.2 Characterization

In order to characterize the phase present and the structure of that phase in the sample, powder XRD data were collected on the PANanalytical Diffractometer (Model: X'Pert Pro) in the diffraction range of 20° - 70°. Magnetic hysteresis measurements in the room temperature (27 °C – 300 K) and fields range of 0 – 1.4 T were performed using a vibrating sample magnetometer (VSM) to identify the magnetic behavior established in each sample.

3 Result and discussion

3.1 Structure of the sample

The summary of our XRD results is shown in Figure 1. The amounts of BFO phase present in all the samples and their corresponding lattice parameters were analyzed and calculated by Rietveld method using X'Pert HighScore plus software with ICSD 98-018-1983 as a reference database (see Table 1). The XRD patterns confirm the presence of strong (012), (104), (110), (202), (024), (214) peaks in all samples. These peak pattern were consistent with the BFO phase structure (rhombohedral structure system with space group R3c) from the previous investigations [3], [4]. The XRD obtained for the BFO and BLFO compound were indexed with lattice parameters $a = b = 5.580$ Å and $c = 13.874$ Å , as mentioned in Table 1. However, doping Zn into Bi-site in BFO-Z were causes the lattice parameter declined because the atomic radius of Zn (142 pm) is smaller than that Bi atom (143pm). This content had resulted in the decrease of lattice parameters with $a = b = 5.576$ Å and $c = 13.863$ Å, comparing to $BiFeO_3$ sample (see Table 1). These occasion coincided with the reveal of $Bi_{20}FeO_{40}$ (reference: ICSD 98-004-1937) which is cubic structure and space groups I23, as the second phase accompanying the BFO phase present in BFO-Z sample (Figure 1).

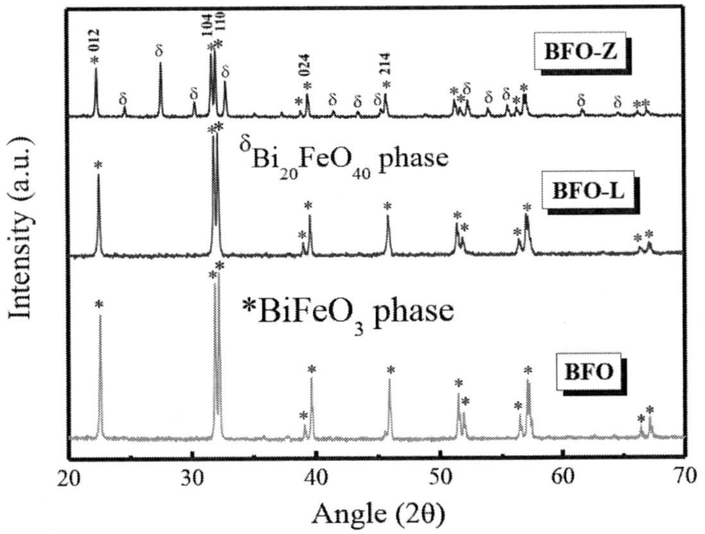

Figure 1. XRD pattern of $BiFeO_3$ (BFO), $Bi_{0.96}Li_{0.02}FeO_3$ (BFO-L), and $Bi_{0.95}Zn_{0.05}FeO_3$ (BFO-Z). All peaks patterns of BFO phase in all samples were consistent with rhombohedral structure with space group R3c.

Table 1. Lattice parameters and percentage of $BiFeO_3$ phase present in $BiFeO_3$, $Bi_{0.96}Li_{0.02}FeO_3$, and $Bi_{0.95}Zn_{0.05}FeO_3$.

Parameter	Sample		
	BFO	BFO-L	BFO-Z
$a=b$ (Å)	5.580	5.580	5.576
c (Å)	13.874	13.874	13.863
V (Å^3)	374.157	374.109	373.318
ρ (g cm^{-3})	8.330	8.330	8.350
D (nm)	193.870	73.840	136.930
Micro strain (%)	0.017	0.114	0.038
$BiFeO_3$ (%)	100.0	100.0	63.5
$Bi_{25}FeO_{40}$ (%)	0.0	0.0	36.5
R_{wp} (%)	4.878	4.71	6.530
GOF	1.599	1.372	1.380

In our circumstance, the refinement *goodness of fit* (*GOF*) parameters of all sample were <1.6 with *R-factors* not more than 7 (<10). As mentioned in Table 1, a broad peaks of the powder XRD sample is due to the crystallite size of BFO phase. Scherrer formula was used to calculate the crystallite size (D) of the BFO nanoparticles,

$$D = \frac{0.9\lambda}{\beta cos\theta} \quad (1)$$

where λ is the wavelength of Cu-Kα radiation, and B is them full-width at half-maximum (FWHM). Nevertheles, the B-unit should be converted into radian. We found that the crystallite size of our sample is varied (193.87 nm, 73.840 nm, and 136.930 nm) which may be caused by particles coagulation.

Three dimensional of $BiFeO_3$ and $Bi_{25}FeO_{40}$ phase structure were illustrated from refinement data results using VESTA software as shown in Figure 2. From the picture, we can be seen the oxygen atoms were forming a polyhedral shape with the Fe-atom as the center. On the $BiFeO_3$ phase structure, we can see that doping Li can replace some of Bi-site precisely. Meanwhile, we can see in the $Bi_{25}FeO_{40}$ phase structure that Bi-atom arranged polyhedral shape much more than the Fe-atom.

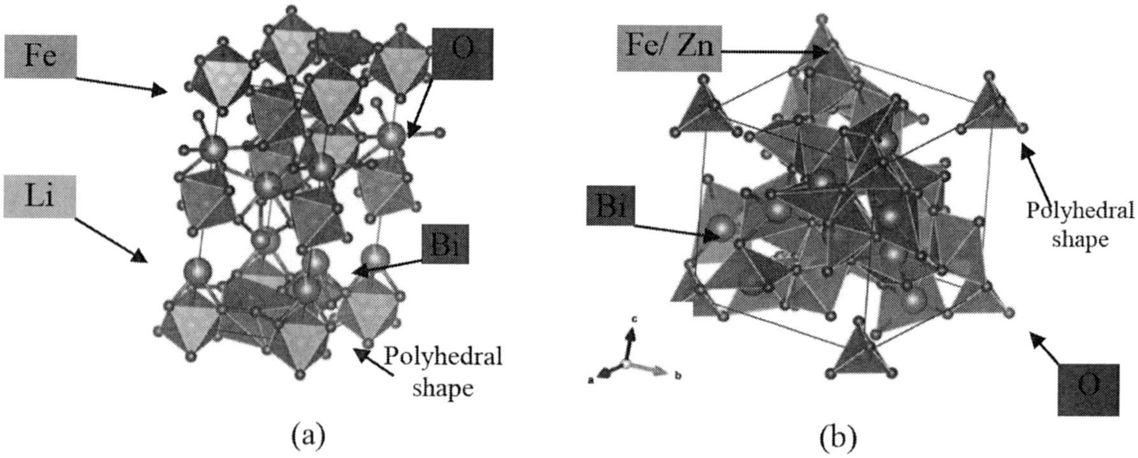

Figure 2. Three dimensional of two phase present in the sample, a) $BiFeO_3$ phase structure and, b) $Bi_{25}FeO_{40}$ phase structure. In bismuth ferrite structure, Li could replaced on Bi-site well. On the other side, the Fe-site in the $Bi_{25}FeO_{40}$ structure could replaced by Zn atom, alternately.

3.2 Magnetic properties

Figure 3 shows room temperature hysteresis loops of BFO, BFO-L, and BFO-Z samples, respectively. There are some factors that can affect the magnetic behavior of all sample such as distortion of crystallite size. The saturation magnetization (M_s), coercivity (H_c), and remanence magnetization (M_r) of all samples are mentioned in Table 2.

The BFO and BFO-Z sample shows a low saturated hysteresis loop M_s (0.08 emu/g and 0.14 emu/g, respectively) which is M_r of both samples is identical value (0.02 emu/g). In comparison, the M_s of BFO-L is the larger (4.89 emu/g) more than BFO and BFO-Z with the M_r is become 1.42 emu/g, as compensate for the Li present in the structure. Refer to Table 1, the increment of M_s in BFO-L and BFO-Z were inline with the increment of micro strain percent value of both sample, compare to BFO value. Contigously, the present of Zn in BFO-Z sample subtituting on the Bi-site had broaden the H_c value larger than two others sample, BFO and BFO-L which both samples have a closed coercivity value. Further, the broaden of H_c value in BFO-Z is associated with the rising of atomic density and the reveal of $Bi_{20}FeO_{40}$ phase structure incorporated with the BFO phase present. This change associated with the increment of stored magnetic energy during hysteresis circle.

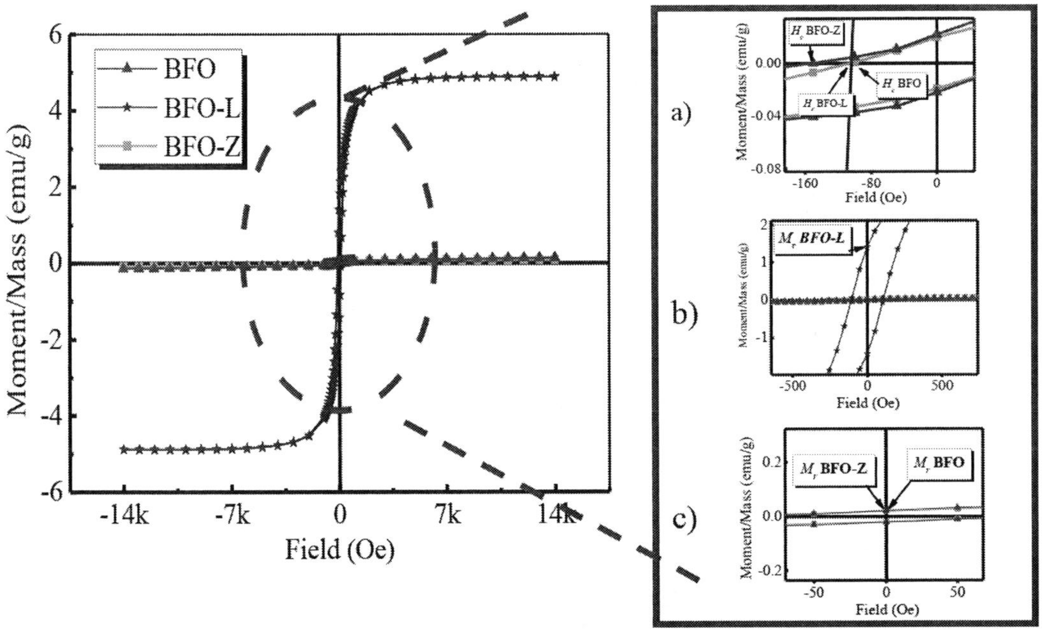

Figure 3. Magnetic properties of BFO, BFO-L, and BFO-Z samples: a) Magnetic Coercivity (H_c) of all the samples, b) magnetic remanent (M_r) of BFO-L sample, c) M_r of BFO and BFO-Z samples.

Table 2 The magnetic saturation (Ms), coercivity (Hc), and remanence magnetization (Mr) of BFO, BFO-L, and BFO-Z samples.

Parameter	Sample		
	BFO	BFO-L	BFO-Z
M_s (emu/g)	0.08	4.89	0.14
M_r (emu/g)	0.02	1.42	0.02
H_c (Oe)	105.39	104.24	152.18

4 Conclusion

The BFO, BFO-L, and BFO-Z samples were synthesized by the sol-gel process for engineering the magnetic behavior of that sample. All the samples were crystalized in rhombohedral with space group R3c, precisely. The lattice parameters of $BiFeO_3$ in BFO and BFO-L samples exhibit identical value, while the lattice parameters of $BiFeO_3$ phase decreased with doping of Zn content. The magnetic saturation and magnetic remanent increased with doping Li. Further, magnetic coercivity broadens coincide with the reveal of $Bi_{25}FeO_{40}$ phase structure after doping Zn in the sample. With this results, the behavior of magnetic properties in bismuth ferrite had been successfully engineered.

Acknowledgement

The author acknowledge with gratitude the financial support from the Ministry of Research, Technology and Higher Education of Indonesia (RISTEK-DIKTI), under grant Penelitian Dasar Unggulan Perguruan Tinggi (PDUPT), No. 366/UN2.R3.1/HKP05.00/2018).

References

1. R. A. M. A. M. Gotardo *et al.*, "Improved magnetic properties and structural characterizations in Mn doped 0.9BiFeO3–0.1BaTiO3 compositions," *Scr. Mater.*, vol. 130, pp. 161–164, 2017.
2. S. Zheng, J. Wang, J. Zhang, H. Ge, Z. Chen, and Y. F. Gao, "The structure and magnetic properties of pure single phase BiFeO3 nanoparticles by microwave-assisted sol-gel method," *J. Alloys Compd.*, vol. 735, pp. 945–949, 2018.
3. S. Godara, N. Sinha, G. Ray, and B. Kumar, "Combined structural, electrical, magnetic and optical characterization of bismuth ferrite nanoparticles synthesized by auto-combustion route," *J. Asian Ceram. Soc.*, vol. 2, no. 4, pp. 416–421, 2014.
4. M. Kumar, K. L. Yadav, and G. D. Varma, "Large magnetization and weak polarization in sol-gel derived BiFeO3ceramics," *Mater. Lett.*, vol. 62, no. 8–9, pp. 1159–1161, 2008.
5. Y. Du, Z. X. Cheng, M. Shahbazi, E. W. Collings, S. X. Dou, and X. L. Wang, "Enhancement of ferromagnetic and dielectric properties in lanthanum doped BiFeO3 by hydrothermal synthesis," *J. Alloys Compd.*, vol. 490, no. 1–2, pp. 637–641, 2010.
6. V. A. Khomchenko *et al.*, "Effect of Sm substitution on ferroelectric and magnetic properties of BiFeO3," *Scr. Mater.*, vol. 62, no. 5, pp. 238–241, 2010.
7. J. Liu *et al.*, "Influence of Zn doping on structural , optical and magnetic properties of BiFeO 3 fi lms fabricated by the sol – gel technique," *Mater. Lett.*, vol. 133, pp. 49–52, 2014.
8. A. Billah, "INVESTIGATION OF MULTIFERROIC AND PHOTOCATALYTIC PROPERTIES OF Li DOPED BiFeO3 NANOPARTICLES By," Bangladesh University of Engineering and Technology, 2017.
9. H. Baqiah, Z. A. Talib, A. H. Shaari, N. Tamchek, and N. B. Ibrahim, "Synthesis, optical and magnetic behavior of (BiFeO3)1−x(α-Fe2O3)x nanocomposites," *Mater. Sci. Eng. B*, vol. 231, no. August 2017, pp. 5–10, 2018.
10. H. Liu, Y. Guo, B. Guo, and D. Zhang, "nanopowders by a sol e gel method," vol. 19, pp. 69–72, 2013.
11. Q. Rong, W. Xiao, G. Xiao, A. Hu, and L. Wang, "Magnetic properties in BiFeO3 doped with Cu and Zn first-principles investigation," *J. Alloys Compd.*, vol. 674, pp. 463–469, 2016.
12. V. Iurchuk, B. Doudin, J. Bran, and B. Kundys, "Electrical Writing of Magnetic and Resistive Multistates in CoFe Films Deposited onto Pb[ZrxTi1-x]O3," *Phys. Procedia*, vol. 75, pp. 956–966, 2015.
13. B. Ramachandran and M. S. R. Rao, "Low temperature magnetocaloric effect in polycrystalline BiFeO3 ceramics," *Appl. Phys. Lett.*, vol. 95, no. 14, pp. 2007–2010, 2009.
14. M. Guo *et al.*, "Enhancement of multiferroic properties in Bi0.92Ho0.08Fe0.97Mn0.03O3/ Zn0.5Ni0.5Fe2O4 bilayered thin films by tunable schottky barrier and interface barrier," *J. Alloys Compd.*, vol. 741, pp. 420–431, 2018.
15. V. Kumar and S. Singh, "Optical and magnetic properties of (1-x) BiFeO 3 - xCaTiO 3 nanoparticles," vol. 732, pp. 350–357, 2018.
16. R. Safi and H. Shokrollahi, "Physics, chemistry and synthesis methods of nanostructured bismuth ferrite (BiFeO3) as a ferroelectro-magnetic material," *Prog. Solid State Chem.*, vol. 40, no. 1–2, pp. 6–15, 2012.

Increasing The Efficiency of The Cub Engine Assembly Lines In The Automotive Industry Using Ranked Positional Weight

Inaki Maulida Hakim[1,a], Syarafi Auzan Mu'min[2], and Rolina Oktapiani Zaqiah[3]

[1a,2,3]*Department of Industrial Engineering,Faculty of Engineering, Universitas Indonesia, Depok, Indonesia*

Abstract. In this modern era, the competition among the manufacturing industry, especially in the automotive sector will become increasingly tight which causes companies need to innovate so that satisfaction of the consumer can be maintained. The production process will be an important aspect in the automotive industry to maintain the quality of products and ensure consumer demand can be fulfilled. The problems that often occur in the production process is in the form of production flow constraints caused by workload unbalanced in the assembly lines. The imbalance causes the assembly lines do not run in a cycle time that is determined, so that consumer demand can not be meet in the right amount and companies need to spend more to mitigate them. Therefore, this study was conducted to balance workload on the assembly line by using line balancing form Ranked Positional Weight (RPW) with a subsequent increase in the efficiency and productivity of assembly line that affect production process runs without any contraints.

1 Introduction

Nowadays, people tend to use their own vehicles such car and motorcycle instead of public transportation. For improvement in automotive industry, must be have strategy each company to prepare Industry 4.0. Currently in Indonesia, the type of motorcycle vehicle is the transportation that is most widely used by community due to ease of use and the most important is cost and time friendly. Motorcycle become relatively important of transportation for Indonesian society and resulting in an increase number of demand for motorcycle each year. The trend of an increasing number of motor vehicles in Indonesia every year, this has resulted in automotive companies, especially in the manufacturing of the motor so it needs to be improvement impacting the company to meet the market demand continues to increase. Optimal use of available resources and also improving the quality of the product will provide cost reduction and increase the level of customer satisfaction so that the profit for the company can continue to increase. The balance of production line starts from mass production, where production process must have a balance of time between each work station. Line balancing learn how balancing in assembly lines

[a] Corresponding author : inakimhakim@eng.ui.ac.id

can occur with leveler existing workload for each operator. The term line balancing is a method to assign jobs to workstations interrelated in a single production line so that each station has a time not exceeding the cycle time of work stations. Linkage number of jobs in an assembly lines need to be considered in determining the division of jobs into each work station. Relationships or interconnections between work and other work is described in a preliminary precedence diagram or chart, while it is called precedence relations job or precedence network (Bedworth, 1982).

The main problems that often occurs in assembly lines such as the occurrence of constraints in the system that is closely related to maintenance. Balancing workload on multiple workstations with the aim of achieving high efficiency and meet production planning that has been made. There are some symptoms of imbalance production line that includes differences in the performance of a work station with high levels of idle time and a work in process at several work station.To improve efficiency and productivity, can be achieved by reducing the manufacturing time required to produce the final product by reducing the number of workstations or both (Hamza, 2013). Therefore, the aims of this study is to increase the productivity and efficiency of the assembly line using line balancing method which focused on the cub engine line and used some parameters such as line efficiency, idle time, and smoothness index.

2 Methodology

Line balancing is usually performed to minimize the imbalance between machines or personnel in order to meet the desired output of the assembly line. Finish line balancing problem, industry management should be able to know about the working methods, equipment, machinery, and personnel used in the work process. In addition, information about the time required for each assembly line and precedence relationship between activities is the arrangement and sequence of the various tasks that need to be done. There are some terms that are commonly used in the line balancing. Here are the terms in question :

2.1 Precedence diagram

Precedence diagram used before stepping on the completion method of line balancing. Precedence diagram is actually a graphic picture of the sequence of work operations, as well as dependence on other working operation whose purpose was to facilitate the control and planning of activities associated therewith, while the mark is used in precedence diagram is as follows:

a. The circle symbol with the letter or number on it to facilitaties identification on a process operation.
b. The arrow shows the dependence and sequence of process operations. This operation is in the base of the arrow means working operations precedence that of the arrowhead.
c. The figure above the circle symbol is the standard time it takes to complete each operation process.

2.2 Assemble Product

Assemble Product is a product that passes order in which workstation, each workstation provides a specific process to complete final product at the final assembly.

2.3 Idle Time

Where operator or worker waiting to perform work processes or operations that will be done next. The difference or distinction between Cycle time (CT) and Station Time (ST), or CT reduced Station Time (ST).

$$Idle\ Time = n.Ws - \sum_{i=1}^{n} W_i \tag{1}$$

N = Number of workstation.
Ws = Highest workstation time
Wi = Time on workstation
I = 1, 2, 3,...n

2.4 Balance delay

Balance delay is measure of inefficiency resulting trajectory of the time unemployed actually caused the allocation is less than perfect work stations. Balance delay can be formulated as follows:

$$D = \frac{n.C - \sum t_i}{(n.t_i)} x\ 100\% \tag{2}$$

D = Balance delay (%)
N = Number of workstation
$\sum t_i$ = Sum of all operation time
t_i = Operation time
C = Cycle time

2.5 Workstation Efficiency

Workstation efficiency is ratio between operating time per work station (Wi) and the largest work station operating time (Ws). The efficiency of the work station can be formulated as follows:

$$Workstation\ Efficiency = \frac{W_i}{W_s} x\ 100\% \tag{3}$$

2.6 Line Efficiency

Line efficiency is ratio of total time divided by the work station cycle multiplied on the number of workstations or the number of workstations efficiency divided by the number of workstations. Line efficiency can be formulated as follows:

$$Line\ Efficiency = \frac{\sum_{i=1}^{k} ST_i}{k.\ CT} x\ 100\% \tag{4}$$

ST_i = Time of workstation i
k = Number of workstation
CT = Cycle time

2.7 Minimum Work Station Required

Workstation is a place on the assembly line where the assembly process is done. After determining the cycle time interval, the number of work stations that can efficiently determined by the formula:

$$kmin = \frac{\sum_{i=1}^{k} t_i}{CT} \quad (5)$$

2.8 Smoothness Index

Smoothness index (SI) is an index that indicates the relative smoothness of a specific assembly line balancing.

$$SI = \sqrt{\sum_{i=1}^{k} (ST_{max} - ST_i)^2} \quad (6)$$

ST_{max} = Maximum workstation time
ST_i = Time of workstation i

Steps in this method used Ranked Positional Weight Method is as follows (5) :

1. Make a precedence diagram.
2. Determine ranking position (positional weight) for each element of the work (a ranking position of a touch operation on the longest flow time of start of operation until the end of network).
3. Sort elements of work based on ranking position in step number 2. Elements of work with the highest ranking position most sorted first.
4. The process of placement of elements of work at workstation, where elements of work with the rank and position of the highest order who placed first.
5. If the workstations are the remaining time after the placing of an operation, place the next sequence operation at workstation during the operation do not violate in precedence relationships, time do not exceed on workstation cycle time.
6. Repeat the steps 4 and 5 until all elements of the work has been already placed at workstation.

3 Case Study of Automotive Industry

This case study is based upon a real problem from the automotive industry which is one of the largest motorcycle manufacturers. The manufacture has a problem in their production line which impact on the company's looses by not able to fulfill customer orders also increased overtime.

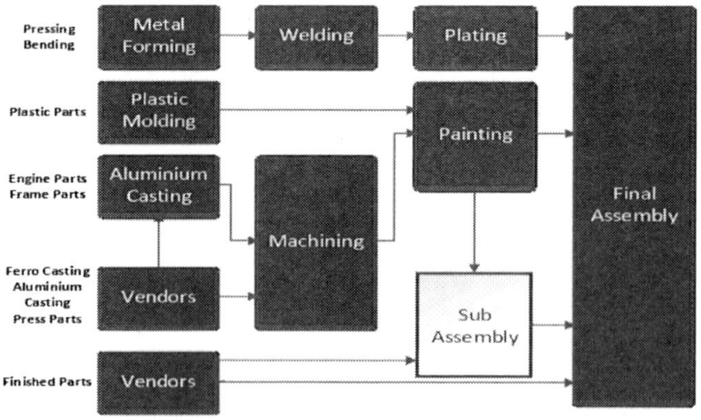

Fig. 1. Production flow of the manufacture

Although the entire section that is in production systems is an important factor for the sustainability of production lines, there are two sections is a major factor that is sub assembly and final assembly. Sub assembly is responsible for generating products such as engine and tire while the final assembly section was responsible for assembling all products from previous section into final product that is motorcycle. Both sections which can be considered important due to the high levels of liability and usability compared to other sections as a result if there is a delay in the sub assembly or final assembly, then the entire production system will also experience delays due to the minimum time available for both sections to be able to repair and fulfill desired order. Particularly in sub assembly section, delay often caused by imbalance production assembly lines that can be seen from the unemployed workers in several work stations, while some other work stations are work optimally. This resulted in the production system is running not efficiently and effectively

3.1 Results

Table 1 shows a comparison of the level of performance of the assembly line with the actual conditions of assembly lines that have been carried out line balancing using parameters that are in line balancing.

Table 2. Comparison of performance levels actual condition with assembly line balancing line results

Parameter Line Balancing	Before Improvement	After Improvement
Line Efficiency	83,12%	91,43%
Number of Work Station	22	20
Smoothness Index	41,59	14,99
WS Efficiency	76,52%	91,71%
Idle Time	168,29	49,58
Balance Delay	16,88%	8,57%

Based on comparison of the level of performance of Table 1 shows that the balance of the assembly lines by using line balancing is generally better than the current condition of the assembly line. Every parameter of line balancing will be further analyzed as follows:

1. *Line Efficiency*

   Line Efficiency is the level of efficiency of the assembly line, expressed as the ratio between the time spent by the assembly line to the available working time is the cycle time. By having a high level of efficiency, it can be said that each station has a working time approached predetermined cycle time so that the higher level of efficiency of assembly line, performance of assembly lines will be more optimal. Based on the calculations, efficiency improvement trajectory of the actual condition from 83.12% increased until 91.43% using the method of line balancing.

2. The number of workstation

   The number of workstations used will affect the planning of expenditures made by the company so as to more reduction in the required workstations will have an impact on cost reduction for companies and profitable for them financially. After the line balancing on assembly line, there is a reduction in number of workstations needed from 22 to 20 workstations resulting cost reduction for the company.

3. Smoothness Index

   Smoothness Index is an index that has a relatively smooth running of the assembly lines of certain balance or equilibrium level of fluency of assembly line has been established. Value smoothness index increasingly close to zero suggest that the level smoothness of assembly line will get better and this contributes to index of delay on production process. The value of the smoothness index decreased assembly line where value on current conditions that amounted to 41.59 to 14.99 so that it can be concluded that the smooth running of assembly lines is increasing and prevent the occurrence of delay production.

4. Workstation Efficiency

Workstation efficiency is a high level of efficiency is expressed by the ratio of a workstation with the greatest time working station for the entire workstation. Workstations can describe balancing efficiency of allocation of workload that is by increasing the value of the workstation efficiency, the labor time required for each workstation will be close to the largest working time from one workstation so that the workload of each work station can be said to be more balanced. Condition of assembly lines currently have a workstation efficiency value of 76.52% to 92.71% after the line balancing that with increase in this parameter states that the allocation of existing workload on assembly line machines cub become more balanced.

5. Idle Time

Idle Time is total waiting time required by work stations to carry out specified work operation. By decreasing value of idle time, waiting time required by the operator will also be getting a little bit so working process can be run faster and increase productivity on workstation. Based on the calculation result, adecline in the value of idle time on assembly lines of 168.29 seconds became 49.58 seconds.

6. Balance Delay

Balance delay is ratio between idle time on assembly line with time available. The value of balance delay starting from percentage of idle time owned production process since the time of components entering the first workstation to workstation last. If value of balance delay to zero then the level of performance of assembly line can be expressed better as time delay on all wor stations will be smaller than time available. The value of balance delay obtained from the calculation amounted to 8.57% where value is decreased from value balance delay in actual conditions which amounted to 16.88%.

4 Summary and Conclusion

By performing line balancing methods using Ranked Positional Weight on the assembly line, there was an increase in the level of performance of assembly lines with an efficiency of 91.43%, number of workstations required as many as 20, the index of smoothness index of 14.99, the level of efficiency workstation at 91.71%, the value of idle time of 49.58 seconds, and the value of balance delay amounted to 8.57%. The increase in line balancing parameter indicates that assembly line will run better than current assembly lines where the level of efficiency has a value of 83.12%. Number of workstation required as many as 22 and index of smoothness index of 41.59, the level of efficiency of workstations amounting to 76.52%, the value of the idle time of 168.29 seconds, and balance delay value amounted to 8.57%. With the increased performance of assembly line so can make increase of t efficiency and productivity which will surely impact production system as a whole so that the company can meet demand of consumers constantly as well as reduce cost. Investigate further associated with other aspects such as layout of production process, maintenance of machinery and equipment also psychologically assessment for employees.

5 Acknowledgements

This work is supported by Hibah PITTA 2018 funded by DRPM Universitas Indonesia, No. 2453/UN2.R3.1/HKP.05.00/2018.

References

1. Barnes.Ralph.M, *Motion and Time Study: Design and Measurement of Work,* (John Wiley and Sons, 1980)
2. D. Bedworth, *Integrated Production and Planning Control* (John Wiley and Sons, 1982)
3. Hong-Jinh Cang,Tung-Meng Chang. (Tamkang Journal of Science and Engineering,2010)
4. R. Hamza, A. Jassim, G. Per. Eng. Mng. E **2**, 70 (2013)
5. S. N. Chapman, *The Fundamentals of Production Planning and Control* (Pearson, London, 2005)
6. S. Ghutukade, S. Sawant, I. J. Ad. Eng. Res. E **5**, 1 (2013)
7. A. Elsayed, *Analysis and Control of Production System* (Prentice Hall, New Jersey, 1994)

Planning to Increasing Production Flexibility on Highest Unmatching Ratio Products in Automotive Company with DMAIC Approach

Inaki Maulida Hakim[1,a], *Rina* Wahyuningtyas[2] and *Narpawandawi* Kusuma[3]

[1a,2,3]Department of Industrial Engineering, Faculty of Engineering, Universitas Indonesia

Abstract. This study aims to increasing production flexibility on highest unmatching ratio products in an Indonesia company. This study focuses on the problem of an inability to absorb orders fluctuations by dealer or end customer at one of Indonesia's leading automotive companies. If it is not solved, it will potentially causes cancelation order. Therefore, the necessary improvements in order to increase the production flexibility of the company. Method used in this research is DMAIC (Define, Measure, Analyze, Improve & Control) in way to get the right solution. Improvements will have an impact on opportunity cost and unmatching ratio. In this study, absorption scheme fluctuations order on a weekly basis in the form of WOC (Weekly Order Change) with adding a safety stock of unique part by using Q-model calculations. The results obtained from this study is the profit as the opportunity cost IDR 431,635,851 and unmatching ratio decrease from 4.9% in before condition become 1.2% in after condition.

1 Introduction

Nowadays competition in the industries or companies requires a companies to improve performance or to innovate on an ongoing basis in order to survive. Therefore industries must be focus to face up industry 4.0 became challenge each industry to improve performance and output company must be reached. So companies must be innovate in all areas to improve competitiveness. If they are not able to adjust to the situation will decline or allow the company will go bankrupt. Likewise with automotive companies need to increase their competitiveness.

Business processes within an automotive company can be divided into three major, such as sales and marketing, manufacturing, ordering and shipping. The first area is the sales and marketing functions to market the product and deal directly with the dealer, before product to end customer. Area ordering and shipping which starts from suppliers to the process of unpacking. Automotive companies are in manufacturing area where the production. In this area, often encountered major problems associated with the production flexibility of

[a] Corresponding author : inakimhakim@eng.ui.ac.id

automotive companies. This problem is an inability to absorb fluctuations in orders placed by a dealer or end customer. This caused a difference or gap (unmatching order) between what is available on dealers with what is required by end customers.

Case study this research at one of big automotive industry in Indonesia. This company has the five models of cars produced for local markets, Model 2 has the highest unmatching order of 4.9%. So this research will focus on the Model 2 only. If it can't be overcome, it would potentially occur cancelation order. In the competitive automotive industry increasingly stringent, customers can easily move to another brand. With the problems described above and through the journals sources, this study offers an increased production flexibility of automotive companies that can also be used as a competitive weapon in manufacturing firms (Awwad, 2008). It will increase production flexibility and can decrease unmatching ratio so that the potential cancelation orders can be minimized or even eliminated. If the value of unmatching ratio decreases, it will be able to improve customer satisfaction in terms of waiting time to get the product, it would also have an impact on the company profitability.

This research focused on finding solutions to these problems. The method used in this research is DMAIC (Define, Measure, Analyze, Improve, Control) to get the right solution. DMAIC is a methodological approach to problem solving that includes key improvement measures sequentially Therefore, this study was entitled "Planning to Increasing Production Flexibility on Highest Unmatching Ratio Products in Automotive Company with DMAIC Approach".

2 Literatur Study

2.1 Toyota Production System

Toyota Production System is triggered by Saichi Toyoda, Kiichiro Toyoda and Taiichi Ohno of Toyota Motor Corporation as a physical embodiment, or way, to achieve the ultimate goal of the six requirements with cost reduction (Monden, 1998). Basically, the main purpose of the TPS is speeding up time by way of the elimination of waste. TPS philosophy used throughout Toyota Manufacturing and affiliates worldwide. In addition, the term TPS is also known by another name, Lean Manufacturing. The main features are reflected in the two pillars of JIT and Jidoka and at the base of the standardized work and kaizen.

2.2 Flexibility

Flexibility plays a significant and important role in the preparation, implementation, and monitoring both for manufacturing and operations strategy (Awwad, 2008). This is why flexibility becomes a strategic weapon in various manufacturing companies. Flexibility is applied in a company (manufacturing flexibility) will make a company more efficient and effective in the face of rapid environmental change. Manufacturing flexibility emerged as a competitive weapon that should be considered the result by many users flexibility, such as the need to change the design of products, shortening product life cycle, using the strategy of change, and cope with uncertainties such as machinery, technology, customer demand, and expectations of the customer (Awwad, 2008).

2.3 Inventory

Inventory is a general term that indicates anything or organizational resources stored in anticipation of the fulfillment of the request (Rangkuti, 2004). In inventory system demand can be divided into two: deterministic and probabilistic. In this study, demand is probabilistic quantity is not known with certainty. Probabilistic demand encourages companies to conduct proper planning inventory control based on the total cost of inventory is the smallest of total accumulation purchase costs, the cost of the message, save costs, shortage costs, and transportation costs. In practice the logistics system, transportation costs include fixed and variable costs (Burhan, 2010).

2.4 Q-model

Q-Model method known method with fixed order quantity models. An inventory control model in which the requested amount is fixed and the purchase was triggered by the decline of inventories to certain inventory level. The method is widely used in Define how many economical to do for ordering is a method of EOQ (Economic Order Quantity). EOQ and its variations are still widely used in industry for inventory management for this type of demand-free. Formulation Q in which an Economic Order Quantity (EOQ):

$$T = \sqrt{\frac{2\ x\ D\ x\ S}{H}} \tag{1}$$

The formulation of safety stock is determined by calculation:

$$SS = \ Z\ X\ \sigma\ x\ \sqrt{LT} \tag{2}$$

The formulation of the reorder point is determined by calculation:

$$ROP = (DxLT) + \left(\ Z\ X\ \sigma\ x\ \sqrt{LT}\right) \tag{3}$$

2.5 Six Sigma & DMAIC

Six Sigma and the DMAIC methodology looks like the last generation of approaches to generate improvements, the addition of the concepts, methods, tools, and remove restrictions on the identification (Snee, 2010). DMAIC approach can be defined as a structured repair procedure (Cunha & Dominguez, 2015). Basically, the steps in the DMAIC is all about improving processes with a key measurement in business. The key is to use the tools, techniques, and calculations can help. So the steps DMAIC is offered as a road map but also flexible in its application. DMAIC improvement in the cycle is the current primary tools we will use Six Sigma projects (Gejdos, 2015).

3 Research Methodology

This study was developed with the DMAIC method that consists of five stages, namely Define, Measure, Analyze, Improve, Control (Carretero, Rosa, & Sanchez-Rodas, 2016). DMAIC phases in this study are described in Table 1.

Table 1. DMAIC Phases

Phase	Input	Process	Output
Define	Problems in automotive company	Analyze journals and reports	Research topic
Measure	Unmatch order	Measuring before condition	- Unmatching ratio - Sigma level
Analyze	Prim and secunder data	- Fishbone - Pareto chart	Customer importance rate
Improve	- Part list - Demand - Lead time - Cost (Unit cost, ordering cost, holding cost)	- Making scheme to increasing production flexibility - Grouping unique part - Calculating Q-model - Interview and observe	- Scheme of WOC (Weekly Order Change) - Unique part - Calculating safety stock - Calculating order quantity - Calculating reorder point
Control	Evaluate an impletion of improvement will execute	Making an action plan	- Flexibility Check Part - Flowchart WOC (Weekly Order Change)

4 Results and Discussion

Safety stock were prepared only for unique part. With the aim to reduce the risk of part shortages if there are fluctuations in the order that resulted in changes in variant Model 2. Before WOC (Weekly Order Change) implementation, there is no safety stock, including for unique part Model 2. The draft increased flexibility of production use WOC (Weekly Order Change) scheme the company must procure safety stock as an implementation of the WOC (Weekly Order Change) have an impact on procurement of safety stock. Each part has safety stock with varying amounts. It depends on parts used and level of uniqueness of parts.

There are two components of costs between inventory costs (total costs) and cost of opportunity lost on circumstances before and after. The first cost is cost of inventory sum from costs of purchase, costs of the message, and save costs. The decision was taken inventory in this study on conditions in which procurement of safety stock only for unique parts of which were obtained from suppliers import, so has ordering and delivery lead time is longer than the local supplier. So if local supplier decision could be made by considering storage costs, in this study a comparison decision of inventory is obtained from the comparison of the total cost of procurement on safety stock with conditions of procurement for parts that have a lead time of booking and delivery quite long, so if there is a shortage it can be ascertained that the procurement costs specifically for the part will be much more expensive compared to cost of storage, because it must make deliveries by air to shorten the lead time.

When viewed from the cost of inventory, before circumstances indicate better conditions, where there is no inventory cost at all. When compared with the situation after that once there is a fee of IDR 2.812.191.099 due to the impact of their procurement of safety stock to 82 unique parts. At an opportunity cost which is the cost incurred for selecting a particular business decision than any other business decision. In the state before there is opportunity lost where the chances of company's inability to meet fluctuations in demand. Opportunity lost value obtained by IDR 3.243.826.950 every month, which is derived from multiplication unmatching ratio, average demand, assuming profit. The average value of demand in July-October 2015 and the assumption of profit earned from the selling price of the cheapest sources of Autobild magazine in February 2016. By contrast, in the state after there is an opportunity to make a profit as the opportunity cost of IDR 431.635.851 every month. This value is obtained from the reduction of opportunity lost on the circumstances before cost of inventory on the state after.

In terms of cost, increased flexibility of production by applying the WOC (Weekly Order Change) is a good decision. Although the company must pay for the provision of safety stock, but it is more profitable than company choose do not to procurement of safety stock. This case will happen that automotive companies are not able to meet fluctuations in demand. Even feared if this situation continues to be able to lower customer confidence towards automotive company.

Value unmatching ratio before and possible values unmatching ratio after to get improvement stage by researchers. Implementation of the WOC (Weekly Order Change) will be held in the month of June 2016, so value of the condition after unmatching ratio can not be seen. Used assumption that by applying the WOC (Weekly Order Change) will be reduced unmatching ratio amounted to 75% of situation before in which 75% was found that the scheme on WOC, dealers have the right to be able to change order in the first week, second, and third of every month. So that in every last week, fluctuations in demand can not be absorbed by the auto manufacturers. Obtained assuming that the WOC (Weekly Order Change) can reduce unmatching ratio of 75%. Figure 4. Here is a ratio that indicates.

In circumstances after unmatch order value can be lowered as much as 1,689 units to 423 units in the state after. It's also affecting depreciation unmatching ratio to 1.2% in the state after that was worth 4.9% on the circumstances before. Based on the analysis in terms of unmatching ratio, increased flexibility of production by applying the WOC (Weekly Change Order) is a good decision because it can decrease the value unmatching ratio.

5 Conclusion

After performing data processing using DMAIC approach, the conclusions obtained from this research are automotive companies increase production flexibility characterized by the falling value of unmatching ratio of 3.7% by obtaining the value of safety stock for a unique part Model 2 local market products. Before the repair, the company has opportunity lost by IDR 3.243.826.950 and after repair companies get opportunity cost of IDR 431.635.851.

6 Acknowledgment

This work is supported by Hibah PITTA 2018 funded by DRPM Universitas Indonesia, No. 2453/UN2.R3.1/HKP.05.00/2018.

References

1. A. d. Awwad, Jord. Al-H. B. T. Uni., (2008)
2. Burhan, Jur. Argo., **4**, (2010)
3. C. Cunha & C. Dominguez, Proc. Comp. Sci., 885-893, (2015)
4. A. Erdogan & H. Canatan, Conf. on Tech. Innov. and Entrepr., 695-704, (2015)
5. P. Fattahi, V. Hajipour, A. Nobari, Els. Jour., 211-223. (2015)

6. P. Gejdos, BEM2015, 566-572. (2015)
7. Y. Gupta, USA Prod. And Ops. Manage., (1996)
8. S. Jevgeni, S. Eduard, R. Zahharov, DAAAM 2014, 511-519, (2015)
9. Laureani, Bready, Anthony, Lead. Heal. Serv., 322-337, (2006)
10. C. Leaphart, T. Gonwa, M. Mai, M. Prendergast, H. Wadei, J. Tepas, C.B. Taner, Jour. Of Surg. Research 177, **13**, (2012)
11. Y. Monden, Georg: Eng. & Manage. Press, (1998)
12. A. B. Sin, S. Zailani, M. Iranmanesh, T. Ramayah, Int. J. Prod. Econ., **168**, 105-117, (2015)
13. Snee, Int. Jour. Of Lean Six Sigma, 9-29, (2010)

MATEC Web of Conferences **218**, 04030 (2018) https://doi.org/10.1051/matecconf/201821804030
ICIEE 2018

Influence of sheath gas flow rate in Ar induction thermal plasma with Ti powder injection on the plasma temperature by numerical calculation

Yulianta Siregar[1,2*], *Yasunori* Tanaka[1] , *Tatsuo* Ishijima[1], *Yoshihiko* Uesugi[1]

[1]Division of Electrical Engineering and Computer Science, Kanazawa University, Japan. Kakuma, Kanazawa 9201192, Japan.
[2] Departement of Electrical Engineering, University of Sumatera Utara, Indonesia. Jalan Dr. T. Mansur No. 9 Kampus USU, Medan, Indonesia 20155. +6262 8211045

Abstract. Numerical calculation was conducted to study the gas flow fields and temperature distribution in Ar inductively couple thermal plasma (ICTP) with titanium (Ti) powder injection. Influence of sheath gas flow rates at 90 L/min, 80 L/min and 70 L/min was investigated on the thermal plasma temperature in the torch. Results indicated that higher sheath gas flow rate raises the axial temperature and gas flow velocity in the thermal plasma

1 Introduction

Inductively coupled thermal plasma (ICTP) in high-pressure and high-power has been widely used for different materials processing such as nanoparticle synthesis [1], thermal barrier coatings [2], fabrication of diamond films [3], surface modification [4], synthesis of fullerene [5]. The ICTP can be an adequate chemical and heat source to different materials processing with little contamination. The ICTP often have high gas temperature and high enthalpy, which can be adequate to evaporate solid feedstock in materials processing. In case of nanoparticle synthesis, the ICTP gives one-step direct processing with rapid evaporation of injected feedstock using the above high gas temperature and high enthalpy. In addition, the ICTP offers rapid cooling of evaporated materials because of high temperature gradient in the downstream portion of the ICTP. This rapid cooling of the evaporated material can enhance nucleation and produce nanoparticles.

The ICTP is usually sustained with sheath gas along the inner plasma torch. The sheath gas is used to supply the plasma gas itself and also to prevent the plasma from contacting the plasma torch from the thermal damage of the torch. It is well known that this sheath gas influence the generation of the ICTP itself, and the temperature and gas flow fields in the ICTP. However, it is not yet enough understood to study the influence of sheath gas flow rate on the temperature field in the ICTP and also on the evaporation rate of feedstock power injected into the ICTP.

* Corresponding author: julianta_srg@yahoo.co.id, tanaka@ec.t.kanazawa-u.ac.jp

In this report, the influence of sheath gas flow rate on thermal interactions between argon (Ar) ICTP and titanium (Ti) powder injection was studied using our previously

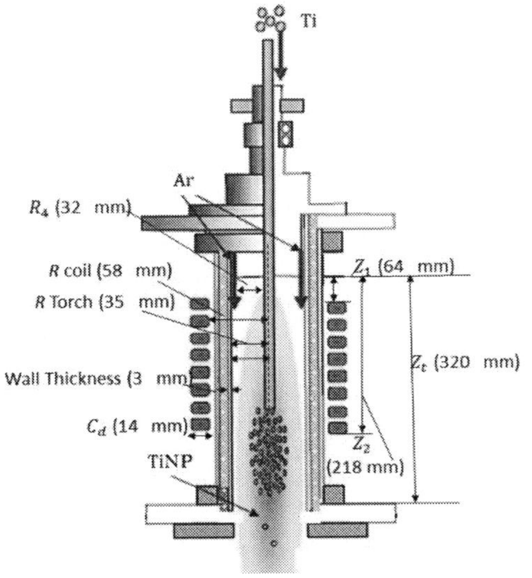

Fig. 1. Configuration of Ar/Ti ICTP torch

developed numerical calculation model [6,7]. Results indicated that higher gas flow rate in the sheath gas enhances evaporation of feedstock by increased thermal plasma temperature along the axis of the plasma torch.

2 Configuration of inductively coupled thermal plasma torch

The configuration of the ICTP treated here is shown in Fig. 1. The plasma torch is composed of two coaxial quartz tubes with a length of 320 mm. The inner quartz tube has an internal radius of 35 mm. Around this tube, an 8 turn-coil is wounded with a radius of 58 mm. To this 8 turn-coil, rf current is supplied to produce electromagnetic field inside the tube. In the tube, the ICTP is sustained by electromagnetic field generated from the coil current. Argon is provided as a sheath gas flow along the inner quartz tube wall. The sheath gas is supplied with a swirl to avoid the plasma from contacting the inner quartz tube.

Titanium feedstock powder is assumed to be injected with Ar carrier gas through a water cooled pipe into the ICTP as shown in Fig.1. For numerical calculation of the temperature and gas flow fields in the ICTP with titanium feedstock powder, a two-dimensional cylindrical r - z cross section of this plasma torch was set as a calculation space.

3 Modeling and calculation of ICTP

3.1 Assumptions

We assumed the following conditions for the ICTP[7]: (i) Local thermodynamic equilibrium (LTE) is established in the ICTP. Thus, all the temperatures including electron temperature and heavy particle temperature are the same to one another. Boltzmann distribution is assumed to the excited state at that temperature. In addition, chemical equilibrium conditions were also set for all reactions. (ii) The optically thin assumption was adopted at wavelengths above 200 nm. (iii) Steady state, axis-symmetric and laminar flow were assumed with no viscous dissipation. (iv) The particle-particle interactions were not considered for injected particles. (v) The mean free path of plasma components was assumed to be much smaller than the particle radius. (v) The particle radius was assumed to be much higher than the mean free path of plasma components.(vi) In evaporation, particle always has spherical shape. (vii) The particles are heated uniformly from the around plasmas. Therefore, evaporation uniformly occurs at the particle surface. (viii) Influence of electronic charging on the particle were not considered. (ix) Rocket force was neglected.

Table 1. Titanium powder of thermodynamic properties

Parameter	Unit	Value
Injection load	g/min	0.5
Mass density	kg/m^3	4506
Melting temperature	K	1953
Boiling temperature	K	3535
Latent heat for melting	J/kg	391000
Latent heat for evaporation	J/kg	8294000
Specific heat of solid	J/(kg.K)	528
Specific heat of liquid	J/(kg.K)	700
Thermal conductivity	W/(m.K)	14
Emissivity of particle source	(-)	0.4

3.2 Numerical calculation condition

In this calculation, the operating frequency for the rf coil current was fixed at 450kHz. The pressure condition was set at 300 Torr (=40 kPa) inside the plasma torch. The input power was set at 40 kW. Sheath gas flow rate was set to three values of 90 L/min, 80L/min, and 70 L/min. To this ICTP, titanium feedstock powder is injected through a water-cooled pipe inserted on the torch axis from the plasma torch head. The injected titanium feedstock powder has a mean diameter of 2 μm and a deviation of 0.5 μm. In the calculation, 35 trajectories for feedstock particles with 7 different diameters with 5 different initial positions were assumed to be injected. The titanium feedstock feed rate was set to 0.5 g/min. In addition, transport properties and thermodynamic of argon and titanium vapor were in advance calculated as a function of temperature. For solid and liquid thermal properties of feedstock, the values were used in the literatures as indicated in Tab.1.

4 Calculation results

4.1 The temperature distribution of the ICTP

Figure 2(a), (b) and (c) show the temperature distribution of the ICTP with an sheath gas flow rate of 90 L/min, 80 L/min and 70 L/min, respectively. In case of Fig.2(a) with a sheath gas rate of 90 L/min, the ICTP has high temperature area about 11000 K in wide region around radial positions of 10 mm $< r <$ 30 mm and axial positions 140$<z<$235mm. In this high temperature area, the rf eddy current flows inside the ICTP to heat it. At a sheath gas flow rate of 80 L/min as shown in Fig.2(b), the high temperature area at temperatures above 11000 K becomes smaller around radial positions of 9 mm $<$ r $<$ 29 mm and axial positions of 135 mm $<$ z $<$ 225 mm than that in Fig.2(a). Furthermore, lower sheath gas flow rate at 70 L/min case as indicated in Fig.2(c) presents a further smaller high temperature area at temperatures above 11000 K around radial positions of 15mm$<$r$<$28mm and axial positions of 130 mm $<$ z $<$ 220 mm. This means that the high temperature area is shifted to upstream with decreasing sheath gas flow rate. This is due to reduced gas flow convection transport by decreasing gas flow rate. In addition, the high temperature area is shrunk in axial direction with decreasing gas flow rate. These features can be seen for mid-temperature area of 6000-6500 K.

On the axis of the torch, the temperature is higher at higher gas flow rate of 90L/min than at lower gas flow rate of 70 L/min. This may be because at higher gas flow rate of 90 L/min, the high temperature area is present near the center axis by strong gas flow around the torch wall. That stronger gas flow make the high temperature plasma present near the axis. This higher temperature around the axis is related to the evaporation rate of feedstock as described later.

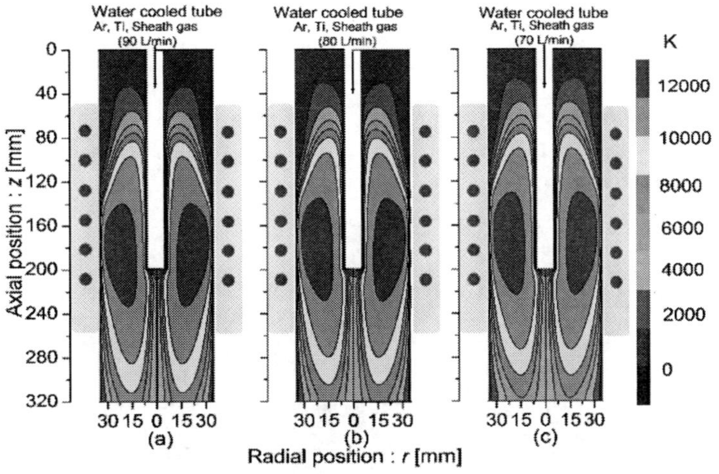

Fig. 2. The temperature distribution of Ar and Ti powder injection with sheath gas flow rate at 90L/min, 80 L/min, 70 L/min.

4.2 Titanium feedstock particle evaporation

In this temperature fields of the ICTP, titanium feedstock particles were assumed to be injected. Figure 3 indicates variation in diameters of 15 particles with 5 different initial diameters injected from 3 different initial positions into the sheath gas as a function of axial position. Figure 3(a) and 3(b) corresponds to the results for sheath gas flow rates of 90L/min and 70 L/min, respectively.

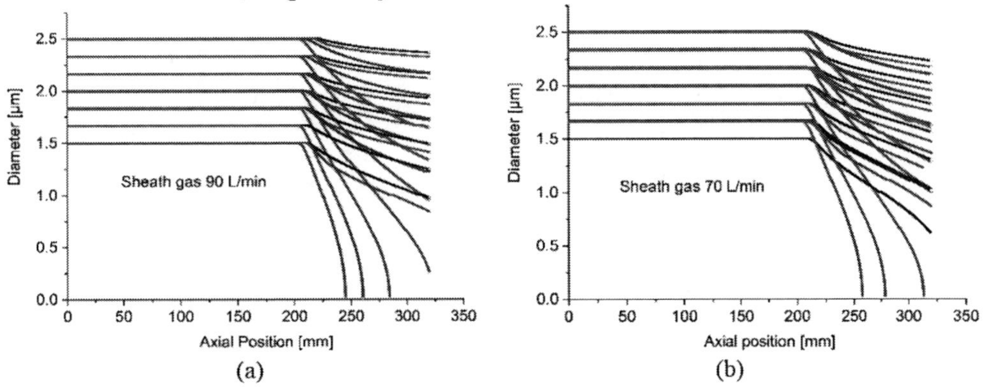

Fig. 3. Diameter variation of titanium particles injected for 35 kinds with 5 different initial positions and different 7 initial diameters as a function of axial position in argon ICTPs with sheath gas flow rate at 90 L/min (a) and 70 L/min (b).

The diameters do not change at axial position from 0 to 200 nm because all the particles injected are present inside the water-cooled pipe. From axial position of 200 mm, the particles start flying into the ICTP, and then they are heated by the ICTP. After some fly to be heated, the Ti particles have the diameter decreased because of its evaporation. In this calculation, the particles with initial diameter of 1.5 μm are strongly heated from the ICTP, and then the particle diameter decreases to 0.01 μm. As shown in Fig.3(a) and (b), higher sheath gas flow rate of 90 L/min involves rapid decrease in diameter than 70 L/min. This is attributed to the higher temperature along the torch axis as indicated in the previous section. This evaporation of the feedstock particles produces titanium vapor to the ICTP, which can change the properties of the Ar ICTP.

4.3 Mass fraction of titanium vapor

Influence of sheath gas flow rate on titanium vapor creation in the ICTP can be seen in Fig. 4. As shown in this figure, the mass fraction of titanium (Ti) vapor is distributed from the tip of the water-cooled pipe.

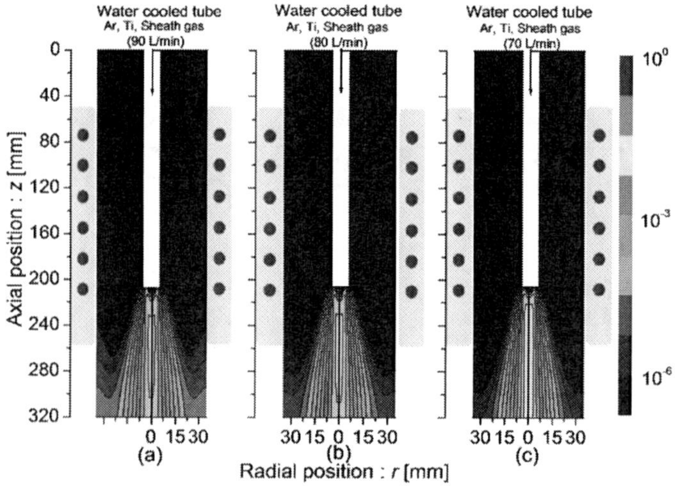

Fig. 4. Mass fraction of Ar and Ti powder injection with sheath gas flow rate at 90 L/min, 80 L/min, 70 L/min

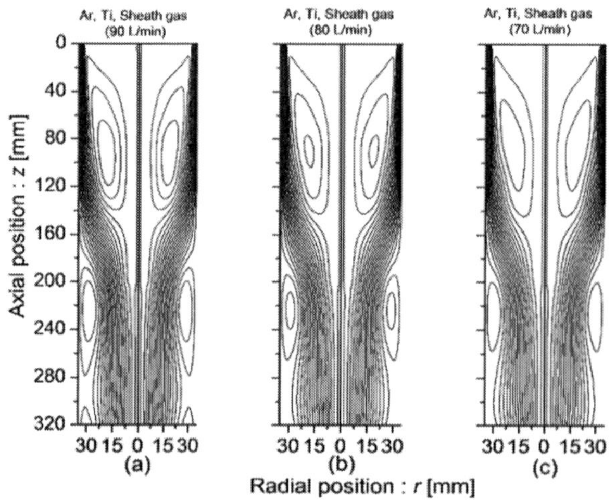

Fig. 5. Streamline of Ar and Ti powder injection with sheath gas flow rate at 90 L/min, 80 L/min, 70L/min

The injected Ti feedstock particles are injected to the ICTP through the feeding pipe. The Ti particles are evaporated during their flight to produce Ti vapor as contamination to Ar ICTP. This contaminated Ti vapor is transported by gas flow convection of Ar carrier gas and by diffusion. This Ti vapor is present downstream of the feeding pipe along the axis. Titanium vapor concentration just slowly increases with decreasing sheath gas flow rate. This is due to lower gas flow velocity along the axis, and thus the Ti vapor concentration is higher at 50 L/min of sheath gas flow rate.

Another difference in Ti vapor concentration is that Ti vapor is present around the wall at higher gas flow rate of 90 L/min. This arises from the vortex near the wall. Figure 5 shows the streamline inside the ICTP at different sheath gas flow rate at 90 L/min, 80L/min and 70 L/min. As indicated in Fig.5, increasing sheath gas flow rate gives strong vortex at the upper side of the plasma torch at axial position 40 mm $< z <$140 mm. Furthermore, another vortex present at axial position 200 mm $< z <$ 280 mm around the torch wall is strengthened by increasing sheath gas flow rate. Because of this vortex, the mass fraction of Ti vapor is transported to upstream direction near the torch wall around axial position 240 mm $< z <$ 320 mm.

5 Conclusion

Numerical simulation was made to study influence of sheath gas flow rate on the temperature distribution in argon (Ar) induction thermal plasma with titanium (Ti) feedstock powder. The calculated temperature distributions of argon thermal plasma and titanium feedstock powder were compared with different sheath gas flow rate at 90 L/min, 80 L/min and 70 L/min. Results indicated that higher sheath gas flow rate raises the axial temperature distribution and gas flow velocity in the thermal plasma torch.

6 Acknowledgment

The authors thank to Kanazawa University, Japan and Ministry of Research, Technology and Higher education (KEMRISTEKDIKTI) of Indonesian Government for the existing bilateral academic-cooperation. An appreciation is also granted to Indonesia Endowment Fund for Education (LPDP) that provides fund through a BUDI-LN Scholarship Scheme.

References

1. L. Tong and R.G. Reddy. Mater. Res. Bull, **4**, 2303 (2006)
2. J. Mao, M. Liu, C.G. Deng, C.M. Deng, K.S. Zhou, Z.Q. Deng, J.Eng. Mater. Technol, **139**, 041003 (2017)
3. S. Gottlieb, N. Wohrl, S. Schulz, V. Buck, SpringerPlus, **5**, 568 (2016)
4. W. Chen, Y. Xu, S. Shi, Y. Cao, M. Chen, X. Zhou, Scien. Report, **8**, 2279 (2018).
5. K.S. Kim, A. Moradian, J. Mostaghimi, G. Soucy. Plasma Chem, **30**, 91-110 (2010)
6. Y. Tanaka, T. Sakuta, J. Phys. D: Appl. Phys, **58**, 2149 (2002)
7. Y. Siregar, N. Kodama, Y. Tanaka, T. Ishijima, Y. Uesugi, IOP Conf. Series: Materials Scence and Engineering, **309**, 012090 (2018)

MATEC Web of Conferences **218**, 04031 (2018) https://doi.org/10.1051/matecconf/201821804031
ICIEE 2018

The Comparison of Genset Machine CS 1000 L Performance Between LPG and Methane Gas Usage as The Fuel

Caturwati Ni Ketut[1,a], *Agung* Sudrajad[1], *Mekro* Permana[1], *Heri* Haryanto[2] and *Rizka Khaerani* [1]

[1]Mechanical Engineering Department- Universitas Sultan Ageng Tirtayasa,
[2]Electrical Engineering Department- Universitas Sultan Ageng Tirtayasa,

Abstract. Increasing demand of the energy along with the depletion of the world oil reserves had forced us to utilize the alternative energies in our daily life. One of those is that we should develop its utilization on biogas which a source of renewable energy and be able to overcome the problems that caused by the abundance of organic waste around us. Methane is a major element which found in biogas, therefore a preliminary study of the biogas utilization as resource of energy in a power plant can be preceed by the use of methane as a fuel gas in genset. This study compares the performance of CS 1000 L generator set that use methane gas fuel and LPG gas fuel. The results shows that the efficiency of thermal generator using methane fuel reaches 26.22 %. This value is higher than thermal efficiency genset that using LPG gas which only reaches the value of 8,26 %.

1 Introduction

Fuel oil is the main energy source used in human life today. This is because this fuel is easier to find and also easier to transportation handle. However, the oil prices are increasing along with the depletion of world oil reserves. This condition forces us to switch to using the other types of fuels such as coal and gas as much as possible to replace the use of fuel oil. In 2012 the Indonesian government issued an energy conversion policy by revoking oil price subsidies and carrying out a national campaign to use gas as the main fuel for cooking needs.

The types of natural gas used today are Liquid Petroleum Gas (LPG) and Compressed Natural Gas (CNG). LPG is a gas produced from refining petrolleum with the main content of propane (C3H8) and butane (C4H10) and a few other light hydrocarbons such as ethane (C2H6) and pentane (C5H12). While CNG is a natural gas with the main composition of methane (CH4). In addition, at this time various organic waste processing technologies were developed to produce biogas fuels which contained the main element of methane gas. Biogas is an alternative energy source that is environmentally friendly and renewable, which is produced from anaerobic fermentation by microorganisms in organic materials,

[a] Corresponding Author : n4wati@untirta.ac.id

such as agricultural waste, livestock manure, human waste and household waste [1][2][3][4][5][6] .

Methane gas can also be produced naturally from piles of waste using landfill methods. The thing that must be considered is that methane gas is one of the gases included in the group of greenhouse gases which release freely into the atmosphere can cause an increase in global warming. To reduce this problem, methane gas can be used as fuel first before being released freely into the atmosphere.

Based on the description above, it is necessary to intensify the use of biogas as an alternative fuel in addressing our energy needs. For this reason, this experiment was carried out by testing the performance and exhaust emissions of LPG-fueled generators using methane as its fuel.

2 Method of Experiment

Experiments to test the performance and the emissions of type 1000 L generators using LPG as fuel compared to the performance of the same engine using methane as fuel are carried out with the composition of the test equipment as shown in Fig. 1.

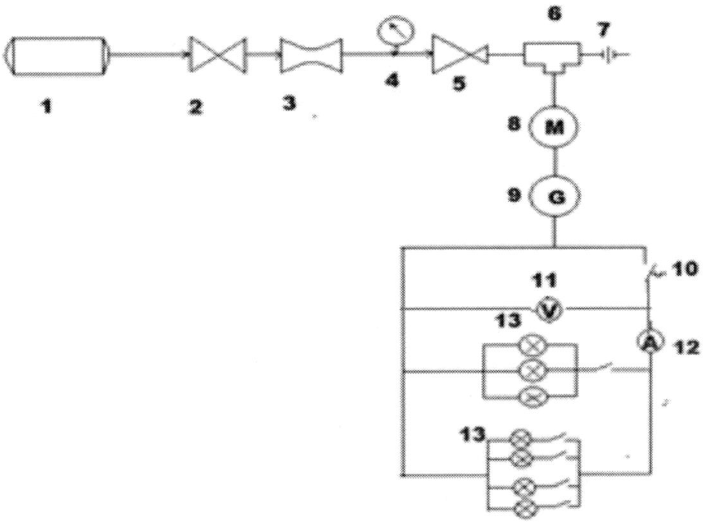

1. gas vessel
2. valve
3. flow meter
4. manometer
5. Reducer
6. carburator
7. orifice (air inlet)
8. Engine
9. Generator
10. circuit breaker
11. Voltmeter
12 Amperemeter
13 lamp loads

Fig. 1. Genset performance test equipment scheme

Genset engine performance test is carried out with variations in lamp load of: 0, 200, 400, 600, 700,725 and 750 Watts by adjusting the number of lights installed. The amount of fuel flow is measured by a rotameter while the amount of air entering the engine combustion chamber is measured by an orificemeter.

3 Results and Discussion

Tests were carried out on the HL 1500 LX generator set using LPG and methane fuels, the installed lamp loads varied by 200, 400, 600, 700, 725, and 750 Watts, but the methane fuel generator test could only run up to 700 Watt.

3.1 Indicated Horse Power (IHP)

IHP is the total amount of power developed by an engine which is resulted from multiplying the measured voltage value and the measured electric current.

$$IHP = VxIxcos\ \varphi \tag{1}$$

Fig. 2 shows the value of the IHP generator with LPG fuel and methane gas fuel. In general the value of the IHP is above the installed lamp load. In addition, the test results show that the use of methane fuel in generators produces a higher IHP than LPG, but the maximum light load that can be installed on a generator with methane fuel is lower than LPG.

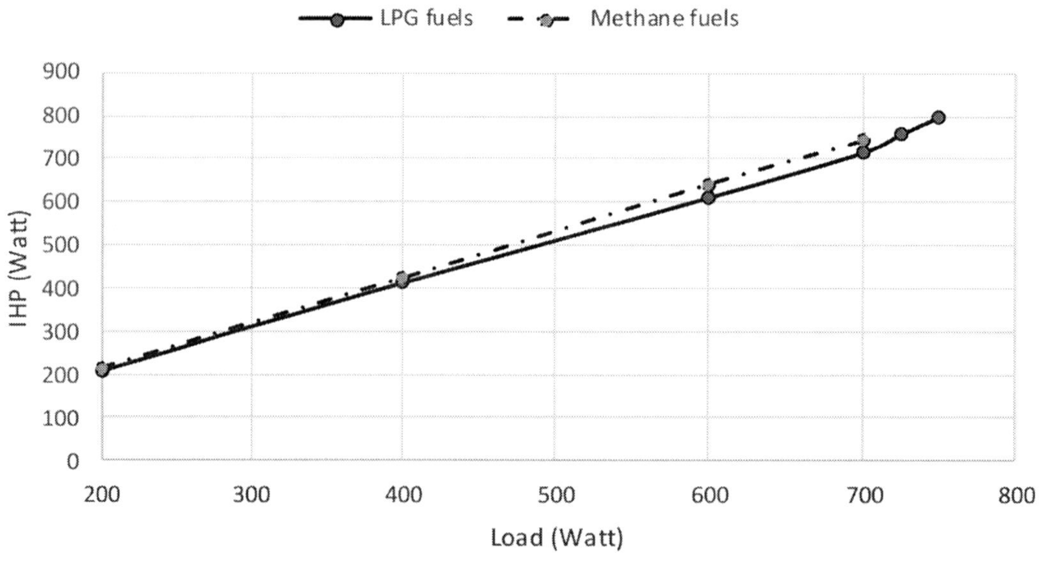

Fig. 2. The value of IHP for various lamp load on the generator.

3.2 Specific Fuel Consumption (SFC)

SFC is the fuel efficiency of an engine with respect to the output.

$$SFC = \frac{m_f}{IHP} \tag{2}$$

Where : $\dot{m}_f$ is mass flow rate of fuel (kg/s) and it can be determined from the flow rate of gas volume

$$\dot{m}_f = \rho_f \cdot \dot{Q}_f \quad (3)$$

Fig. 3 shows that in general the value of the SFC generator is lower with the increase in the load of the installed lamp. In particular, the use of methane fuel in generators is much lower than LPG fuel.

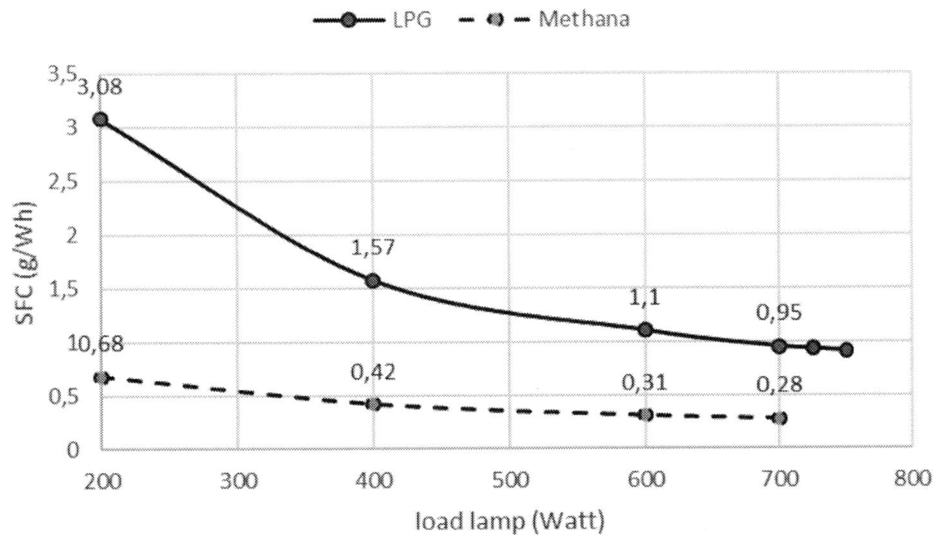

Fig. 3. The value of SFC for various load lamp.

3.3 Thermal Efficiency

Thermal efficiency is a dimensionless performance measure of a device that uses thermal energy, such as an internal combustion engine. Or the thermal efficiency is the size and magnitude of the heat energy contained in a fuel that can be converted to be a useful power, theoretically can be formulated as follows :

$$\eta_{th} = \frac{IHP}{m_f x LHV} \text{ x } 100\ \% \quad (4)$$

Where LHV is lower heat values of the fuel. LHV of LPG (50 % Propana; 50 % Butana) is 46.050 kJ/kg and LHV of methane (CH_4) is about 50.000 kJ/kg. So the value of the thermal efficiency of the generator with LPG and methane gas is shown in Fig. 4.

The value of thermal efficiency in general increases with the increasing value of the load installed on the generator. In addition, the thermal efficiency of methane gas is higher than the thermal efficiency of LPG gas.

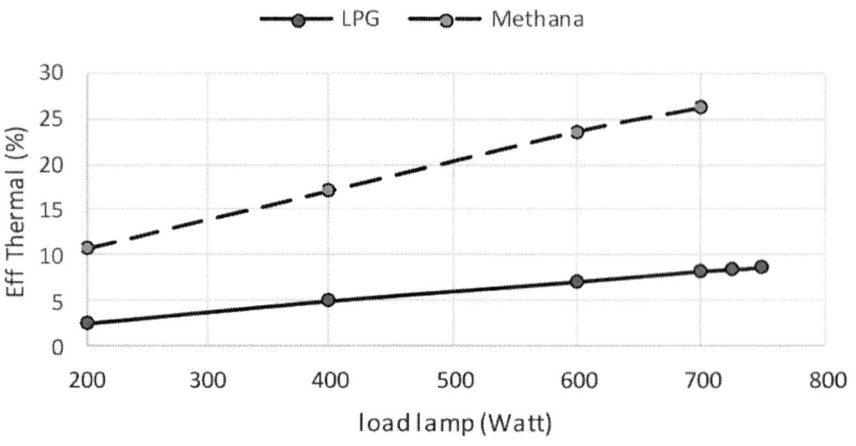

Fig. 4. Thermal Efficiencies of genset with methana and LPG as fuel

4 Conclusions

The use of methane gas as fuel for the LPG gas generator set can improve engine performance which is marked by the value of the IHP generated by the generator is always above the installed lamp load. The difference in the value of the IHP and the lamp load installed on the methane gas-fueled generator set reaches an average of 6.4% while the average LPG-fueled generator set is only 3.47%. So that in general the use of methane fuel in the generator produces greater output power than LPG. So that in general the use of methane fuel in the generator produces greater output power than LPG.

Fuel consumption for generator sets with methane gas fuel is lower than LPG gas fuel which is indicated by the SFC values. The value of SFC genset with methane as the fuel and load installed of 700 W is 0.28 g / WH, while the value of SFC generator with LPG fuel at the same load reaches 0.95 g / WH. This shows the use of methane gas as fuel generator is 30% of the amount of LPG gas needed.

The thermal efficiency of the generator with methane gas fuel can reach 26.22% when the installed load is 700 Watts, while the use of LPG fuel only produces a maximum value of thermal efficiency of 8.62% when the load is installed at 750 Watts of the several advantages above, the use of methane as a fuel in a generator set reduces the generator load capacity. The test results show that the installed load capacity for methane gas-fueled generators reaches 700 Watts, this value is lower than LPG gas which reaches 750 Watts.

5 Acknowledgment

The authors would like to thank the Ministry of Research and Higher Education of Republic of Indonesia for the financial support under PTUPT Research Grant.

References

1. S. Amini, P. Ordoney, and I. Khuzestan, "Investigation of biogas as renewable energy source.," *Int. J.* ..., pp. 1453–1457, (2013).
2. R. Arthur, M. F. Baidoo, and E. Antwi, "Biogas as a potential renewable energy

source: A Ghanaian case study," *Renew. Energy*, vol. 36, no. 5, pp. 1510–1516, (2011).

3. L. G. a Potts and D. J. Martin, "Anaerobic digestion, gasification, and pyrolysis," *Waste Manag. Minimization*, (2011).
4. J. Fierro, "Best Available Techniques in Pyrolysis and Anaerobic Digestion Introduction : Pyrolysis," no. September, pp. 29–30, (2011).
5. K. Jonasson, "Project 3 . Biomass to energy and chemicals through anaerobic digestion and gasification Kevin Jonasson Sebnem Madrali NRCan - Bioenergy Development Group."
6. C. N. Ketut, A. Sudrajad, M. Permana, and H. Haryanto, "Experimental study of anaerobic digester biogas method using leachate from landfill municipal waste," *Int. J. Appl. Eng. Res.*, vol. 12, no. 19, pp. 9112–9115, (2017).

AUTHOR INDEX

AUTHOR INDEX

AUTHOR INDEX